Electromagnetic Interference and Electromagnetic Compatibility

Electromagnetic compatibility is concerned with the generation, transmission, and reception of electromagnetic energy. This book discusses about the basic principles of electromagnetic interference (EMI) and electromagnetic compatibility (EMC) including causes, events, and mitigation of issues. The design procedures for EMI filter, the types of filters, and filter implementation methods are explained. The simulation of printed circuit board designs using different software and a step-by-step method is discussed in detail. This book addresses the gap between theory and practice using case studies with design, experiments, and supporting analysis.

Features:

- Discusses about the basic principles of EMI/EMC including causes and events.
- Makes readers understand the problems in different applications because of EMI/EMC and the reducing methods.
- Explores real-world case studies with code to provide hands-on experience.
- Reviews design strategies for mitigation of noise.
- Includes MATLAB, PSPICE, and ADS simulations for designing EMI Filter circuits.

The book is aimed at graduate students and researchers in electromagnetics, circuit and systems, and electrical engineering.

Electromagnetic Interference and Electromagnetic Compatibility

Principles, Design, Simulation, and Applications

L. Ashok Kumar and Y. Uma Maheswari

CRC Press
Taylor & Francis Group
Boca Raton London New York

CRC Press is an imprint of the
Taylor & Francis Group, an **informa** business

First edition published 2024
by CRC Press
6000 Broken Sound Parkway NW, Suite 300, Boca Raton, FL 33487-2742

and by CRC Press
4 Park Square, Milton Park, Abingdon, Oxon, OX14 4RN

CRC Press is an imprint of Taylor & Francis Group, LLC

Library of Congress Cataloging-in-Publication Data
Names: Kumar, L. Ashok, author. | Maheswari, Y. Uma author.
Title: Electromagnetic interference and electromagnetic compatibility : principles, design, simulation, and applications / L. Ashok Kumar, Uma Maheswari Y.
Description: Boca Raton : CRC Press, 2024. | Includes bibliographical references and index.
Subjects: LCSH: Electromagnetic interference. | Electromagnetic compatibility.
Classification: LCC TK7867.2 .K86 2024 (print) | LCC TK7867.2 (ebook) | DDC 621.382/24–dc23/eng/20230608
LC record available at https://lccn.loc.gov/2023012948
LC ebook record available at https://lccn.loc.gov/2023012949

ISBN: 978-1-032-41976-3 (hbk)
ISBN: 978-1-032-42478-1 (pbk)
ISBN: 978-1-003-36295-1 (ebk)

DOI: 10.1201/9781003362951

Typeset in Times
by codeMantra

Contents

Preface

The term EMI, which refers to electromagnetic interference, can cause interference to the signal and it crosses with other electronic devices. It encompasses the entire electromagnetic spectrum, which can be applicable to modern electronic devices over the frequency range of 10 kHz to 10 GHz. It can be formed through intentional or unintentional sources over the continuous or discontinuous or intermittent at a single frequency or across a broad range of frequencies.

The unintentional EMI sources include switch mode power supplies (SMPS), digital devices, brushed DC motors, high-voltage ignition systems, and fluorescent lighting. SMPS are the most common unintentional EMI sources, since they are now used almost exclusively in LED light bulbs, digital devices, and battery chargers for cell phones and laptops. EMI sources are most commonly radio frequencies and transmitters, whose emissions are often referred to as radio frequency interference (RFI). This includes AM radio, FM Radio, television, cell phones, Wi-Fi, Bluetooth, and many other fixed and mobile radio communication systems used by aviation, emergency services, police, and the military. Transients may also include EMI that can cause catastrophic damage to electronics including the electrostatic discharge, lightning, inductive kickback, and the electromagnetic pulse event.

EMC ensures that multiple electronic components can coexist within the same electromagnetic environment by not interfering with each other. EMI is one of the biggest challenges faced during the production of any electronic device. If EMI profile doesn't meet the required standards, then it is necessary to reduce the unwanted interference so that the equipment can be used in the real world. In other words, we can say that any external effects that disturb electric circuits by way of induction or radiation are called as electromagnetic interference.

MATLAB® is a registered trademark of The MathWorks, Inc. For product information, please contact:

The MathWorks, Inc.
3 Apple Hill Drive
Natick, MA 01760-2098 USA
Tel: 508-647-7000
Fax: 508-647-7001
E-mail: info@mathworks.com
Web: www.mathworks.com

Acknowledgements

I (Dr. L. Ashok Kumar) would like to thank everyone who assisted me in finishing this book. I am grateful to all of my research scholars and students who collaborate with me on projects and research. However, this book would not have been possible without the help of my family members, particularly my parents and sisters. Most importantly, I am grateful to my wife, co-author of this book, Y. Uma Maheswari, for her unwavering support throughout the writing process. All of this would not be possible without her. I'd like to thank our daughter, A.K. Sangamithra, in particular for her cheerful face and support. I'd like to dedicate this piece to her. Also, a very special thank you goes out to COVID 19 for releasing us from our regular obligations and providing us with the opportunity to rethink some aspects of our job in order to make this book a reality, as we started working during that period!!!!

The co-author (Y. Uma Maheswari) would like to thank our daughter A.K. Sangamithra for being patient and giving her all the love, time, and space to finish this work. She wishes to acknowledge her husband Aski's support and guidance in the successful completion of this book. She dedicates this book to her father Mr. S. Yuvaraj and mother Mrs. Y. Kalavathy who laid the foundation for all her successes and special thanks to her brother Y. Dhayaneswaran and to everyone of the publishing team.

About the Authors

L. Ashok Kumar was a Postdoctoral Research Fellow from San Diego State University, California. He was selected among seven scientists from India for the BHAVAN Fellowship from the Indo-US Science and Technology Forum and also, he received SYST Fellowship from DST, Govt. of India. He has 3 years of industrial experience and 22 years of academic and research experience. He has published 173 technical papers in international and national journals and presented 167 papers in national and international conferences. He has completed 26 Government of India-funded projects worth about 15 crores and currently 9 projects are in progress worth about 12 crores. He has developed 27 products, out of which 23 products have been technology transferred to industries and for government funding agencies. His PhD work on wearable electronics earned him a national award from ISTE, and he has received 26 awards at the national and international levels. He has guided 92 graduate and postgraduate projects. He has produced 10 PhD scholars, and 12 candidates are doing PhD under his supervision. He has visited many countries for institute–industry collaboration and as a keynote speaker. He has been an invited speaker in 345 programmes. Also, he has organized 102 events, including conferences, workshops, and seminars. He completed his graduate programme in Electrical and Electronics Engineering from University of Madras and his postgraduate degree from PSG College of Technology, India, and Master's in Business Administration from IGNOU, New Delhi. After completion of his graduate degree, he joined as a project engineer in Serval Paper Boards Ltd., Coimbatore (now ITC Unit, Kovai). Presently he is working as a Professor in the Department of EEE, PSG College of Technology. He is also a Certified Charted Engineer and BSI Certified ISO 500001 2008 Lead Auditor. He has authored 24 books in his areas of interest published by Springer, CRC Press, Elsevier, Nova Publishers, Cambridge University Press, Wiley, Lambert Publishing, and IGI Global. He has 11 patents, 1 design patent, and 2 copyrights to his credit and also has contributed 23 chapters in various books. He is also the Chairman of Indian Association of Energy Management Professionals and Executive Member in Institution of Engineers, Coimbatore, Vice President of Sustainability and Energy Practitioners Association, Executive Council Member in Institute of Smart Structure and Systems, Bangalore, and Associate Member in CODISSIA. He is also holding prestigious positions in various national and international forums, and he is a Fellow Member in IET (UK), Fellow Member in IETE, Fellow Member in IE, and Senior Member in IEEE. He has been appointed as Nodal Officer for Technology Enabling Centre, Anna University to strengthen industry–academic interaction and to promote technology development ecosystem.

Y. Uma Maheswari is a PhD scholar in Karpagam Academy of Higher Education. Currently working in Cognizant Technology Solutions, Coimbatore, she has around 22 years of experience in the field of PCB design, simulation software, and quality assurance engineering. She has completed her graduation programme in Electrical and Electronics Engineering, Amrita Institute of Technology, Coimbatore and her postgraduation programme in Embedded System Technologies, Anna University, Coimbatore. She has authored three books published by Elsevier, UK, Cambridge University Press, UK, and Nova Science publishers, US. She has also published many papers in national and international conferences and in reputed journals. She is also a Japanese Language Proficiency Test (JLPT) N3 Certificate holder.

ORGANIZATION OF THE BOOK

Chapter 1 describes about EMI, EMC, and their techniques. It is necessary to have knowledge on electromagnetic spectrum, time domain and frequency domain, EMI effects on humans and other electronic appliances, differential-mode current and common-mode current, EMI mitigation techniques, full-wave simulation techniques, impedance, and material and EMC testing. In this chapter, briefing of all the required is addressed.

Chapter 2 discusses about EMI/EMC qualification testing. The four key tests in an EMI/EMC qualification test phase are conducted emissions, conducted susceptibility, radiated emissions, and radiated susceptibility. The purpose of this chapter is to go over the theory and explanation of the aforementioned test and its methodologies.

Chapter 3 gives an insight into EMI filters. The filter installation is more important than the insertion loss performance for radiated frequencies. Some of the EMI/EMC filter components are installed on the power supply's input, some on the power supply's output, and some between the power supply's input and output. This chapter discusses about EMI filters, including their design and types, as well as their components and types.

Chapter 4 provides a detailed explanation about EMI/EMC design for boards with printed circuits. EMI has consequences for the human nervous system, visual system, immunological system, and so on. The purpose of this chapter is to give design principles for creating printed circuit boards for EMI and EMC investigation.

Chapter 5 goes through several simulation software used for EMI/EMC study. The clear insights into complicated systems provided by simulation software are invaluable in a wide range of fields and industries. There are numerous simulation software available for EMI, and various software simulation tools are discussed in this chapter.

Chapter 6 looks into EMI measurements. EMI reduction strategies are provided in four sections, with specific emphasis on electromagnetic shielding, and various methodologies used by the scientific community to test the shielding efficacy of a material or microwave absorber and its use in EMI reduction. Following that, EMI filters, circuit topology modifications, and spread spectrum are also discussed in this chapter.

Chapter 7 goes through how EMI works in the MATLAB environment. The programme is used to simulate electromagnetic coupling between systems and to anticipate electromagnetic radiation emission from high-voltage systems outside the system.

Chapter 8 goes through EMI and EMC simulation in the PSPICE environment. This chapter provides a comprehensive overview of how to utilize all PSpice A/D capabilities.

Chapter 9 is used for analysing EMI and EMC using the advanced design system (ADS) software. This chapter clearly showcases how to get started with the software, the simulation schematic window, component placement, connection techniques, several parameters required for simulation, and how to build and run the design to get waveforms on the output data window.

1 Introduction to Electromagnetic Interference and Electromagnetic Compatibility

1.1 INTRODUCTION

Electromagnetic interference and electromagnetic compatibility (EMI/EMC) became a problem for the first time in the 1940s and 1950s, when motor noise travelled along power lines and damaged sensitive equipment. EMI/EMC was mostly important to the military during this time and through the 1960s because it made sure that things were electromagnetically compatible (EMC). In a few well-known accidents, radar emissions accidentally set off weapons or EMI caused navigation systems to stop working. Because of this, military EMI/EMC was mostly concerned with EMC, especially within a weapons system like an aeroplane or ship. During the 1970s and 1980s, computers became more and more common. As a result, interference from computers became a big problem for broadcast TV and radio reception, as well as radio reception for emergency services. The government decided to set rules about how much electromagnetic (EM) radiation these products give off. The Federal Communications Commission (FCC) made a set of rules about how much emissions each type of computer makes and how to measure those emissions. In a similar way, governments in Europe and other places started to control the emissions from electronics and computers. But at that time, EMI/EMC control was only available for computers, computer accessories, and computer communication products. EMI/EMC became a lot more of a concern in the 1990s. In fact, many countries have put in place import controls that require products to meet EMI/EMC regulations before they can be brought into the country. All devices and equipment must be able to work together well in the electromagnetic environment (EME) as a whole. Emissions, susceptibility to emissions from other equipment, and susceptibility to electrostatic discharge are all controlled. All of these things can come from either radiated or conducted media. This control is no longer just for computers; now, any product that could possibly give off EMI or be vulnerable to other emissions must be carefully tested. It is necessary to have knowledge on electromagnetic (EM) spectrum, time domain and frequency domain, EMI Effects on humans and other electronic appliances, differential-mode (DM) current and common-mode (CM) current, EMI mitigation techniques, full-wave simulation techniques, impedance, and material and EMC testing. In this chapter, briefing of all the required is addressed.

1.2 EM SPECTRUM—TIME DOMAIN—FREQUENCY DOMAIN

1.2.1 EM RADIATION

EM radiation from electronic devices has become a common and unnoticeable fact. EM radiation was first predicted using Maxwell's equation. Their existence as EM waves was first transmitted and detected by Heinrich Hertz in 1888. According to Maxwell's theory, EM radiation, which includes both electric and magnetic fields, is the result of accelerated charges. The two fields here are at right

DOI: 10.1201/9781003362951-1

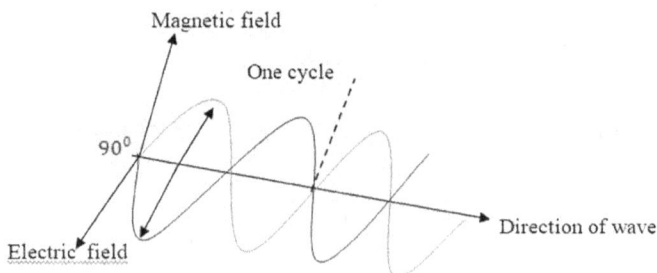

FIGURE 1.1 Electromagnetic radiations.

angles to one another. Photons, the fundamental units of EM radiation, combine the properties of particles and waves. This power travels at the speed of light, as shown in Figure 1.1.

Let us now assume that the charge is pulsating at some frequency. This will cause a space-wide electric field to oscillate, which in turn will cause a magnetic field to oscillate, and so on. The electric and magnetic fields mutually regenerate as waves that go through the void. Thus EM radiation can be described as a wave occurring simultaneously in electric and magnetic fields.

1.2.2 EM WAVES

There are two main categories of waves: mechanical and EM. It is possible to observe the propagation of mechanical waves by throwing a stone into a body of water and absorbing the resulting circular structure. EM waves are distinct from mechanical waves in that they do not need a medium in which to spread. This implies that EM waves are capable of penetrating not only air and solids but even the vacuum of space. According to Maxwell's theory, electric and magnetic fields can pair to generate EM waves. Radio waves were first created and discovered by Heinrich Hertz, who later used Maxwell's theories. Consider a pair of square copper and zinc plates spaced at least 50 feet apart. Using thin copper wires, these metal discs are linked to two shiny metal orbs. A high voltage is applied across the spheres. Small sparks are created when the air between the spheres becomes ionized as a result of the large potential difference across the sphere. Due to this jolt, EM waves were released. The wave propagated through the air and sparked the metal coil 1 m away. Therefore, this is the experimental method by which EM waves are first detected. This experiment confirmed two of Heinrich's hypotheses: first, that radio waves are kind of light, and second, that the velocity of radio waves is the same as the velocity of light. To continue, he discovered how EM waves are formed when electric and magnetic fields break apart and travel through space.

1.2.2.1 Wave Property

Wavelength (λ) and frequency (v) are used to describe the waves. The distance between the consecutive crests (or troughs) is said to be wavelength. Thus, the wavelength in the EM waves are expressed in metres (m), millimetres (1 mm = 10^{-3}m), micrometres (1 μm = 10^{-6}m), or nanometres (1 nm = 10^{-9}m), as shown in Figure 1.2.

Frequency characteristics of EM radiation (waves) are defined as the number of complete cycle per sec (cps). Hertz (Hz) is the unit of frequency (one cycle per second) named after Heinrich Hertz, as shown in Figure 1.3.

1.2.2.2 Relationship between Frequency, Wave Length, and Energy

According to the definition of frequency and wavelength, both are inversely perpendicular to each other. Also, the energy of the quantum (photon) of EM energy has a direct link with the frequency of radiation.

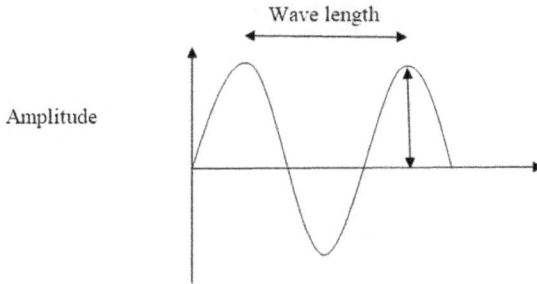

FIGURE 1.2 Wave property.

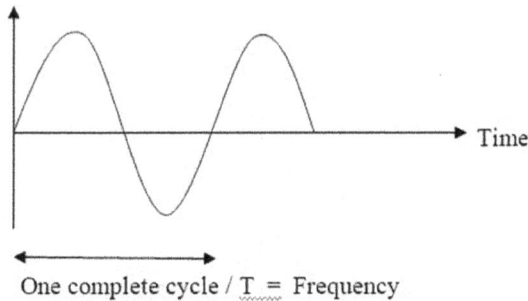

FIGURE 1.3 Wave frequency.

$$v = \frac{c}{\lambda} \rightarrow 1$$

where:

v=frequency (Hz)

λ=wave length (cm)

$c=3\times 10^{10}$ cm/sec (speed/velocity of light)

$$E = hv \rightarrow 2$$

where: h=Planck's constant (6.63×10^{-34} J/s)

From 1 and 2, we can say energy (E) as

$$E = \frac{v}{\lambda}$$

1.2.3 EM SPECTRUM

After the observations of the Hertz experimental discovery, many other different EM waves were discovered. Thus, EM radiation produces different kinds of EM waves. These EM waves differ according to their distinct wavelength and frequency properties. These distinct property groups are collectively said to be the EM spectrum.

This EM spectrum is arranged in increasing or decreasing order of frequency and wavelength and energy values. The spectrum of EM radiation is shown in Figures 1.4 and 1.5. We know frequency and energy are directly proportional, whereas wavelength and energy are inversely proportional to each other. Thus, the higher the frequency of radiation, the higher the energy, and the

Electromagnetic Spectrum

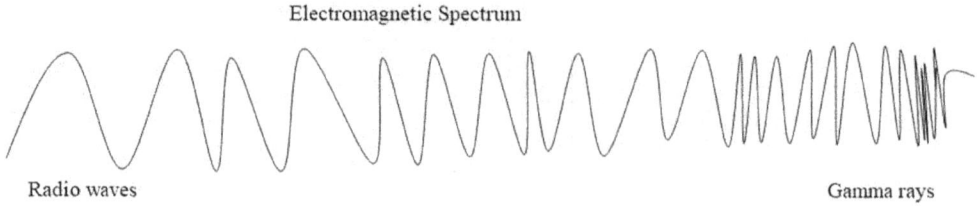

Radio waves   Gamma rays

FIGURE 1.4   Electromagnetic radiations.

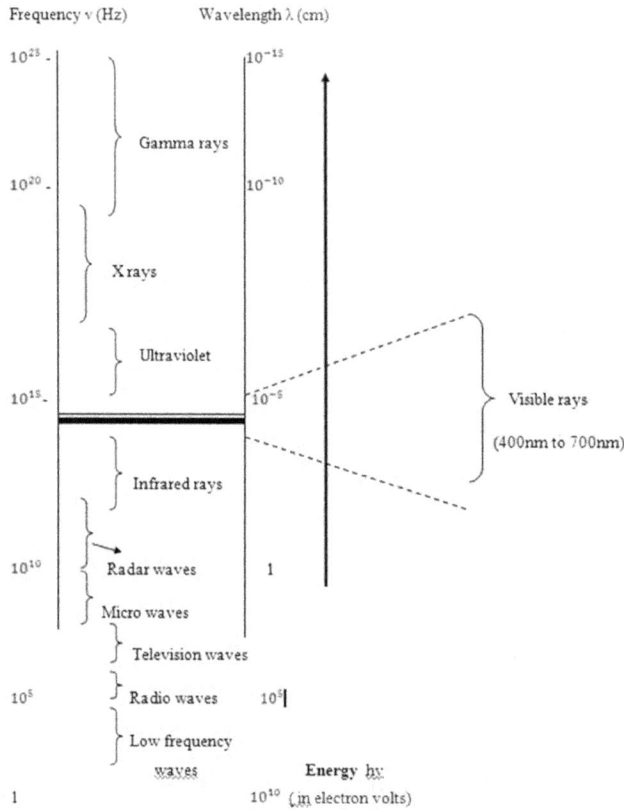

FIGURE 1.5   Spectrum of electromagnetic radiation.

longer the wavelength, the lesser the energy. This is said to be the spectrum, which has the arrangement of radio waves to cosmic ways in increasing order of its energy.

1.2.3.1   Radio Waves

This wave is produced by the oscillating electric circuit having an inductor and capacitor. The most important aspect from this concept of radio waves is that Hertz discovered or detected the EM waves. The frequency range of radio waves is from 5×10^5 to 10^9 Hz. These radio waves were at the bottom of the EM spectrum. Thus, these waves have longer wavelength and higher frequency and energy. Because of this lower frequency phenomenon, these radio waves have good penetrative capability; hence, these radio waves are used for a wide range of applications.

1.2.3.2 Radio Wave Applications

Our present situation mainly depends on wireless network/wireless communication. Examples for this are mobile phones, laptops, etc., which are collectively called cellular communication system. Radio waves play a vital role in this. They are also used to transmit radio and TV signals.

1.2.3.3 Microwaves

The increased frequency of radio waves, increases energy level and travel speedfv. These are said to be short-wavelength radio waves. Special vacuum tubes generate microwaves (called klystrons, magnetrons, and Gunn diodes). Electronic circuits can also be thought of as the source of these waves. The frequency range of microwaves are in gigahertz (1 –300 GHz). Because of its shortest wavelength, it also has a wide range of applications, especially for a larger transmission purpose.

1.2.3.4 Microwave Applications

The most widely used microwave application in the world is the microwave oven. In normal cooking, the vessel gets heated first, and then heat is transferred from the vessel to the food. In a microwave oven, the cooking technique is just a reversed version of normal cooking. The water molecules present in food material get heated first; the heat is then transferred to the food to heat it up. It is important to note that porcelain containers are used for cooking; if we use metal containers, the accumulated charge carriers result in shock, and metal can also melt. Microwaves are also used for radar system communication and long-distance telephone communications.

1.2.3.5 Infrared Waves

Infrared radiation is created by objects and molecules at high temperatures. Another name for these waves is heat waves. Almost all materials have water molecules, and these IR wavelengths can be absorbed by water. The molecules present in the material undergo rotational and vibration transitions. We can't see these infrared rays, but we can sense these rays when they heat up our body. The frequency ranges from 3×10^{14} to 3×10^{11} Hz. This region of the EM spectrum is immediately close to the long-wavelength or low-frequency (LF) end of the visible spectrum.

1.2.3.6 Infrared Wave Applications

This infrared radiation is crucial to keeping the planet's average temperature from dropping below freezing (the greenhouse effect). Some greenhouse gases absorb this light. In comparison to other rays, we can easily take photographs in foggy conditions with the help of these infrared rays. Thus, these infrared photographs are used for weather forecasting. For physical therapy, infrared lamps are used. Because electronic devices (such as semiconductor LEDs) emit infrared rays, they are widely used in household applications such as remote controls for TV sets and video recorders.

1.2.3.7 Visible Ray

Light in this range is the only entity that can be viewed by the human eye, as shown in Figure 1.6. This range of EM waves is also the most widely recognized. The range of frequencies is from

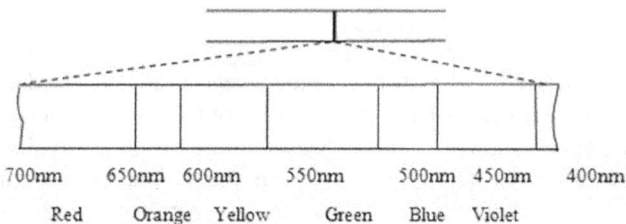

FIGURE 1.6 Visible ray spectrum.

4 to 7 Hz, while the range of wavelengths is from around 700 to 400 nm. The energy released from the oscillation of negatively charged particles within an atom is responsible for creating short wavelengths (visible rays), while electrical current is responsible for creating longer wavelengths (infrared and ultraviolet (UV)). These visible rays are said to be visible lights. Light has several properties, including reflection, refraction, dispersion, absorption, polarization, and the photoelectric effect. The dispersion of light property states that the refractive indices differ for different types of wavelengths; this phenomenon is demonstrated by passing white light through a prism, which produces visible rays and a broad range of the rainbow spectrum.

1.2.3.8   Visible Ray Applications

A close examination of this spectrum reveals some dark lines known as absorption lines. The absorption property is used to study the external shells of atoms, such as their molecular structure and electron arrangements. It gives a sensation of vision. Remote sensing by using visible lights is implemented in laser altimetry. National Aeronautics and Space Administration (NASA) uses GLAS (Geoscience Laser Altimeter System) to examine the amount of water stored as ice particles on Earth, and this technology is also used to give height measurements and cloud characteristics in a detailed view.

1.2.3.9   UV Rays

The Sun is the major source of these UV rays, or, in other words, these UV rays are produced by very hot bodies. Light is often created when electrons in an atom's inner shell drop from a higher to a lower energy level. These UV rays are most harmful for the melanin in our human skin because they cause skin tanning. We most likely have our production layer in the atmosphere, known as ozone layer, which has the ability to absorb UV rays. Unfortunately, scientists discovered that the presence of holes in the ozone layer results in chlorofluorocarbons (CFCs) gas depletion, which is critical for our future. These UV rays are invisible to the human eye, but some insects can detect them. The wave length is shorter than visible spectra having the ranges of about 400 nm (4×10^{-7} m) to 0.6 nm (6×10^{-10} m).

UV rays received from the Sun have three categories, namely, UV-A, UV-B, and UV-C. UV-C is the most harmful, and 100% of it is fully absorbed by atmosphere. UV-B is dangerous to DNA of living organisms, and 95% of it is absorbed by the ozone layer. Based on the wavelength of UV spectra, it is classified as near (NUV), middle (MUV), far (FUV), and extreme (EUV). These UV rays were first detected by Johann Ritter in 1801. He observed the spectrum of visible light made to fall on white paper. He observed that the paper turned black on the blue colour side than on the red colour. He then exposed the paper to more light beyond violet, and the paper got burnt. This is how UV rays are discovered.

1.2.3.10   UV Ray Applications

Laser-assisted in situ keratomileusis (LASIK) eye surgery, for example, benefits from UV radiation's ability to be focused into narrow beams. Seeing this as a source of brilliant light allows us to investigate planets, stars, and galaxy properties, and it also allows us to detect new and older stars in the atmosphere using UV imagery.

1.2.3.11   X-Rays

In 1895, a German physicist named Wilhelm Conrad Roentgen first detected and reported x-rays. According to the results of his experiment, since bones are more solid than skin and absorb more x-rays, their shadows appear on x-ray images. X-rays are more commonly discussed in scientific circles in terms of their energy than their wavelength. Since their wavelengths are significantly shorter and their energy much higher than UV light, some portions of x-rays are not even bigger than a single atom of some elements, as their wavelength is between 0.03 and 3 nm. The Sun's corona is significantly hotter and emits primarily x-rays. To generate x-rays, high-energy electrons are often directed at a metal target.

X-rays originate from things that are millions of degrees celsius, like pulsars, the accretion discs of black holes, and the leftovers of galactic supernovae. Electrons in Earth's magnetosphere are energized when auroras, created by fast-moving charged particles from the Sun, hit the planet. These x-ray emissions are caused by electrons that travel with Earth's magnetic field and crash into the ionosphere. The tremendous energy and penetrating quality of x-rays means that they are not reflected back.

1.2.3.12   X-Ray Applications

Although x-rays are most commonly associated with their usage in examining bones, they have also found utility in the diagnosis and treatment of several forms of cancer due to their capacity to damage and kill live organisms and tissues. Other than this, the NASA Mars Exploration Rover employed x-rays to analyse the spectral signature of elements like zinc and nickel. The Alpha Proton X-ray Spectrometer (APXS) can be used to analyse the structure and composition of metals, which are much heavier than other elements. X-rays can also be used to observe the aurora on Earth.

1.2.3.13   Gamma Rays

Gamma rays are the shortest wavelength EM waves and carry the highest energy. They are created at the most intense and powerful places, like the centres of pulsars and stars, the aftermaths of supernova explosions, and the space surrounding black holes. Simply we can say that gamma rays are produced when the radioactive atom decays a ray with high amount of energy. Gamma rays on Earth were generated by nuclear explosions (fission and fusion), lightning, and the radioactive decay with less dramatic activity.

Let us consider the example of the nuclear fission process. When the uranium atom is bombarded with the slow-moving neutron, uranium the radioactive atom decays to produce thorium and positively charged rays called alpha rays. A new atom, that is, thorium, has to rearrange its protons and neutrons to reach a lower energy state. The transition from a higher to lower energy state results in the emission of gamma rays, which carry a tremendous amount of energy.

Then, when the thorium atom is bombarded with a slow-moving neutron again it produces another new atom with negatively charged rays called beta rays. A newly produced atom will also reach its lower energy state so that we will get gamma rays. This is how a radioactive atom decays and produces gamma rays.

Gamma rays cannot be captured and reflected by mirrors or prism like visible rays. These gamma rays can pass through the crystal and collide with it. This is said to be Compton scattering process shown in Figure 1.7. Here, the gamma rays collide with the electron, depleting its energy; the resulting collisions produce charged, motionless particles, which are picked up by the sensor.

Gamma rays released by atomic nuclei on mercury's surface after being struck by cosmic rays can be measured by the mercury surface, space environment, geochemistry, and ranging (MESSENGER) and gamma ray spectrometer (GRS).

1.2.3.14   Gamma Ray Applications

In the medical field, gamma rays are used to kill cancer cells. New elements on other planets are discovered with the help of gamma rays. Magnesium, oxygen, hydrogen, silicon, iron, sodium,

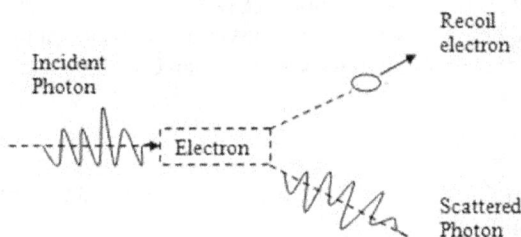

FIGURE 1.7   Gamma rays.

calcium, and titanium are all very significant in the Earth's geology, and the data obtained by the GRS helps scientists discover and estimate these elements and their abundances.

1.2.4 EMI/EM Spectrum

Electric circuits can be disrupted by electromagnetic interference (EMI) if they are subjected to either EM induction or EM radiation from sources outside the circuit. Interference from one electrical or electronic system to another due to the EM fields produced by its operation is known as EMI. With the use of more sophisticated and complex electronic devices, as well as the increasing use of the EM spectrum, EMI issues have become a significant factor. In other words, we can say that whenever a device makes use of electronic circuitry, it can be influenced by EMI. Adopting EMC is becoming a threat to the control of EMI.

EMI can happen anywhere in the EM spectrum, but it is most common at radio and microwave frequencies. Interference from EMI disrupts the operation of other electronics. EM emissions are produced by any equipment with rapidly varying electrical currents; these emissions can "interfere" with those produced by other objects. EM distortion occurs when one EMI interacts with another. EM waves can cause interference and disturbance even if they are not in phase. When radio frequencies are switched, this interference may be heard, and when the signal is distorted on television, the picture becomes warped. Thus, EMI is now also known as "radio-frequency interference" when it occurs in the radio-frequency (RF) range.

1.3 SPECTRUM ANALYSIS: TIME DOMAIN AND FREQUENCY DOMAIN

Time domains and frequency domains are commonly used to represent the properties of EMI signals. As an example, when converting signals from one domain to another, Fourier analysis is often employed. To achieve optimal EMC, it is necessary to consider the electrical dimensions of an electric circuit or EM radiating structure. Electrical dimensions are a more important consideration in design and construction than physical dimensions of a radiation structure, such as an antenna. Wavelengths are important electrical dimensions and are more significant in determining EM energy.

1.3.1 Time-Domain Analysis

Time becomes the independent variable denoted by "t" in the time domain, where we examine waves in amplitude versus time along the Y and X axes, respectively. Most events in the real world occur only in the time domain. An oscilloscope is used for time-domain analysis.

1.3.2 Oscilloscope

For the purposes of studying and fixing broken circuits, it is an invaluable tool. They measure factors like amplitude, frequency, and transient signals that a multimeter can't. It displays the signal wave with respect to time versus amplitude, or, in other words, in the time domain. Here we output a frequency signal from the oscilloscope waveform generator to channel one of the oscilloscopes, and while visualizing that, we observe as a sinusoidal waveform appears on the oscilloscope. Basically, the higher the frequency, the more waves we see in the same span on our oscilloscope.

By examining two different frequencies in the oscilloscope, we will get the following figures as output, where Frequency = 1/ time.

For case 1: Given frequency as 10 MHz, we get the time period for one complete cycle as 100 ns, as shown in Figure 1.8.

For case 2: Given frequency as 20 MHz, we get the time period as 50 ns, as shown in Figure 1.8.

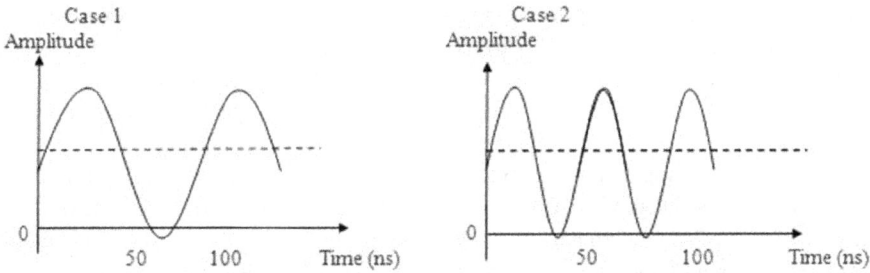

FIGURE 1.8 Case 1: 10 MHz frequency waveform. Case 2: 20 MHz frequency waveform.

The voltage between a pair of vertical deflection plates in a cathode-ray oscilloscope changes in response to the electrical signal being examined, causing the electron beam to rise and fall as it scans a phosphorescent screen. Current digital scopes (oscilloscopes) sample the signal and display the wave shape on the screen by acquiring measurements using an analogue to digital converter.

1.3.3 DISADVANTAGES OVER THE SIGNAL ANALYSER

In an oscilloscope, it is not possible to examine the signal when there's noise or harmonics. It is possible to output sine wave. Signals in the real world, on the other hand, are not pure sinusoidal; they are a combination or sum of signals with different frequencies, making them easier to examine in a signal analyser.

1.3.4 FREQUENCY-DOMAIN ANALYSIS OF SPECTRA

In the frequency domain, frequency becomes the independent variable denoted by "f" in Hz; it is possible to examine signals in amplitude versus frequency along the Y and X axes, respectively. With the help of this frequency domain, it is possible to analyse the spectrum of waves in a wide range, where frequency-domain analysis is carried out with the help of a signal analyser, which we can also refer to as a "spectrum analyser" because of its wide range of applications.

1.3.5 SIGNAL ANALYSER

In the context of wave data analysis, it is employed in the amplitude and frequency domains. As demonstrated in Figure 1.9, a spectrum analyser shows the amplitude of a signal as it varies with frequency, but it does not quantify the signal's phase. Some characteristics, such as those involved in digital communications like LTE (long-term evolution) on our cell phones, cannot be measured without first determining the phase of the incoming signal. In contrast, a signal analyser only looks

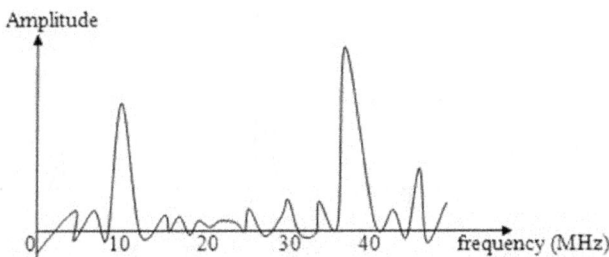

FIGURE 1.9 Demonstration of waveform in signal analyser.

at one frequency at a time to determine its amplitude and phase. A signal analyser may take an RF signal transmitted from your device and convert it to a digital signal, which can then be displayed in a variety of ways, including as a constellation plot, a spectrum, a power plot, etc. You can't measure the amplitude, frequency, or phase of an interesting signal using a spectrum analyser, but you can with a signal analyser.

Let's say you're designing something that can only function in one frequency range and can't emit in any of the others. Then, we need to figure out what additional frequencies the other signal is generating that are interfering with the signal we need to receive from the gadget. And here is where signal analysers come in; they take a variety of sinusoidal signals and break them down into their component frequencies so you can see them all in one place. We may then apply filters to isolate the most obnoxious parts of the signal in this way, as shown in Figure 1.9.

A signal analyser is primarily used to test the strengths of RF and audio signals, among others. It is possible to examine the signal and circuitry of a device with the help of this signal analyser. Time-domain spectrum and signal analysers may perform a wide range of tests, from checking for harmonic distortion in transmitters to analysing wireless networks, creating spectral masks to display the boundaries of undesired emissions, and checking the performance of RF power amplifiers. Signal analysers are so useful that they are almost standard equipment in all research and development centres for engineers. In-depth spectrum analysis and visibility into the frequency composition of your signals are both made possible by modern signal analysers.

The signal analysers have lower noise floors, which means they are much more sensitive to frequency components, and they are also better at showing the low-level signals of the waveform spectra. Signal analysers also have better dynamic range (i.e., spurious free dynamic range, or SFDR) than oscilloscopes.

1.3.6 Time-Domain Analysis versus Frequency-Domain Analysis

Harmonics are the result of a waveform's summation, and they are typically present when there is a disturbance of some kind. The time- and frequency-domain representation is displayed in Figure 1.10. Using frequency-domain analysis, we can quickly identify spurious signals, which we can then filter out using appropriate circuitry to obtain the desired waveform spectra.

1.3.7 Propagation of Time

EM phenomena are truly a distributed parameter process. The distributed parameter spells out the properties of the structure such as capacitance and inductance; these are distributed throughout space rather than being lumped parameters at discrete points. The EM field's distributed nature can be ignored, when constructing electric circuit models with lumped parameters.

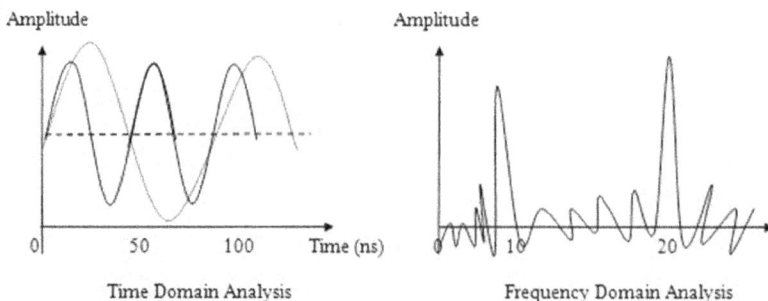

FIGURE 1.10 Time- and frequency-domain analysis.

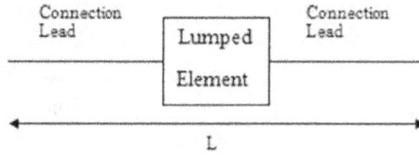

FIGURE 1.11 Propagation of time.

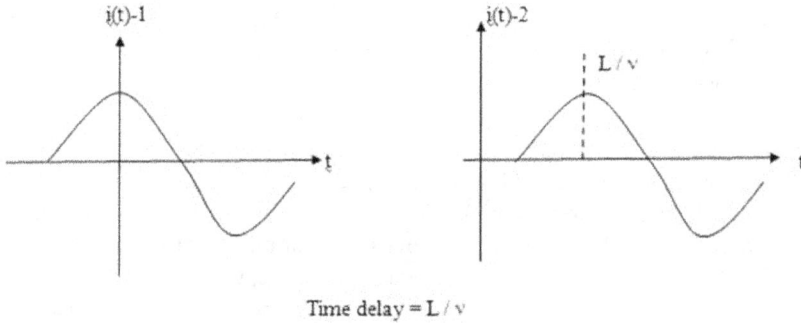

FIGURE 1.12 Time propagation delay.

For example, consider connecting the resistor Figure 1.11 (lumped circuit element) with lead wire connection. A sine wave is passed from one end to another end of the lead wire through a lumped element as the resister. It is said that the element with lead connection results with no consequence and most probably their effects may get ignored. It is observed that the time lag of the input wave forms at the output side (Figure 1.12). This happens because of the surrounding medium of air.

Actually, the current is the wave propagation property of velocity (v). The medium surrounding the lead is air; it has the velocity of propagation approximately equal to 3×10^8 m/s. Because of this finite time delay is required for current wave to transit. In today's world, this propagation time delay becomes one of the major constraints. For example, today, clock speed in personal computers is about 3 GHz and the time transitions are about 100–500 ps, whereas in the 1980s clock speed was about 10 MHz and transitions were about 20 ns. Therefore, while designing a printed circuit board (PCB), these are certain points most important for consideration.

1.3.8 RF and Microwave Frequency Ranges

Every aspect of our lives has been altered by the lightning-fast development of electronics, electro-optics, and computer science. They have also paved the way for a flurry of activity in the fields of medical device research and development that has never been seen before. In particular, developments in RF and microwave technology and computation techniques, among others, have cleared the way for promising new therapeutic and diagnostic approaches. Researchers are currently exploring the therapeutic and diagnostic potential of a wide range of frequencies, from RF as low as 400 kHz to microwave frequency (MW) as high as 10 GHz, in fields as diverse as cardiology, urology, surgery, ophthalmology, cancer therapy, and others.

Researchers have voiced worries over the potential harmful consequences of EM radiation on living organisms, especially at higher frequencies. Different types of waves and signals, such as digital transmissions used in radio, television, and mobile phone networks, must be taken into account. It now takes a high level of competence to set safety guidelines or standards and to take sufficient measurements, as the sector has got increasingly complex.

Focus solely on the consequences and uses of RF and microwave fields should be happening. This includes frequencies up to and including 10 GHz. Though the effects at RF and microwaves MW are distinct, this option is still suitable. Effects at low and extremely low frequencies (LF and ELF) are not included because they do not involve radiation. Ionizing radiation, such as UV and x-rays, is also blocked off. The RF and MW spectrum discussed here is sometimes called the "non-ionizing" range.

The RF and MW ranges are distinguished by a phenomenon known as radiation. As is often known, the smaller a structure is in relation to the wavelength, the worse it radiates. Power distribution frequencies of 50 and 60 Hz, for instance, have wavelengths of 6,000 and 5,000 km, respectively, which are vast compared to the objects we use every day. Actually, a structure needs to be sufficiently massive in relation to the wavelength in order to radiate effectively. Radiation, antennae, and the far and near fields must all be studied.

Conversely, at RF and MW ranges, both electric (E) and magnetic (H) fields coexist; if there is an electric field, there is also a coupled magnetic field, and vice versa. The value of the other can be deduced from the knowledge of the former: The famous Maxwell's equations serve as a bridge between them. At a later point in this book, we'll be able to disentangle the biological effects of one field from those of the other. However, we must keep in mind that we are thinking about the full field, which is the EM field, as a general instance. Therefore, we are not taking into account the penetration of alternating current (AC) and LF electric or magnetic fields into living tissue.

If we limit our attention to the RF/microwave range, the interaction of electric and magnetic fields with living matter can be classified as non-ionizing radiation biological impacts. It is important to take note of the fact that when the materials and systems of interest are subjected to a source of sinusoidal fields, we choose to characterize the processes in what is called the frequency domain. Changing the source's frequency allows us to examine its qualities over a broad or small frequency range. A sinusoidal source is not "physical," as it began existing infinitely in the past and will continue to exist infinitely into the future, making the frequency domain an unphysical concept. Furthermore, the frequency-domain representation as a whole implies non-physical complex variables (real and imaginary parts). Even so, many sources are (nearly) monochromatic; therefore, the frequency-domain description is very helpful.

However, one must work in the so-called time domain, in which events are characterized as a function of time, in order to explore the actual effect of physical causes, which is real and physically quantifiable. Time-domain operations may be more challenging than frequency-domain ones. Typically, the frequency domain is used to study the effects of RF and MW fluxes on living tissues, with sinusoidal sources assumed. However, time-domain analyses and measurements may be required for modern numerical signals used in telecommunications, television, and frequency-modulated (FM) radio.

Microwaves have the following notable features: They span the frequency range where the wavelength is on the order of the size of everyday things, from metres down to decimetres, centimetres, and millimetres (within reason). One might therefore speculate as to whether or not such wavelengths can trigger resonance in biological tissues and systems.

1.3.9 FIELDS

EM fields and their interactions with biological tissues call for a solid physics and math foundation. It is the association between a spatial location and a physical phenomenon that is called a field. Example: A room's temperature is a temperature field made up of the individual temperature readings taken at various locations within the space. The same may be said, for instance, about the distribution of body heat. Although unseen, the field is real, allowing us to imagine things like isothermal or constant-temperature surfaces. There are various types of fields.

As a first point, fields might be either stationary or dynamic. Taking the previously mentioned temperature field as an example, the room can be heated or cooled, making the temperature field time-dependent. The distribution of body heat can also be affected by a number of factors, both internal and external, that the human body may be subjected to. In this situation, the forms of the isothermal surfaces will evolve with the passage of time.

Second, the field's characteristics may be such that only a single parameter, like magnitude, is connected to it. Then, the field is defined as a scalar. An example of an internal scalar field is the temperature of a room or a human body. If the structure is specified in great detail or if the observer demands a precise description of the field in space, it becomes clear that mapping the field may require skill and considerable memory space. When the field is scalar and static, this is true even in the simplest of situations.

However, in a vector field, a vector represents both the magnitude and the direction of the physical quantity of interest at locations in space, and this vector field may be either static or time-dependent. Work in visualization is already required when plotting a static scalar field, that is, one quantity in points in space. However, considerably more care must be used when drawing a time-dependent vector field, which consists of three time-varying values at discrete locations. Sets of direction lines (sometimes called stream lines or flux lines) characterize a vector field. The direction line is the curve that is built so that the field is tangential to the curve at all places along the curve.

1.3.10 ELECTRIC FIELD AND FLUX DENSITY

The electric field E is derived from Coulomb's law, which expresses the interaction between two electric point charges. Experimentally, it has been shown that:

- A charge produces a field of force, and charges of opposing polarity attract one another while charges of the same polarity repel one another.
- Force = product of charges.
- A vectorial force field is manifested because the force is directed down the line connecting the charges.
- When charges are close together, they exert a greater force, and when the charges are spread out, the force is affected by the electric characteristics of the medium in which they are located.
- If it occurs at regular intervals, it is typically represented by the symbol q and is scaled in coulombs.
- Linear spatial distribution is possible. The unit of measurement is the Coulomb per metre, with the symbol rl commonly used (cm^{-1}).
- This material has the ability to be spread out in all directions across a surface (material of not). Typically, it is indicated by the symbol rs and quantified in coulombs per square metre in such cases (cm^{-2}).
- It can also be depicted in volume and typically indicated by the symbol r and quantified in coulombs/m³ (cm^{-3}).

An electric field causes a material to become polarized and the direction and magnitude of this polarization is known as the polarization vector. This is because of the material's dielectric characteristics, which are triggered whenever electric dipoles are introduced into, or altered inside, the material. As a result, the polarization is the electric dipole moment per unit volume, expressed in coulombs per square metre.

In a dielectric substance, the total electric field is equal to the sum of the applied electric field and the induced electric field due to polarization. A perfect electric conductor, for instance, is an equipotential substance. If all locations in a material have the same electric potential, then the electric field must be zero and there must be no electric charges in the material. Applying a field to a perfect electric conductor causes that field to permeate the entire substance. To have an electric field that vanishes everywhere in the material, the material must generate an induced electric field with a magnitude smaller than the applied field. It is necessary to consider the boundary conditions and the geometry of the issue in order to compute the induced field, both of which might add complexity to the calculation. The human body, for instance, generates an induced electric field in response to an externally applied electric field, with the result that the combined value of the applied and generated fields meets the boundary requirements at the body's surface. When you add up the fields that are applied to and caused by the body, you get the total field. This term is completely inclusive and can be used to refer to any material, including biological ones. As a matter of fact, it is valid for those items in which

- An isotropic material is one in which the polarization vector and the electric field vector do not point in the same direction.
- Polarization can be slowed down in lossy materials relative to the change in electric field. As a universal trait, the lossy nature of physical materials applies to everything. However, it's ignored when the losses are negligible, which isn't always the case in biological tissues.
- Having a polarization that is independent of the electric field indicates that the material is non-linear.

1.3.11 EM FIELD

According to the laws of electrostatics, the electric field is appropriately estimated in DC conditions. This is also true for very low and low frequencies, though to a lesser extent as frequency increases. In DC circumstances, the magnetic field is also determined by applying the rules of magnetostatics. The same is true at extremely low and low frequencies; however, as frequency increases, approximation quality decreases. The geometric, electrical, and magnetic features of the issue determine the values of the frequency at which the approximation no longer holds.

It becomes impossible to isolate electric and magnetic fields as frequency increases. One field's existence implies the other's existence. No single discipline can be seen as producing the results seen in another. Maxwell's equations describe the universal nature of this relationship. The British physicist J. C. Maxwell formalized the rules of electromagnetics about 1880; however, he did not create the equations we use today.

He made two critical contributions on this occasion:

- It was proposed that the previous rules, which were mostly derived from the experimental observations of Gauss, Ampere, Lenz, and others, should be accepted but that they should be viewed as a system of equations.
- The other was to mention that there was a missing term from these laws as a whole. It was decided to add the displacement current to the law, which had previously been absent.

A connection between electricity and magnetism results from the first contribution linking the electric and magnetic fields together. Because of the second input, there is now a previously unrecognized shape that electric current can take. (The electric current passing through a given area is obtained by the surface integrating the corresponding current density.)

1. A convection current density, due, for instance, to a density of electric charges moving in vacuum, is described by

$$\bar{J}_{conv} = \varphi \bar{v}$$

2. A conduction current density, due to the conductivity of some materials, is described by a relationship based on the electric current in the conductor

$$\bar{J}_{cond} = \sigma \bar{v}$$

3. A displacement current density, due to the time variation of the electric field, is equal to the time derivative of the displacement field (the electric current density)

$$\bar{J}_d = \frac{\partial \bar{D}}{\partial t}$$

4. It is the displacement current that maintains continuity between the conduction current cycling in the wire and the electric phenomenon occurring between the two capacitor plates when an AC source is used to charge the capacitor, which consists of two parallel plates in vacuum. Integration of the displacement current density across the plate area and time, that is, multiplication by jw, yields the same value as the conduction current in the wire.

5. A source current density: Both the vector and scalar equations that make up Maxwell's equations are first order. These rules are typically viewed as a broader application of earlier laws defining electricity and magnetism. It's generally accompanied by the remark that no EM phenomenon has been identified so far that doesn't agree with Maxwell's equations. However, it has been shown that, with the observer's speed of light held constant, Maxwell's equations can be obtained from a relativistic translation of the Coulomb law. In the simple situation of linear motion, when special relativity may be used, the proof is rather straightforward. But a good deal of vector calculus is required. However, when rotation is included, general relativity must be taken into account, which greatly complicates the calculation. We might therefore draw the conclusion that re-evaluating Coulomb's law, relativity, and the constancy of the speed of light would be necessary if an EM phenomenon were to be discovered that did not conform to Maxwell's equations.

Boundary conditions are required since Maxwell's equations do not work there. Both the time and frequency domains can be used to create them. Both differential and integral forms are acceptable. For that reason, they're adaptable in four distinct ways. When describing EM processes in continuous media, the differential forms are used, either in the time or frequency domain. They hold true at each and every location along a continuous medium subjected to EM fields without fail. However, integral forms can be used to link average quantities and can be applied to volumes, surfaces, and contour lines.

In most cases, they are utilized to examine any kind of circuit, mobile or not. Measurements of physical quantities are evaluated using time-domain descriptions. For example, they are required for the study of some optoelectronic phenomena in which the effect does not immediately follow the cause. In addition, they are required for the study of broadband phenomena. They hold true regardless of the rate at which they are applied. When looking into occurrences that are monochromatic or have a restricted band, a frequency-domain description is invaluable. Although they may have a true physical component, they are not actual, physically quantifiable quantities. They can be used whenever you like. It may be quite challenging to solve Maxwell's equations under generic conditions. The system of equations that needs to be solved consists of 16 equations once the vector equations are decomposed into scalar ones. There are also 16 quantities that are not known: the 15 scalar quantities, the scalar charge density, and the 5 vector quantities (electric field, electric flux density, magnetic field, magnetic flux density, and current density). Fortunately, there are numerous circumstances where the system is far simpler, for example, as a result of symmetry, simplicity of the materials involved, slowly varying phenomena, or low frequency. Particularly, some symmetry can be very helpful in making calculations easier to handle.

EM phenomena are typically quite slow in biological tissues, especially when compared to the enormously diverse range of phenomena that must be assessed in the fields of physics and engineering. Most biological reactions take far longer than the shortest response time, which is on the order of 10^{-4}s. Therefore, Maxwell's equations are rarely employed to assess the impact of EM fields on physiological tissues and systems. When compared to biological tissues and systems, such as the human body as a whole, the wavelength in vacuum is quite enormous at RF frequencies. In a vacuum, the wavelength at 1 MHz is 300 m long. Conversely, at microwave frequencies, the period of oscillation is tiny, corresponding to 10^{-9} s at 1 GHz, which is significantly smaller than the fastest biological responses. This means that in most circumstances, Maxwell's equations are not required for the evaluation of biological effects in live tissues and systems caused by EM stimulation, at RF as well as at microwaves. Therefore, quasi-static techniques work well enough in biological material, and electric and magnetic forces are typically considered independently, even at RF/microwave frequencies.

1.3.12 EM Wave

Around the year 1888, Hertz theoretically and empirically discovered that the special coupling between the electric and magnetic fields due to the particular form of Maxwell's vector equations incorporated the notion of propagation of EM waves inside its structure. The motion of waves is the outcome of the interaction between the spatial and temporal components of the electric and magnetic fields. The speed of light, which is finite, was the speed at which the waves were moving. The term "wireless communications" wasn't coined until 1895, when Marconi began testing transmissions at increasing distances.

All the field components and associated physical quantities, such as current and charge densities, have a z dependence expressed as the factor in a cylindrical coordinate system or a r dependence expressed as the factor in a spherical coordinate system, and this is the mathematical definition of propagation. An EM wave is a collection of fields. Thus, there is a tight relationship between the words "propagation" and "wave."

The equations formulated by Maxwell have a first order of complexity. If we remove one field from these equations, we get the wave equation or Helmholtz equation, which describes the remaining field in the second order. It may be shown that a moving EM wave is the universal solution to the wave equation. The wave equation can be satisfied by a wide variety of waves in even the simplest medium, vacuum, with transverse and/or longitudinal components of the fields. Whether the propagation medium is infinite or finite has an effect on the wave's field structure. Both a coaxial wire and a waveguide are limited in width. The term "free space" is commonly used to describe a medium with no physical boundaries. Reflection occurs at the boundaries of a limited medium. Refraction can occur at edges and corners.

However, the propagating field's orientations may change spatially depending on the type of wave and the medium. This polarization is the defining feature of this difference. It's typically associated with electric fields, but magnetic fields could be involved as well. However, in this case, it is necessary to state this explicitly.

As a matter of first principles, we all know what an EM wave looks like at its most fundamental level. It possesses a magnetic field perpendicular in space to the electric field, both being perpendicular in space to the direction of transmission. The fields are stationary neither in space nor in time: the wave displaces itself in space as a function of time at constant speed, and the field amplitudes change as a function of time, sinusoidally in a monochromatic scenario. In this case, the polarization is said to be linear since the fields are always aligned in a straight line. It is called circular polarization when the fields spin with constant amplitude in a plane that is perpendicular to the direction of propagation. Circular polarization may be right- or left-handed. The polarization is said to be elliptic if its amplitude and direction both change as it travels. Many RF and microwave sources actually use linear polarization. That doesn't mean that's always the case, though. Circular polarization is often employed in technical applications.

Geometrical considerations, such as the geometry of the wave's constant-phase surface, can be used to categorize waves. The wave can be planar, cylindrical, or spherical, depending on whether the surface it's travelling over is flat, round, or oblate. When the field's amplitude is the same at all points along the surface of constant phase, the wave is uniform; otherwise, the wave is non-uniform. A uniform plane wave is one in which the amplitude of the field is always the same over all of its surfaces, or planes of constant phase. Using a uniform plane wave as an example, we can see that the aforementioned structure is the most basic type of wave.

The number of field components is another criterion for grouping waves into distinct categories. Transverse EM waves are those that lack any longitudinal components in their electric or magnetic fields and instead solely have transverse field components (TEM). It is not possible to find more than four parts to a transverse electromagnetic (TEM) wave. When there is only one longitudinal component to a wave, it is referred to it as transverse electric (TE) if the longitudinal component is magnetic and transverse magnetic (TM) if it is electric. A total of five parts make up both TE and TM waves. There are six parts to the most fundamental wave. It is a linear combination of TEM and transverse mechanical (TM) waves, with or without boundary conditions.

1.3.13 ANTENNAE AND NEAR FIELD

From the reactive near-field limit out to a distance of 2D2/l, where D is the maximum dimension of the antenna, lies the radiating near-field (Fresnel) zone. D must also be large compared to the wavelength for this formula to hold. This field region might not be present if the antenna's largest overall dimension is much smaller than the wavelength. Because of its resemblance to the Fresnel zone in optical theory, the near field of an antenna pointed at infinity is sometimes called the Fresnel region. The radial field component may be sizable and the field pattern is generally a function of the radial distance in this region (Figure 1.13).

At distances greater than 2D2/l from the antenna, the far-field (Fraunhofer) area is assumed to exist. The maximum allowable phase error for this criterion is p/8. Fraunhofer is mentioned because the situation is analogous to optical terminology: the antenna is focused to infinity. Here, the fields are almost entirely perpendicular to the direction of the observer, and the angular distribution does not change with changes in the radial position of the instrument.

As a general rule, in RF and microwave communications, long-range circumstances are of primary interest. As a result, approximate solutions can be obtained for the fields, allowing for closed-form solutions to be obtained, which are very useful. It is crucial to differentiate between near-field and far-field exposures when assessing biological impacts, yet this is not always the case. Typically, far-field conditions are used for assessing the risks of RF/microwave exposure to humans or animals. The antenna of a mobile phone is so small in comparison to the wavelength that the user's head is in the far field of the antenna, and transmitting stations are typically located far enough away from residential and commercial areas to prevent interference.

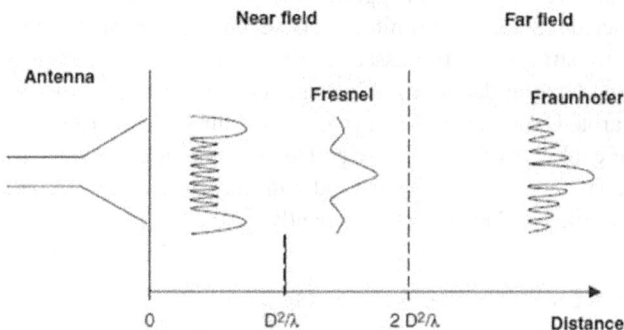

FIGURE 1.13   Angular distribution of fields around an antenna in the zone of near fields (Fresnel zone) and far fields (Fraunhofer zone).

TV and FM radio transmitters and also receivers account for near-field circumstances. Medical uses for antenna implantation in living tissue and organisms are possible. Because of this, there is a need to precisely regulate the temperature of the affected tissue; antenna placement is crucial when microwave power is used to heat tissue. As a result, there is a wide range of antenna types and forms to choose from, each best suited to a particular task. Choosing the right tool for the job is a major challenge. On the other hand, the tissue's absorption qualities have a major impact on the efficiency of EM energy transfer. It's also dependent upon how often it occurs. Millimetre-wave sources, for instance, provide outcomes that are analogous to those of infrared wavelengths.

1.4   EMI EFFECTS ON HUMANS AND OTHER ELECTRONIC APPLIANCES

EMI is a key factor in the operation and performance of a product. Hence, its measurement is essential for ensuring its quality. Unwanted EM emission from a product, often known as EMI, can cause disruptions in electronic systems. The evaluation of EMI is known as EMC testing. Whenever one system breaks down or stops working entirely, it always causes EMI to spread throughout the surrounding area. It's crucial to detect EMI in a product because it reduces the product's useful life and can even destroy it. Also, the human nervous system, eyes, immune system, etc. are all vulnerable to EMI's damaging impacts. In order to comprehend and quantify the EMC of the device, its performance impact from these inevitable interferences must be assessed accurately. In order to make the system usable in the real world, it may be essential to take steps to mitigate the effects of EMI if the system's EMI profile does not satisfy the required criteria. Although EMI and EMC play a crucial role in electronic engineering; the field has received comparatively less attention than other areas of electronic engineering and product development. This book provides a high-level summary of EMI, its causes, EMC, the EMI standards that must be met by any given device, the consequences of EMI on electrical appliances, and the impacts of EMI on people.

Radio-frequency interference (RFI) is another name for EMI. Although EMI refers to the unintended frequency range of all electrical disturbances, and RFI is typically understood to refer to a smaller subset of all electrical noises absorbed by an electrical system, the two terms are often used interchangeably. Any source of EM radiation or conduction can cause RFI in an electrical circuit. In the worst case, this could cause malfunctions and decrease the efficiency of the vulnerable circuits.

In short, EMI is dangerous because it disturbs nearby equipment like radios, televisions, navigation systems, radar systems, switch-mode power supplies (SMPS), and other electronic communication devices from operating in a common EME.

Radiated EMI can have both immediate and far-reaching effects on the efficiency of electronic equipment and systems. Direct effects of EMI on system performance include, but are not limited to, false targets and missed targets in radar display systems, incorrect navigation data or landing system errors in aircraft, lost or garbled messages in communications systems, false commands to a missile or electro-explosive device, and triggering the heart pacemaker's demand-mode operation. Its indirect effects include, but are not limited to: false alerts in an air-defence system due to false targets; surprise enemy attacks due to missed targets; mid-air collisions due to navigation errors; aircraft crashes during landing due to altitude or glide-slope errors; ineffective control of riots or fires due to lost or garbled emergency fire or police communications; accidental launching of missiles or detonation of explosives due to lost or garbled communications; and mid-air collisions due to navigation errors. They can happen again, and with the consecutive growth in EMI generators and receivers, they are likely to occur more frequently.

1.4.1   EMI's EFFECTS ON ELECTRONIC APPLIANCES

- One of the most difficult parts is figuring out the switching frequency and switching time of the power switches. Providing an optimized system right from the start of the design process necessitates in-depth familiarity with EMC and power electronics components.

- Reduced losses and EMI noise are the outcomes of a well-designed power converter layout.
- An understanding of EMI issues in power converters, the analysis of switching transients in a power device, taking EMI into account from the start of the design process and compensating for it with low-order harmonics and losses, and choosing the right pulse-width modulation (PWM) technique and topology are all important steps in developing a functional system.
- The magnitude of the ripple on the output voltage, the total losses, the thermal resistance between the power switches the ambient temperature arfve all factors to consider when deciding on the switching frequency and the PWM modulation, which are influencing the cost and size of the passive filters.
- Switching time is accounted for by picking the right factors, such as switching losses, maximum dv/dt and di/dt values of power switches, overvoltage, and leakage of current magnitudes that affect the EMI filter size and cost.

1.4.2 IMPACT ON THE HUMAN BODY

- There is a corpus of literature on the subject of the human body's reaction to EM fields, particularly those of very low frequency (ELF) and in particular MW. The harmful effects of RF radiation on various tissues and organs are dangerous. Researchers are still digging into claims that low-intensity radiation fields have both thermal and non-thermal effects on the body.
- The data presented here suggest that effects of RF radiation are unlikely to occur at the power densities and absorbed energies relevant to GWEN fields, although more research is needed to confirm this. Since the waveforms and frequency spectra of these fields are distinct from those of the GWEN fields, the possibility of energy effects, such as the corresponding effects of low-intensity, amplitude-modulated RF fields on Ca2+ binding in nerve tissue, does not change that result. Typically, tremendous field intensities and massive induced-current densities in tissue are associated with the physical effects of ELF fields, whose frequencies are about one-tenth those of GWEN fields. Physical effects of ELF fields, such as changes in pineal hormone concentration, that may come from low evoked current in tissue have not been proven to create immediate damage to human health. Overall, there is scant evidence from physical studies of ELF fields to suggest that public exposure to LF fields from GWEN antennae poses any sort of health danger.

1.4.3 NERVOUS SYSTEM

- During the 1950s and 1960s, scientists in the Soviet Union published their findings on the nervous system, including their investigations into topics like animal behaviour modification and the electrical properties of nerves. The effects of radiation on neural tissues are reportedly the subject of an ongoing study. Isolated nerve preparations, brain chemistry, the blood–brain barrier, and brain histology have all been documented to be altered by RF radiation. Changes in Aplysia neuron firing rates and isolated frog sciatic nerve refractoriness have been found after exposure to 2.45 GHz microwaves at SAR values greater than 5 W/kg in in vitro nerve preparation tests. Multiple research teams have found that Ca2+ binding to the surfaces of nerve cells in isolated brain hemispheres and neuroblastoma cells cultivated in vitro is unaffected by RF fields at thermal or subthermal levels. In vivo studies of pulsed and continuous-wave (CW) RF fields' effects on brain electrical activity have shown that temporary impacts can occur at SAR levels greater than 1 W/kg. Voluntary human subjects in recent investigations revealed no consistent effects on the electroencephalogram (EEG) when exposed to electric and magnetic fields at 50 Hz at levels comparable to those of high-voltage power lines. Reaction times and heart rates did fluctuate, but only slightly outside of normal.

1.4.4 VISUAL SYSTEM

- It has been investigated for over 30 years whether high-intensity RF radiation causes cataracts in the eye. Near-field exposure at 1–10 GHz has been found to induce cataracts at the lowest thresholds, with a power density greater than 100 mW/cm^2 administered for at least an hour. There is a preponderance of evidence suggesting that both thermal and pulsed CW microwave fields have identical thresholds for causing cataracts. When the RF field's wavelength is not well suited to the eye's size, even at extremely high power densities nearing deadly levels, cataracts are not created. Cataract formation threshold power densities are believed to be similar in rabbits and people due to structural similarities and the equivalent size of the eyes, even if it is difficult to extrapolate data from laboratory animals to humans. The effect has been seen to manifest at a power density as low as 1 mW/cm^2, which is equivalent to an intraocular SAR of 0.26 W/kg. However, the RF fields produced by GWEN antennae in publicly accessible places are still significantly stronger than the threshold power density at 2.45 GHz. In addition to phosphenes, changes in visually evoked potential (VEP) have been reported in response to extremely low-frequency (ELF) magnetic fields with flux densities 5–10 times greater than those required to produce phosphenes. Since these phenomena are observed only with fields below 100 Hz, it is not expected that they will occur in response to the LF fields associated with the GWEN antenna.

1.4.5 IMMUNE SYSTEM

- Both in vitro and in vivo test methods have indicated that exposure to RF fields can have an effect on immune system cells. Several studies have indicated lymphoblast transformation and altered response to mitoses, but the effects seen have varied widely between laboratories. According to the data we have, the SAR threshold for pulsed and CW microwaves that can affect lymphocyte responses to mitogens is greater than 4 W/kg. Lowered natural killer cell activity and increased macrophage activation have both been linked to thermogenic exposure. The changes observed in immune system components at RF power densities that cause tissue heating are consistent with the expected consequences of increased steroid hormone release into the circulation. In another investigation, including a 2-year exposure of rats to a non-thermal dose of pulsed 2.45 GHz microwaves (SAR, 0.4 W/kg), no significant permanent alterations were identified in the concentrations of lymphocytes or their responses to mitogen stimulation. Numerous studies have been conducted to assess the effects of ELF electric and magnetic fields on components of the immune system.
- In general, sinusoidal ELF fields were found to have no significant effects on immunological competence in laboratory animals after in vivo exposure. However, lymphocytes exposed in vitro to pulsed magnetic fields or 50 Hz amplitude-modulated RF fields have shown reduced mitogen responses and target-cell toxicity. These effects might have stemmed from the unusually high current densities created in the cell suspensions. In a study involving 60 Hz electric and magnetic fields with a sinusoidal waveform and intensities comparable with those of the fields near high-voltage power lines, no effects were observed on the immunologic functions of peripheral human and canine lymphocytes obtained from donors that either were normal or had been challenged with specific antigens.

1.5 BASICS OF DM CURRENT AND CM CURRENT

1.5.1 DEFINITION OF DM

- Temporary differences in voltage between a live power line and its matching neutral or return line are all that constitute normal-mode noise. Potentially, the power grid's neutral wire is not grounded. These two wires symbolize the conventional flow of electricity in electronic circuits, allowing every normal-mode transient a direct route into sensitive cir-cuits and the potential to damage system performance. When referring to normal mode, the term "differential mode" is more common. Power supply or load switching is typically the culprit behind differential-mode (DM) noise. AC transients, surges, and interruptions on any line with regard to ground can also be blamed. Power supplies and motor controls that use PWM control to rapidly switch loads are major sources of DM noise. Due to strong ripple currents in the DC link capacitors, the switching action generates DM noise at the source.

1.5.2 Definition of CM

• CM noise is typically manifested as a momentary voltage difference between ground (which does not have to be the neutral line) and the two normal-mode lines. This means that the interference is shared by all lines relative to some sort of ground or neutral point. In sensitive analogue and digital circuits, CM transients are a major reason for alarm since they can cause the circuits to perform erratically, possibly leading to the breakdown of the system. The frequency of the impulses in CM noise is typically greater than that of the corresponding DM noise. Coupling between the line, neutral, and ground occurs at higher frequencies. In general, CM noise is far more detrimental to electronic circuits than DM noise. In conclusion, the CM noise voltage is seen between the lines (usually two) and ground, while the DM noise voltage is impressed between the lines.

1.5.3 Origin of CM Noise

CM noise is a signal that emerges on two or more lines at the same time with respect to ground. This noise or signal travels in the same direction. For this discussion, we will assume that a DC-DC con-verter, also known as a PWM switching supply, is in use and examine its CM effects. To convert DC input voltage to AC for use across a transformer, a flyback DC-DC converter features an input power switch. When it reaches the output, the AC is corrected to DC. Because the primary and secondary windings are constructed from concentric windings that sit on top of each other, the transformer has high inter-winding capacitance as well as capacitance between the input and output. When the power switch is conducting, an impressive dv/dt is impressed across the transformer's input–output capacitance by the input stage. As a result, the capacitance of the transformer induces a current to flow from input to output. During each switching cycle, twice as much current flows and needs a way to return to its "source." Since the current can flow through any of the inputs and outputs, either independently or simultaneously, the term "common-mode current" has come to be used to describe it. The primary and secondary windings of a transformer are able to exchange electrical energy with the presence of perfect magnetic coupling, making the transformer an excellent circuit element. Transmitting solely alternating, DM electricity is the only acceptable operation for a per-fect transformer. Transfer of CM current is prevented because there is no magnetic field generated in the transformer windings when the potential difference across them is zero. A small but non-zero capacitance will connect the primary and secondary windings of any practical transformer. The capacitance arises from the dielectric between the windings and the physical separation between them. Increasing the distance between the windings and filling the space between them with a low-permittivity material can help lower the magnitude of the inter-winding capacitance. Parasitic capacitance provides a conduit across the transformer for CM current; the impedance of this path is a function of both the magnitude of the capacitance and the signal frequency. Depending on the values of the parasitic inductance and capacitance, the CM current will be of varying magnitudes and may also contribute to the DM noise. The CM current is confined within the converter and flows through the stray capacitance from the input to the output when there is no external current channel from input to output (i.e., the converter is driving an isolated load that has very minimal capacitance back to the input).

If the converter is not isolated and an external current path is inserted, such as a PCB trace, or if there is a direct electrical connection between the input and output grounds, the current will preferentially travel through the input ground. As long as the parasitic inductance of this connection is small, it is generally fine to do so. The current in this conductor varies with time (*di/dt*) due to power switching (*dv/dt*). If there is significant inductance in the external path, a voltage *V=L*(*di/dt*) will be created between the grounds due to the quick change in induced current, *di/dt*. This will be evidenced by erratic voltage readings at the input or output connectors. To mitigate these effects, grounding must be either solid or low inductance, and loop areas must be kept to a minimum. Currents flow between the isolated (primary and secondary) sides of the transformer because of the inter-winding capacitance, which might result in a high-frequency (HF) component switching the secondary side ground voltage (Vsg). As another example of CM effects, think about the potential damage caused to utility power distribution equipment by an indirect lightning strike. Lightning strikes produce a powerful magnetic field that couples into all power lines in relation to ground potential. This induced voltage impresses between all conductors and ground, resulting in a potentially very large transient current flow in all conductors in the same direction. Without some way to dampen the excessive voltage and current, the machinery will almost certainly collapse catastrophically. Multiple bursts, or multiple high-voltage pulses at around 50 kHz, are possible in such an attack. Lines can have close or loose spacing depending on the use, such as in a power wire. If the distance between the conductors is large enough, the magnetic field coupling will be slightly different, resulting in a little different induced voltage in the two or more power lines. Occasionally, the sum of all these voltages will equal to zero. Any AC power line voltage during a strike will have its induced voltage increased algebraically. To the extent that the magnitudes of the two impressed voltages differ, the secondary of the transformer will take on the value of the difference between the two line voltages feeding it. As DM noise, this variation is now translated (stepped down) to the secondary plane. Since the induced noise or voltage pulses are at higher frequencies, the core losses of the transformer will be amplified, leading to greater losses. Pulse losses are exacerbated by the "skin effect," which occurs both in the transformer windings and on the lines. The CM current flows through the transformer's primary to secondary capacitance. However, the secondary transformer has a number of pathways to ground through the winding capacitance to the centre taps of the transformer. Adding a Faraday shield or screen to the transformer further decreases the capacitance between the primary and secondary windings. The interwoven secondary capacitance of the transformer further attenuates the pulse's high frequency. The capacitance between the primary and secondary will divert most of the voltage to the various service grounds. The pulse will primarily reach utility customers via the voltage difference across the transformer.

This voltage, however, travels around the two outside legs in a DM. Two-phase is a common term for utility voltage that is 180 degrees out of phase. The filter must be able to withstand a CM pulse if the appliance is wired across the two external lines (220 V). A safety green wire connects back to the service ground, and both wires from the power panel are black (occasionally two different colours are used). It is common for a filter to contain three transzorbs one between the lines and another between the lines and the equipment grounded through the green wire. This category of Transzorb gadgets will be available in two different power levels. All line-to-line Transzorbs will have an RMS voltage rating of 250 or 275 V, while the two line-to-ground Transzorbs will have an RMS voltage rating of 150 V. The RMS value is used to rank the Transzorbs. Harris Transzorb V150LA20Bs are able to discharge between 212 and 240 V, and their RMS voltage rating is 150 V. An arrester is installed to stop these high-voltage surges. Line-to-ground arresters help with DM noise suppression while also eliminating the CM.

The pulse is in DM if it is supplied from either side to the common ground. The hot (usually black) cable is connected to the circuit breaker at the service, the neutral (usually white) is connected directly to the service common ground, and the safety ground (usually green) is also connected directly to the service common ground. The unwanted pulse travels along the heated black wire to the apparatus, and the white common ground wire returns it to the ground. This DM noise pulse must be processed by the filter. In most cases, a Transzorb will be connected from the hot wire to the filter case or piece of equipment ground (the green wire), and another will be connected from the hot wire to the return. Even if lightning strikes between the transformer and the user, the magnetic field will still couple to all conductors in the same way; however, the central wire is grounded many times at the services and the transformer to prevent this from happening. The CM pulses in this example travel along the two outside lines. When connected to the exterior lines and the green ground wire, the filter must deal with CM noise problems. Two of the Transzorbs will be rated at more than 120 V RMS, while the line-to-line unit will be rated at more than 220 V RMS, usually 250 V RMS. The hot and return wires, commonly black and white, carry the noise energy in DM if the equipment is connected to only one line and ground. Transzorb would have a voltage rating higher than 120 V RMS.

1.5.4 GENERATION OF CM NOISE—LOAD

A storage capacitor is typically connected to the diode outputs and ground in most power supplies. Here, we have the service ground green wire, which connects to the chassis. The incoming power lines rise and fall in phase with the storage capacitor's voltage as measured in relation to ground. This results in CM noise propagating from the equipment side or the load side back to the source. Similar results can be achieved by switching converters, which couple the CM noise from the transformer's primary and secondary through the latter's parasitic capacitance. The *dv/dt* on the switch will create *di/dt*, and the CM current will flow to switch ground due to the relatively high parasitic capacitance of the power switch to ground. A power factor correction coil wired from the line-to-ground or an input transformer can both be used to get rid of CM. The CM noise can also be eliminated by first isolating the input supply from ground by placing the storage capacitor across the diode bridge and then using an isolated switching converter. The storage capacitor, switch and load resistor make up the isolated supply. The leakage or reactive current on the ground (green) wire is reduced as the transformer helps to lessen CM noise. Leakage current will be limited if the EMI filter in front of this supply is balanced by connecting differential capacitors from line to line rather than to the green ground line. If the system is supplied with power from the 220 V side, greater CM loss is required; however, bigger capacitors to ground may be employed because the system is balanced.

1.5.5 ELIMINATION OF CM NOISE—LINE AND LOAD

To measure CM noise, capacitors to ground must be used as low-impedance shunts between both or all lines and ground. Also, we'll talk about CM chokes, which are used to offer high impedance to CM current. A leakage current is the current flowing from a reactive capacitor to ground. The voltage drop between the lines is equal to the current flowing through the ground capacitors. An isolation transformer eliminates the leakage current and considerably reduces the CM noise. Balanced lines of 220 V AC are depicted with two capacitors to ground. Feed-through capacitors or leaded capacitors could be used here. At the line frequency, the ground current is zero if the voltages are equal and opposite (180 degrees out of phase) and the capacitors are also equal. This works for capacitors with leads as well as for feed-through capacitors.

There is no such thing as perfect DM balance, as we discussed earlier in this chapter, and as a result, some current will always flow to ground. For instance, 115 V line-to-ground, 230 V line-to-line, 5% capacitor tolerance, 5 mA ground-to-line current, and 60 Hz. Please note that if a capacitor measurement is done from either line to the common ground and the capacitance is outside the range allowed by the criterion, this approach will fail. Isolating the filter and load from the ground and then measuring the current flowing through the green wire is a foolproof way to meet the current limit if one is provided.

Therefore, two feed-through or leaded capacitors of 1 F will work well if the capacitor tolerance is 5% or less. Keep in mind that the values employed above are only applicable in a line-to-line system, and that a far smaller single value of capacitance to ground would be required in a 120 V-to-ground arrangement to satisfy this low current value.

In addition, this method should never be used on any piece of medical equipment that could potentially have a patient linked to it. On wire failure, the remaining capacitor will be subjected to the full line-to-ground voltage (120 V) of the single remaining line, and its current to ground will be 45 mA, which is far higher than the limit set by the patient. When employing a capacitor to ground in place of ground at the common point, the two-phase balanced system can avoid having to use feed-through capacitors. This is used for CM attenuation.

1.5.6   Generation of DM Noise

Whenever inductive devices are switched on and off, DM noise can be generated on the line side. Specifically, this is a voltage spike between the wires. Similarly, transformers connected to the filter's output and the device generate DM noise. Line-side MOVs, also known as Transzorb devices, help clamp the higher voltage pulses from the line and the equipment, and the differential filter section is responsible for the rest.

1.5.6.1   Three-Phase Virtual Ground

Only three-phase virtual programmes running in the same container can take advantage of this method. This method can't be used with devices that need separate insert filters with the entire capacitance connected to ground. With a capacitor connected between each phase and a central point, a virtual ground is created. If all three capacitors have the same value and the phase voltages are equal, the junction voltage will be zero. In the absence of a current source, a fourth capacitor connected between the junction and ground will have no effect. The current on the ground wire should be far below the specification limit if the unit is tested for ground current rather than capacitance to ground by isolating both the equipment and the filter from ground. The best overall performance is achieved when the capacitor values are the same. Find solutions to equations that are analogous to the ones just solved. If one phase fails, the line voltage remains constant at the junction, and the phase angle will be 120 degrees between the two surviving phases. Therefore, this poses a threat to the safety of medical devices that must be applied to a patient. While three-phase high-power equipment may be useful in some industrial settings, it is highly unlikely to ever be employed in a setting where human lives are at stake. Again, a current sensor can be used to keep tabs on the flow of electricity, and a relay can be opened to cut power to the device, the patient, or both. The design engineer is doing more than just designing the EMI filter and hence, has authority to order the installation of a ground in fault device. The three capacitors in series for the CM voltages and the three in parallel for the three-phase voltages are the nicest components of this setup. In order to get rid of CM noise, this provides very low impedance to the ground. Any CM issue can be avoided with the method described here, which involves the use of a well-designed CM choke. Keep in mind that the ferrite CM inductor needs to have three identical windings for the three phases plus one additional winding for the neutral wire if there is a neutral wire. In a CM core, the overall magnetic flux is zero because any imbalance in the phase currents is carried by the neutral. According to this hypothesis, a well-balanced system has a negligible neutral current. To the extent that there is a third harmonic in the system, any multiple of that harmonic (such as 3, 6, 9, etc.) adds back in phase to the neutral. Accordingly, the sum of all third-order currents and the current that flows in the opposite direction adds up to zero. A minimum wire size equal to the peak phase currents should be used. Overheating of neutral filters is prevalent because of current imbalance and third-order harmonics, making them much hotter than the other phase filters. See if the neutral current is lower than the other three by measuring it across all four wires. In conclusion, to get rid of CM noise, a CM choke, capacitors to ground (feed-through or Y caps), transformers, and arresters are the correct components to use.

1.5.6.2　Analysis of CM and DM in AC Drives

It is common practice for modern electromechanical energy conversion systems to employ AC motor drives. In today's power electronic inverters, PWM techniques are used, and the switching periods of the associated switches are extremely short. Because of this, a great deal of HF voltage is applied to the motor windings. These HF voltage components lead to HF leakage currents and conducted EMI in the power mains and ground system due to parasitic winding capacitances (i.e., turn-to-turn and turn-to-ground capacitances). The resulting HF currents are separated into CM and DM subcomponents based on their circulation patterns. Power electronics often focus on conducted EMI in the megahertz-to-kilohertz range. Since MOSFETs and IGBTs switch at such high frequencies, the power converter's less-than-ideal behaviour has a significant impact on the HF component prediction. Therefore, parasitic parameters in semiconductor devices and passive elements of the power circuit must be considered while modelling the HF behaviour of the converter. The low-voltage induction motor is the most common motor design in the low and medium power ranges. Here, tightly wound coils connected in sequence represent the stator windings. A single-turn model, such as that used for forming wound coils, cannot be used for an analytical evaluation of the coil model parameters due to the random distribution of the turns in each coil. Therefore, a suitable three-terminal network may be used to determine the lumped equivalent circuit of a tightly wound coil in terms of equivalent impedance. To accurately simulate the coil's behaviour, one must take into account both the real and imaginary parts of the impedance. From the single-coil model, the multi-coil stator winding model can be generated. With one terminal per phase and another for the motor frame, a three-phase induction motor can be thought of as a "black box" (ground). Phase-to-phase and phase-to-ground impedances are used to express the equivalence between the model and the stator windings. To isolate the DC source from the inverter, two-line impedance stabilization networks (LISNs, 50/50 H) can be used. Software tools like PSpice can be used to perform a numerical analysis of the HF current interferences on both the input and output sides of the inverter. This HF coupling between the DC link and the motor terminals is emphasized by taking into account both CM and DM current components. To determine the HF current harmonic spectrum, a time-domain analysis combined with a frequency-domain study can be performed.

1.5.6.3　Analysis of CM and DM in Three-Phase Cables

In many systems, the cable between the PWM drive and the motor is one of the most expensive parts. In addition to controlling CM leakage current, which can cause EMI issues in the system, and DM reflected wave overvoltage, which can cause cable/motor failures at cable end termination, special cable geometry and voltage rating concerns must be addressed. A three-phase simplified R-L-C per-phase lumped parameter cable model is driven by the inverter, which induces EMI via line-line charging current via line-line capacitance Cll and line-ground CM current via line-ground capacitance Clg. A typical line-ground current waveform at the fast-rising edge of inverter switching current waveform may be large enough to trigger inverter over-current trips. The cable surge impedance, $Zo = L/C$, and the motor surge impedance both define a transient magnitude of twice the bus voltage for a single drive output pulse. With cable sizes smaller than #8AWG, increasing AC cable resistance at high frequencies due to skin and proximity effects may have a significant impact on minimizing peak over. The HF oscillation of a cable depends on its length, geometry, and electrical qualities. The multi-pulse spacing of the inverter modulator and the cable damping time = 2 Ls/Rs to steady-state levels interact in such a way that the transient magnitude spike can be three times the DC bus voltage. Due to HF self-inductance skin effects and mutual-inductance coupling based on cable design, the Ls value is much smaller than the DC quantity. Given the wide range of possible impedance resonance and anti-resonance points, cable oscillation frequencies and the equivalent frequency of drive voltage rise time $f_{rise} = 1/t_{rise}$ may further excite the system's natural response, potentially causing unexpected problems. Therefore, a method is required to accurately retrieve RLC parameters per metre at both low and high frequencies. Due to oversimplified geometric assumptions, the capacitance calculations are found to have a large discrepancy with measured values. When it comes to bundled cables, the typical capacitance formulas are inadequate since they require wire spacing that is significantly larger than the conductor radius. There are close equations for self and mutual inductance, but they cannot be used to predict HF skin effects. Past efforts used LCR meters to physically test cable properties from low to HF. Following this, similar parameters for a single-line transmission line model with transport delay were derived from the per-phase parameters. Since (1) most inverter switching patterns couple two output lines to the (+ or −) DC bus and the third phase is coupled to the opposite bus polarity and (2) only single-line transmission line models were available in power electronic simulators at the time, it is reasonable to assume that this is how real-world inverters work. The fundamental problem of these transmission line models is the absence of interaction with different three-phase modulation schemes that could impact cable or motor waveforms under a multi-pulse train, despite the fact that they provide more accurate results than a lumped parameter or multi-segment RLC model. In addition, single-line models are being replaced by new three-phase motor transfer function models. There is also the issue of cable variances, which makes it more difficult to obtain precise LF and HF RLC values per metre. Finally, a three-phase transmission line model and a three-phase motor model, activated by three-phase inverter modulation techniques, are required for the analysis of issues related to asymmetrical cables. Maxwell 2D Finite Element Analysis (FEA) is used to extract and estimate cable R-L-C parameters throughout a broad frequency range, as it has the potential to obviate the need for time-consuming and costly cable testing. To do this, the frequency dependence RLC parameters can be predicted by FEA. Predicted FEA parameters can be used by a drive-cable-motor power electronic simulator tool to calculate the ringing voltage and current of a DM motor.

1.6 EMI ISSUES AND MITIGATION

EMI occurs when a signal from one electronic equipment interferes with that of another. It covers the full range of frequencies from 10 kHz to 10 GHz, where many current electrical gadgets operate. The source may be deliberate or accidental, and the output may be continuous or discontinuous, intermittent or centred on a single frequency or a broad spectrum of frequencies.

SMPS, digital electronics, brushed DC motors, high-voltage ignition systems, and fluorescent lighting are examples of unintended EMI sources. Since SMPSs are now ubiquitous in LED light bulbs, digital devices, laptop and mobile phone battery chargers, they are the most prevalent unintended EMI source. AM radio, FM radio, television, cell phones, Wi-Fi, Bluetooth, and many other fixed and mobile radio communications systems used by aviation, emergency services, police, and the military are all examples of RF transmitters whose emissions are commonly referred to as RFI. In addition to electrostatic discharge, lightning, inductive kickback, and EM pulse events, EMI is another type of transient that can cause catastrophic damage to electronics.

The coupling mechanisms play a significant role in EMI and EMC properties and may manifest problems, including conducted radiated emissions. Communication channels between devices are formed via coupling mechanisms in which two or more devices are connected. In order to function properly in the same EME, various electronic components must be EMC-compliant.

1.6.1 EMI EMISSIONS SOURCES

In order to cut down EMI, it is best to focus where it starts. There are several potential sources for such signals, but in most cases, they are generated by the rapid switching currents seen in integrated circuits. The vast majority of EMI emissions are caused by CM currents that flow through the product. All of these CM currents originate from intended currents or currents in integral circuits (ICs) that are essential to the product's operation. The potential for HF harmonics to cause undesirable emissions can be mitigated by limiting the intentional signal to only those harmonics required for the product's correct operation. It is likely that the designed signal's return path is the origin of these CM currents. Engineers designing PCBs need to pay close attention to how a trace is routed from its origin to its final destination (the driver to the receiver), but they fail to give the same consideration to the circuit's return current path. At lower clock rates (10 MHz), the return current path was not crucial. Signal traces today must be thought of as microwave transmission lines due to on-board clock speeds of 200–400 MHz and data bus speeds above I GHz. If you want good EMC performance and reliable signal tracing functionality, you need to pay close attention to the return current path at high frequencies.

The best place to begin when designing for EMI is by pinpointing and isolating each potential source of EMI emissions. Most sources of EMI on PC boards can be isolated and addressed independently without affecting the performance of other components. All the signals on the board must be sorted into two groups first: those that are meant to be seen and those that aren't.

The engineers designing the board will naturally take the intended signals into account. The signals they "expect" to be present on the board are routed along specifically planned traces. The design process frequently overlooks the potential for unintended messages. They aren't usually on our minds since designers don't plan for them to be on the board. More than 90% of a PC board's EMI emissions come from these unintended signals. It doesn't matter how meticulously the board is constructed; there will always be some of these accidental signals. Unintentional signal emissions can come from two places: the loop model and the CM. CM, crosstalk linking power planes, and above-board structures are all potential sources of emissions.

1.6.2   EMI PROBLEM SOLUTIONS

There are many ways to reduce the instrument's EMI and make it fit in with the real EME.

- Software solution
- Hardware solution
- EMI shielding technique
- Filter
- Grounding
- Fabric foam
- Ferrite, etc...
- System-level solution
- Optimize assembly code, there is no impact on the physical redesign of the system.

1.6.3   EMI SHIELDING

EMI shielding is one of the methods used to reduce EMI. This is the method for regulating and coupling radio waves, electrostatic fields, and EM fields. The effectiveness of shielding depends on three main factors, namely, absorption, reflections, and multiple reflections. Shielded cable, unshielded cable, and fibre cable are the main categories in the shielding of cables.

Braided copper, spiral copper tape, or polymer conductive materials are make-up shielded cables. Typically, they are thicker and more stringent than standard unshielded cables. When working with them, comprehensive care is required. Unshielded twisted wires have no interior shielding to prevent interference from magnetic fields. However, they are deliberately bent to reduce EMI. Because of its low mass and minimal profile, this category of cables serves a practical purpose. These cables are typically used for LAN cable systems or network connections inside buildings, such as offices and homes. To safely transmit private information, fibre optics has emerged as the preferred medium of transmission. Unlike copper cables, these are susceptible to interference from electrical noise.

The use of EM shields is a common way to lessen the impact of EMI on electronic devices. To shield an electronic circuit, a conductive surface is placed around its most vital components to

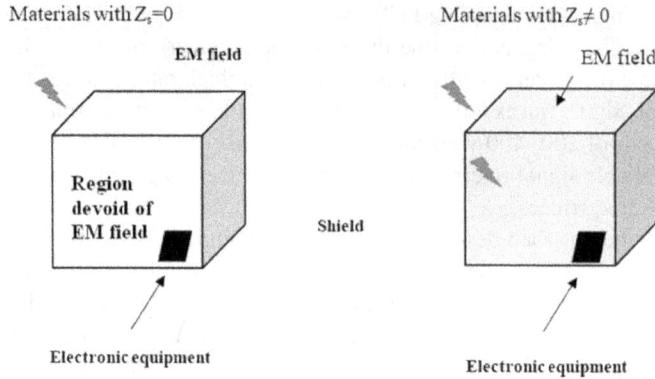

FIGURE 1.14   (a) Perfect shielding. (b) Real shielding.

reduce EMI through a process of combined reflection and absorption. At low frequencies, all metals can be used in the shield, whereas at high frequencies, a thin conductive layer is deposited.

Choosing a material with zero surface impedance ($Zs = 0$) while interacting with the surrounding environment provides flawless shielding. A medium with such a shield would be fully unaffected by fields, as shown in Figure 1.14a. However, there does not appear to be any material with zero surface impedance that could be used to achieve perfect shielding. Therefore, EM waves will always be able to pass through the walls, as seen in Figure 1.14b.

1.6.4   EFFECTIVENESS OF SHIELDING

Some of the EMI field will be transmitted, while the majority will be reflected or absorbed by the shield. The structure's EMI shielding effectiveness (EMI SE) is defined as the decibel-ratio of the EM power before and after the shielding effect. Permeability, conductivity, and permittivity are the properties of the frequency at which the measurement has to be done; the angle and polarization of incidence of the impinging wave, near-field or far-field application, govern its EMI SE.

1.6.5   FILTERING

HF currents can emit radiation if allowed to flow freely through a circuit; thus, filters are used to block them. Unshielded products depend on filtering and careful PCB layout to prevent interference. CM and DM noise make up the majority of conducted emission noise (CM). The regular flow of current through the loops formed by components in a circuit is the source of DM noise. Such loops will serve as antennae, emitting EMI interference. The resistance introduced into the circuit as a result of unwanted voltage dips is the root cause of CM noise. If there is a leak in the power supply due to stray capacitance or inductance, the current will flow back to the source. It is usual practice to use an EMI filter to mitigate CM and DM disturbances. Filtering CM and DM noise signals is accomplished with dissimilar methods; first, a noise separator is utilized to isolate the two unwanted signals, and then filters are implemented. Filters for reducing EMI have the advantages of being simple in concept and implementation, as well as being more cost-effective for smaller systems. The primary benefit of utilizing a filter is that it can identify and eliminate interferences whose properties are distinct from those of the system.

1.6.6   CABLE BONDING

The shielded enclosure of the product needs to have a strong bond with the conductive shell of the power connector. With the ability to prevent cables from radiating, a full 360-degree bond is ideal.

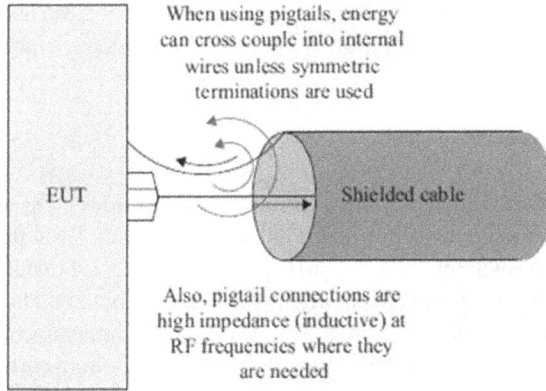

When using pigtails, energy can cross couple into internal wires unless symmetric terminations are used

EUT

Shielded cable

Also, pigtail connections are high impedance (inductive) at RF frequencies where they are needed

FIGURE 1.15   Cable shield termination using a pigtail ground to chassis.

Cable shield termination using a pigtail ground to chassis has high impedance at RF, and energy can cross couple, resulting in magnetic field coupling to the signal wires, as shown in Figure 1.15

1.6.7   Circuit Topology

There will be some level of conducted and radiated emissions from any switching power supply topology.

For instance,

1. Forward converters are superior to flyback converters, as their lower peak currents and much lower secondary AC currents make them the superior choice.
2. Topologies that operate at a duty cycle greater than 50% (like pulse-width modulation or push–pull) also result in lower peak currents. By utilizing input power factor correction, input spikes on AC power can be minimized. If the circuit's components are not laid out properly, spurious noise signals, such as loop inductances and parasitic capacitances, may be produced. As a result, it is crucial to provide each part of the electronic system with its own designated spot.

1.6.8   Spread-Spectrum Technique

Spread-spectrum clock generation (SSCG) is a technique recently developed for lowering radiated emission from digital electronic devices that uses frequency modulation of the clock. This novel method is similar to the popular spread-spectrum method of transmitting data.

As a result of the clock frequency, the emission spectrum is broadened by the presence of side bands. The effectiveness of switching power circuits lies in lowering the fundamental frequency level, particularly if the switching frequency is less than 150 kHz and the frequency for modulation is selected to be greater than 200 Hz.

1.6.9   Hard- and Soft-Switched Power Converters

In the soft-switched approach, all power electronic devices, such as switches and diodes, are activated and deactivated at zero voltage and zero current.

EMI emission levels are measured from both hard- and soft-switched buck and boost converters with the same power rating. The findings point to the efficacy of the soft switching technique in power converters for lowering EMI emissions.

Conventional hard-switched power electronics generate a lot of EMI because of the high dv/dt and di/dt involved in the switching operation. Soft switching techniques have been proposed to:

- Lower switching losses
- Reduce the SMPS switching stress

Through the use of soft switching techniques, the instantaneous power loss associated with turning on and off a power switch is greatly diminished or even eliminated. Even though there hasn't been much research on how soft switching affects EMI, early results suggest that it can lower EMI output.

Hard and soft switching were experimentally investigated for both conducted and radiated EMI. During testing, the converter was hooked up to the grid via a line impedance stabilization network's nominal RF impedance (LISN). Place the converter 40 cm from a 4 Sqm earthed conducting surface and maintain a 100 cm separation. A 0.5-m cable was used to link the converter to LISN. Noise measurements were taken using the 50-port voltage generator.

Second, a 20×12×7-foot semi-anechoic room was used to measure the radiated emission. Emission levels were measured using a calibrated and adjusted HP 11955A biconical antenna. It takes 3 m to get from the converter to the antenna. This was accomplished with the use of a special turntable setup. Emitted and conducted fields were evaluated with an EMC analyser. The desired frequency range is swept while the converter is disabled in order to collect data about the natural frequency levels present in the surrounding environment. When the converter is turned on, both the surrounding environment and the converter's own emission signals are logged. By subtracting the ambient signal from the measured signal, the true emission from the converter can be determined.

Each set of hard- and soft-switched converters is tested with the same load, and their fundamental converter components (L and C) are interchangeable. With two of each type of converter, the inductor current and load current are nearly equivalent in magnitude. The converters are not insulated and lack EMI filters.

1.6.9.1 Hard- and Soft-Switched Buck Converters

Soft-switched converters also have a significantly lower diode reverse recovery current compared to their hard-switched counterparts (Figures 1.16 and 1.17). Soft-switched converters have substantially lower *dv/dt* and *di/dt* because transient ringing is drastically reduced. The converter's totally soft-switched nature has been verified by these tests. We can see the combined effects of the conducted and radiated EMI emission from the hard-switched buck converter, as well as the ambient noise. Figure 1.18 displays the results of measurements circuit of both conducted and radiated EMI made with a soft-switched buck converter. Since the waveforms of the soft-switched converter's auxiliary switch are recorded, we may deduce that the former has a higher level of EMI emission,

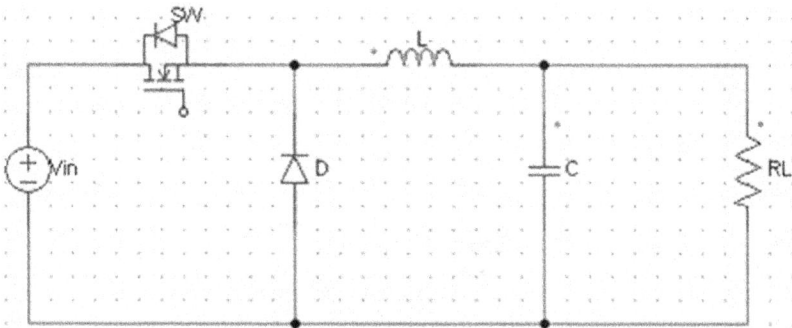

FIGURE 1.16 Hard-switched buck converter. $V_{in}=55$ v, $V_{out}=20$ V, $f_s=115$ kHz, $D=0.4$, $L=2.5$ mH, $C=220$ μF, $R_L=7.5$ Ω.

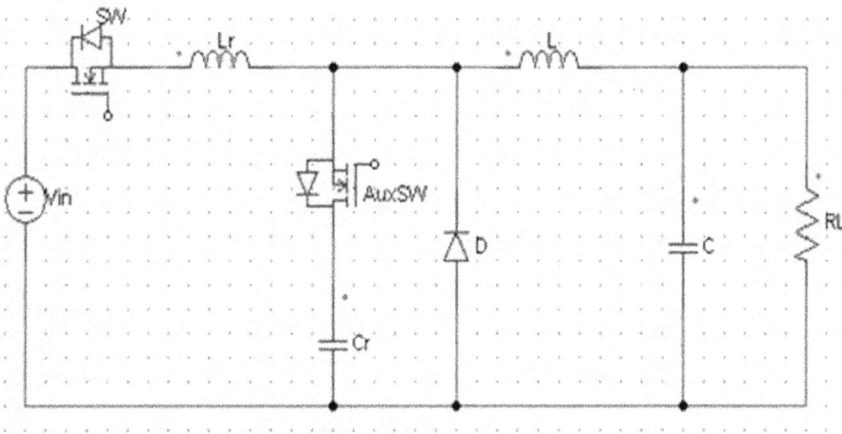

FIGURE 1.17   Soft-switched buck converter. $V_{in}=55$ v, $V_{out}=20$ V, $f_s=115$ kHz, $D=0.4$, $L=2.5$ mH, $C=220\,\mu$F, $R_L=7.5\,\Omega$, $L_r=3.1\,\mu$H, $C_r=46.5$ nF.

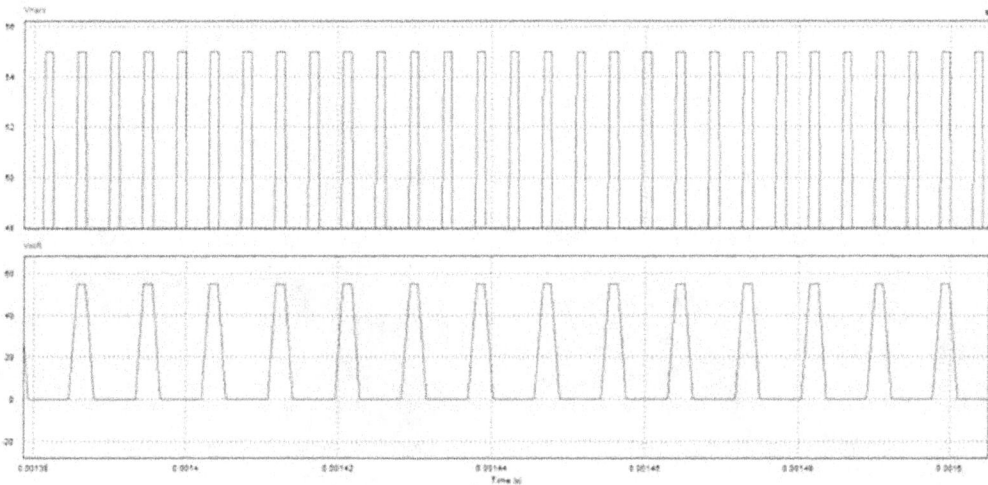

FIGURE 1.18   Voltage across the main switch for soft switching.

both conducted and radiated. All of the converter's soft-switched properties are verified by these measurements.

1.6.9.2   Hard- and Soft-Switched Boost Converters

Figure 1.19 depicts the EM interference caused by the hard switch boost converter. Soft-switched boost converter EMI measurements are provided in Figure 1.20. Comparing the two sets of data reveals that the soft circuits produce less EMI than the hard-switched circuit. Figure 1.21 shows the waveforms of the soft-switched converter's auxiliary switch Q. These findings verify that the converter truly operates as a soft switch.

1.6.10   Sources of Noise and Switching Noise

High-performance digital circuits require fast, time-varying currents, which are a common source of switching noise. Radiation is produced when this current flows between the PCB layers.

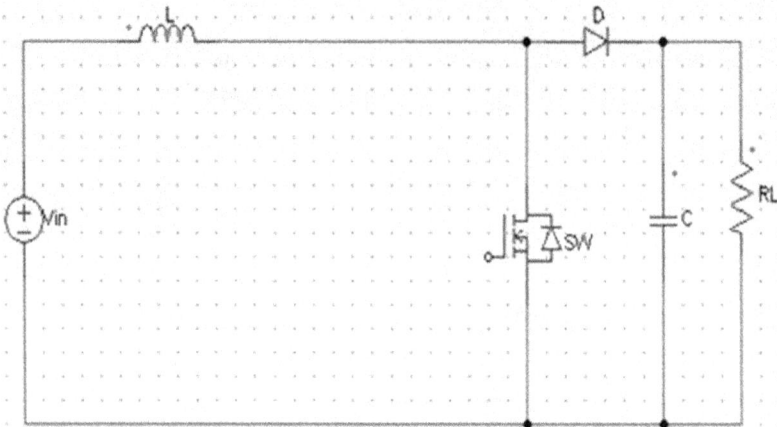

FIGURE 1.19 Hard-switched boost converter. $V_{in}=34$ v, $V_{out}=48$ V, $f_s=115\,\mathrm{kHz}$, $D=0.26$, $L=2$ mH, $C=220$ F, $R_L=7.5\,\Omega$.

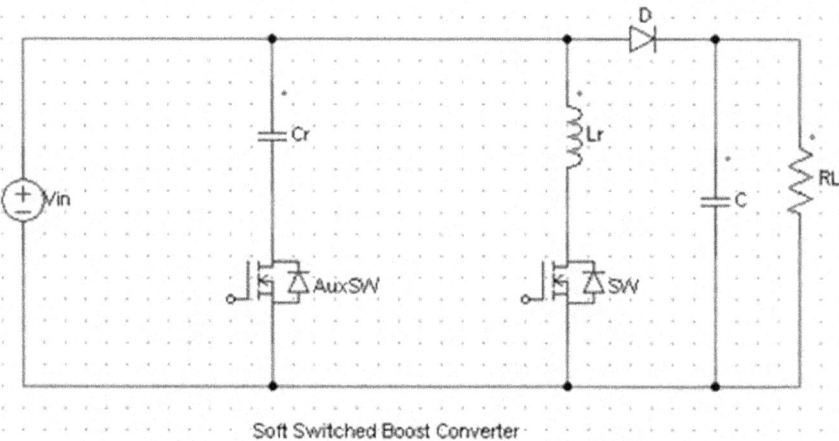

Soft Switched Boost Converter

FIGURE 1.20 Soft-switched boost converter. $V_{in}=34$ v, $V_{out}=48$ V, $f_s=115\,\mathrm{kHz}$, $D=0.26$, $L=2$ mH, $C=220$ F, $R_L=7.5\,\Omega$, $L_r=3.2\,\mu\mathrm{H}$, $C_r=20$ nF.

Power planes act as parallel plates for the radiated wave to travel along. Inductive noise, also known as "simultaneous switching noise" (SSN), is produced when several digital circuit outputs are switched simultaneously. SSNs can be taken into account by

$$V_{noise} = NL_{eq}$$

where V_{noise} is the magnitude of the noise voltage, and N is the number of outputs (drivers) switching simultaneously. L_{eq} is the equivalent inductance through which the current must pass, and current passes through each driver while switching.

As soon as you plug L_{eq} into the wall outlet and switch on the current, multiple signals are switched at the same time. When there is inductance in the path of the current, it causes voltage

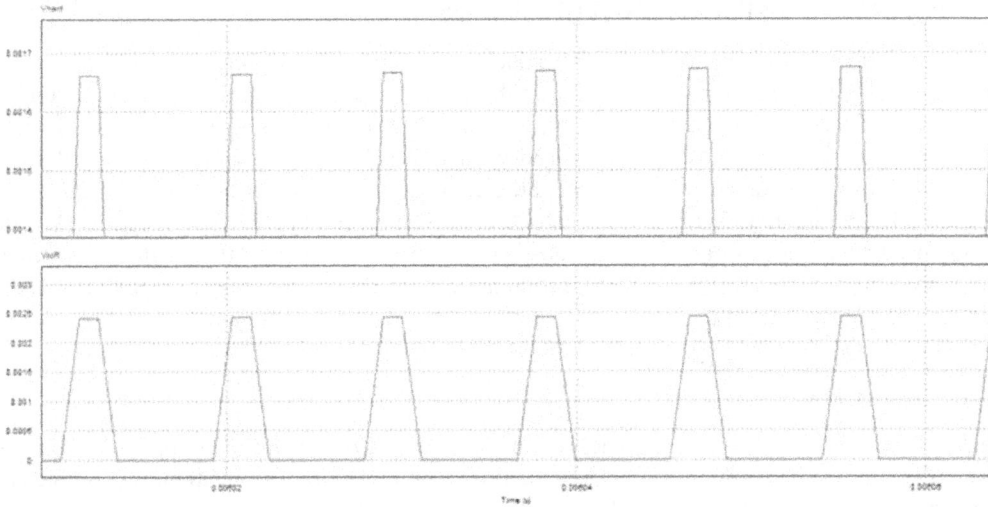

FIGURE 1.21 Voltage across the main switch for hard switching.

changes on the lower planes. These changes affect the driver's outputs and other signals on the board, which leads to wrong switching.

1.6.11 OPTICAL CIRCUIT BOARDS

Different needs, especially those that are level-dependent, must be taken into account when deciding whether an electrical or optical connecting method is appropriate. There's no denying that the PCB is a crucial part of any piece of electronic machinery.

This optical layer consists of integrated dielectric waveguides for high to medium bandwidth interconnections that can be manufactured by hot embossing. The passive elements can be manufactured using the aforementioned technology. The layer must be located inside the PCB due to manufacturing and assembly temperature and space limitations. The optimal size for PCB compatibility is between 5 and 100 mm or high-bandwidth interconnections that can be manufactured by hot embossing. The passive elements can be manufactured using the aforementioned technology. The layer must be located inside the PCB due to manufacturing and assembly temperature and space limitations. The optimal size for PCB compatibility is between 50 and 100 mm. This sizing is also due to the passive tuning required by optical transmitters and receivers. Light is in the 650–850 nm range since that's the minimal local attenuation required for polymer waveguide materials. The optical converters and passive components that allow light to be coupled into and out of the waveguides are just as crucial. Photodiodes with built-in amplifiers will form the basis of the optical receivers. Voltage at the receiver inputs is of primary relevance, so the first step is to build on-board linkages between electrical and optical components on the board.

1.6.12 CABLING AND INTERCONNECT

If the product's enclosure is not shielded or if the cable connector is not correctly bonded, then the cable is likely the source of EMI emissions due to faulty circuit or PC board design. The penetration of shielded enclosures by cables allows HF CM currents to radiate outside the shield.

1.7   FULL-WAVE SIMULATION TECHNIQUES

Electrical circuits can be disrupted by EM induction, electrostatic coupling, or conduction when exposed to EMI, also known as RFI, while in the RF spectrum. It's possible that the disruption would reduce functionality and performance, which could lead to a complete loss of data. EMI is caused by fluctuations in electrical current and voltage, which can be generated by either human activity or natural phenomena (EMI). While all electronic devices can be affected by EMI, some gadgets and parts are more likely to cause interference than others. In electronic warfare, EMI can be employed on purpose to jam radio signals. Interference, both deliberate and inadvertent, has been a problem from the beginning of radio communications and the necessity to regulate the radio spectrum has always been clear.

1.7.1   SIMULATION

Mathematical models are used in electrical circuit simulation to create an environment that is identical to the real-world operation of an electronic device or circuit. The stimulation programme can mimic circuit operation and serve as a beneficial analysis tool. There is a wide variety of simulation software available on the market. Electronic simulation software encourages user participation and active learning. The user or students may become interested in analysing, organizing, and evaluating the outcome if such attractions are used.

A system's behaviour can be simulated before it is actually built. By spotting design flaws and offering insight into how electronic circuits function, it can significantly boost productivity. In order to allow users to observe the effects of their changes to the simulated circuit, several electronics simulators combine a simulation engine with an on-screen waveform display.

1.7.2   THE IMPORTANCE AND METHODS OF FULL-WAVE SIMULATION VALIDATION

Validating simulation and model results are more important than ever. Widespread EMI/EMC issues are solved with the help of full-wave simulation tools that employ a wide range of modelling approaches. Software tools may provide an output based on a model given to them. Sadly, there is no assurance that the model was constructed correctly, that all relevant aspects of the situation were taken into account, or that the results were correctly interpreted. There is some more checking that needs to be done to make sure the results are accurate for that simulation.

1.7.3   FULL-WAVE METHOD

The Maxwell equations are solved by the full-wave technique for any material or conductor within the computational space. It is possible to write the Maxwell equations as a set of hyperbolic partial differential equations. As a result, this opens the door to a potent method for calculating numerical solutions. Full-wave techniques are more difficult to implement than quasi-static ones, but they are more flexible in application. Each full-wave modelling approach has its own limitations in terms of the models it can't adequately simulate. Each full-wave modelling approach excels at simulating certain classes of models, and this is where the approaches diverge. The model's limited frequency validity is a major drawback. The partitioning of the issues and the proximity of the elements to the boundaries of the computing space primarily impose the constraints. When stimulation is carried out outside of the valid range, errors can be introduced.

The use of full-wave techniques can be implemented in one of two ways. If you look hard enough in either the frequency domain or the time domain, you'll eventually find the answer. Frequency-domain codes must be executed for each frequency of interest, while time-domain methods use a Fourier transform to deliver output data as a function of frequency.

1.7.4 TIME-DOMAIN TECHNIQUE

When employing a time-domain approach, a band-limited impulse is used to excite the simulations over a broad frequency range. The model's reaction to the impulse is the output of the time-domain code. A Fourier transform is performed on time-domain data to provide frequency-domain information.

Caution should be taken when selecting the excitation method to achieve optimal performance. While many different kinds of driving wave shapes exist, the Gaussian pulse is by far the most common. An elementary Gaussian pulse stores energy from direct current up to a maximum frequency. A symmetrical spectrum is produced when a signal is modulated onto a sine wave (carrier), and both the maximum and minimum excitation frequencies are determined. The differentiated Gaussian pulse is another type; it lacks a DC component, has a value that decreases by 6 dB per octave as frequency decreases, and exhibits the typical quick roll-off at high frequencies. Specialized pulses can be made up of a sequence of cosines whenever highly specific frequency characteristics are required.

Resonances within the model are highly possible due to the time-domain techniques that excite the model over a broad frequency range. When resonances take place, the currents and fields connected with them will continue to ring for a while within the computational domain. Full impulse response can only be obtained after the field has stabilized. When ringing lasts for a lengthy period of time, the simulation time may need to be very long. Numerous solutions exist that can drastically reduce or do away with this issue altogether in practical models.

1.7.5 FREQUENCY-DOMAIN TECHNIQUES

The frequency-domain approach addresses one frequency at a time. Typically, this is sufficient for antenna repair and examining the particular challenges. As a result, multiple frequency-domain simulations can typically be executed at a time, compared to the time it would take to do a single time-domain simulation. Furthermore, since larger meshes may be used for the lower frequencies with frequency-domain codes, computation times can be reduced.

Frequency-domain codes require multiple simulations to cover a large frequency range. It's worth noting that you can reduce the volume of simulations needed by using interpolation techniques. However, these interpolation methods require caution in order to avoid missing the resonance effect.

1.7.6 FULL-WAVE SIMULATION TECHNIQUE FOR EMI

EMI is rapidly becoming a critical design concern for cutting-edge electronic devices. Current trends in integration are squeezing entire systems into unprecedentedly high circuit densities, while the frequency of operations is always rising. EMC concerns are often dealt with after the prototype has been built in a traditional design. Adding hand-held components, metal plans, or even re-designing the entire system can happen. It has substantial influence on both cost and time to market. EMC considerations should be made early in the design process rather than later. It's possible that the EMI simulation, which allows for the examination of circuits, leads to reducing the time between the design phase and manufacturing.

Common EMC remedies typically require costly and complex electrical machinery for efficient management. Electronic consumer devices with a lesser market share, automobile electronics, and the majority of embedded systems cannot afford these high-priced but effective alternatives. Because these devices require whole-board capabilities and no tool is provided for this purpose, they are more vulnerable to EMI issues and additional post-prototype EMC costs.

1.7.7   Full-Board EMI Simulation

Therefore, having this capacity early on in the design process for high-speed PCBs and IC packages would be crucial for ensuring EMC compliance. For the sake of the research into integral equation methods:

1. Need to concentrate on reducing the number of parameters needed to model each conductor. For the same level of precision in the end result, it is needed to find a different set of basic functions that requires 16–20 times fewer unknowns than the traditional thin filament discretization. For the current density inside the conductors, employ a basis function that is a subset of the physically allowed solutions to the diffusion equation.
2. Using the new fundamental functions, the Krylov subspace iterative methods are used to solve the liner system. An O(N2) matrix vector product operation is the primary contributor to the algorithm's complexity. For such items, quick algorithms with a complexity of O(Nlog(N)) exist. In order to solve the static capacitance extraction problem, such algorithms have been effectively applied to the corresponding tools. For full-wave EMI analysis problems, it is intended to adapt and improve upon one such algorithm, pre-corrected Fast Fourier transform (FFT).

An EMI analysis should be able to do things like describe the spectrum emissions measured on a measurement sphere at 10m and make a model of the interconnect structure that can be connected to a non-linear time-domain simulator. Using an adjoint approach to construct a set of transfer functions from the thousands of circuit inputs (e.g., the pins of the ICs) to the tens of observation locations on the measurement sphere. Thereby efficiently addressing the first specification. On generating dynamic linear systems with same behaviour as the original interconnect structure with a state vector to meet the second specification.

1.8   EQUIVALENT SERIES RESISTANCE AND EQUIVALENT SERIES INDUCTANCE

Any series-connected perfect capacitor will have a resistance and inductance equal to the equivalent series resistance and inductance, respectively. The maximum power ratings are calculated by coupling the equivalent series resistance with double-layer capacitors, which is also known as the effective series resistance (ESR) or the equivalent series inductance (ESL). The inductive portion of a component's impedance can be calculated by calculating its ESL. A vendor's datasheet will detail the parameters and ratings. The RF characteristics are calculated using the equivalent series resistance (ESR) and ESL. As a function of the device's characteristics, they can be used to quantify a variety of frequencies, and Q-factors are often associated with ESR. Finding the total resistance and inductance of the circuit is also incorporated with the equivalent series resistance (ESR), which is useful for performing the series resistance both in-circuit and out-of-circuit. Both the ESR and ESL represent a form of AC resistance. The system's heat dissipation is primarily targeted by concentrating on the ESR and ESL. Consequently, the ESR and ESL are essential to the functioning of the system.

1.8.1   Equivalent Series Resistance

All the circuits involve three main components: resistors, inductors, and capacitors. Most of the circuits are connected with parasitic components, which led to some distortions. When the ideal components, such as capacitance, are coupled with resistance in the series connection, they lead to some internal resistance; then they are called "EQUIVALENT SERIES REISTANCE." The values will range from few milliohms to several ohms; it leads to power losses, reduced efficiency, etc.

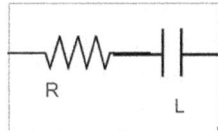

FIGURE 1.22   Circuit representation of the ESR.

This property is present in all capacitors in varying degrees depending on the construction, dielectric materials, and quality. This method is useful for testing EMI and EMC, as shown in Figure 1.22.

1.8.2   Need for ESR

The increased ESR, heat, and ripple current of electrolytic capacitors can lead to malfunctioning of machinery. A capacitor with a high ESR can cause equipment to malfunction or create permanent damage, needing maintenance, often by causing power supply voltages to grow extremely high. This is in contrast to previous technology, which produced hum. However, electrolytic capacitors are frequently utilized due to their low cost and high capacitance per unit volume or weight; typical electrolytic capacitors have capacitance ranging from a few microfarads to the hundreds.

Faulty capacitors with high ESR tend to overheat, bulge, and leak when the electrolyte chemicals disintegrate into gases, making them very easy to recognize visually. However, capacitors that appear visually fine may still have a high ESR, identifiable only by measurement.

In most cases, a less-than-exact measurement of ESR will suffice for troubleshooting. Because ESR fluctuates with frequency, applied voltage, and temperature, precise measurements must be done under precisely specified conditions. When taking measurements in the lab, a general-purpose ESR meter that just has one frequency and one waveform will not do.

1.8.2.1   ESR Components

- Metallic resistance.
- Electrolytic and paper resistance.
- Frequency dependent.

1.8.2.2   Metallic Resistance

Metallurgical resistance is the impedance a conductor presents to an electric current. Since the temperature coefficient of conductors is positive in nature and the resistance is directly proportional to the temperature of the devices, the resistance of the conductor is optimized for metallic conductors as the temperature rises. To some extent, the weak current will flow through it.

1.8.2.3   Electrolytic and Paper Resistance

Measuring the resistance of a solution to the flow of current between two separate electrodes over a specific distance yields information on the solution's electrical conductivity. The term "conductive paper" is used to describe paper that has been impregnated or coated with a conductive substance to make it resistant throughout the medium. When a voltage is applied to tele deltos paper, for example, the paper does not conduct electricity; however, the voltage rises, and the paper eventually breaks down due to a weak point or moisture.

1.8.2.4   Dielectric with Frequency Dependence

Materials "dielectric constant" and "permeability" are frequency-dependent sources. This does not alter for single-frequency modes, but it does lead to some dispersion if we have superposition.

This material is more efficient and keeps electrical charges in the medium range while also stopping the flow of electric charges and losing energy as heat. This is called dielectric losses.

Unlike metals, dielectrics have tighter bonds between their electrons, that might otherwise drift through the substance instead get polarized when placed in an electric field. The electric property of the medium is greatly enhanced by the dielectric substance, therefore limiting the amount of current that can be passed through the substance.

1.8.3   Factors That Increase the ESR Value

1.8.3.1   Bad Electrical Connections

Welding or mechanical crimps are typically used to attach the capacitor's copper leads to the aluminium plates. Due to the inability to solder aluminium, this form of connection necessarily involves some series resistance. As the liquid component of the electrolyte dries out due to elevated temperatures, the electrical resistance increases.

Increase in temperature and frequency: In supplying high currents, the power dissipation associated with the ESR may further increase the temperature and lead to capacitor failure.

1.8.4   Effect of Frequency on ESR

The ESR is the portion of the capacitors' impedance that causes the overall, real power losses. The frequency-dependent equation can be shown below with real power losses:

$$\text{ESR} = DF_R/2\pi fC + DF_L/2\pi fC + DF_D/2\pi fC = R_c + 1/R_L\left(2\pi fC\right)^{2.}$$

Where:
   DF_R = dissipation factor associated with contact resistance.
   DF_L = dissipation factor with leakage losses.
   DF_D = dissipation factor with dielectric losses.

The leakage and dielectric losses decrease with an increase in frequency until the contact resistance dominates up to a specific point. ESR becomes very high at higher frequencies, largely due to the skin effect of the AC signal.

1.8.5   Minimizing ESR in Circuits

Low equivalent series resistance (ESR) capacitors, such as low ESR solid polymer capacitors and multilayer ceramic capacitors (MLCC), are used in high-performance applications. Parallel connections are used to connect capacitors, such as in low-power supply smoothing circuits. In addition, lowering of the ripple voltage allows the capacitor to handle greater currents with fewer losses by decreasing the effective series resistance (ESR).

1.8.6   Measuring Equivalent Series Resistance

1.8.6.1   ESR Meter

Both in-circuit and out-of-circuit series resistance measurements can be taken with the use of ESR meters. The equivalent series resistance (ESR) is measured with an ESR meter, which is a two-terminal electronic measuring instrument typically connected to capacitors. Since most ESR meters are non-inductive, low-value resistance devices, they can only be used with capacitors.

1.8.6.2 Principle

If a true electrolytic capacitor is discharged and an electric current flows through it for a very short period of time, the ESR will be activated. This will cause a voltage across the device proportional to the product of the current and the equivalent series resistance (ESR) plus a small contribution from the charge on the capacitor; this voltage is measured, and the value is divided by the current (i.e., ESR) displayed in ohms or milliohms on a digital display or by the position of a pointer on a scale. This happens hundreds of times every second.

AC can also be employed, but only if the frequency is high enough such that the capacitor's reactance is significantly smaller than the ESR. Typical aluminium capacitors have an ESR that becomes unacceptable over a microfarad; hence, the settings of the circuit are often chosen to yield significant results up to that value.

1.8.6.3 Uses of ESR Meters

The internal resistance of batteries can be measured using an ESR meter outfitted with a pair of protection diodes across its input. ESR meters can also be used to measure the resistance of parts of the printed circuit track, the resistance of switch contacts, and so on, depending on the specific circuit being tested.

An ESR meter is helpful since it can measure low resistances without skewing the readings by turning on semiconductor junctions, but there are dedicated instruments for detecting short-circuits between neighbouring PCB tracks. Short-circuits can be located with the help of an ESR meter, and it is possible to pinpoint the faulty component among a group of capacitors or transistors connected in parallel via printed circuit tracks or wires. Many standard ohmmeters and millimetres can't handle testing circuits with extremely low resistance and the ones that can use too much voltage, potentially damaging the circuit.

When the test points on a piece of equipment are tightly spaced, as is often the case with devices constructed using surface mount technology, tweezer probes come in handy. Testing tools can be held steady or moved with one hand while the tweezers are held in the other.

The important property of capacitors is that they can block DC current and allow only AC current to flow through them, so regular ohm meters cannot be used to determine resistance values; instead, special meters are available. It is important to note that the capacitor's inductance can be ignored in the circuit and not taken into account in the circuit. The transfer function to be calculated for the system is given below:

$$H(S) = R_2/R_2 + R_1 * \left(S + 1/R_2 * C/\left(S + 1/(R_2 + R_3) * C\right)\right)$$

In the above equation, it is shown that the high-pass feature of the circuit is reflected, an approximation is made in the transfer function of the system, and it can be further evaluated as:

$$H(s) = R_2/(R_2 + R_1) = R_2/R_1$$

This shows the transfer function of the system, which can be represented in the circuit analysis. The above operations are suitable for HF operation. At this point, the circuit starts to attenuate. The attenuation factor can also be expressed as a condition. $\alpha = R_2/(R_2 + R_1)$. This module, which increases the resistance value of the system, can be resistive to the circuit of the system.

The inductive qualities that lie inside a capacitor prevent it from behaving in the system in the same way as pure capacitors; hence, it is common practice to use both resistive and inductive capacitors together.

Therefore, ESR may be estimated when paired with ideal components; no real component should be included. When designing, it is critical to consider the circuit elements that are ideal components,; in this case, capacitance as well as their parasitic components (inductive and resistive).

1.8.7 MUTUALLY COUPLED INDUCTORS IN PARALLEL

When two or more inductors are linked in parallel, the total inductance will either increase or decrease due to the effect of mutual inductance, depending on the strength of the magnetic coupling between the coils. The result of this mutual inductance is conditional on the coil's separation and orientation with respect to one another.

Compared to coils with zero mutual inductance, the equivalent inductance of two mutually linked inductors in parallel can be thought of as either "aiding" or "opposing" depending on which direction the current flows. Polarity dots or markers can be used to indicate whether a pair of parallel coils are mutually coupled in a cooperative or antagonistic fashion.

When inductors are connected in parallel, the voltage across them is the same as that across a resistor. When N inductors are linked in parallel, the equivalent inductance of the circuit is the reciprocal of the sum of the reciprocals of the individual inductances, thereby decreasing the effective inductance of the circuit. The direction of the coupling between the coils determines whether the inductors in a parallel circuit are "aiding" or "opposing" this total inductance, just as it does for series-connected inductors (in the opposite direction).

1.8.8 EQUIVALENT SERIES INDUCTANCE

High equivalency current can be affected by the distance between the lead termination and the circuit joining point. Thus, the appropriate capacitors are crucial.

1.8.9 MEASURING ESL

A simple method is to analyse the impedance versus frequency plot. Capacitor impedance shifts as frequency is varied; at the "knee point" where capacitive and inductive reactance cancels out, the capacitor self-resonates. The capacitor's equivalent series resistance (ESR) helps to flatten the "knee point" or the frequency at which it begins to resonate, and this is where the ESL begins to rise.

If any one of the knee points is identified, ESL can be calculated with the below equation:

$$\text{Frequency} = \frac{1}{\left(2\pi\sqrt{\text{ESL} \times C}\right)}$$

1.8.10 ESL EFFECTS ON CAPACITOR OUTPUT

When the equivalent series resistance (ESR) of a capacitor increases, the capacitor's output suffers. This is because increased ESL also increases the flow of unwanted current and EMI in HF applications, which can cause malfunctions and a high ripple factor in output variables like voltage. For a given capacitor's ESL value, the ripple voltage will be in direct proportion. The ring waveform can be induced by a capacitor with a high enough ESL value.

1.8.11 IMPORTANCE OF ESL

Modern techniques for making capacitors lower the ESL values of capacitors used as output filters in designs. This has a wide range of effects on the performance of electronic circuits, especially in HF and high-current applications.

A switched-mode power supply's (SMPS) input and filter capacitors must have low ESR values so that ripple doesn't affect the power supply's performance.

It features a low-noise supply, which is very useful in situations where there is a requirement in noise suppression high-quality output filter and reduction of ripples in the power supply. Capacitors' equivalent series resistance (ESR) should be kept low so that they don't interfere with the switching frequencies of the power supply.

It can be that the impedance of a capacitor falls with increasing frequency at 20 dB/decade, which is useful for smooth output filter stages and stable power supplies to the load. The noise levels are low, the noises have been suppressed, and the number of output filter stages should have been minimized. Partially as a result of the inductive property of the connections and the non-ideal characteristics of the capacitor material, real capacitors also have inductive properties, whose impedance rises with frequency at 20 dB/decade. The sum of both is minimal at the resonance frequency. Above this, the parasitic series inductance of the capacitor dominates. These non-ideal inductances can be easily accounted for in-circuit analysis by expressing the physical components as a combination of an ideal component and a small inductor in series, with the inductor's value being equal to the inductance present in the non-ideal physical devices. The transient time of any inductive element is determined by the relationship between the inductance and the resistance,

- with the fixed value of the resistance being the same throughout,
- number of turns in the coil,
- diameter of the coil,
- coil length,
- type of materials,
- coil length and number of layers in the winding in the materials.

As a result, this module provides detailed information about the ESL, as well as the actors that affect the normal inductance.

1.9   IMPEDANCE AND ITS EFFECT ON EMI

Every electronic circuit causes EMI and is affected by it. EMI is defined as "a disturbance caused by something outside the circuit that affects the circuit through electromagnetic induction, electrostatic coupling, or conduction." EMI, whether radiated or conducted, couples to other circuits. The circuit's efficiency may drop or it may stop working altogether if the disturbance persists. Standards for EMI at the international level define permissible emission limits. Products should be resistant to EMI above acceptable levels and should not create EMI over the values defined by standards. To effectively design approaches to reduce the impact of EMI, it is necessary to investigate the influence of impedance in addition to identifying the noise source.

In order for a noise problem to exist, the following components are required:

- EMI noise source to generate the noise.
- Receiving device (victim), which is affected by the noise.
- Coupling channel between the source and victim.

The goal of EMC is to reduce or get rid of one of these three things that are needed to make noise.

1.9.1   THE EFFECT OF IMPEDANCE ON EMI

Resistance, capacitance, and inductance are attributes shared by all elements (conductors) of a grounding system, whether it is used for power grounding, signal grounding, or lightning protection. It is the material, the length, and the cross-sectional area of a ground path conductor that determine its resistance. A ground conductor's capacitance depends on its geometry, its closeness to other conductors, and the dielectric material between them. The inductance depends on the dimensions, shape, and length of the coil, as well as the relative permeability of the metal. As a result of the interplay of resistance, inductance, capacitance, and frequency, grounding impedance is calculated.

1.9.2 CABLING IMPEDANCE

When two pieces of equipment are connected by an electrical cable, the existence of the wire has a direct impact on the EMI experienced by the devices. Cables have a major impact on circuit parameters, including high-speed digital data transfer and compliance with conducted EMI. The signal travels from the driver to the receiver along the upper conductor of the generic single-ended circuit depicted in Figure 1.23 and then back along the lower conductor.

Short Cables: Close proximity between the vehicle's driver and its passenger may make it possible to disregard the cable's conductors. The impedance of the conductors is almost zero; therefore, they don't have much of an impact on the functioning of the circuit. Exactly the same waveform that was sent out by the driver has been received by the receiver.

Long Cables: Due to the increasing separation between the transmitter and receiver, the cable is modelled as a transmission line and the receiver's impedance, as seen by the transmitter, undergoes a transformation. There may be problems with the driver's ability to maintain appropriate logic levels if the impedance shift is substantial. In addition, data mistakes may occur due to the intensity of reflections brought on by a mismatch between the cable's characteristic impedance and the circuit's impedance. When this happens, the received waveform degrades, and you may experience ringing and other issues with the signal quality. The capacitance between the cable conductors, which is parallel with the receiver, causes the impedance to drop as the cable length increases from zero. Further length increases cause the impedance to show peaks when the impedance is very high and nulls when the impedance is very low.

When the length of the cable is an odd multiple of a quarter wavelength, null occur. At a certain frequency, the impedance of each null is equal to the resistance of the cable conductors. Increasing the frequency causes a small rise in resistance.

If the driver is not able to provide enough current to sustain proper signal levels at frequencies close to the nulls, the waveform voltage at the receiver will be distorted. Cable lengths that are even multiples of a quarter wavelength produce the greatest peaks. Each peak can have an extremely high impedance. It is related to the impedance of the circuit at the cable's termination. Near-peak frequencies, the driver voltage can be twice as high as normal. Changes in impedance due to the wires result in ringing and standing waves. This noise causes issues with the cable's signal integrity and also increases the cable's conducted and radiated emissions.

1.9.3 CONTROLLED IMPEDANCE (CI)

These trends have led to smaller and more intricate PCB designs and components. Important nets and traces, impedances, and an analysis of the board's effect on signal performance must be initiated. Signals in the gigahertz range are the norm in modern electronics as circuit speeds continue to rise. The signal traces on a PCB behave like transmission lines at high frequencies, with impedance at each node along the signal trace's path. Whenever there is a change in this impedance from one

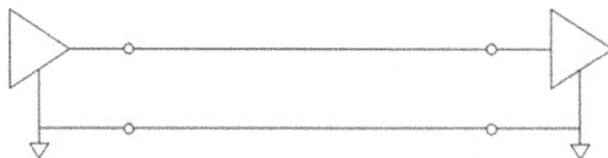

FIGURE 1.23 Generic single-ended circuit.

location to the next, a reflected signal will occur, the strength of which is proportional to the disparity between the two impedances. As the disparity widen, reflection occur. Because of the signal's direction changes, the reflected signal will eventually overlap with the original one.

As a result, the signal that was originally intended to be conveyed from the transmitter side will be altered by the time it reaches the receiver side. If the level of distortion is high enough, the signal might not be useful anymore. The regulated impedance of the PCB signal traces must be uniform to reduce reflection-induced signal distortions and allow for clean signal transmission. CI of traces is important for the integrity and performance of signals on circuit boards. This is because a uniform transmission line on a PCB has a definite width and height and is at a uniform distance from the return path of the conductor, which is usually a plane at a certain distance from the signal trace.

PCB traces and their corresponding reference planes form a transmission line with CI. This becomes important when PCB transmission lines are being used for the transmission of HF signals. Circuit impedance is based on the size of the PCB and the dielectric materials used. Because of this, CI is very important for addressing signal integrity issues, which are problems with how signals travel without getting messed up. The ohm is the unit of measurement. Single-ended microstrip, stripline, differential pair, embedded microstrip, and co-planar (single-ended and differential) PCB transmission lines all need CI and are affected by factors such as resin content percentage and physical PCB tolerances such as trace height and width at the top and bottom of the trace.

- Impedance is inversely proportional to trace width and trace thickness.
- Impedance is proportional to the laminate's height, and it is inversely proportional to the square root of the laminate's dielectric constant (Er).

1.9.4 INPUT AND OUTPUT IMPEDANCE

An electrical network's input impedance is its static resistance to current and its dynamic reactance to current flowing from the power source into the load network. The load's propensity to draw current is represented by the input admittance (1/impedance). As signals travel between various circuit blocks, they face some apparent impedance. The source network is the part of the network that transmits power, while the load network is the part of the network that absorbs power. How much of a signal's power is reflected by a circuit and how much is transmitted into it is controlled by the input impedance observed by the signal, which is a function of the total number of network parts.

- Applying the series and parallel rules to a simple electrical network made up of resistors, inductors, and capacitors makes it possible to figure out the input impedance.
- Each and every circuit has a closed-form solution to this equivalent impedance. Non-linear devices (transistors and diodes) in a circuit cause its input impedance to vary with the strength of the signal voltage.
- The way a waveguide or transmission line is built determines what its input impedance is.

In Figure 1.24, a source (Vs) outputs a digital signal.

- The source is connected to the electrical network and has some equivalent output impedance (Z_{out}).
- When the signal reaches the input port of the network, the signal might reflect off of Z_1. The behaviour of the signal does not just depend on Z_1. Instead, it depends on the value of *zinc*. The elements in the network (Z_1–Z_4) combine in some way to create the input impedance Z_{in}.
- Z_1–Z_4 could be transmission lines, passive elements, non-linear components, or any other component with a defined impedance.

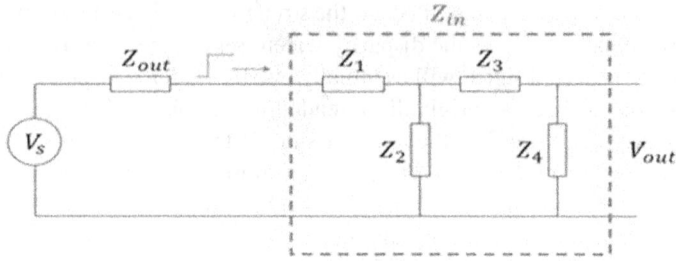

FIGURE 1.24 Example electrical network and its input impedance.

- Signal reflection does not depend only on Z_1 because the various components in the network couple together through the EM field. Components Z_2–Z_4 also add some impedance to Z_1, and when all four are added together, they make an input impedance called Z_{in}.
- The input impedance is determined so that the sending and receiving networks have identical impedances. When the source's output impedance is equal to the complex conjugate of the network's input impedance, power transfer is at its highest and reflection is at its lowest.

$$\text{Reflection coefficient at input}: \tau = \frac{Z_{in} - Z_{out}}{Z_{in} + Z_{out}}$$

$$\text{Power transfer}: P = \frac{v_s^2 Z_{in}}{Z_{in} + Z_{out}^2}$$

Analytically calculating a network's input impedance can be challenging. Internal circuit voltage and current distribution can be intricate. The circuit will have a defined output impedance that is not necessarily equal to its input impedance because waves can still reflect between components, different components might be connected with transmission lines that have length-dependent impedance, and all this behaviour can be a function of frequency (i.e., for reactive components). Assuming the circuit is perfectly symmetrical, the input impedance is equivalent to the output impedance when viewed from the output side (i.e., when looking into the circuit from the Z_4 side of the network).

In order to analyse the signal in both directions along the transmission line, we need to calculate the S parameters, namely, the $S11$ and $S22$ parameters. Designing high-speed digital channels with enough bandwidth for the signals requires knowledge of the $S11$ spectrum. A precise model of the characteristic impedance of the transmission line is necessary for obtaining precise spectra of the S parameters.

1.9.5 CHARACTERISTIC IMPEDANCE

In parallel circuits and power planes, the characteristic impedance is the resistance to the passage of AC and is typically denoted by the symbol Z_0. A uniform transmission line's characteristic impedance can be calculated as the voltage amplitude divided by the current amplitude of a single wave travelling along the line. The transmission line's characteristic impedance is set by the line's geometry and materials. Impedance is measured in ohms, and there are two types of characteristic impedance: single-ended and differential. There are three types of differential impedance: the odd mode, the even mode, and the CM. Both microstrip and strip line impedances have distinguishing characteristics.

In terms of characteristic impedance, the width of the circuit trace, the thickness of the circuit trace, and the dielectric thickness (the thickness of the core, prepreg, or solder mask) around the circuit trace are of paramount importance. The dielectric thickness depends on the type of core and prepreg that are used in the lamination process. The layer design, copper percentage, and copper thickness all have a role in regulating the dielectric thickness. Plotting, imaging, developing, and etching all contribute to the thickness of a line. This means that the transmission characteristic impedance is affected by the following variables:

- **Dielectric Losses and Dispersion:** Each type of dielectric has its own unique dielectric loss spectrum and dispersion that must be accounted for when determining its characteristic impedance.
- **Copper Roughness and Skin Effect:** Because of the skin effect, the characteristic impedance includes both resistive and inductive components. When compared to smooth copper, rough copper has higher skin effect impedance and a more nuanced enhancement factor.
- **Load Capacitance:** Real integrated circuits have some parasitic shunt impedance at the input pin. For bi-directional channels, the input also has some load impedance.

1.9.6 DAMPING AND RESONANCE

One of the most crucial aspects of inducing or being subjected to noise is resonance. An unintentional resonant circuit in a circuit can cause noise interference by producing a significant current or voltage at the resonance frequency. Reduce or get rid of any circuit resonances that you can. Damping resistors are used to dampen resonance.

A resonant frequency is the frequency at which a circuit's inductive reactance and capacitive reactance cancel out. An impedance's minimal value (preferably zero) is achieved at a series resonance, whereas its maximum value is during a parallel resonance. A value known as Q quantifies how powerful this resonance is (the quality factor). When Q increases, resonance becomes more powerful. In addition, the performance of a capacitor or inductor can be represented by the index Q. The parasitic components of a capacitor or inductor produce a resonance at a given frequency when the component is employed in the HF region. One term for this phenomenon is "self-resonance."

1.9.6.1 EMC Measurement Issues for Resonant Circuits

- A resonant circuit has the ability to boost voltage. Unwanted resonance in an electric circuit causes an increase in current or voltage at the resonance frequency, which can be a source of unwanted noise.
- The unintentional creation of a resonant circuit is possible. Consider the case of a digital IC with a few picofarads of electrostatic capacitance floating at its input port. Furthermore, the inductance of the wiring is roughly 1 uH/m. For this reason, a resonant circuit can be formed if a cable of roughly 1 m is linked to the input terminal of a digital IC (e.g., to connect it to an external sensor). It is possible for this location to become a source of noise if a conductor is attached carelessly to it.

1.9.7 LINE IMPEDANCE STABILIZATION NETWORK (LISN)

As required by several EMC and EMI test standards, a LISN is used to measure radiated and conducted RF emissions and susceptibilities.

A low-pass filter, or LISN, is often installed between an AC or DC power source and the equipment under test (EUT) to produce known impedance and offer a RF noise measurement port, effectively isolating the EUT from any interference that may be present in the power supply. Another application for LISNs is predicting conducted emissions for diagnostic and pre-compliance testing.

The functions of LISN are as follows:

1. Stable line impedance.
2. Isolation of the power source noise.
3. Safe connection of the measuring equipment.

1.10   MATERIALS

The most important EMI qualities are the materials and shielding used. It is the functional and multifunctional structural shielding materials that play the most crucial role. Materials like metals and carbon are effective. Except when paired with functional materials, ceramics, cement, and polymers are often ineffective. Metal carbon, ceramic carbon, cement carbon, and polymer carbon combinations have been accepted as shielding materials because of the wide availability of micro- and nanocarbons.

Structural shielding materials typically consist of continuous carbon fibre composites or cement-based materials. By analysing the underlying science and material structure, this chapter explains the fundamentals of shielding material design. Common pitfalls encountered when investigating shielding materials are also discussed.

1.10.1   TYPES OF MATERIALS

Materials used to block EMI can be divided into two categories: structural and non-structural. The materials used for shielding are not the same as those used for building. A shielding component in a mobile phone, for example, could be made of a functional material. It could be built inside or mounted on a structure. Functional materials, rather than structural materials, have been the focus of most EMI shielding research.

Multifunctional structural materials are better than adding functional materials to a structure to make it shielding-capable because they are cheaper, last longer, have a larger functional volume, and don't change the structure's mechanical properties.

1.10.2   POLYMER-BASED COMPOSITES

1.10.2.1   Metallic Fillers
Polymer matrices have been infused with metallic fillers of varying physical shapes, such as fibres or nanoparticles, to enhance interaction with incident EMI radiation.

1.10.2.2   Injection Modelling
It provides a direct method to disperse metallic fillers into the polymer matrix.

1.10.2.3   Stainless Steel Fibres
Injection moulding is used to incorporate stainless steel fibres (SSF) into a polycarbonate matrix, and the moulding parameters that result in maximum electrical conductivity are crucial to reduce EMI.

1.10.2.4   Syntactic Foams
A lightweight composite material is syntactic foam, which fills hollow spheres in a matrix. This method for improving the EMI SE of syntactic foams includes the following steps:

    i. Hollow particles made of a conductive material
   ii. Coating a conductive layer onto the surface of hollow particles
  iii. Adding a second conductive filter to the syntactic foam matrix

1.10.2.5   Carbon Foams
Carbon foams are a type of porous, three-dimensional architecture. They are made up of linked networks of carbon atoms, much like a sponge. Carbon foams, thanks to their strong thermal and electrical resistance, low density, and resilience to chemical corrosion, have been put to use as EMI shielding materials.

1.10.2.6   Graphene Foams

This is a novel, lightweight EMI shielding material that may be manufactured under relaxed manufacturing conditions and is in high demand. Polymer processing charge yield is directly proportional to the stringent circumstances under which it is produced.

1.10.2.7   Graphene Aerogels

Aerogel, a synthetic, porous, ultralight substance produced from gel in which the liquid component utilized in gel is replaced with air, is a good example of this. Because of its increased electrical conductivity after graphitization, GA is better able to reduce EMI. Strong polarization effects were created in GAC by nitrogen doping and by side polar groups. Lesser numbers of side polar groups were left when the graphene sheets were reduced further in GAT. Internally, numerous reflections of the EM waves within the cavities of the highly porous GA absorbed the waves' energy.

1.10.2.8   Metals

Customers of Leader Tech typically employ pre-tin-plated steel in bright and matte finishes, as well as the copper alloy 770, also called nickel silver or German silver, for EMI shielding.

1.10.2.9   Pre-tin-Plated Steel

Pre-tin-plated steel is a low-cost, high-performance option from the low kHz range all the way up to the low GHz range. Due to its low permeability value in the hundreds, carbon steel can act as a LF magnetic shield, a feature not present in alloy 770, copper, or aluminium. The tin plating makes it possible to connect the shield to the traces on the surface board during assembly. It also keeps the steel from rusting or corroding.

1.10.2.10   Copper Alloy 770 or Nickel Silver

Alloy 770, also known as "copper alloy 770," is a copper, nickel, and zinc alloy that is commonly used in EMI shielding applications due to its high corrosion resistance. The UNS C77000 code is the universally accepted identification for this particular alloy. The base material already has these desirable qualities

—sturdiness, resistance to corrosion, and the ability to be soldered.
—no further plating is necessary.

Using the material as an EMI shield is effective from about 500 kHz all the way up to 1 GHz. A permeability of 1 makes it is good for the use in MRI machines, which prohibit the use of any magnetic materials.

1.10.2.11   Copper

Because of its tremendous efficiency in dampening magnetic and electrical waves, copper is the most dependable metal in EMI shielding. Because MRI machines in hospitals are linked to standard computer hardware, copper RFI shielding is a practical solution. This metal and its alloys, including brass, phosphorous bronze, and beryllium copper, are extremely versatile and may be easily produced. Although these metals often cost more than pre-tin-plated steel or copper alloy 770, which are two common shielding alloy alternatives, they may offer better conductivity. Phosphorous bronze and beryllium copper are often used in contact applications, such as batteries and springs, because they are very flexible.

1.10.2.12   Aluminium

Due to its non-ferrous characteristics, high strength-to-weight ratio, and high conductivity, aluminium is still a great material choice for a wide range of applications, despite a few manufacturing difficulties. Due to its high conductivity relative to copper (about 60% higher), aluminium requires careful consideration of its galvanic corrosion and oxidation properties before use. Over time, an oxide will form, and the material will be difficult to solder.

1.10.3   EMI SHIELDING TYPES

1.10.3.1   Board-Level Shielding

BLS or board-level shielding is a method of preventing EMI that uses metal cages made of either one or two pieces. By isolating components on the board, minimizing crosstalk, and decreasing EMI susceptibility, these BLS keep the system running at full speed.

Cold-rolled steel (CRS), a metal commonly used in manufacturing, and aluminium are both options for the BLS. The use of aluminium reduces the product's total mass while still meeting the EMI suppression requirements, and the material's thermal conductivity is five times that of CRS, making it a great choice for any application that necessitates board-level EMI suppression.

1.10.3.2   Shielding of BLS
- plated with stainless steel
- German silver
- copper beryllium

1.10.3.3   Product Forms
- one-piece shields
- two-piece shields (frame and cover)
- Surface mount device (SMD)-compatible shielding clips and separate covers

1.10.3.4   One-Piece Shields

Those metal walls have five sides for maximum protection. The PCB's electrical grounding connector defines the sixth side.

1.10.3.5   Two-Piece Shields

The fence and the cover are separated into six equal sections. The cover can be installed at a later time, which is a benefit. One-piece and two-piece shields are placed on the PCB with THT or SMD technology, depending on the circumstances (e.g., following component alignment). In addition to our standard board-level shields, MTC also offers custom-designed shielding enclosures. Shielding covers are a low-cost prototype for mass manufacturing, offering a direct shield for your PCB to prevent radiation. It is recommended that air holes be installed in the cover to prevent overheating and the diseases associated with its components.

1.10.3.6   SMD-Compatible Shielding Clips

Putting them on the PCB one by one has the benefit of creating a custom fit for the shielding cover. Using one or more clips on either side is suggested, with the number depending on the size of the area to be protected. Belt-mounted shielding clips are made from tin-plated steel. MTC's product line-up includes gap filler that conducts heat well; it can be used even if you need a high level of heat sinking. Adding microwave absorbers to the shielding cover can increase the shielding effect in the HF band (Figure 1.25).

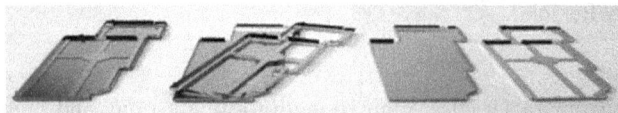

FIGURE 1.25   SMD-compatible shielding clips.

1.11 EMC TESTING

The goal of EMC is to ensure that various pieces of equipment work together successfully under similar EM conditions. Also, any unintended consequences should be minimized or eliminated. When it comes to electrical and electronic devices, the EMC directive makes sure nothing interferes with or causes them to produce any unwanted EM fields.

For your electronic or electrical device to continue working as intended in the presence of multiple EM phenomena, it must pass EMC testing, which verifies that it does not produce excessive EMI in the form of radiated and conducted emissions. Hence, it is important to test the equipment for EMC compatibility.

EMC can be grouped into two categories:

1. Immunity Testing
2. Emissions Testing

1.11.1 EMI MEASUREMENT TECHNIQUES

Figure 1.26 shows the different EMI measurement techniques.

1.11.1.1 Emission Testing

Electronic instruments operate as EM polluters due to their conducted or radiated emissions, which originate from power cables, wires, resistors, capacitors, and OpAmps. These emissions can reach the GHz range and can be conducted by AC power systems or antennae. Therefore, emission testing is necessary for all electronic devices to ensure a healthy and safe EME.

An emission test is used to determine how much EM noise a device produces while functioning normally. The goal of this check is to ensure that the emission levels are well within the parameters set for them. This will ensure that the item will not interact with other electronics while functioning normally within its designated operating range.

The most common key inferences are as follows:

EMI is a key obstacle to the improvement of power electronic equipment. The importance of EMC increases when considering the fact that power electronics are frequently found in extremely noisy situations.

EMC is a major factor that needs to be considered early in the design process. Designing for EMC is an involved process that begins early in the product life cycle and continues through testing and even post-production. Engineers must consider EMC at every stage of a system's creation. The system is put through a series of tests to make sure it will function as expected in its intended setting. In addition, there are a variety of methods for regulating the emissions they're responsible for.

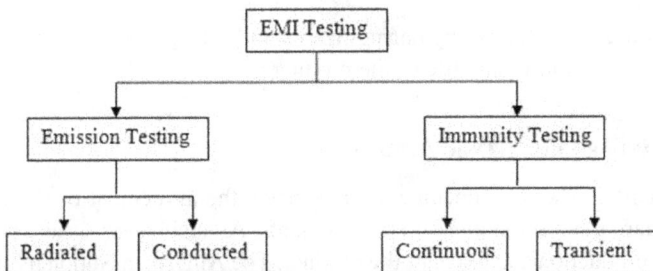

FIGURE 1.26 Electromagnetic interference (EMI) measurement techniques.

1.11.1.2 Radiation Emission Analysis

Radiated emission testing was performed in the frequency range of 30 MHz to 1 GHz. The open-area test site is one of the most frequent types of test setup. It consists of an infinite metallic ground plane, a receiving antenna, an EMI receiver, or a spectrum analyser, and the EUT. The distance is calculated by locating the reception antenna in relation to the closest outside surface of the EUT. In order to ensure that measurements are made in the field region, a considerable distance must be set between the EUT and the receiver.

Measurements made while in flight provide an indirect method of verifying the multiple-input and multiple-output (MIMO) system's radiated emissions. Although measuring the MIMO system's radiation performance is the main goal of OTA tests, the sensitivity of the receiver is actually measured due to noise signals produced by the MIMO system itself. An estimate of the noise released by the equipment owing to a sudden change in voltage or current is obtained by doing this test. The interference from the outside noise could be disastrous for the linked gadget and cause it to fail.

1.11.1.3 Immunity Testing

When compared to emission testing, this is the polar opposite. In emission testing, the noise emitted by the EUT was quantified, whereas in immunity testing, the EUT is exposed to an electromagnetically hostile environment and its responsiveness is evaluated. It is measured how well a device performs when exposed to EM noise and other disturbances. If the tool can't handle the rigours of the testing environment, it can't handle the rigours of the actual world. The goal of these tests is to guarantee that the gadget will function properly under typical conditions.

1.11.1.4 Testing for Continuous Source Immunity

This test is aimed at determining if the EUT will perform as intended when it is exposed to continuous noise sources such as microwave background, Sun radiation, broadcast stations, motor vehicles, and magnetic fields.

1.11.1.5 Temporary Source Immunity

Lightning, EM pulses, electrostatic discharge, voltage fluctuations, quick switching, and relaying are all examples of transitory sources of EM that can have devastating effects on system performance while only lasting a fraction of a second.

1.12 EMC STANDARDS

1.12.1 The Need for an EMC Standard

The following are some of the intentional requirements for EMC standards:

- The EMC standards are required for trouble-free coexistence and to ensure satisfactory operation.
- They are also required for providing compatibility between electronic, computer control, and other systems.
- They are required for establishing harmonized standards to reduce internal trade barriers and improve the reliability and life of the product.

1.12.2 EMC Standard Electronic Appliances

In general, a standard is a written document that reflects the agreement of those having a vested interest in the standard's subject matter and requirements. As such, it serves as a resource for everyone involved in the production and usage of the product. The American National Standards Program is based on this philosophical or mission concept of a standard. Other standards, whether they are for the military, for civilians, for a country, or for the whole world, all have similar goals.

A product's compliance with standards established by organizations like the International Special Committee for Radio Interference (CISPR), CENELEC, ETSI, Institute of Electrical and Electronics Engineers (IEEE), International Organization for Standardization (ISO), FCC, and IEC is essential, whether you're making 5G items, automotive equipment, or a basic table lamp. Delays in product certification due to tight EMC limit enforcement might result in lost revenue, additional cost, and design difficulties.

Every year, more and more EMC standards are released. Standardization can be broken down into a few different categories: minimum requirements, general requirements, and specific requirements for a given product or product family. Fundamental specifications outlining what kind of measuring gear, measurement procedures, measurement uncertainty, and test facilities are needed. In the absence of more narrowly tailored norms for a given product or category of products, the general set of norms will be applied. Standards restricting LF and HF emission and standards outlining the requirements of immunity to EM emission make up the product family standards. In addition, EMC requirements are defined by a set of industry-specific product standards.

The fundamental goal of an EMC standard is to ensure that home appliances may be used safely and effectively in the expected EM environment. Over the past few years, there has been a rise in awareness of EMC and standards organization in home appliances among users and manufacturers. The German government enacted the EMC Directive 89/336/EEC, officially known as the EMVO, in November 1992. At the beginning of 2019, the European Economic Area (EEA) will require EMC and CE certification on all domestically produced goods. A collection of standards that have been harmonized and published in both the EU's Official Journal and the German PTT Ministry's Official Journal. The emission standards to be enforced under 89/336/EEC for household appliances such as vacuum cleaners, washing machines, microwaves, and heating appliances are EN 60555-2, EN555-3, and EN 55014. For TV receivers and audio equipment, the standards are ENW 555-2 and ENSS 013.

Some of the commercial EMC standards and their marking codes are given in Table 1.1.

1.12.3 CISPR STANDARD

CISPR (International Special Committee on Radio Interference) standards are commonly used to determine EMC Emission Test Methods and Limits.

The following are examples of CISPR standards currently on the market.

TABLE 1.1
Commercial Electromagnetic Compatibility Standards

Commercial Standard	CISPR	IEC (Europe)	FCC (USA)	METI (Japan)
Industrial, Scientific, and Medical equipment	11	EN 55011	Part 18, C	J 55001
Vehicles, Boats, and Internal Combustion Engines	12/25	EN 55012 EN55025	SAEJ551 J 1113	JASO D001–82
Electrical Devices, Household Appliances, and Tools	14-1	EN 55014-1	-	J 55014-1
Electrical Lighting	15	EN 55015	-	J 55015
Multimedia Equipment	32	EN 55032	Part 15, B	J 55032
Military Equipment			MIL-STD-461	
Aviation			DO-160	

- **CISPR11:** Radio-frequency Instruments Used in Industry, Science, and Medicine (ISM): Limitations and Measurement Techniques for Electromagnetic Disturbances.
- **CISPR 12:** Limits and methods of measurement for the protection of receivers other than those installed in the vehicle, boat, or device itself or in neighbouring vehicles, boats, or devices from radio disturbances emitted from moving or stationary sources, such as internal combustion engine-powered vehicles and boats.
- **CISPR 14-1:** Electromagnetic Compatibility: Requirements for Household Appliances, Electric Tools, and Similar Apparatus, **Part 1:** Emission.
- **CISPR 14-2:** Electromagnetic Compatibility: Requirements for Household Appliances, Electric Tools, and Similar Apparatus, **Part 2:** Immunity: Product Family Standard.
- **CISPR 15:** Limits and methods of measurement of radio disturbance characteristics of electrical lighting and similar equipment.
- **CISPR 16-1:** Specification for radio disturbance and immunity measurement apparatus and methods **Part 1:** Radio Disturbance and Immunity Measuring Apparatus.
- **CISPR 16-2:** Specification for radio disturbance and immunity measurement apparatus and methods, **Part 2:** Methods of measurement of disturbances and immunity.
- **CISPR 16-3:** Specification for radio disturbance and immunity measurement apparatus and methods—**Part 3:** Reports and recommendations of CISPR.
- **CISPR 16-4: Part 4-1:** Uncertainties, Statistics, and Limit Modelling: Uncertainties in Standardized EMC Tests.
- **CISPR 22:** Limits and methods for measuring radio disturbances in IT hardware and infrastructure The following are some examples of radiated and conducted emissions from ITE:
  1. Data Display, Telephone, CRT, Plasma, LED
  2. Keyboard, Mouse
  3. Magnetic Card Reader, Optical Character Reader
  4. Image Scanner, Pen, Data Printer, Dot Matrix, Laser
  5. Data Processor, Computer, Calculator, LAN
  6. Modem
  7. Automatic Teller Machine
- **CISPR 24:** Information technology equipment; immunity characteristics; limits and methods of measurement.
- **CISPR 25:** "Vehicles, boats, and internal combustion engines—Radio disturbance characteristics—Limits and methods of measurement for the protection of on-board receivers."
- **CISPR32:** Electromagnetic Compatibility of Multimedia Equipment—Emission Requirements.

1.12.4 IEC STANDARD

Many of the International Electrotechnical Commission's (IEC) guidelines for ensuring devices are EMC, which can be found in the IEC 61000 family of standards. Listed below are a few instances of this.

- **IEC/TR EN** 61000-1-1, "**Electromagnetic compatibility (EMC)—Part 1:** General—Section 1: Application and interpretation of Fundamental Definitions and Terms"
- **IEC/TR EN** 61000-2-1, "**Electromagnetic compatibility (EMC)—Part 2:** Environment—Section 1: Description of the environment–Electromagnetic environment for low-frequency conducted disturbances and signalling in public power supply systems"
- **IEC/TR EN** 61000-2-3, "**Electromagnetic compatibility (EMC)—Part 2:** Environment—Section 3: Description of the environment—Radiated and non-network-frequency-related conducted phenomena"

- IEC EN 61000-3-2, "**Electromagnetic compatibility (EMC)—Part 3-2:** Limits: Limits for harmonic current emissions (equipment input current 16 A per phase)"
- **IEC EN 61000-3-3, Limitation of voltage changes, voltage fluctuations, and flicker in public low-voltage supply systems, for equipment with rated current 16 A per phase and not subject to conditional connection.**
- **IEC EN 61000-3-12, Electromagnetic compatibility (EMC)— Part 3–4: Limits: Limitation of harmonic current emission in low-voltage power supply systems for equipment with a rated current greater than 16 A**
- **IEC/TS EN** 61000-3-5, "**Electromagnetic compatibility (EMC)—Part 3:** Limits–Section 5: Limitation of voltage fluctuations and flicker in low-voltage power supply systems for equipment with rated current greater than 16 A"
- **IEC EN** 61000-3-12, "**Electromagnetic compatibility (EMC)—Part 3–12:** Limits: Limits for harmonic currents produced by equipment connected to public low-voltage systems with input current > 16 A and 75 A per phase"
- **IEC EN** 61000-4-2, "**Electromagnetic compatibility (EMC)—Part 4-2:** Testing and measurement techniques—Electrostatic discharge immunity test"
- **IEC EN** 61000-4-3, "**Electromagnetic compatibility (EMC)—Part 4-3:** Testing and measurement techniques—Radiated, radio-frequency, electromagnetic field immunity test"
- **IEC EN** 61000-4-4, "**Electromagnetic compatibility (EMC)—Part 4-4:** Testing and measurement techniques—Electrical fast transient or burst immunity test"
- **IEC EN** 61000-4-5, "**Electromagnetic compatibility (EMC)—Part 4–5:** Testing and measurement techniques—Surge Immunity Test"
- **IEC EN** 61000-4-6, "**Electromagnetic compatibility (EMC)—Part 4–6**: Testing and measurement techniques" –Immunity to radio-frequency fields-induced conducted disturbances
- **IEC EN 61000-4-7**, "Electromagnetic compatibility (EMC)—Part 4–7: Testing and measurement techniques—General guide on harmonics and interharmonics measurements and instrumentation, for power supply systems and equipment connected thereto"
- **IEC EN** 61000-4-8, "**Electromagnetic compatibility (EMC)—Part 4–8:** Testing and measurement techniques—Power frequency magnetic field immunity test"
- **IEC EN** 61000-4-9, "**Electromagnetic compatibility (EMC)—Part 4–9:** Testing and measurement techniques—Pulse magnetic field immunity test"
- **IEC EN 61000-4-11, "Electromagnetic compatibility (EMC)—Part 4-11: Testing and measurement techniques—Voltage dips, short interruptions, and voltage variations immunity tests"**
- **IEC EN 61000-6-1, TC 77**, "**Electromagnetic compatibility (EMC)—Part 6-1:** Generic standards—Immunity for residential, commercial, and light-industrial environments"
- **IEC EN 61000-6-2, TC 77**, "**Electromagnetic compatibility (EMC)—Part 6-2:** Generic standards—Immunity for industrial environments"
- **IEC EN 61000-6-3, CIS/H, "Electromagnetic compatibility (EMC)—Part 6-3: Generic standards—Emission standard for residential, commercial, and light-industrial environments"**
- **IEC EN 61000-6-4, CIS/H,** "**Electromagnetic compatibility (EMC)—Part 6-4:** Generic Standards—Emission Standard for Industrial Environment"

1.12.5 ISO STANDARD

The following are ISO standards on automotive EMC issues:

- ISO 7637, Road vehicles—Electrical disturbances from conduction and coupling
- ISO 11452-1, Road vehicles—Vehicle test methods for electrical disturbances from narrowband radiated electromagnetic energy—Part 1: General and definition

- ISO 11452-2, Road vehicles—Vehicle test methods for electrical disturbances from narrowband radiated electromagnetic energy—Part 2: Off-vehicle radiation source
- ISO 11452-3, Road vehicles—Vehicle test methods for electrical disturbances from narrowband radiated electromagnetic energy—Part 3: On-board transmitter simulation
- ISO 11452-4, Road vehicles—Vehicle test methods for electrical disturbances from narrowband radiated electromagnetic energy—Part 4: Bulk current injection (BCI)
- ISO 11452-5, Road Vehicles—Component test methods for electrical disturbances from narrowband radiated electromagnetic energy—Part 5: Strip line
- ISO 11452-6, Road Vehicles—Component test methods for electrical disturbances from narrowband radiated electromagnetic energy—Part 6: Parallel plate antenna
- ISO 11452-7, Road Vehicles—Component test methods for electrical disturbances from narrowband radiated electromagnetic energy—Part 7: Direct radio-frequency (RF) power injection
- ISO 11452-8, Road Vehicles—Component test methods for electrical disturbances from narrowband radiated electromagnetic energy—Part 8: Immunity to magnetic fields
- ISO 11452-9, Road Vehicles—Component test methods for electrical disturbances from narrowband radiated electromagnetic energy—Part 9: Portable transmitters
- ISO 11452-10, Road Vehicles—Component test methods for electrical disturbances from narrowband radiated electromagnetic energy—Part 10: Immunity to conducted disturbances in the extended audio frequency range
- ISO 11452, Road vehicles—Electrical disturbances by narrowband radiated electromagnetic energy—Component test methods
- ISO 13766, Earthmoving Machinery—Electromagnetic compatibility
- ISO 14982, Agricultural and forestry machinery—Electromagnetic compatibility—Test methods and acceptance criteria

1.12.6   SAE EMC Standard Committee

The following are some of the SAE standards available for EMC issues:

- J1113/1, Electromagnetic Compatibility Measurement Procedures and Limits for Components of Vehicles, Boats (up to 15 m), and Machines (Except Aircraft) (16.6 Hz to 18 GHz)
- J1113/11, Immunity to Conducted Transients on Power Leads
- J1113/12, Electrical Interference by Conduction and Coupling—Capacitive and Inductive Coupling via Lines Other than Supply Lines
- J1113/13, Electromagnetic Compatibility Measurement Procedure for Vehicle Component—Part 13: Immunity to Electrostatic Discharge
- J1113/21, Electromagnetic Compatibility Measurement Procedure for Vehicle Components—Part 21: Immunity to Electromagnetic Fields, 30 MHz to 18 GHz, Absorber-Lined Chamber
- J1113/26, Electromagnetic Compatibility Measurement Procedure for Vehicle Components—Immunity to AC Power Line Electric Fields
- J1113/27, Electromagnetic Compatibility Measurements Procedure for Vehicle Components—Part 27: Immunity to Radiated Electromagnetic Fields—Mode Stir Reverberation Method
- J1113/4, Immunity to Radiated Electromagnetic Fields—Bulk Current Injection (BCI) Method
- J1752/1, Electromagnetic Compatibility Measurement Procedures for Integrated Circuits—Integrated Circuit EMC Measurement Procedures—General and Definitions

- J1752/2, Measurement of Radiated Emissions from Integrated Circuits—Surface Scan Method (Loop Probe Method) 10 MHz to 3 GHz
- J1752/3, Measurement of Radiated Emissions from Integrated Circuits—TEM/Wideband TEM (GTEM) Cell Method; TEM Cell (150 kHz to 1 GHz); Wideband TEM Cell (150 kHz to 8 GHz)
- J1812, Function Performance Status Classification for EMC Immunity Testing
- J2556, Radiated Emissions (RE) Narrowband Data Analysis—Power Spectral Density (PSD)
- J2628, Characterization—Conducted Immunity
- J551/1, Performance Levels and Methods of Measurement of Electromagnetic Compatibility of Vehicles, Boats (up to 15 m), & Machines (16.6 Hz to 18 GHz)
- J551/15, Vehicle Electromagnetic Immunity—Electrostatic Discharge (ESD)
- J551/16, Electromagnetic Immunity—Off-Vehicle Source (Reverberation Chamber Method) —Part 16: Immunity to Radiated Electromagnetic Fields
- J551/17, Vehicle Electromagnetic Immunity—Power Line Magnetic Fields
- J551/5, Performance Levels and Methods of Measurement of Magnetic and Electric Field Strength from Electric Vehicles, Broadband, 9 kHz to 30 MHz

1.12.7 European Standards Concerning Unwanted Electrical Emissions

- EN 50 081 part1 European Generic emission standard, part1: Domestic, commercial and light industry environment, replaced by EN61000-6-3
- EN 50 081 part2 European Generic emission standard, part2: industrial environment, replaced by EN61000-6-4
- EN 55 011 European limits and methods of measurement of radio disturbance characteristics for scientific and medical equipment
- EN 55 013 European limits and methods of measurement of radio disturbance characteristics of broadcast receivers
- EN 55 014 European limits and methods of measurement of radio disturbance characteristics of household appliances and power tools, replaced by EN55014-1, and immunity part is covered by EN55014-2
- EN 55 015 European limits and methods of measurement of radio disturbance characteristics of fluorescent lamps
- EN 55 022 European limits and methods of measurement of radio disturbance characteristics of information technology equipment
- EN 55 032 Electromagnetic compatibility of multimedia equipment—Emission requirements
- EN 60 555 parts 2 and 3 Disturbances of power supply network (part 2) and power fluctuations (part 3) caused by of household appliances and power tools, replaced by EN61000-3-2 and EN61000-3-3
- EN 13309 Construction Machinery—Electromagnetic compatibility of machines with internal electrical power supplies
- VDE 0875 German EMC directive for broadband interference generated by household appliances
- VDE 0871 German EMC directive for broadband and narrowband interference generated by information technology equipment

1.12.8 European Standards Concerning Immunity to Electrical Emissions

- EN 50 082 part1 European immunity standard, part1: Domestic, commercial and light industry environment, replaced by EN61000-6-1

- EN 50 082 part2 European immunity standard, part2: industrial environment, replaced by EN61000-6-2
- EN 50 093 European, immunity to short dips in the power supply (brownouts)
- EN 55 020 European, immunity from radio interference of broadcast receivers
- EN 55 024 European immunity requirements for information technology equipment
- EN 55 101 older draft of immunity requirements for information technology equipment, replaced by EN 55 024
- EN 50 081 part1 European Generic emission standard, part1: Domestic, commercial and light industry environment, replaced by EN61000-6-3
- EN 50 081 part2 European immunity requirements for information technology equipment, replaced by EN61000-6-4

1.12.9   AMERICAN STANDARD

- FCC Part 15 regulates unlicensed radio-frequency transmissions, both intentional and unintentional
  - FCC Part 15 Subpart A contains a general provision that "devices may not cause interference and must accept interference from other sources"
  - FCC Part 15 Subpart B US limits and methods of measurement of radio disturbance, measuring radio waves accidentally emitted from devices not specifically designed to emit radio waves ("unintentional"), both directly ("radiated") and indirectly ("conducted")
  - The rest of FCC Part 15 (subparts C through H) deal with unlicensed devices specifically designed to emit radio waves ("intentional"), such as wireless LAN, cordless telephones, low-power broadcasting, walkie-talkies, etc.
  - Conducted emissions are regulated from 150 kHz to 30 MHz, and radiated emissions are regulated from 30 MHz and up
- **MIL-STD 461** is a US Military Standard addressing EMC for subsystem and components. Currently in revision G, it covers Conducted and Radiated Emissions and Susceptibility
- **MIL-STD 464** is a US Military Standard addressing EMC for systems. Currently in revision G, it covers Conducted and Radiated Emissions and Susceptibility

MIL-STD-469, MILITARY STANDARD: This standard establishes the engineering interface requirements to control the electromagnetic emission and susceptibility characteristics of all new military radar equipment and systems operating between 100 megahertz (MHz) and 100 gigahertz (GHz).

1.13   SUMMARY

In this chapter, it is discussedall uses of EM waves so-called EM spectrum and also viewed on major disturbances of EM. The importance of EMC factor for product development is clearly understood. The real-world applications are mainly based on making use of the EM field. In future, as the signal speed increases, it is necessary to address EM issues on all the electronic products. Therefore, the evolution and uses of EM waves or spectra never end.

BIBLIOGRAPHY

1. A. Azpurua, Marc Pous, and Ferran Silva, "Decomposition of Electromagnetic Interferences in the Time-Domain," *IEEE Transactions on Electromagnetic Compatibility*, vol. 58, no. 2, pp. 385–392, April 2016. doi: 10.1109/TEMC.2016.2518302
2. Clayton R. Paul, Introduction to Electromagnetic Compatibility, A John Wiley & Sons, Inc. Publication, Second Edition, 2006.

3. Florian Krug, and Peter Russer, "The Time-Domain Electromagnetic Interference Measurement System," IEEE Transactions on Electromagnetic Compatibility, vol. 45, no. 2, pp. 330–338, May 2003. doi: 10.1109/TEMC.2003.811303

4. Tongkai Cui, Qishuang Ma, Ping Xu, and Yuchen Wang, "Analysis and Optimization of Power MOSFETs Shaped Switching Transients for Reduced EMI Generation," *IEEE-Access*, vol 5, pp. 20440–20448, October 2017. doi: 10.1109/ACCESS.2017.2758443

5. François Costa, Eric Laboure, and Bertrand Revol, *Electromagnetic Compatibility in Power Electronics*. Electronics Engineering Series. London England: Wiley-ISTE, 2014.

6. William G. Duff, *Designing Electronic Systems for EMC*. SciTech Series on Electromagnetic Compatibility. Scitech Publishing, Inc. 2011.

7. Fabio Pareschi, Riccardo Rovatti, and Gianluca Setti, EMI Reduction via Spread Spectrum in DC/DC Converters: State of the Art, Optimization, and Tradeoffs.

8. Andreas Fhager, Parham Hashemzadeh, and Mikael Persson, "Reconstruction Quality and Spectral Content of an Electromagnetic Time-Domain Inversion Algorithm," *IEEE Transactions on Biomedical Engineering*, vol. 53, no. 8, pp. 1594–1604, August 2006. doi: 10.1109/TBME.2006.878079

9. Jeffrey H. Reed, Jennifer T. Bernhard, and Jung-Min Park, "Spectrum Access Technologies: The Past, the Present, and the Future," *Proceedings of the IEEE*, vol. 100, pp. 1676–1684, May 2012. doi: 10.1109/JPROC.2012.2187140

10. Fatemeh Abolqasemi Kharanaq, Ali Emadi, and Berker Bilgin, "Modeling of Conducted Emissions for EMI Analysis of Power Converters: State-of-the-Art Review," *IEEE Access*, vol. 8, pp. 189313–189325, October 2020. doi: 10.1109/ACCESS.2020.3031693.

2 Electromagnetic Interference Events

2.1 INTRODUCTION

Conducted electromagnetic interference (EMI) is the result of direct contact between the conductors, which are opposed by radiated EMI as generated by induction. Lower frequency and higher frequency EMI are triggered by conduction and radiation. Both compliance and pre-compliance EMI testing are possible, with the latter involving highly accurate simulations of the former in terms of hardware, software, and testing methodology. Emitter/source, which serves as the source of undesirable interferences. Coupling channel, transmits the interference from the source to the receiver. If the coupling channel is conducting, EMI is said to result from conducted emission. Emission via radiated channels only occurs if the coupling channel is of the radiated kind. Intersystem EMI develops when interference occurs between two or more independent systems, while intra system EMI occurs between parts of the same system. EMI can be measured and categorized based on all these criteria.

Most modern electronic applications employ small, compact devices that contain hundreds of thousands or millions of passive or active components to execute numerous functions concurrently, although arranging such a large number of components on a device with dimensions of no more than a few centimetres provides a formidable challenge in and of itself. The greatest obstacle for any electronic hardware engineer or manufacturer is the development of high-speed multifunctional devices that are not only compact but also electromagnetically compatible with their predecessors.

2.2 EMI MEASUREMENT TECHNIQUES

The EMI measurement techniques come under two processes:

- Emission testing
    1. Radiated emission
    2. Conducted emission
- Immunity testing
    1. Continuous sources
    2. Transient sources

We can evaluate radiated emission using many ways, for instance:

- Open-Air Test Sites (OATS)
- Anechoic Chamber
- Gigahertz Transverse Electromagnetic (GTEM) Cell
- Reverberation Chamber

As well as we can test conducted emission using different methods, such as:

- Line Impedance Stabilization Network (LISN)
- 1 ohm Method

DOI: 10.1201/9781003362951-2

- Probes
  1. Current probe
  2. Voltage probe
- Transverse Electromagnetic (TEM) Cell

2.3 RADIATED EMISSIONS

When conducting compliance testing at the test site, radiated emissions pose the greatest potential danger. Harmonics of clock frequencies and other fast-edged devices are easily radiated by today's electronic devices due to the extensive use of digital circuitry.

2.4 RADIATED EMISSIONS CHECKLIST

Very high-frequency (or radio frequency, RF) sources of energy can be generated using only a small current and correspondingly small voltage, leading to radiated emissions. Cross-coupled noise and parasitic energy are often the issue. Any metallic object, especially a cable, can act as an antenna. So, consider the following:

- The chassis may emit radiation at frequencies above 200 MHz. The chassis or, in the case of an open frame or chassis-less machine, the circuit board will be the most probable reason at higher frequencies.
- Taking for granted the low impedance of all shielded wires below 200 MHz, emitted radiation is most likely originating from the cables themselves. At lower frequencies, when the wavelengths are larger, wires or cables make for more effective antennae.
- United at both ends, it ensures that the shields have a direct connection to the chassis or the connector. Use a pigtail only if necessary.
- Make sure the chassis' metal pieces are in perfect contact with one another. Paint or other coatings, grease or dirt, corrosion or oxidation shouldn't be present as they could cause impedance.
- Make sure each line exiting the apparatus is filtered, and that the filter is situated close to the point where the line breaks through the apparatus.

2.4.1 FAILURE MODES

The majority of goods do not pass the radiated emissions test because of leaky chassis enclosures or radiated connections.

2.4.2 CABLE RADIATION

The usage of pigtails is recommended when power cables emit high-frequency harmonics due to poor shield bonding to the chassis or enclosure, insufficient filtering, or being poked through the shielded enclosure. The necessity for actual length to create a good antenna is just what causes lower frequency emissions to originate from the cables. The cables are typically the longest part, making it the major source of low-frequency radiation.

2.4.3 METAL CHASSIS

The equipment's metal chassis is a common source of emission at higher frequencies (usually greater than 200 MHz). The input/output (I/O) cables become more radiated at these higher frequencies because their inductive nature makes them a higher impedance path for RF currents than the chassis. When the equipment is being tested physically large, makes an exception.

Internally mounted circuit boards are capable of generating currents that travel down the inner surface of the chassis. These HF currents escape through seams or openings in the chassis or enclosure and travel around its exterior. This makes the entire enclosure a transmitting antenna. When the majority of the current can be returned to the source extremely close to where it is coupled onto the chassis, this is an exception. This is why it is said to be good to use reference return planes that are well bonded to the chassis as reference return planes on a circuit board.

2.5   RADIATED EMISSION TESTING

A common way is to test for radiated emission electromagnetic interference (RE EMI). Usually, radiated emission tests were typically conducted between 30 MHz and 1 GHz (corresponding wavelengths of 10 and 0.3 m). An open area test site is one of the most typical locations for radiated emission testing on large instruments (OATS). This type of setup consists of a theoretically infinite metallic ground plane, a receiving antenna connected via cables to an EMI receiver or spectrum analyser, and the EUT, which is typically maintained 3 or 10 m away from the receiver. The reception antenna was used as a reference point to determine how far away the EUT was. This huge gap was principally chosen to guarantee that the observations were made in the far-field region, where the radiated field is more stable than in the close field or Fresnel zone. In reality, distances of 30 m were permitted in the early days of EMI testing in an OATS setting. Researchers are considering adopting a distance of 5 m between the EUT and the receiving antenna; however, they have previously only been able to do so at a distance of 3 or 10 m. As the study is still in its development, the OATS measurement setup with a 5-m distance has not yet been standardized. The maximum size of EUTs allowed for OATS testing is determined by the distance between the EUT and the receiving antenna, as stipulated in standards like CISPR EMI. The receiving antenna evaluated the EUT's radiated emissions in both vertical and horizontal polarizations. The success or failure of the EUT was determined by comparing the measured field strength to the field strength specified in the standard.

Multiple input multiple output (MIMO) systems can be directly tested for their radiated emissions utilizing over-the-air (OTA) measurements. OTA tests are performed to evaluate the MIMO system's radiation performance. The resulting measure can also be utilized to determine the receiver sensitivity degradation caused by spurious noise signals created within the MIMO system. The two-stage procedure is one of the most popular and reliable approaches of OTA testing due to its low cost and high accuracy. In the beginning, a regular anechoic chamber would have produced all feasible three-dimensional radiation patterns for all polarizations. When the selected base station emulator and channel model have been applied to all the collected patterns, the resulting matrix is then utilized to calculate the RE EMI associated with the MIMO system. However, these tests were conducted in specialized chambers, including an anechoic chamber, a reverberation chamber, or a gigahertz transverse electromagnetic cell, for smaller prototypes (GTEM cell) (Figure 2.1).

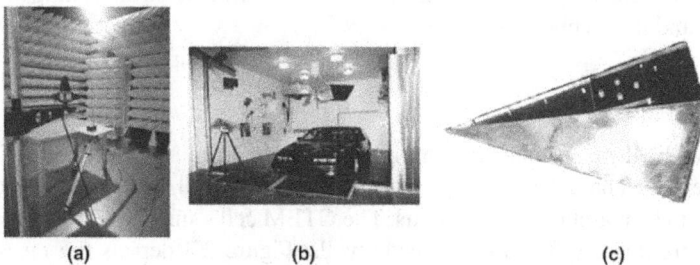

(a)   (b)   (c)

FIGURE 2.1   Radiated emission testing chambers. (a) Anechoic chamber. (b) Reverberation chamber. (c) Gigahertz Transverse Electromagnetic (GTEM) cell.

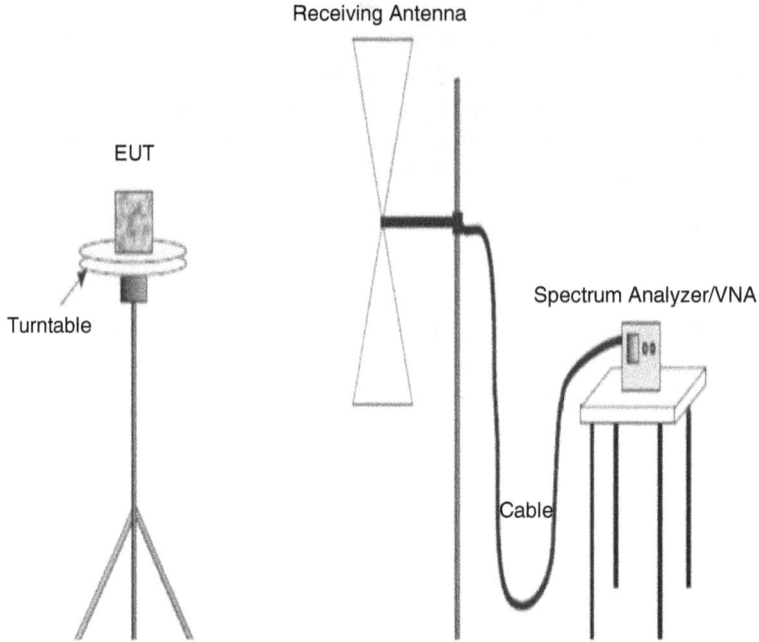

FIGURE 2.2　Radiated emission test setup in an anechoic chamber.

2.5.1　ANECHOIC CHAMBER

In order to take accurate electromagnetic measurements, the walls and floor of an anechoic chamber were covered with microwave absorbers, either entirely (in the case of a full- or half-anechoic chamber) or partially. RE testing was performed in a semi-anechoic chamber (SAC) to measure the cable's common-mode (CM) current and radiated transfer function to determine the interferences caused by the cable connecting a mobile instrument. The authors compared their results to the EN 55022 (CISPR 22) standard using an IEC 61000-4-3 standard measuring equipment. The frequency range for the analysis is 30. In order to determine the directivity of electrically massive EUTs and to correlate the radiated emissions for similar EUTs under different test settings, Wang and Vick conducted the tests in an anechoic chamber. Gao et al. hypothesized that RE EMI emitted by the battery management system (BMS) of an electric vehicle might cause severe damage to the on-board circuitry and the operation of a neighbouring electric vehicle. To find out what was causing the electric vehicle's emission peaks to rise above acceptable levels, they performed a diagnostic analysis. Anechoic chambers are commonly used to conduct emission tests for EMI measurements, and Figure 2.2 shows that a separation distance of 10 m was utilized during testing between the EUT (electric vehicle) and the receiving antenna.

2.5.2　GTEM CELL

Emission tests for EMI measurements can also be performed in a GTEM cell, in addition to an anechoic chamber. The characteristic impedance of the GTEM cell, a tapered two-conductor transmission line, was to be maintained at 50 ohms. The GTEM cell's smaller end serves as the I/O port, while RF absorbers line the flared ends inside walls. Figure 2.3 depicts the measurement setup for radiated emission testing using a GTEM cell, in which the GTEM I/O port was linked to the spectrum analyser or a network analyser. The comparable OATS field strength from the measured emissions from a portable power bank were extracted by using the GTEM cell from the equation

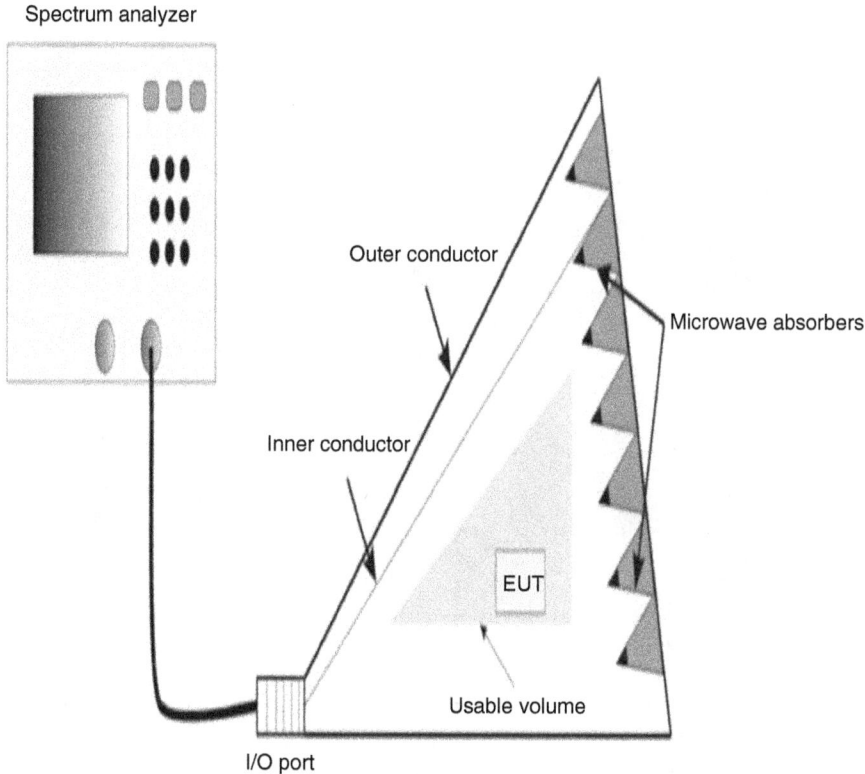

FIGURE 2.3 Measurement setup for electromagnetic interference (EMI) testing using a gigahertz transverse electromagnetic (GTEM) cell.

$$E_{m/dB} 20\log(g_m) + 20\log\left(\frac{n_o.k_o}{2\pi e_{oy}}\right) + 10\log\left(\frac{V_x^2 + V_y^2 + V_z^2}{Z_c}\right) + 120$$

where $E_{m/dB}$, maximum electric field strength; g_m, variable dependent on the antenna and geometry of the EUT to be used in the OATS setup; Ko, wave number; e_{oy}, field in TEM mode; and Z_c, characteristic impedance and some output voltages in three orthogonal directions along which the EUT is sequentially oriented. This was done to guarantee that the criteria were met by the final product. When compared to other options like anechoic chambers or open-air test structures (OATS), the GTEM cell has the benefit of being both smaller and cheaper (hence easy to maintain). In addition, GTEM cells can deliver reasonably precise data between 0 Hz and 20 GHz.

2.5.3 Reverberation Chamber

Reversible chambers (RCs) are metal enclosures with a high-quality factor (Q) that typically house a transmitter antenna, receiver antenna, and end-user terminal (EUT). Standing waves generated at the chamber's metallic boundary generate an uneven electric field inside the chamber. By adjusting the stirrers' orientation, we may change the chamber's boundary conditions and create a more uniform electric field. The image depicts a typical EMC testing laboratory with a reverberation chamber. The greatest benefit of employing a reverberation chamber for EMI testing is that it provides the most precise and dependable findings in the shortest amount of time, especially for radiated immunity testing. This is because, unlike in an anechoic room, the EUT was exposed to a field from all sides at once. Therefore, there is no longer any need to manually rotate the EUT in the correct

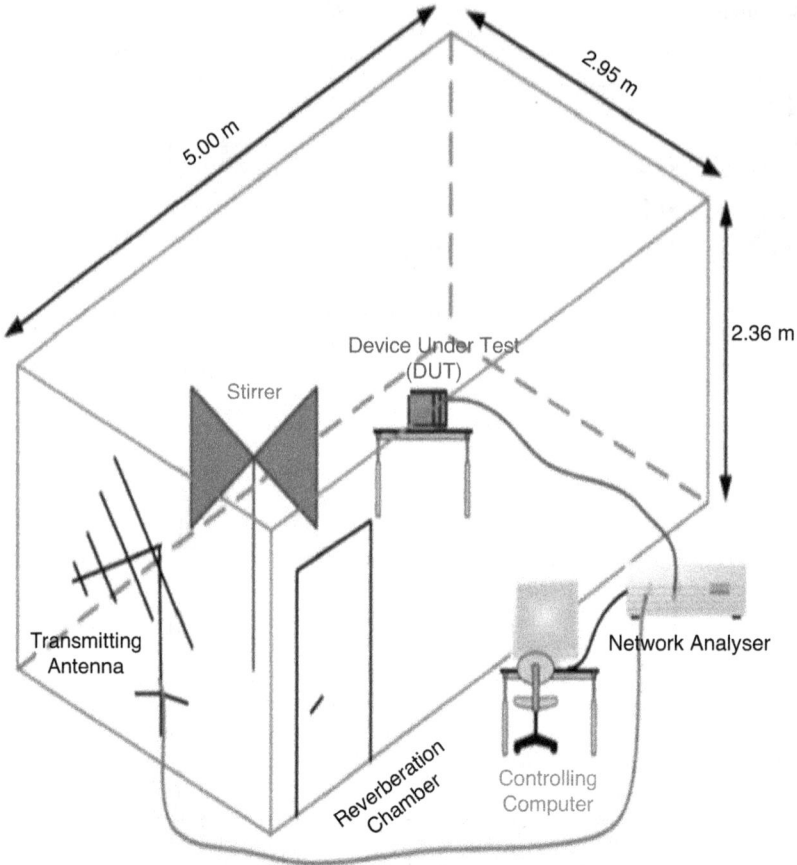

FIGURE 2.4   Measurement setup for electromagnetic interference (EMI) emission testing using reverberation chamber.

directions in order to take the necessary measurements. However, directivity and polarization data are lost during the process. The approach was cost-effective since microwave and RF absorbers weren't required inside the chamber (Figure 2.4).

2.6   TEST SETUP FOR RADIATED EMISSION MEASUREMENT

Radiated emissions from the first generation of televisions can be measured with this test apparatus. The measurements of the original television, without any changes to the chassis or internal components, are referred to as baseline assessment. For attaining maximal emission values, the turntable was rotated from 0 to 360 degrees in 10 degree increments utilizing the max hold mode. Both the original Device Under Test (DUT) and a copy with the metal back cover removed were used in the measurement. By comparing the outcomes of the two tests, we can assess the metal back cover's shielding efficiency. Maximized radiation from 30 to 1,000 MHz over all angles of the turntable rotation for both polarizations is plotted using post-processed data from the two test scenarios.

Observing the difference in EMI levels between the television with its back cover removed and not removed demonstrates the intricacy of EMI in electrically big constructions with numerous holes. The most intense emissions occur at 540 MHz and are therefore of greatest interest. Also, when using vertical polarization, removing the back cover results in a 15 dB reduction in emissions. This seemingly counterintuitive finding is analysed by comparing the radiated patterns for the two

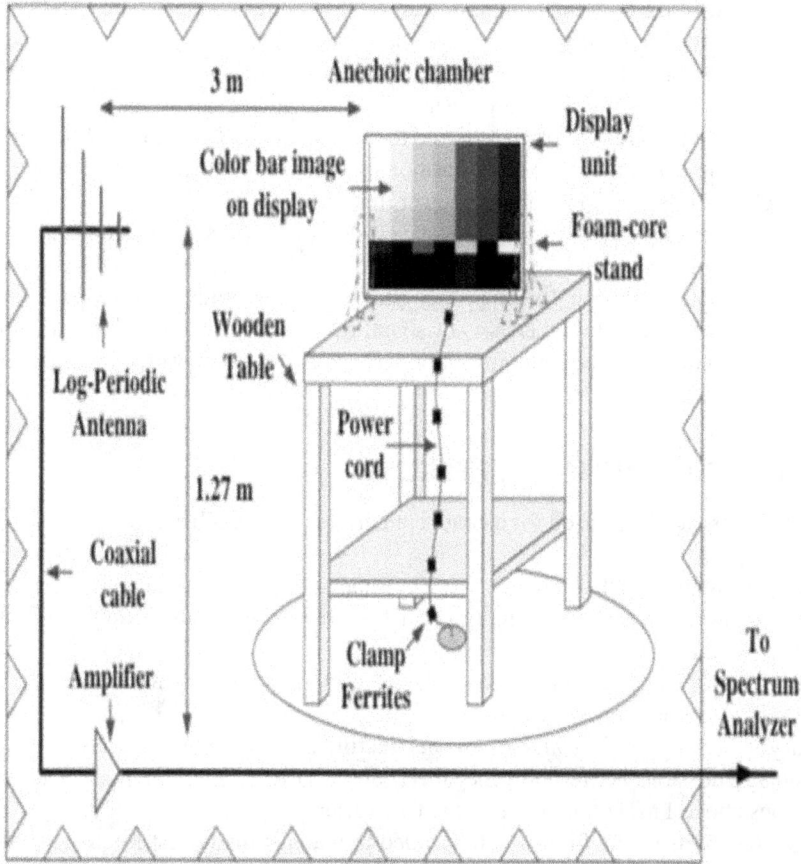

FIGURE 2.5 Test setup for radiated emission measurement.

test cases at 540 MHz, which shows that the back cover increases emissions from the angles 330 to 60 degrees (front side of the TV) and from 150 to 240 degrees (back side of the TV) in the vertical polarization (Figure 2.5).

In a reverberation chamber, the total radiated power of the television is also measured. There is a comparison of the total radiated powers for the two scenarios. The data demonstrate that the back cover decreases the overall radiated power at 540 MHz by approximately 5 dB. In addition to acting as a shield against the radiation from the TV, the back cover appears to boost the vertical field component of the emitted radiation. Even if the total radiation power is lower, the anechoic chamber measurements indicate that the inclusion of the rear cover enhances the radiation in the vertical polarization at 540 MHz and in certain directions.

2.7 CONDUCTED EMISSION

It should be simpler to limit and prevent conducted emissions than radiated. Its lower frequency makes it less vulnerable to parasitic interference than signals operating at higher frequencies. However, they remain a problem that needs attention. As a result, the root causes of and remedies for conducted emissions are more intuitive to grasp than those for radiated emissions. Switch-mode power supplies (SMPS) are the primary source of conducted emissions, and the best power supply designs typically provide effective filtering of the power input.

However, many original equipment manufacturer (OEM) power supplies are poorly constructed, have dreadful emissions, and nonetheless have the Federal Communications Commission (FCC) and Conducted Emission (CE) marks. When these power supplies are loaded with a reactive load, as opposed to the resistive load they were designed for, the power supply may become unstable or noisy, and additional measures are often required to keep it in compliance. In addition, with modern products, it is possible for higher frequency harmonics to contaminate the system power supply and leak out through the filter and back into the power line.

So, while we know from experience that most well-designed filters are adequate, it's important to be on the lookout for cases where the filter is inadequate, either because of its design or because of problems with the system's design, like improper internal cable routing, filter or power supply placement, or a lack of connection to the chassis or signal returns. High emissions can disturb delicate adjacent measuring equipment or communications receivers that are powered by the same circuit or by the same power line as the product itself.

2.8   CONDUCTED EMISSIONS CHECKLIST

The following items should be used for a conducted emissions checklist:

- In the spectrum above 10 MHz, distinct emission spikes (narrowband emissions) may be caused by internal digital clocking or other high-frequency sources coupling around the power input filter.
- The most common cause of broad peaks of closely spaced harmonics below roughly 10 MHz is switching power supplies.
- Primary switching devices and AC power rectifiers are common sources of broadband emissions. Emissions below a few megahertz are usually differential mode in nature.
- Emissions above 1 MHz tend to be of the CM variety.
- The greatest parts of the filter will be needed to manage the lowest frequency emissions. Even so, it's important to take precautions to prevent leakage currents from becoming too high.
- Both line-to-chassis capacitors and CM inductors are effective at dampening CM noise. CM noise cannot be filtered by line-to-line capacitors.
- The filter should be mounted as close as possible to the product's powerline connection.

2.8.1   Failure Modes

Most problems caused by conducted emissions tend to happen at the limits of the frequency range. These events have a bias toward either the greatest or lowest frequency limit. Inadequate filter components are often the root cause of low-frequency emissions. The filter frequently uses big value line-to-line capacitors and large-series CM chokes and series inductors. The difficulties arise when the necessary components to adequately filter the equipment do not fit or are judged too heavy, which is the case when the equipment's size and weight become issues. A portion of this can be regulated through the right location and technology of the filter components. However, the principles of physics must be followed, and the values may still be substantial.

Parasitic and cross-coupled noises are commonly the root of the problem when it comes to high-frequency emissions. The positioning of filters and the twisting of connected cables gain more significance. When designing, it is important to pay close attention to it and manage the placement of any components that will produce magnetic fields. Problems may arise if a filter is not placed close to the power connector to which it is connected.

Other concerns include the misapplying components, such as utilizing electrolytic capacitors to filter frequencies beyond 200 kHz, employing ferrite cores as linear inductor cores, and not filtering the neutral or power return lines because of the misconception that grounds are noise-free. If a discrete filter module is located too far away from the power line entrance location or if the input and output wires of the filter are close together, it is possible for RF noise currents to bypass the filter and exit on the power line.

2.9   EMI CONDUCTED TERMINOLOGY

The phase conducted EMI refers to EMI that is connected between circuits, equipment, or systems because it is conducted along a power or signal wire or cable. Voltage and current are the common units of measurement for conducted EMI.

2.10   DECIBELS

EMI characterization generally necessitates dealing with signal and susceptibility levels spanning many orders of magnitude. For example, receivers often have sensitivities on the order of 1CT13 watts, whereas high-power transmitters have power outputs on the order of kilowatts or megawatts. Usually, signals with a wide dynamic range are shown on a logarithmic scale so that the resolution may be preserved when the signal spans more and more decades. In the EMC field, the decibel is a popular logarithmic representation.

$$Db = 10\log(P_1/P_2)$$

2.11   STANDARDIZED MEASUREMENTS OF CONDUCTED EMI

Originally, radio television (RTV) signal reception integrity has been the primary emphasis of electromagnetic compatibility (EMC) guidelines. Therefore, it seemed reasonable to use equipment in EMC testing that resembles a conventional RTV receiver. The majority of RTV signal receivers are super heterodyne, hence super heterodyne receivers were chosen for EMC testing. Typical practice in electromagnetic emission testing involves using standardized instruments and comparing the results obtained in a standard configuration to predetermined limits associated with the tested device (EUT). Knowledge of specialized EMI measuring techniques, especially in the conducted EMI frequency band, is necessary for the evaluation of EMC from a definitional (not EMC standard) perspective and for the creation of EMI mitigation measures.

2.12   CONDUCTED EMISSION TESTING

Testing was performed to estimate or determine the noise caused by a sudden change in voltage or current in the equipment's circuitry and emitted through the power cable into peripheral or loaded devices. The interference could be fatal to the connected devices and cause them to break down. Common techniques used in emission lab tests include:

- Super heterodyne EMI Receiver
- LISN

2.12.1   SUPER HETERODYNE EMI RECEIVER

Measuring voltages that occur at a receiver's input is typically how conducted EMI is measured. Specific supplemental devices, such as the line impedance stabilization network (LISN), current probes, field probes, and absorption clamps, are utilized to convert other physical values, such as current, electromagnetic field, and radiated power, into voltage. The measured voltage is only one component of a typical conducted EMI measurement result. To fully utilize the information contained in typical, normalized, performed EMI measurement findings, it is required to have a

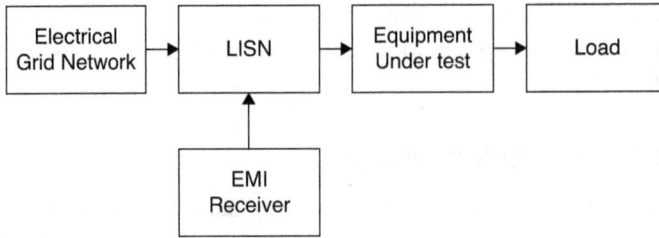

FIGURE 2.6   Block diagram of electromagnetic interference (EMI) receiver.

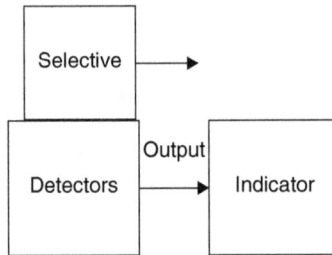

FIGURE 2.7   Main functional element of the electromagnetic interference (EMI) receiver.

fundamental understanding of the applicable measuring technique. EMI's receiver is a super heterodyne selective micro-voltmeter with a specially stated connection between reading and input voltage. Each receiver's characteristics are normalized to ensure reliable and predictable readings. Responses of the receiver to the specified signals are described in the standard (Figure 2.6).

- Sinusoidal (continuous wave) of determined amplitude A,
- Pulse (pulse train of amplitude A, duration ti, pulse repetition rate fr),
- Noise (normal distribution of amplitude probability).

The EMI receiver must adhere to industry standards in the following areas.

- Pulse response,
- Selectivity (band pass in individual frequency ranges, attenuation of intermediate frequency (IF) signals, attenuation of mirror frequencies, and other unwanted responses),
- Intermodulation effects,
- Limitation of noises and internal unwanted signals (internal noises, continuous wave signals),
- Shielding.

The block diagram of the typical EMI receiver comprises the following: (1) attenuator, (2) preselection, preamplifier, (3) mixer, (4) local oscillator, (5) IF bandwidths, (6) detectors, (7) display, (8) loudspeaker, and (9) internal reference generator. The primary functions of an EMI receiver are typically described in detail to aid in the reader's comprehension of the device's working principles like selective element, detectors, and indicator. Figure 2.7 depicts the layout of the primary functional aspects (Figure 2.8).

2.12.2   Line Impedance Stabilization Network

LISN is to ensure constant, stable, and consistent measuring conditions of conducted interferences introduced. LISN is employed in EUT for the conducted emission measurement.

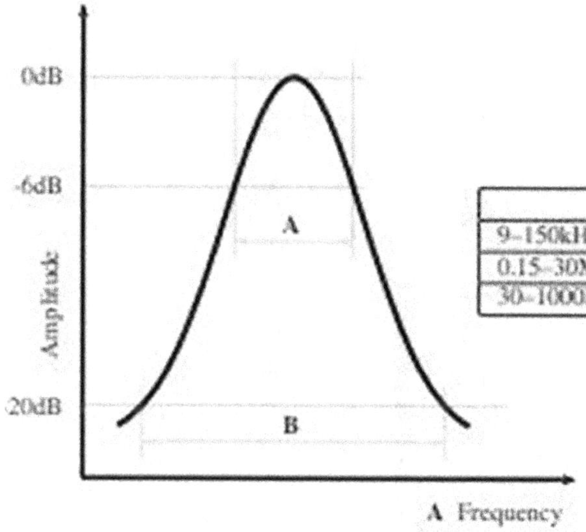

	A	B
9–150kHz	170–220Hz	170–440Hz
0.15–30MHz	8–10kHz	8–20kHz
30–1000MHz	100–140kHz	100–280kHz

FIGURE 2.8 IF bandwidth specification according to CISPR 16-1-1.

The LISN simultaneously serves three basic purposes:

- It is important that the high-frequency components present in the power mains are isolated from the EUT side of the LISN, which is why the LISN supplies the supply voltage at the frequency of the power mains. Power line noise that is linked to the EUT side is mistaken for EMI produced by the EUT.
- LISN guarantees constant impedance between all of the line wires and the PE wire in the CISPR A and CISPR B frequency bands. The impedance of an LISN at various frequencies is shown in the following diagram.
- A high-pass filter built into the LISN couples HF interference signals produced by the EUT to the EMI receiver. An additional benefit of the LISN configuration is that it guarantees impedance matching with a 50 load, which is the EMI receiver's input impedance.

Conducted emission, which is recorded by the detectors as a voltage drop on the EMI receiver's input impedance, is the result of interference currents created by the EUT and flowing through a circuit consisting of the standardized impedance of the LISN. Noises caused by fluctuations in the impedance of the AC power supply can be suppressed by installing an LISN between the EUT and the power supply. The LISN maintains constant impedance at a fixed frequency for the EUT to operate upon. Then, the procedure outlined in CISPR 25, 2010 is applied to measure the noise voltage. The LISN's noise voltage measuring receiver is linked to the network's RF input. A typical measurement setup for conducted emission testing using an LISN is shown in Figure 2.9. The accuracy of this approach is limited to 100 MHz; consequently, it cannot be employed. The voltage drop between the external resistances connected in series with the EUT's pins is used to calculate the conducted emission noise, as described in the 1 ohm noise measurement described in IEC 61967-4. This technique provides accuracy in the gigahertz range.

A probe is a simpler way to calculate the EUT's conducted emission. The power lead of the EUT serves as the primary coil in a current transformer, while the probe itself serves as the secondary coil. The probe consists of a wire coiled around a ferrite core. In order to quantify the volume of conducted emission noises, voltage probes are also employed. A modest resistance is linked to the voltage probe and a sensitive oscilloscope. After determining the voltage across the resistor, the fast fourier transform (FFT) transforms the voltage into the current spectrum (Figure 2.10).

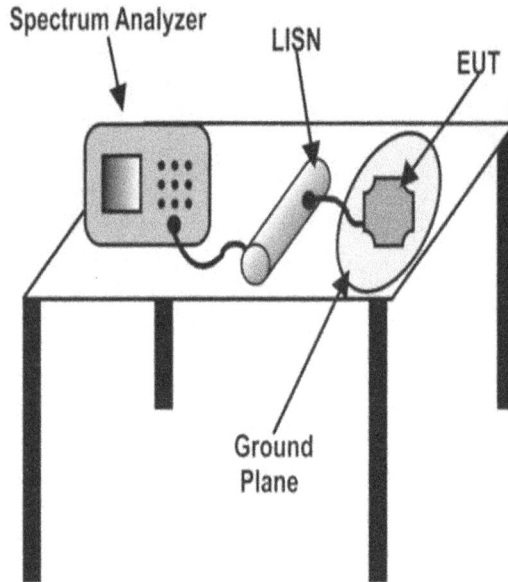

FIGURE 2.9 Measurement setup for electromagnetic interference (EMI) conducted emission testing using line impedance stabilization network (LISN).

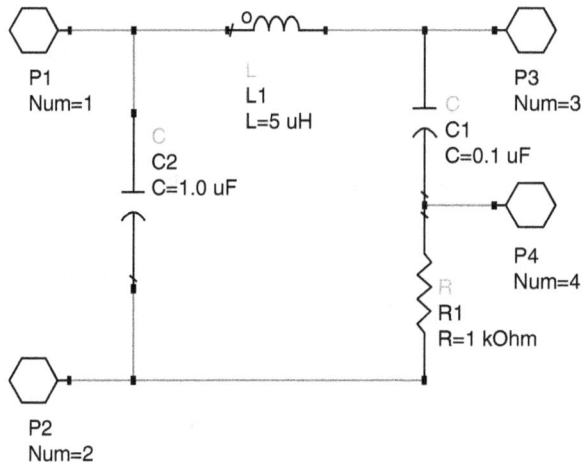

FIGURE 2.10 Model of line impedance stabilization network (LISN) block.

2.13 MEASUREMENT SELECTIVITY

The use of selected elements ensures that interference amplitude measurements have a high degree of resolution. Receiver detectors are supplied with IF signals filtered by Gaussian filters with specified bandwidth. Figure shows specification of the IF BW filter for various frequency ranges of EMI measurements. Measurements of a pure sinusoidal interference signal using an EMI receiver with standard filters of IF BW = 200 Hz and IF BW = 9 kHz for peak and average detectors show that the use of a selective receiver produces an image of the filter rather than the single expected spectral line predicted by Fourier's theorem. The accompanying diagram depicts the steps involved in conducting a standard EMI scan with an EMI receiver. Selective element filters are applied to the input signals of the detectors once the superheterodyne receiver has been set to the desired measuring frequency. Because the measuring step depends on the IF BW of the filter, a faster scan is possible with a filter with a broader IF bandwidth (typically, the measuring step should not be bigger than half of the 6 dB bandwidth). However, the higher the bandwidth, the lower the resolution. This is why the standard defines the B6 bandwidths for various CISPR frequency ranges with great precision.

2.14 MEASURING DETECTORS

EMI receivers are typically equipped with numerous types of detectors. Nevertheless, the fundamental detector is the quasi-peak detector. In order to determine the nature of the interference, many additional detectors (peak, average, RMS, etc.) are used. Performing interference analyses like these calls for a foundational understanding of standard detectors. Different types of detectors are distinguished by their charging and discharging time constants, respectively.

2.14.1 PEAK DETECTOR

The smallest time constant of the RC circuit allows the peak detector to provide the fastest feasible sweep. The above diagram is a block diagram of a peak detector that describes how it works. Since the capacitor is charged to the maximum level of the IF signal envelope, a peak detector can track the quickest changes in the IF signal's envelope, but not its instantaneous value. In their default settings, most spectrum analyser and EMI receivers show signals in peak. Since peak detector readings are always greater than average and quasi-peak detectors' indication, using the peak detector sweep is a fast and straightforward technique to compare produced results with limit lines during pre-compliance, engineering test.

2.14.2 QUASI-PEAK DETECTOR

In order to calculate the so-called "annoyance factor," the quasi-peak detection mode uses an integration circuit with two distinct time constants. Since the quasi-peak detector's charge rate is substantially larger than its discharge rate, the detector's output grows as the signal repetition rate increases. This detector's response to signals of varying amplitude is linear. It's possible that high-amplitude, low-repetition rate signals would have the same effect as low-amplitude, high-rate ones.

2.15 PARAMETERS ASSOCIATED WITH EMI MEASUREMENT TECHNIQUES

2.15.1 RADIATED SUSCEPTIBILITY

The system's radiated susceptibility indicates how well it performs in the presence of an imposed electromagnetic field. Radiation susceptibility testing is also known as radiation immunity testing. In the presence of a fluctuating electromagnetic field, a current is induced in a conductor. The strength of an inducing field is directly correlated with the magnitude of the induced current and the induced voltages at the circuits at both ends of the wire.

A system's susceptibility threshold, or limit, is the field strength that causes a current that is too large for satisfactory operation. Accordingly, the gadget performs satisfactorily at field strengths below the susceptibility threshold. The device's performance degrades beyond acceptability once the applied field strength is beyond the susceptibility threshold. This restriction often changes with changing frequencies in a given system. During radiated susceptibility testing, a device is subjected to electromagnetic fields of a predetermined amplitude and frequency range. The device is regarded to have passed if it continues to function suitably as the field is applied and swept across the designated frequency range. If not, then it was a bust. When the field is no longer there, many devices

TABLE 2.1
IEC-61000-4-3 Test Levels

Level	Test Field Strength (V/m)
1	1
2	3
3	10
4	30
X(1)	Special

that were negatively impacted by all this resumed normal operation. As a rule, radiated EMI occurs when the dimensions of the tested apparatus are of the same order of magnitude as the wavelength of the interfering signal. The applicable test standard is IEC-61000-4-3. The process is carried out in a SAC, same as the radiated emissions setup, with the signal chain swapped around. The range of frequencies scanned is narrower, between 80 MHz and 1 GHz. The modulation is 80% AM with a 1 kHz sine wave. The performance of the system is typically described in terms of the stress field levels at which it continues to function normally. Table 2.1 displays the various prescribed test levels for 80–1,000 MHz.

2.15.2 INTRODUCTION TO RADIATED SUSCEPTIBILITY

Determining if your product is vulnerable to interference from external RF fields is a crucial part of ensuring it is EMC compliant. According to the International Electrotechnical Commission (IEC) 61000-4-3 standard for industrial and commercial products, this evaluation is known by a variety of other names. Depending on the product's environment or intended use, commercial devices undergo testing at frequencies between 80 and 1,000 MHz and E-field strengths between 3 and 20 V/m. This test was done in a SAC with a broadband antenna pointing in the direction of the product being tested. The insulated room ensures no other communication services are disrupted. Testing to voltages of 200–1,000 V/m and frequencies of 18 GHz or higher may be necessary for some military, vehicle, or aerospace applications. The RF signal is commonly modulated at 1 kHz with a square wave or short-duration (sometimes less than 1%) pulsed modulation for military and aerospace testing, and with a 1,000 Hz AM sine wave modulation set to 80% for commercial testing.

Audio rectification concerns can be tested extensively with the help of the modulation (or radar pulse interference for military testing). The low-frequency modulation could produce bias upsets or otherwise disrupt sensitive analogue circuitry, for instance, if the RF signal is rectified by semiconductor junctions or in audio or other analogue circuitry.

2.16 TROUBLESHOOTING AT YOUR FACILITY

Conducting the troubleshooting process in-house may be the most time- and cost-effective option if the issue cannot be resolved at the testing lab. If you want to quickly locate the weak spots in your product, injecting a controlled RF source at certain spots is the way to go. It may be difficult to generate the necessary E-field test levels on a workbench, but by hooking up a small loop probe to an RF generator and probing around, you may readily identify cables or circuitry components that are overly sensitive. A significant number of revolutions of the loop may be required to significantly increase the field generation capability. The ideal RF generator has an output of between −20 and −20 dBm. If not, you may need to install a 10 W broadband amplifier. Make sure your commercial probe can withstand all this power if you plan on utilizing it. The possibility to use an AM modulation of 1 kHz at 80% would be a huge benefit. The ideal setup would include thorough filtering of all

I/O ports as well as any direct current (DC) or line power. I/O ports (e.g., USB, Ethernet) typically require the usage of CM chokes or filters. Otherwise, RF energy could be transferred into your circuitry through the I/O or power wires. At the testing facility, you should have previously established whether the problem lies with the cables or the enclosure (using aluminium foil). The wires could be acting as antennae to couple noise into the machine if the susceptibility frequency is below 200–300 MHz. Try the following once you suspect a susceptible cable:

- To figure out which cable (or cables) is malfunctioning, try removing them one by one.
- The cable should have a ferrite choke installed around it as close to the product connector as is practical.
- One can try putting a basic R-C low-pass filter in front of any input or output ports you have reason to doubt. An equivalent 47–100 W resistor and 1–10 nF capacitor in series could be used for power or signal return.
- If you can get your hands on a bulk current injection probe, clamp it onto the cable in trouble, and then attach it to an RF source. If you suspect leakage from the chassis or enclosure and not the cables, try the following:
- Ensure all the enclosure fasteners are tight.

Use copper tape to secure any questionable joints. Make sure the tape is making multiple points of contact with the metal chassis.

2.16.1 Low-Cost Tools

The use of an RF generator coupled with a compact E-field or H-field loop probe is that of a powerful troubleshooting tool (Figure 2.11). An extremely strong RF field (up to 10 V/m or more) will be generated, allowing for probing of areas around cables, connectors, and internal circuitry. You'll need to have a system in place to ensure your product is functioning as intended. Take care to avoid any interruptions while you're investigating. If the product is not vulnerable to the RF from the short loop probes, you can try wrapping a longer piece of wire loosely around and along the length of each I/O or power connection to couple in the RF more efficiently. Finally, link this cable to the radio frequency generator. A wideband power amplifier of 10 to 20 watts may be required to increase the generator's RF output in the worst case scenario. Ensure to use bulk injection probe when testing cables. A basic dipole antenna, like TV rabbit ears or a DIY dipole built from two lengths of wire cut to the approximate quarter-wavelength (each side), with one side connected to the shield and the other to the centre wire of a coax cable would also work .

Solving the issue could also be done at the compliance testing facility. Use of a licence-free portable Family Radio Service (FRS) FM two-way radio to transmit close to sensitive sections of the product is yet another low-cost method. The 465 MHz frequency range is covered by these half-watt radios. Despite its limited frequency range, this basic equipment has proven important in the detection and resolution of numerous RF susceptibility issues. A portable CB radio and a cellular (or PCS) mobile phone with transmission capabilities are two other tools that can be used without a licence. A valid licence is needed to operate the GMRS transmitter.

Contact opening creates high voltage across the contacts, which can cause arcing if not properly grounded. Repeated short bursts of this happen until the magnetic field energy stored in the relay coil is completely drained. To utilize it as an antenna, the wire must be stretched out far from the coil to the contacts. Now you may secure the unit's various cords and enclosure using this wire. The voltage depends on the inductance of the relay winding and can be as high as 1,000 V for most relays.

$$V = -L\frac{dI}{dt}$$

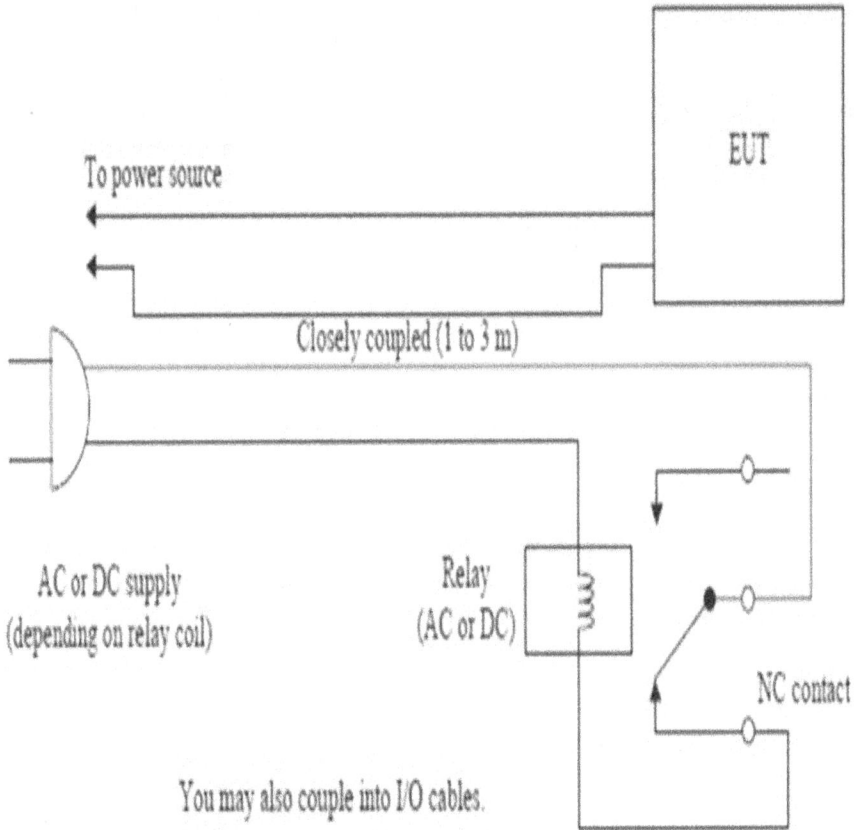

FIGURE 2.11　Schematic drawing of a simple chattering relay.

Where:
　V is the voltage produced
　L is the relay coil inductance
　dI/dt is the change in relay current versus time

If you wish to limit the maximum voltage produced, you can connect two zener diodes across the relay coil in series. Each power and I/O cable should be subjected to the test to determine its potential vulnerability. The EMI created by a short antenna in close proximity to the analyser is shown in Figure 2.11 (upper trace). Until roughly 1 GHz, the average amplitude is around 75 dBuV. The 85 dBuV line on the screen is the loudest reading. The background noise level is indicated by the lower trace. Upper and lower traces are both set to peak hold .

2.17　SUSCEPTIBILITY MODELS FOR WIRES AND PCB LANDS

In order to bring a digital electronic product to market, it is necessary to ensure that the device complies with the regulatory restrictions on radiated (and conducted) emissions. However, as was previously said, being able to comply with regulation emission limitations does not constitute a complete product design from the aspect of EMC. Unreliable performance and customer satisfaction can occur if a product is vulnerable to environmental disturbances such as radiated fields from radio transmitters and radars, or to lightning, or electrostatic-discharge (ESD)-induced transients.

If a product is sensitive to external disturbances such as radiated fields from radio transmitters and radars or lightning or ESD-induced transients, its performance will be unstable and customer satisfaction will suffer.

By adjusting the two parameters depicted in Figure 2.12 (a), the model can be applied to the situation when a PCB has two parallel lands. Load resistances RS and RL are present on the wires, which are separated by a distance. Putting the two wires in the xy plane with RS at $x=0$ and RL at $x=L$ will help us quantify our findings. The wires lie in a straight line with the x-axis. Our goal is to foretell terminal voltages given the size, polarization, and propagation direction of a sinusoidal, steady-state electric field of a homogeneous plane wave. The induced voltages are caused by two components of the incident wave. These are the transverse component of the incident electric field and the normal component of the incident magnetic field, as depicted in Figure 2.12b. The line will have per-unit-length inductance l and capacitance c parameters. For the parallel-wire line with wires of radius rw, these parameters per unit length are applicable.

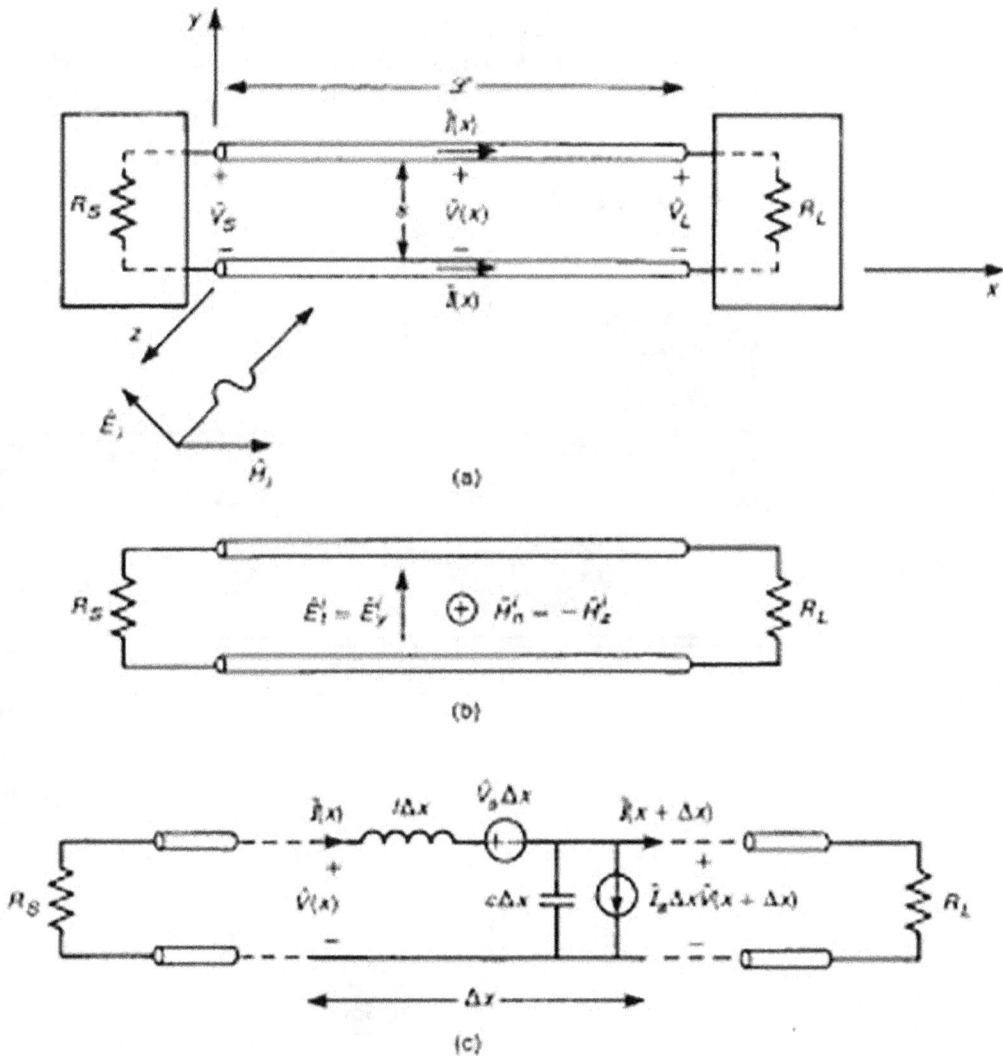

FIGURE 2.12   Modelling a two-conductor line to determine the terminal voltages induced by an incident electromagnetic field: (a) problem definition; (b) effects of the transverse electric field component and the normal magnetic field component. (c) a per-unit-length equivalent circuit.

$$l = \frac{\mu_o}{\pi} \ln \frac{s}{rw}$$

$$c = \pi \varepsilon_o \varepsilon_r / \ln \frac{s}{rw}$$

The relative permittivity of the surrounding medium is denoted by εr. For the following model to apply to two parallel lands on a PCB, it is necessary to utilize the appropriate per-unit-length capacitance and inductance characteristics. Figure 2.12c depicts a model of a Dx section of the line, where the per-unit-length parameters are multiplied by the section's length, Dx. The incident wave generates the per-unit-length-induced sources Vs and Is according to the following arguments.

2.17.1 Radiated Susceptibility Tests

The Figure 2.13 RS test setup is used to validate the DUT and its associated cabling's resistance to electric fields applied to the DUT from a predetermined distance, through an antenna. An electric field is generated and applied to the DUT with the help of the antenna signal generator, amplifier,

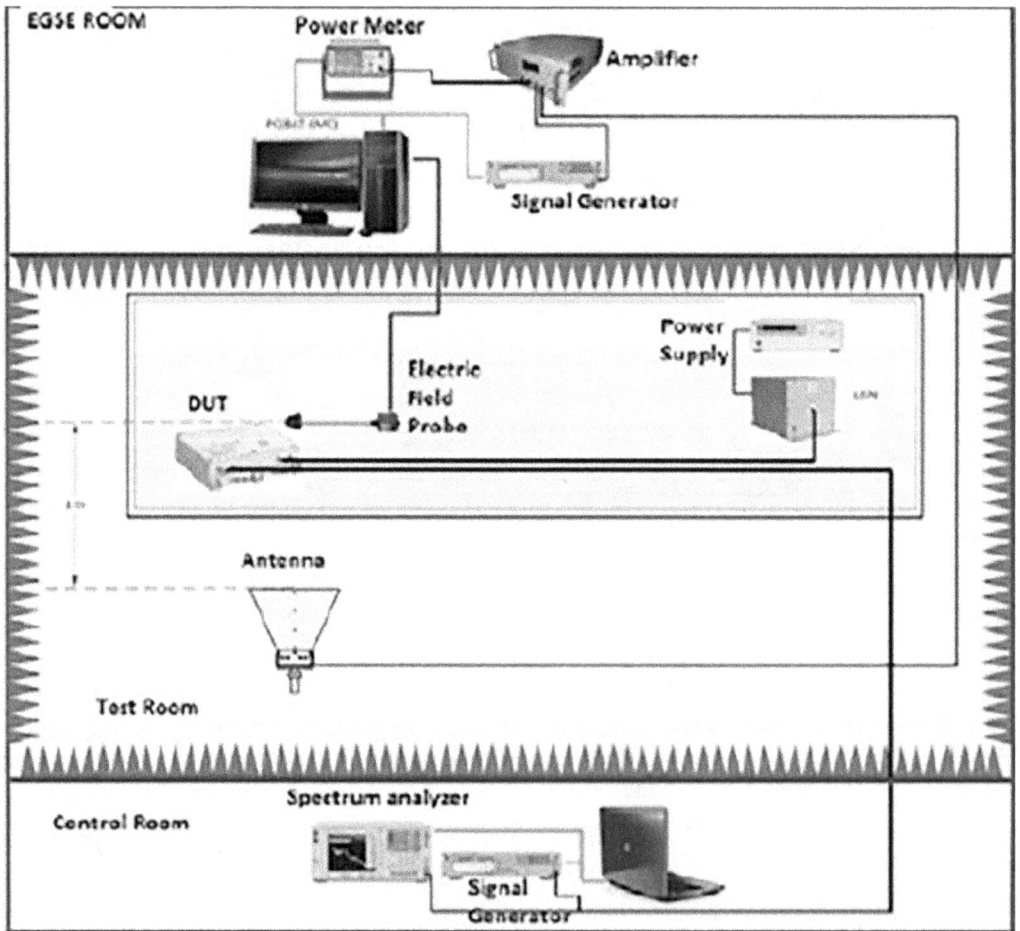

FIGURE 2.13   Radio susceptibility test setup.

and antenna, within the prescribed frequency and amplitude ranges. In order to fine-tune the generated electric field level, the signal generator and power amplifier are fed back the measured data from the power metre and electric field probe.

2.17.2 ANTENNA FOR RADIATED SUSCEPTIBILITY TEST

The double-ridged horn antenna has the potential to act as a transducer, converting an electrical signal into a radio frequency electromagnetic wave that is uniform across the whole testing spectrum. As a result, the antenna might be used to characterize and calibrate a test site's field uniformity. In addition, the antenna can be used to measure the field's actual strength for a more accurate assessment of its consistency across several locations. Careful design and analysis of the antenna's impedance matching band, antenna gain, and radiation beam width are required for these uses. Achieving the required level of field uniformity at higher frequencies is difficult otherwise.

IEC 61000-4-3 details the standards that must be met for an anechoic chamber's field homogeneity. The radiation pattern, radiation efficiency, and impedance matching quality of the double-ridged horn antenna all play a role in the antenna's ability to produce a consistent field. Changing the feed structure, waveguide dimensions, ridge spacing, and ridge curvature can all improve the antenna's performance. The antenna has a brand-new coaxial feed construction designed to increase the impedance matching band. As a result, the suggested antenna makes it simple to conduct field uniformity tests in a variety of locations. We built and analysed a prototype of the double-ridged antenna to prove the efficacy of the new design. At a 3-m certified SAC with absorbers laid on the ground, this prototype antenna is measured using the standard radiated susceptibility test equipment. The measured field homogeneity between 1 and 18 GHz is quite favourable and suggests the antenna's potential for EMC applications.

2.18   ANTENNA DESIGN AND ITS PERFORMANCE

The suggested double-ridged horn antenna's feed structure and prototype are seen in Figure 2.14a and b, respectively. As the coaxial feed construction, the ridges protruding inward from both sides of the antenna are joined to a rectangular stub 8.43 mm behind the waveguide terminal plate. It will alter the current distribution on the antenna and significantly enhance the homogeneity of the higher frequency field.

Due to its symmetry, it is also possible to produce an ideal current distribution across a wider frequency spectrum. It can not only provide a broader impedance matching bandwidth for the fed signal but also alter the radiating structure's capacitance and inductance to improve the main-beam direction and high-frequency half power beam width (HPBW). To conduct a steady and repeatable radiated susceptibility test, the parameters of the field-generating antenna are essential. This antenna's radiation pattern and impedance matching property are enhanced with the addition of a tapering flared section and a new coaxial feed arrangement.

The voltage standing wave ratio (VSWR) of the prototype antenna. According to our experience, the maximum value of VSWR throughout the entire frequency spectrum must be less than 2:1. This condition is rigorous and impossible to fulfil. From 1 to 18 GHz, the VSWR of the proposed antenna is substantially below 2:1. Modifying the feed structures, the curvature of the ridges, and the waveguide size optimizes the antenna pattern.

At frequencies between 6 and 18 GHz, it is very challenging to generate the ideal antenna pattern for greater field uniformity. Radiation patterns in both E-plane and H-plane are measured. Based on our above-described design, the HPBW of the primary beam gets broader, and this beam can be guided along the line of sight at higher frequencies. The outcomes illustrate the expanded coverage of HPBW and the consistency of antenna gain.

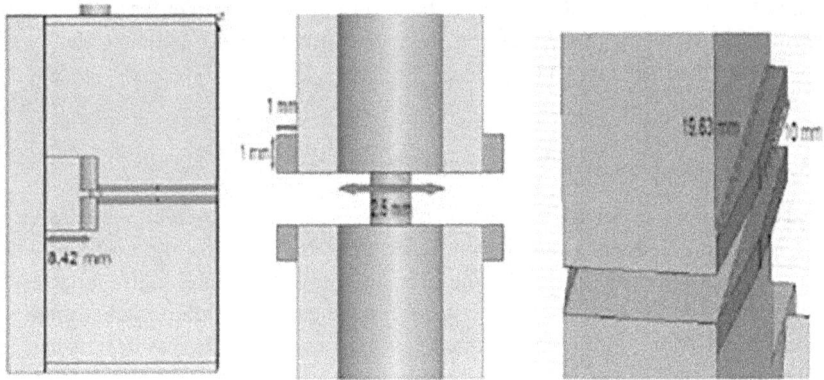

(a)

(b)

FIGURE 2.14   (a) Feed structure and (b) prototype of the proposed antenna.

2.19   CONDUCTED SUSCEPTIBILITY

The capacity of a system to function normally when exposed to radio frequency voltage or current on connecting conductors is known as its conducted susceptibility (CS) or immunity. It is common for electrical conductors to be subjected to radio frequency noise created by a range of sources via a number of coupling methods. Examples include load switching, electromagnetic fields, and conducted emissions from devices that share conductors.

There are many different types of interference signals that might enter a device through its power cable. Transients caused by lightning are a good illustration of this. Transmission lines and substations are routinely damaged by lightning. Breakers in a circuit are designed to open briefly to clear faults and then close again once a certain number of cycles of the AC waveform have passed. The product must function normally despite transient spikes and other power line fluctuations. While a total power "blackout" is out of the manufacturer's control, customers typically assume the equipment will continue functioning during brief spikes in voltage. Most companies put their products through these kinds of tests by sending spikes into the AC power cord to mimic lightning-induced transients.

In addition, the AC voltage is temporarily lowered and/or interrupted to make sure the product can keep running even if the power goes out. Sensitivity tests in this context are analogous to those that have been really carried out. While the FCC doesn't require it, the EU and MIL-STD-461E standards call for actual tests of susceptibility to interference.

2.19.1 INTRODUCTION TO CONDUCTED SUSCEPTIBILITY

An essential EMC compliance test consists of determining if externally emitted low-frequency RF fields can couple with your product through I/O or power cords. This test is commonly referred to as conducted immunity or CS and is defined by IEC 61000-4-6 for commercial products. The test frequency can range between 150 kHz and 230 MHz, and the voltage can be 1 V, 3 V, or 10 V RMS, depending on the product's environment or intended application. For usage in the military, on the road, or in space, stricter testing requirements are necessary.

In contrast to the 1 kHz square wave pulsed modulation used for most military and aerospace testing, the more common 1,000 Hz AM modulation is used for commercial testing. This modulation is useful for diagnosing problems with audio rectification. Modulation that is rectified (as it often is in audio and other analogue circuitry) can cause bias upsets and other problems for these systems. Due to the difficulty of producing a uniform field at such high frequencies in a shielded chamber, RF is typically connected directly to the I/O or power wires of the product. For commercial equipment, the test is only performed on I/O cables that are typically longer than 3 m (such as Ethernet or power cables). It was found that the military and aerospace low-frequency sources were dispersed across a large area and were positioned at great distances (the whole ship, the whole aircraft). As a result, antennae with wire runs of tens to hundreds of metres proved effective.

Due to the extensive lengths and distances required to perform the test correctly, this could not be repeated in the testing chamber. However, it was discovered that 1 V/m generated around 1.5 mA of current, and this was employed. Consequently, the limit lines for these standards tend to peak at ratios of 1.5 to 1 for radiated standards. Below 500 kHz, where ships and aircraft become electrically short, the limit lines begin to decrease (they become poor antennae).

2.19.2 TYPICAL FAILURE MODES

As stated in the book's radiated susceptibility chapter, induced signals from this test can cause a broad variety of issues.

- Rebooting the system
- Disruption of analogue or digital circuitry
- False readings on displays
- Loss of data
- Halting, slowing, or disruption of data transfer
- High bit errors (BER)
- Change of state of the product (e.g., mode, timing)
- Disruption of the switch-mode power supply

2.20 TROUBLESHOOTING AT YOUR FACILITY

If the problem cannot be fixed in the testing lab, bringing the troubleshooting process in-house may be the most efficient use of time and resources. The frequencies at which the EUT is failing the test should be known in advance so that troubleshooting can be focused on just those frequencies. To make sure you haven't just changed the frequency at which the issues manifest, it's important to retest throughout the whole frequency range after making any changes to address the known immunity issues. Since stray inductances and capacitances in the final build configuration can have a dominant effect on the RF behaviour of the circuit, it is important to replicate the test setup and final

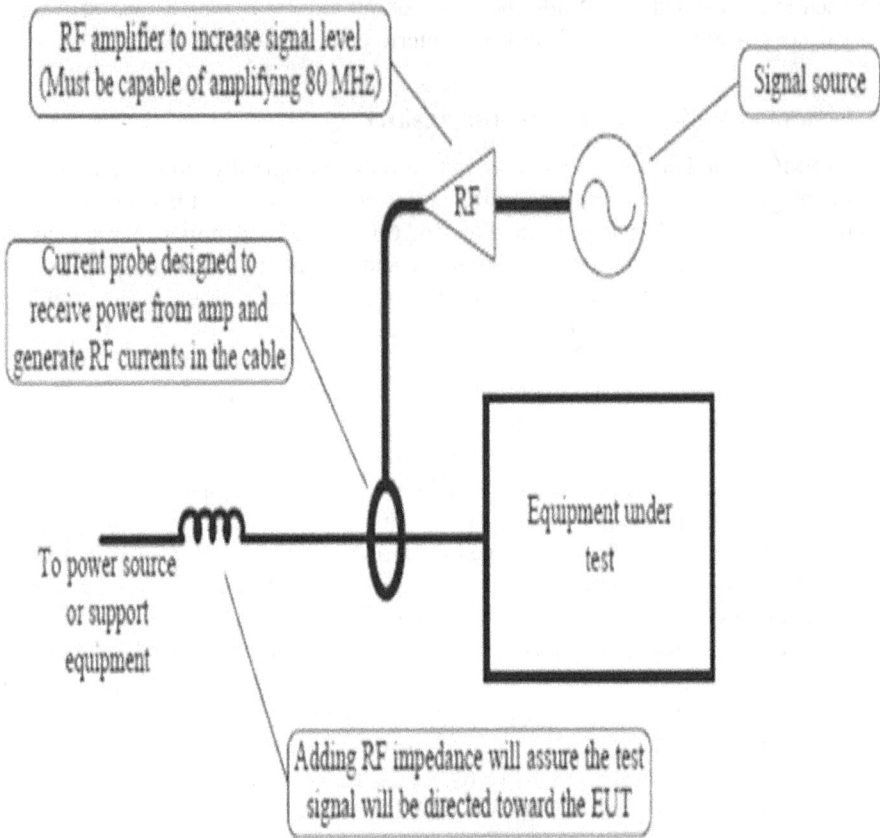

FIGURE 2.15   Simplified test setup for conducted immunity testing.

build of the product under test as closely as possible (e.g., shielding, earth bonding, proximity to metal objects or structures). Clamping a current probe around the cables to be examined makes bulk current injection (BCI) a convenient testing method (Figure 2.15). Some EMC standards, including IEC 61000-4-6, MIL-STD-461, DO-160 for civil aviation, SAE J1113–4, and ISO 11452-4:1995, allow for this form of testing. It is used in the aerospace, military, and automotive industries.

By introducing a controlled RF source into a portion of a product's cables, it is frequently easier to identify its weak points. Even if you cannot execute the required RF injection test levels on the workstation, connecting a clamp-on current probe to an RF generator will help you discover sensitive wires.

If a power amplifier is used, an RF BCI probe built to tolerate higher power levels must be employed. If you are using a commercial probe, ensure that it can withstand this level of power. Most current sensors designed to detect tiny RF currents are incapable of withstanding high levels of power. However, some probes incorporate a low-power internal resistance used to adjust the probe's impedance.

You'll need a high-wattage RF BCI probe if you're going to be utilizing a power amplifier. Make sure a commercial probe can handle this much power if you plan on using it. Sensing probes for weak RF currents often cannot withstand strong currents. Some of them have a low-power internal resistance that can be used to change the probe's impedance.

Filtering should be present on all I/O ports and DC or line power. CM chokes or filters are often used in I/O ports (such as USB and Ethernet) to prevent interference. Otherwise, RF energy could potentially enter your circuitry through the I/O or power lines. To know whether the issue is with the cables or the enclosure, you should have determined that before arriving at the testing facility (using aluminium foil). Once a suspect cable has been identified, you can try the following:

- By checking each cable separately, you can pinpoint the bad one or ones.
- Ensure the problematic cable is as close as possible to the product connector before installing the ferrite choke.
- Make sure that the protected cable is properly fastened into the shielded housing.
- Put in a simple low-pass R-C filter at every problematic input and output port.
- CM chokes or a clamp-on ferrite choke should be included in products that lack shielded enclosures.

2.20.1 Low-Cost Tools

Assembling an RF generator with a current probe that can be clamped around each cable is a reliable method of troubleshooting. As a manufacturer, you need a way to check that your product is performing as expected. As you inquire, be aware of any distractions. If there is no discernible change, possibly not enough energy is being applied. A possible solution is to wind the cord around the current probe several times. When the cable is looped around the injection probe, the configuration becomes a transformer with a higher output voltage.

It's vital to test the power amplifier in a shielded area to prevent interference with surrounding communications or broadcasts. As an alternative, you might conduct the probe at the same site as the compliance check. Inductively linking the product's power or I/O line to a clattering relay is another option for generating wide-range noise (Figure 2.16). Since the relay coil and the connecting wire have high inductances, the current will tend to remain flowing indefinitely once it has been set in motion.

An arc forms when a high voltage builds up across the contacts after they have been opened. This process occurs in rapid succession until the magnetic field energy stored in the relay coil has been depleted. The coil and contacts can only be used as an antenna if the wire is very long. Having done that, you can wrap the wire securely around the device's housing and all its individual wires. The highest allowable voltage is based on the inductance of the relay winding and can reach 1,000 V.

$$V = -L\frac{dI}{dt}$$

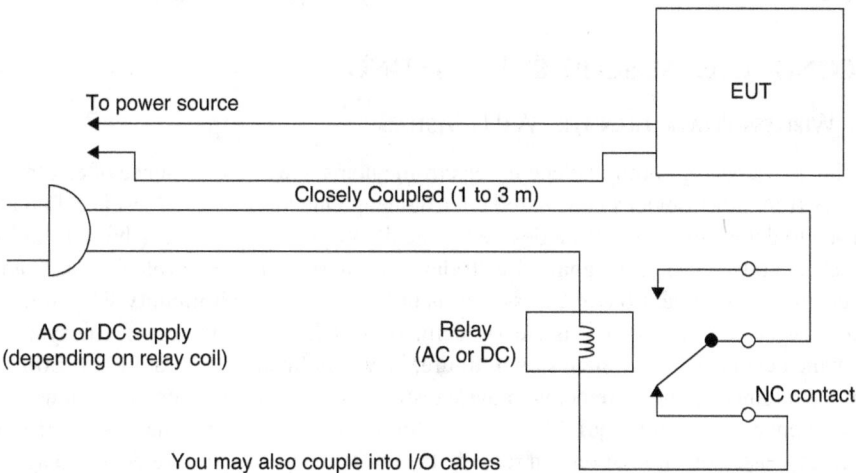

FIGURE 2.16   Alternative, try coupling a chattering relay to the cable under test. This serves as a broadband noise source.

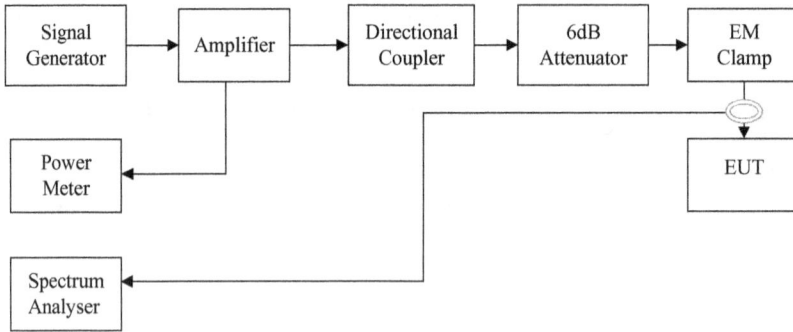

FIGURE 2.17   IEC-61000-4-6 test setup.

Where:
  V is the voltage produced
  L is the relay coil inductance
  dI/dt is the change in relay current versus time

If you want to make sure that your power and I/O cables are secure, you should put them all to the test. The fields can grow to great heights if left unchecked. Even at a distance of 20 feet, the coil was able to cause damage to one especially fragile item. To learn more about what happens during a conducted immunity test, read the IEC-61000-4-6 standard. The two most critical aspects of RF testing are the frequency range and the test level, as depicted in Figure 2.17.

We offer a frequency range from 150 kHz to 230 MHz, with a typical value of 80 MHz and a maximum value of 230 MHz; this should be more than adequate for most applications. Each product standard has its own required frequency range for testing, and that range is based on the product being evaluated. Choosing a frequency range depends on the EUT and the cable's diameter. The term "test level" can be taken in several directions when discussing actual exams. In the standards, signal levels measured into a completely resistive load and unmodulated are used as references. How test results look at the EUT can be heavily influenced by its impedance characteristics. As long as it is indicative of the electromagnetic environment in which the equipment is expected to work normally, the user is free to select any nominal test level, where rms is defined as volts into a 50 matched load. Voltages of 1, 3, and 10 V are recommended per the standards.

2.21   CONDUCTED SUSCEPTIBILITY TESTING

2.21.1   WIRELESS POWER TRANSFER (WPT) SYSTEMS

EMC testing is becoming commonplace for electronic items due to the presence of electromagnetic disturbances from other devices in a shared environment. One kind of EMC test is called a CS test, and it's used to determine how well a device can deal with interference that's brought in by a conductor, such as power cord or a signal wire. Today, the International Electrotechnical Commission (IEC) standard measurement-based assessment is utilized to test for immunity. The immunity test for system-integrated circuits (ICs) is clearly defined by IEC 62132. Moreover, the system immunity test, which evaluates the system's own failure, is well defined within the IEC 61000 standard. However, the reliance on measurements in these testing methods makes them inefficient in terms of resources expended and time required to prepare for and carry out the actual tests. Utilizing a CS testing strategy based on simulation can remedy the inefficiency. Figure 2.18 depicts a system-level CS test environment that complies with IEC 61000-4-6 and uses a BCI. The testing infrastructure consists of a brain–computer interface (BCI) probe, an end-of-trailer (EUT), a decoupling device, and some other tools.

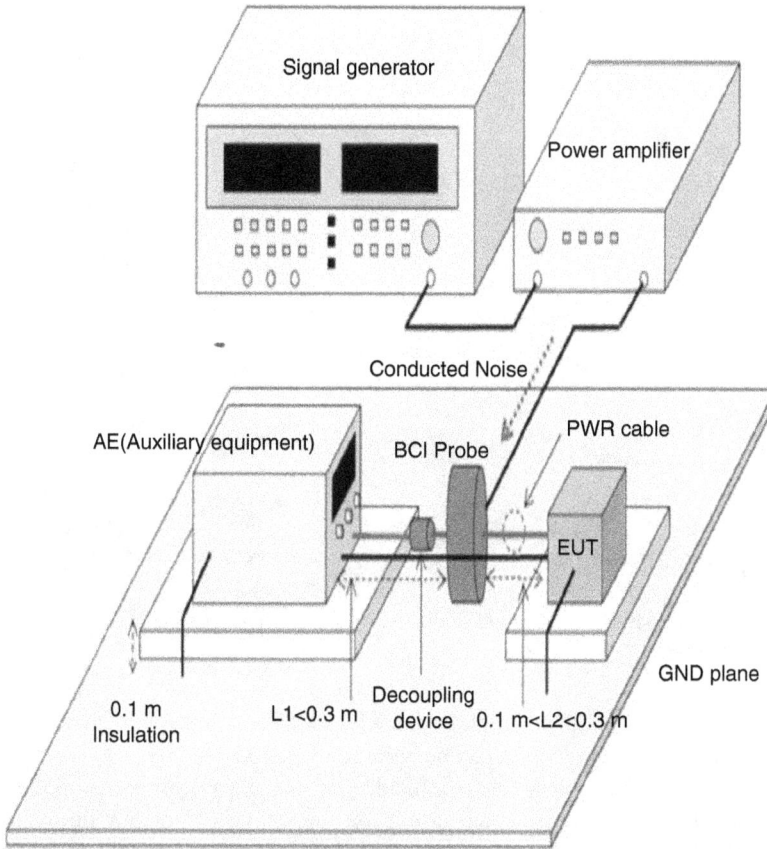

FIGURE 2.18   A system-level conducted susceptibility (CS) test setup using bulk current injection (BCI) method based on IEC 61000-4-6.

The signal generator must be capable of producing a signal with a 1-kHz sine wave as the modulation source and a frequency range of 150 kHz to 230 MHz and an amplitude modulation depth of 80%. You may additionally need a power amplifier to amplify the RF signal if the output power of the RF generator is low. Because it uses wires, which are themselves susceptible to conductive noise, the EUT might be thought of as a victim system. A decoupling device or network is used in the BCI method to reduce the influence of the EUT's peripherals. Due to the precise requirements for the testing apparatus and the testing arrangement, even a single test according to the measurement-based CS testing standard necessitates a great deal of planning and preparation (which includes cabling and insulation).

2.21.2   MITIGATION TECHNIQUES

There are a number of ways to reduce the EMI and bring the instrument into line with the electromagnetic environment in the real world if it fails any of the EMI tests.

2.21.3   ELECTROMAGNETIC SHIELDING

Often used to mitigate the effects of EMI on sensitive electronics, electromagnetic shields have become an integral part of the industry. The most important parts of an electronic circuit can be protected from EMI by surrounding them with a conductive surface and using a combination of

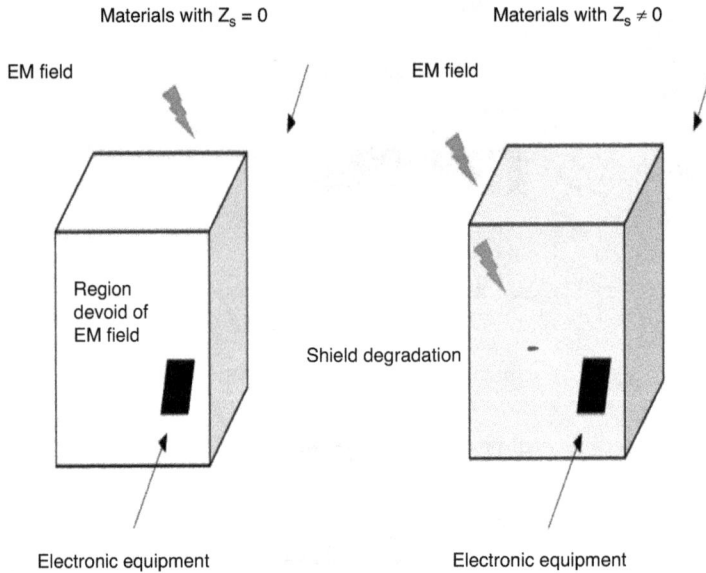

FIGURE 2.19   (a) Perfect shielding. (b) Real shielding.

reflection and absorption. Shielding at high frequencies requires a thin conductive layer to be deposited, while shielding at low frequencies can be built of any metal.

In theory, the best way to achieve perfect shielding is to use a shielding material with a surface impedance of $Zs=0$. A material protected in this way would be completely immune to fields, as depicted in Figure 2.19a. The lack of a material with zero-surface impedance makes it impossible to provide flawless shielding at the present time. So, EM waves will always be able to penetrate the enclosure, as seen in Figure 2.19b.

2.21.3.1   Effectiveness of Shielding

The shield is effective in blocking or reflecting most of the EMI field while allowing only a negligible amount to pass through. A structure's EMI shielding efficiency (EMI SE) is defined as the decibel-valued ratio of the EMI

$$SE = 10\log\frac{P_{rx}}{P'_{rx}}$$

Where:

P_{rx} is the power intercepted by the receiver in the absence of the shield

P'_{rx} is the power intercepted by the receiver when the shield is placed between the receiver and transmitter

Permeability, conductivity, permittivity, etc., are the characteristics of the frequency at which the measurement must be conducted; the angle and polarization of incidence of the impinging wave, as well as the near field or far-field application, govern its EMI SE.

2.21.4   Filtering

A filter can stop high-frequency currents, which can produce radiation if not filtered, in their tracks. To prevent interference, unshielded products rely on filtering and careful PCB layout. The two major types of conducted emission noise are differential mode (DM) and CM noise. The natural flow of current in the circuit's loops produces DM noise. These loops will function as antennae that emit EMI. CM noise is caused by resistance injected into the circuit as a result of unintended voltage dips. When electricity flows back into the source via capacitance or inductance that is not connected to the source. When dealing with CM and DM disturbances, it is typical to employ an EMI filter. To eliminate CM and DM noise signals, distinct techniques are required. Initially, a noise separator is used to separate the two undesired signals. The primary benefit of utilizing a filter to mitigate EMI is that they are straightforward in terms of design, particularly for smaller systems, and implementation. An important benefit of utilizing a filter is that it can identify and eliminate interferences whose properties are incompatible with those of the system.

2.21.5   CABLE BONDING

A strong connection must exist between the shielded enclosure and the conductive shell of the power connector. A 360-degree bond should be used to prevent cables from radiating for optimal results (Figure 2.20).

2.22   CIRCUIT TOPOLOGY

Any switching power supply topology will emit conductive and radiated emissions, but some are better than others. It has been discovered that topologies that operate at a duty cycle greater than 50% (such as pulse width modulation or push–pull) result in lower peak currents, and that forward converters are superior to fly back converters due to their lower peak currents and significantly lower secondary AC currents. By utilizing input power factor adjustment, AC power input spikes will be reduced. Loop inductances, parasitic capacitances, etc., may be formed if the placement of the circuit's components is not optimal, resulting in spurious noise signals. Therefore, it is of the utmost importance to correctly position each component in an electronic system.

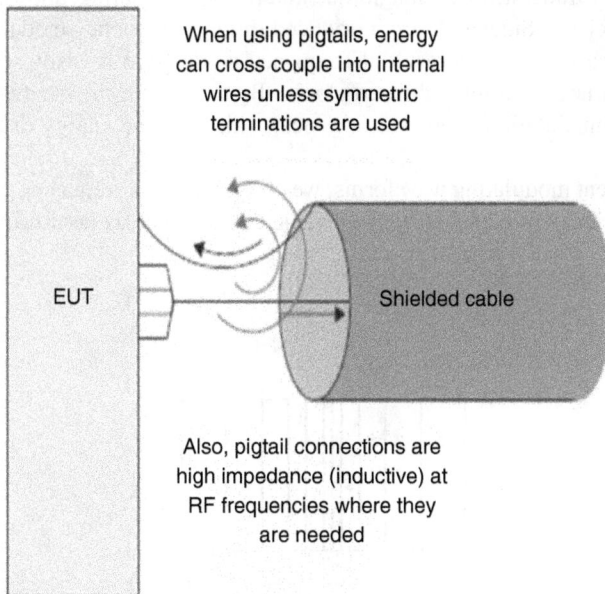

FIGURE 2.20   Cable shield termination using a pigtail ground to chassis and the resulting magnetic field coupling to the signal wires.

2.23   SPREAD SPECTRUM TECHNIQUE

A new technique for reducing radiated emission from digital electronic equipment is the spread spectrum clock generation (SSCG) technology, which uses frequency modulation of the clock. This cutting-edge technique is remarkably comparable to the standard spread spectrum approach

2.24   SPREAD SPECTRUM CLOCK

The clock frequency causes side bands, which expand the emission wavelength. This method is useful for decreasing the fundamental frequency in switching power circuits if the switching frequency is less than 150 kHz and the frequency for modulation is selected to be greater than 200 Hz.

Consider a clock signal represented by $f(t)$ in Figure 2.21. The pulse shape depicted in this image conveys the idea that the pulse width of the clock signal fluctuates across the modulating waveform's period T.

The Fourier Coefficients of $f(t)$ shown in Figure 2.21 are given by

$$a_n = \int_{-T/2}^{T/2} f(t)\cos(n\omega_o t)dt$$

$$b_n = \int_{-T/2}^{T/2} f(t)\sin(n\omega_o t)dt$$

Where $n = 0, 1, 2 \ldots$ and $\omega_o = \left(\dfrac{2\pi}{T}\right)$ where T is the modulation period.

From the above two equations, the magnitude of nth harmonics can be

$$|Cn| = \sqrt{a_n^2 + b_n^2}$$

Clock pulses were used to create a numerical sequence of pulse widths, which were then used to perform a Fast Fourier transform with magnitude and frequency scaling, allowing for the computation of Fourier coefficients. Sidebands are introduced through frequency modulation. The lowering of the peak level of harmonics as a result of frequency modulation is easily seen in a comparison of spectra from modulated and unmodulated clocks. By expanding the bandwidth of the harmonics, clock frequency modulation suppresses the peak values of the energy distribution across the frequency spectrum.

To simulate different modulating waveforms, we chose the clock frequency to be 65 MHz, which is typical of system clocks in many digital electronic devices such as laser printers. We also chose

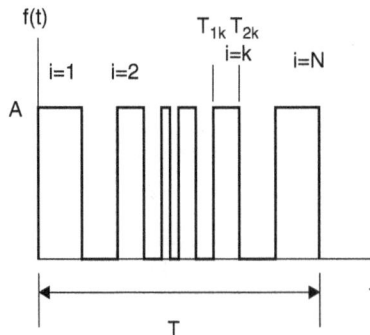

FIGURE 2.21   Time domain representation of the frequency modulated clock.

the modulating waveform period to be 10nS and we restricted the maximum frequency deviation of the modulated clock to be below a specified level in order to avoid a system failure.

2.25   SOFT SWITCHING TECHNIQUES

Comparing the EMI emissions of hard-switched and soft-switched buck and boost converters of the same rating, both in terms of conducted and radiated fields, is demonstrated. The results suggest that the use of soft switching in power converters can reduce EMI emission. Traditional hard-switched power electronics are thought to be a primary source of EMI emission in power electronics circuits due to the high dv/dt and di/dt involved in the switching process.

The soft switching techniques have been proposed to:

- Reduce Switching Losses
- Reduce the Switching Stress of Switched Mode Power Supply (SMPS)

In order to minimize or eliminate the instantaneous power loss associated with turning on or off a power switch, "soft switching" methods establish circumstances of zero voltage or zero current. The research into how soft switching affects EMI has been lacking, but preliminary findings suggest it can lessen EMI output.

2.26   HARD-SWITCHED AND SOFT-SWITCHED POWER CONVERTERS

All power electronic devices, such as switches and diodes, are activated and deactivated at zero voltage and zero current in the soft-switched approach.

2.26.1   EXPERIMENT INVESTIGATION

Both Hard and Soft switching were tested experimentally to determine their levels of conducted EMI and radiated EMI. A line impedance stabilization network was used to link the converter to the network and do the measurements at a nominal radio frequency impedance (LISN). The converters should be set up 40 cm away from a 4-square-metre earthed conducting surface and maintained 100 cm away from such a surface. A 0.5 m cable was used to link the converter to LISN. The 50-port voltage developed noise measurements were taken. Second, the measurements of radiated emission were performed in a semi-anechoic room that was 20 feet long, 12 feet wide, and 7 feet high. Emissions levels were measured using a calibrated and adjusted HP 11955A biconical antenna. An antenna is located 3 m from the converter. This was accomplished with the use of a special turntable set up. Radiated and conducted emissions were measured with an EMC analyser. The desired frequency range is swept while the converter is turned off to gauge background noise. When the converter is turned on, both the surrounding signal and the signal emitted by the converter must be recorded. If the measured signal is subtracted from the ambient signal, the true emission from the converter can be determined. Each set of hard-switched and soft-switched converters is tested with the same load, and their fundamental converter components (L and C) are interchangeable. When comparing two of the same type of converters, the inductor current and load current are virtually comparable in magnitude. The converters are not insulated and lack EMI filters.

2.26.2   HARD- AND SOFT-SWITCHED BUCK CONVERTER

Soft-switched converters also have a significantly lower diode reverse recovery current compared to their hard-switched counterparts. Soft-switched converters have substantially lower dv/dt and di/dt because transient ringing is drastically reduced. Figure 2.22 depicts the waveforms of the soft-switched converter's auxiliary switch Q. These findings verify the converter's totally soft-switched

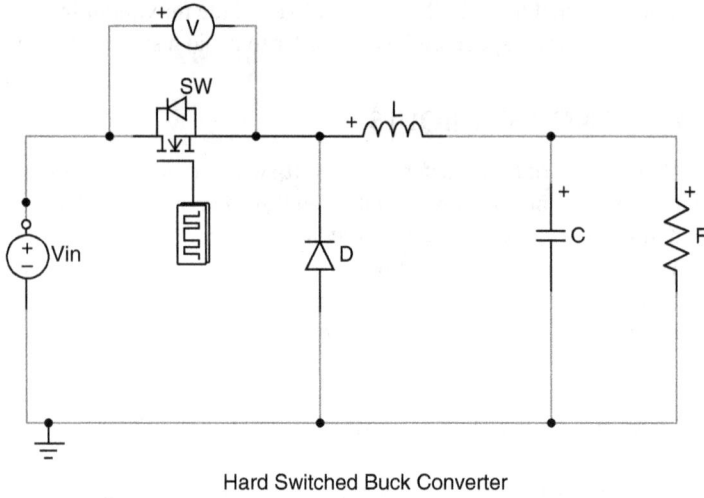

Hard Switched Buck Converter

FIGURE 2.22 Hard-switched buck converter. $V_{in}=55$ v, $V_{out}=20$V, $f_s=115$kHz, $D=0.4$, $L=2.5$ mH, $C=220$ F, $R_L=7.5$.

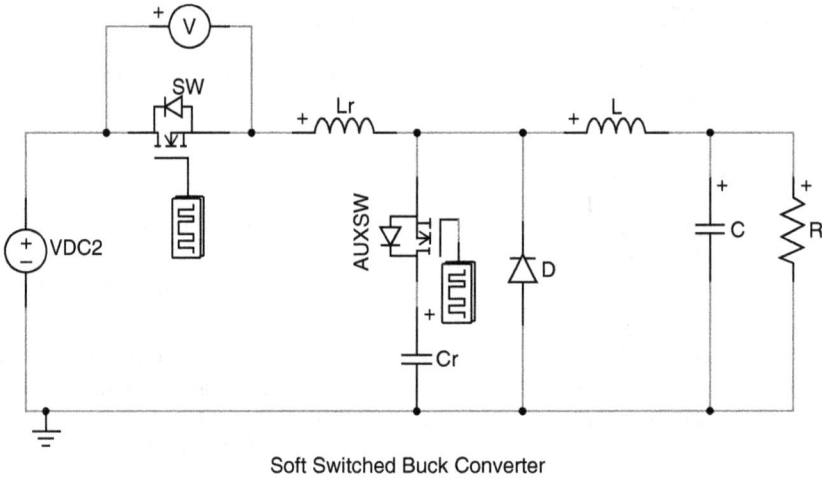

Soft Switched Buck Converter

FIGURE 2.23 Soft-switched buck converter. $V_{in}=55$ v, $V_{out}=20$V, $f_s=115$kHz, $D=0.4$, $L=2.5$ mH, $C=220$ F, $R_L=7.5$, $L_r=3.1$, $C_r=46.5$ nF.

nature. In Figure 2.22, we can see the combined effect of the conducted and radiated EMI emission from the hard-switched buck converter, as well as the ambient noise. Figure 2.23 shows the Measurements of Conducted and Radiated EMI in a Soft-Switched Buck Converter. The waveforms of the soft-switched converter's auxiliary switch Q are recorded in Figure 2.24; these findings validate the converter's fully soft-switched properties.

Figure 2.25 depicts the conducted and radiated EMI from the hard-switch boost converter. The equivalent EMI readings from the soft-switched boost converter are depicted in Figure 2.26. Compared to the hard-switched circuit, the soft circuits exhibit less EMI. The waveforms of the auxiliary switch Q in the soft-switched converter are depicted in Figure 2.27. This demonstrates that the soft-switched boost converter has greater EMI emission than the hard-switched one. These data demonstrate that the converter is fully soft switched.

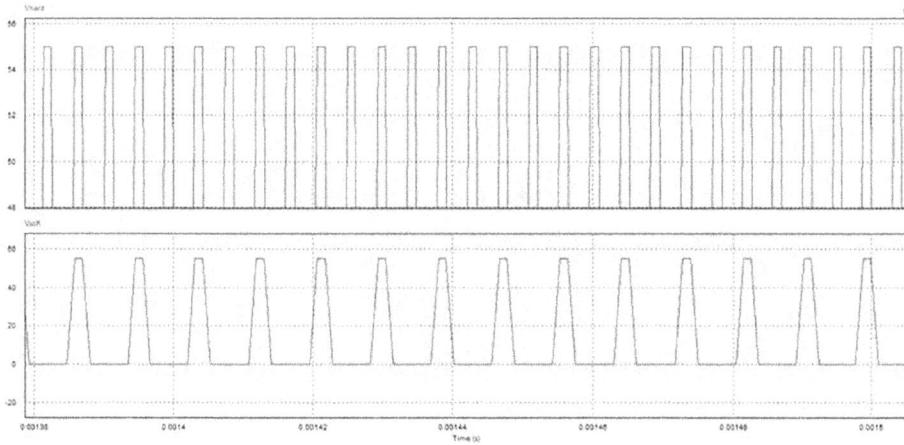

FIGURE 2.24 Voltage across the main switch for hard and soft switching hard- and soft-switched boost converter.

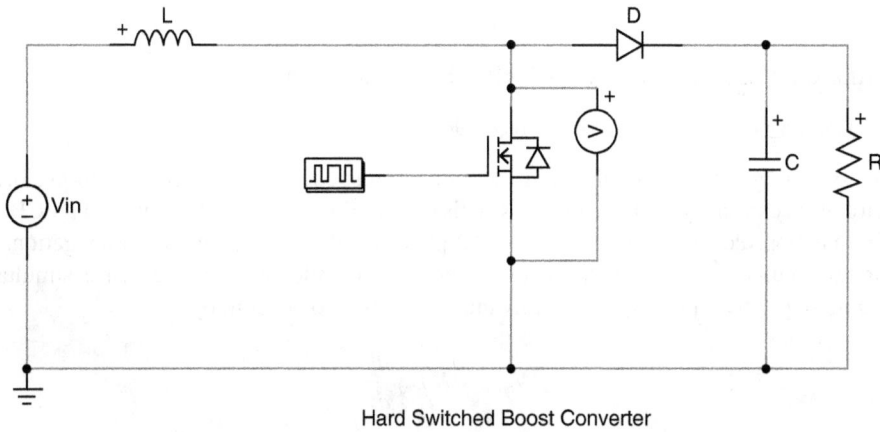

Hard Switched Boost Converter

FIGURE 2.25 Hard-switched boost converter. $V_{in}=34$ v, $V_{out}=48$ V, $f_s=115$ kHz, $D=0.26$, $L=2$ mH, $C=220$ F, $R_L=7.5$.

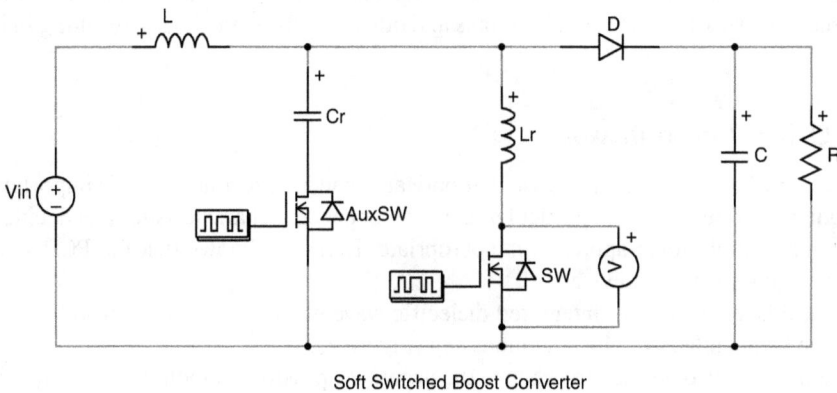

Soft Switched Boost Converter

FIGURE 2.26 Soft-switched boost converter. $V_{in}=34$ v, $V_{out}=48$ V, $f_s=115$ kHz, $D=0.26$, $L=2$ mH, $C=220$ F, $R_L=7.5$, $L_r=3.2$, $C_r=20$ nF.

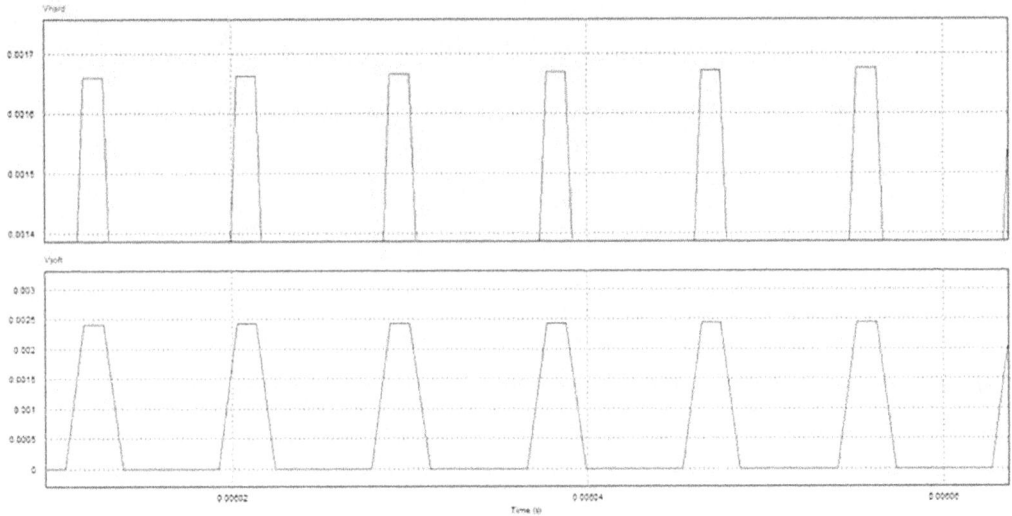

FIGURE 2.27   Voltage across the main switch for hard and soft switching.

2.27   REDUCTION IN PRINTED CIRCUIT BOARD (PCB)

2.27.1   Sources of Noise and Switching Noise

Typically, switching noise is brought on by the time-varying currents required by high-performance digital circuits at extremely high frequencies. This current flows between the layers of a PCB, which results in radiation. Power planes act as parallel plates for the radiated wave's propagation. When many digital circuit outputs are switched simultaneously, an inductive noise known as simultaneous switching noise (SSN) is produced. An SSN may be taken into account by

$$V_{\text{noise}} = N L_{eq} \frac{di}{dt}$$

In which V_{noise} is the noise voltage and N is the total number of switching outputs (drivers), Current flows through each driver during switching; this inductance, denoted by L_{eq}, must be overcome in order for the switches to close. Connected to a power source, current flows to L_{eq} whenever multiple signals change state simultaneously. This inductance causes voltage fluctuations on the lower planes, which in turn affect the driver's outputs and other signals on the board, resulting in incorrect switching.

2.27.2   Optical Circuit Boards

When deciding whether to use an electrical or optical connecting technology, it is important to take into account the various needs; in particular, the level-dependent characteristics that decide whether a single-mode or multimode approach is appropriate. Everyone knows that the PCB is the most crucial part of any electrical device.

This optical layer consists of integrated dielectric wave guides for high bandwidth interconnections that can be manufactured by hot embossing (Figure 2.28).

This method can also be used to produce the necessary passive components. The layer is placed inside the PCB due to manufacturing and assembly temperature limits. Sizes in the 50–100 m range are recommended for optimal PCB compatibility. The requirement that the optical transmitters and receivers be adjusted passively also contributes to this range's size. To accommodate the least local

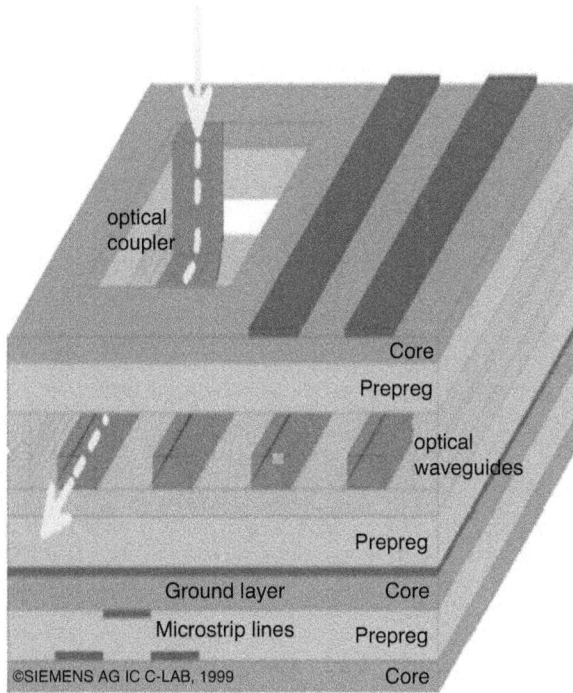

FIGURE 2.28 Concept of an optical circuit board.

attenuation of polymer material employed as a wave guide, the light's wavelength is kept within the range of 650–850 nm. The optical converters and passive components that allow light to be coupled into and out of the waveguides are just as crucial as the waveguides themselves. To make optical receivers, integrated photodiodes and amplifiers will be used (Figures 2.29 and 2.30).

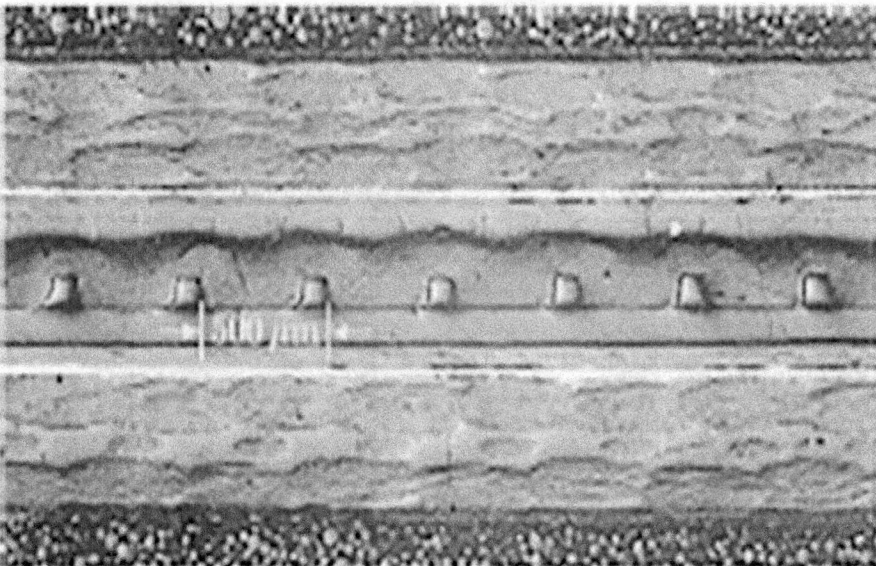

FIGURE 2.29 Cross section of an electrical/optical circuit board.

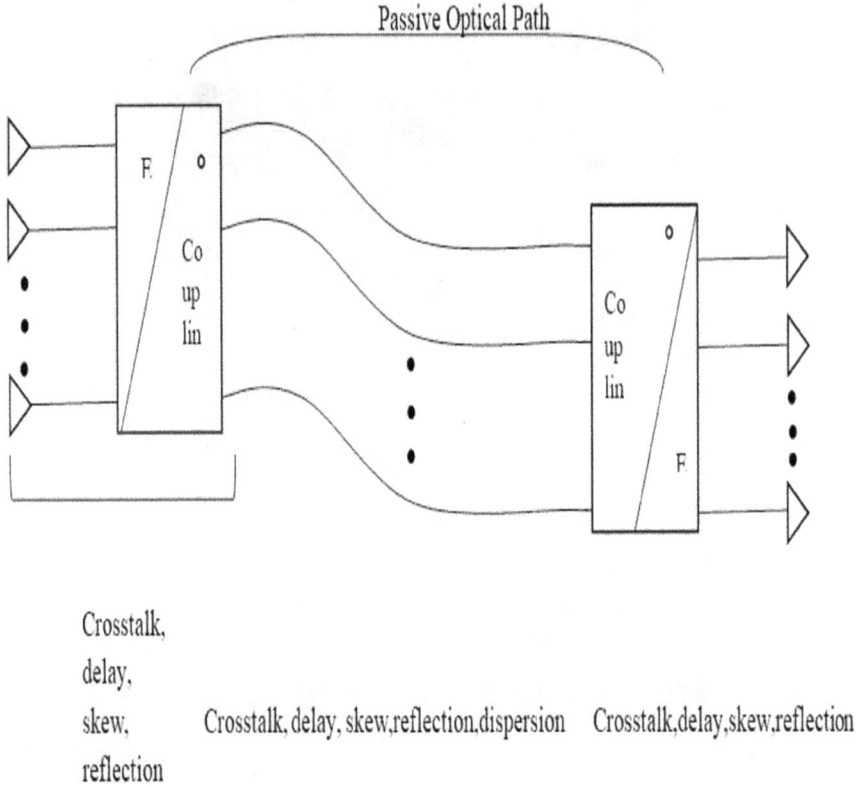

FIGURE 2.30   Optical interconnection system.

Voltage at the receiver inputs is of primary relevance, thus the first step is to build on-board linkages between electrical and optical components on the board.

2.27.3   CABLING AND INTERCONNECT

The most common source of EMI is the cable itself, which can occur when the connector is not correctly bonded or when the cable itself penetrates the product's shielding (see Figure 2.32).

As seen, a single USB connector was used to attach the protected enclosure, although numerous bonding points would have been preferable. The illustrated connectors will be PCB-mounted and inserted through the enclosure's cutouts as in Figure 2.31.

2.28   SUMMARY

The procedures commonly employed to measure EMI fall into two major categories: emission testing and immunity testing. For EMI measurements, two types of emissions are considered: radiated emission and conducted emission. Radiated emission testing is conducted in an open-area test site configuration or in chambers such as anechoic chambers, GTEMs, and reverberation chambers. All of these EMI measurement methods are described in detail, and the applicable globally recognized standards are referenced for additional study. The review paper discusses EMI in its entirety, including measurement methodologies. Thus, both established and novice researchers could benefit from acquiring a thorough understanding of EMI. Techniques such as electromagnetic shielding, EMI filters, switching techniques, cable bonding, and spread spectrum techniques are employed to reduce EMI caused by electronic devices. This is advantageous for the advancement of electrical devices.

FIGURE 2.31 Cable and interconnect.

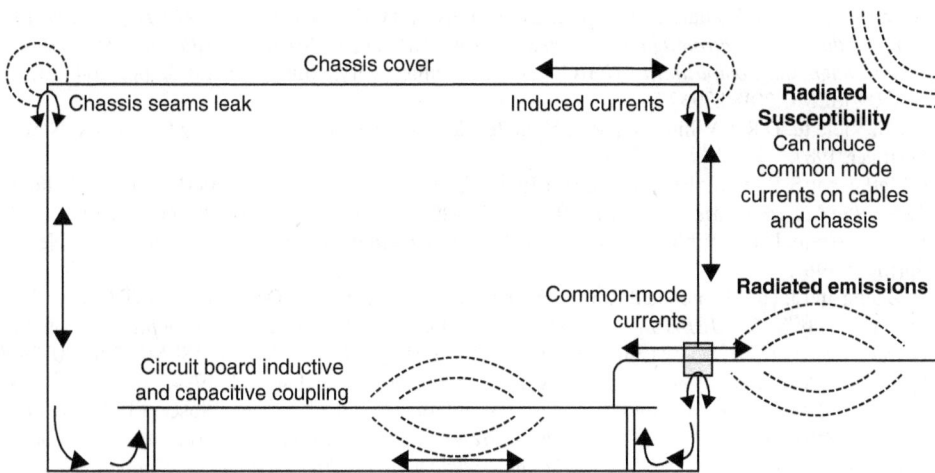

FIGURE 2.32 Cables that penetrate shielded enclosures defeat the shielding and allow high-frequency common-mode (CM) currents to radiate on the outside of the shield.

This chapter addresses the complexity of a radiated susceptibility test linked with the directivities of the EUT using straightforward simulations. If the sampling is inadequate, the uncertainty in determining the maximum coupling to a EUT from an anechoic chamber test could be substantial. Consideration must be given to reverberation chambers for optimal testing. As the electronic components aboard spacecraft operate in severe electromagnetic conditions, RS testing for the aerospace industry is of utmost importance and entails rigorous measuring processes, thereby increasing physical understanding of electromagnetic coupling mechanisms, lowering EMC test complexity, and associated costs. This is a question that will be addressed in future investigations. In conclusion, it is important to note that none of the alternative tests presented in this article require operation in an anechoic environment. The consequent benefits in terms of lower cost and faster testing time are largely attributable to this factor and can be easily evaluated if certain setups and testing procedures are selected and compared.

BIBLIOGRAPHY

1. Patrick G. André and Kenneth Wyatt, *EMI Troubleshooting Cookbook for Product Designers*. Science Tech, 2014. Doi: 10.1049/SBEW510E
2. David A. Weston, *Electromagnetic Compatibility: Methods, Analysis, Circuits, and Measurement*. CRC Press, 2016. https://doi.org/10.1201/9781315372013
3. H. Chung, S. Y. R. Hui, and K. K. Tse, "Reduction of Power Converter EMI Emission Using Soft-Switching Technique," *IEEE Transactions on Electromagnetic Compatibility*, vol. 40, no. 3, pp. 282–287, August 1998. doi: 10.1109/15.709428
4. Shahrooz Shahparnia, "Electromagnetic Interference (EMI) Reduction from Printed Circuit Boards (PCB) Using Electromagnetic Band Gap Structures," *IEEE Transactions on Electromagnetic Compatibility*, vol. 46, no. 4, pp. 580–587, November 2004. doi: 10.1109/TEMC.2004.837671
5. Yoonjae Lee and Raj Mittra, "Electromagnetic Interference Mitigation by Using a Spread-Spectrum Approach," *IEEE Transactions on Electromagnetic Compatibility*, vol. 44, no. 2, pp. 380–385, May 2002. doi: 10.1109/TEMC.2002.1003404
6. Clayton R. Paul, *Introduction to Electromagnetic Compatibility*. A John Wiley & Sons, Inc. Publication, 2022.
7. Sarangan Valavan, Understanding Electromagnetic Compliance Tests in Digital Isolators. Texas Instruments. November 2014.
8. G. Yavaş and S. Akgül, "Göktürk-1 Satellite System Level Radiated Emission and Radiated Susceptibility Tests," *2019 9th International Conference on Recent Advances in Space Technologies (RAST)*, Istanbul, Turkey, 2019, pp. 851–853, DOI: 10.1109/RAST.2019.8767824.
9. T. Lin, C. Lee, J. Dong, C. Chiu, D. Lin, and H. Lin, "A New Uniformity-Enhanced Double Ridged Horn Antenna for Radiated Susceptibility Test from 1 GHz to 18 GHz," *2018 IEEE International Symposium on Electromagnetic Compatibility and 2018 IEEE Asia-Pacific Symposium on Electromagnetic Compatibility (EMC/APEMC)*, Suntec City, Singapore, 2018, pp. 264–267, DOI: 10.1109/ISEMC.2018.8393779.
10. J.L.N. Violette, D.R.J. White, and M.F. Violette, *Electromagnetic Compatibility Handbook*. New York: Springer, 1987.
11. P. Mathur and S. Raman, "Electromagnetic Interference (EMI): Measurement and Reduction Techniques," *Journal of Electronic Materials*, vol. 49, pp. 2975–2998, 2020, DOI: 10.1007/s11664-020-07979-1.
12. J.L.N. Violette, D.R.J. White, and M.F. Violette, *Electromagnetic Compatibility Handbook*. New York: Springer, 1987.
13. B. Zheng and J. He, "Conducted Emission Analysis and Suppression Design of a DC/DC Semi-isolated Power Supply," *2017 IEEE International Symposium on Electromagnetic Compatibility & Signal/Power Integrity (EMCSI)*, Washington, DC, 2017, pp. 383–387, DOI: 10.1109/ISEMC.2017.8077900.
14. D. Morgan, *A Handbook for EMC Testing and Measurement*. London: IET, 1994.
15. J. Joo, E. Song, S. Kwak, and J. Kwon, "Simulation-based Conducted Susceptibility Testing for Wireless Power Transfer (WPT) Systems," *2018 IEEE International Symposium on Electromagnetic Compatibility and 2018 IEEE Asia-Pacific Symposium on Electromagnetic Compatibility (EMC/APEMC)*, Suntec City, Singapore, 2018, pp. 568–571, DOI: 10.1109/ISEMC.2018.8393843.

3 EMI Filter

3.1 INTRODUCTION

Electromagnetic interference (EMI) comes in two forms: conducted and radiated, and it is on the rise due to the demand for high-speed electronics at high frequencies. EMI filters are the most effective tool for reducing conducted interference, which is a major problem in modern technology. The use of an EMI filter is essential for the safety of any electrical device (EMI). The development and selection of filters is influenced by EMI regulations, electrical specifications, and other design requirements. An off-the-shelf filter will usually suffice, but a custom EMI filter solution may be required when other filters won't work. Passive, active, analogue, and digital types of electrical filters all exist. It is a piece of equipment that can be positioned anywhere between two circuits, networks, or pieces of equipment or systems in order to amplify, attenuate, or otherwise manipulate the frequency components of a signal.

3.2 EMI FILTER SOURCE IMPEDANCE OF VARIOUS POWER LINES

For optimal stability, EMI filter design necessitates a good approximation of the transfer function H(s) of an EMI filter. The filter Q, half-power bandwidth, and frequency magnitude attenuation can all be calculated with an accurate equivalent two-port transfer function. Insertion loss must be calculated using both the line and load impedances for accuracy. It is desirable to have the similar level of filtering as specified by selecting any one filter. The undesirable noise cannot be removed due to the presence of impedance that is present on either side of the filter. A filter's performance is not the same when measured in isolation from the circuit as it is when the filter is installed, and this fact must be taken into account during the selection and design phases. Since the impedance of the final circuit design typically differs from that of the measurement system. Input and output impedance for radio frequency (RF) test gear is typically 50 ohm. The impedance of digital and analogue circuits in electronic systems rarely averages 50 ohm and varies significantly with frequency.

3.3 SKIN EFFECT

A line's depth of conduction decreases as frequency rises. As the radius of conduction decreases, the cross-sectional area of the wire also decreases. More of this excess energy is lost as alternating current (AC) resistance increases. The power lines' higher frequency energy can be degraded by skin effect. This serves to dissipate the electromagnetic pulse, often known as an electromagnetic pulse (EMP), as other higher frequency noise. It is designed to carry electricity for lower frequencies but is now suffering from power loss due to higher frequencies. Hence, they were constructed accordingly. Even though the line may have a characteristic impedance of 50 ohms, the loss that the line experiences along each unit of length rises as the frequency of the signal increases. Copper's skin-effect depth, measured in centimetres, can be calculated using the following equation:

$$D = \text{depth}(\text{cm}) = \frac{6.61}{\sqrt{F}}$$

Frequencies over the skin depth cause the wire's cross section of the conducting area (CA) to be

$$CA = \left| R^2 - (R - D) \right| \pi$$

$$= D(2R - D)\pi$$

DOI: 10.1201/9781003362951-3

where R is the radius of the wire in centimetres, and D is the skin depth in centimetres. As the frequency goes up, D goes down, and eventually, the D term becomes much less significant than the $2R$ term. There is little impact, if any at all, that the skin-effect term has on the characteristic impedance. The inductors, transformers (which are employed), and the EMI filter's wiring are all susceptible to the skin effect.

In order to function properly, EMP applications require a filter input end that has high impedance. Energy is lost along power lines not just because of the characteristic impedance of the lines themselves but also because of resistance elements that are present along the lines. Some of the elements that contribute to these losses are the skin effect, the DC resistance (DCR), and the conductance across the line. When several elements are used in a filter, the elements take on properties that are analogous to transmission lines. At much higher frequencies, the shorter power lines begin to exhibit their variable characteristic impedance. This occurs because of the shorter power lines' lower length. Characteristic impedance for the open wire type can be anywhere from 50 to 180 ohms, depending on the size of the wire and the distance between its conductors. The resistance of twisted wires housed in conduit, or the paired kind, is typically between 50 and 90 ohms. Due to its thin construction, conduit provides only marginal protection. Around 95% of the time, multiple power line segments will be operating in tandem with one another. As the lengths of the various power lines get closer to their resonant values, the impedances of these lines become back and forth. Slow propagation can be expected because the wire was intended to transmit load factor (LF) and not RF. These wires' apparent length is roughly eight times their true length. When a pulse hits a power line, some of the energy is dissipated in the resistive elements. At the slower speed of propagation in these wires, the pulse will make its way toward the filter end. Lightning has a basic frequency of about 50 kHz and quickly adopts the 377 ohms of open space. As the impedance at the input side is large, the voltage is also doubled. If the filter impedance is low or is seen as a short to this pulse, the voltage drops to zero and the line current increases by a factor of two. Connections between nodes are apparently formed by a series of short lines. Below 10 kHz, the lines appear to be resistive, with values near to the DCR value. The output impedance of the lines rises with frequency to 4 ohms at 10 kHz. Impedance swells to 50 ohms at around 100 kHz for longer lines and 250 kHz for shorter lines, before decaying back to zero. Inductors utilize wires with appropriate amperage carrying capacity at the LF. The wire gauge must be chosen with consideration to the voltage loss and inductor temperature rise.

3.4   DIFFERENCES AMONG POWER LINE MEASUREMENTS

Variation in line impedance is commonplace among power lines. The ratio is eight times the actual length of some lines. At higher frequencies, the line is cut off by devices like generators or transformers. Power for 50, 60, or 400 Hz is all that can be safely transmitted across these wires; anything greater would damage the equipment.

CM energy will be produced by a lightning strike among the line. If this occurs further, the characteristic impedance between the line and ground will equal to V/I. The capacitance between this line and ground is equal to the square root of the inductance. Each metre has an inductance of about 1.5 H and a capacitance of less than 1 pF. Also, the impedance will be extremely high.

3.5   EMC FILTER DESIGN

An EMI filter for mains power is a low pass filter that allows only the intended input (often DC or 50/60/400 Hz power frequency) through rejecting all other frequencies. A perfect EMI filter will attenuate signals above and below the filter's cut-off frequency. When a signal is attenuated by 3 dB below the acceptance line, the transition between the passband and the reject bands is at the cut-off frequency. Insertion loss, often known as attenuation, is the degree to which a filter attenuates a signal. The installation of a power line or mains EMI filter blocks any noise from either leaving or entering the device.

It is important to identify the product-specific characteristics of the device, like real estate constraints, unusual environmental conditions, and electrical needs, when developing a bespoke EMI filter specification in addition to the basic electrical, function, and mechanical requirements. It is also necessary to determine whether or if there are any application-specific demands, such as meeting stringent medical or military design standards. An EMI filter's job is to substantially attenuate the electrical noise or disturbance while minimizing the loss or distortion of the required signal or power across a wide range of frequencies. The filter must be housed and positioned so that the unwanted electrical noise or disturbance cannot get past it. Components like capacitors, inductors, and resistors make up passive low-pass filters. Below are some examples of straightforward filter topologies. Most practical EMI filters consist of many stages, damping networks, and a magnetically connected inductor. Source and load impedances should be considered while designing an EMC filter. A capacitor linked between the line and ground typically produces better results in high impedance circuits, while a series inductor put within the line typically produces better results in low impedance circuits.

3.6 ADAPTATION PRINCIPLE OF EMI FILTERS

Power supply filters typically employ either passive or active filter circuits. Capacitor filters, inductance filters, and complicated filters (such as inverted L-type, LC filters, RC filters, etc.) are the most common types of passive filtration. Most active filters are active RC filters (electronic filters). The pulsation coefficient S quantifies the size of the DC alternating current's pulsing component. A higher value has a weaker filtering effect. Pulsating coefficient (S)=fundamental maximum of output voltage AC component/DC component of output voltage.

Filtering's job is to take the rectifier's fluctuating output voltage and smooth it out to where the DC power is practically constant throughout the system. Transmission of AC power to the power supply is unaffected by the EMI characteristics of the power port when an EMI filter is used. Furthermore, the EMI noise sent with the AC is significantly attenuated, and the EMI noise generated by the devices in the power supply is successfully suppressed. The power terminals' EMI is taken into account when developing the EMI. Inductor, capacitor, resistor, or ferrite device make up the bulk of this two-terminal network that is selective in its operation. Within the filter's stopband, its high series impedance and low parallel impedance create a significant mismatch with the noise source's impedance and load impedance, resulting in the transmission of the noise source's undesirable frequency components.

There are a number of limitations that must be:

1. Equipment application
2. EMC performance requirements
   a. CISPR, DO-160, MIL-STD-461, other
3. Mechanical constraints
   a. Form factor
   b. Weight
4. Business constraints
   a. Cost
   b. Use of standard/commercial-off-the-shelf (COTS) parts
   c. Schedule
5. Input power source
   a. AC single or three phase
   b. DC single ended or floating (differential)
   c. Inrush requirements
   d. Lightning requirements
   e. Power-interrupt requirements

6. Output power load requirements
7. Switching frequency in the case of switching converters
8. Differential-mode design goals
9. CM design goals

Numerous technical indicators such as order (the number of stages), complete spectrum, relative download speed, cut-off frequency, standing wave, out-of-band rejection, ripple, absolute group delay, group delay fluctuation, power capacity, phase consistency, amplitude uniformity, and operating temperature range are used to evaluate a filter's electrical performance.

3.7   ELECTRONIC FILTER PARAMETERS

- Order (Number of Stages): The order of a high-pass or low-pass filter is equivalent to addition of capacitor and inductor count. The order of a bandpass filter is equivalent to the sum of the numbers of parallel resonators, while the order of a bandstop filter is equal to the sum of the numbers of series and parallel resonators.
- Absolute Bandwidth/Relative Bandwidth: Bandpass filters often utilize this indicator to describe the range of frequencies that can be sent through the filter. The ratio of absolute bandwidth to the fundamental frequency is known as the relative bandwidth.
- Cut-off Frequency: High pass and low pass filters often make use of the cut-off frequency. The cut-off frequency of a low-pass filter defines its upper frequency limit, whereas the cut-off frequency of a high-pass filter defines its lower frequency limit.
- Standing Wave: The vector network's S11 measurement reveals how well the port impedance of the filter corresponds to the system's desired impedance. Shows how much of the incoming signal was reflected back through the filter's input.
- Loss: A filter's "loss" is the amount of energy that is lost as a result of the filter's operation on the input signal.
- Passband Flatness: Maximum loss minus minimum loss in the filter's passband, expressed as an absolute value. Identifies the variance in filter power usage between signal frequencies.
- Out-of-band Rejection: The "attenuation amount" is the amount of signal loss experienced by frequencies that are outside the passband of the filter. Defines the filter's capacity to remove signals at particular frequencies.
- Ripple: The filter's passband S21 curve's slope difference between its highest and lowest points.
- Phase Linearity: The phasing of the passband frequencies of the filter minus the phasing of the transmission line at the centre frequency. Describes the filter's dispersion capabilities.
- Absolute Group Delay: The duration of time it takes for a signal to travel from the filter's input to its output port.
- Group Delay Fluctuation: Absolute group delay passband range is the range between the highest and lowest values of the group delay. Describes the filter's dispersion capabilities.
- Power Capacity: The maximum amount of power that can be fed into the signal being passed via the filter's passband.
- Phase Consistency: The phase shift of the transmitted signal that occurs when comparing various filters of the same indication to one another in the same indicator. Discuss the differences (consistencies) that exist amongst batch filters.
- Amplitude Consistency: The variation in the amount of signal that is lost during transmission between the various filters of the same indication in the same indicator. Explain the difference (in consistency) between batch filters and other filters.

3.8 FILTER CONFIGURATIONS

In its most fundamental form, an EMI filter can be broken down into two primary categories of component: capacitors and inductors. The most basic kind is referred to as a first-order filter, and it just has one reactive component to its make-up. Inductors block or reduce noise, whereas capacitors redirect noise current away from the load they are connected to. The attenuation of these single component filters only grows at a rate of 6 dB per octave or 20 dB/decade, which means that they are not very effective in most situations. In order to attain a higher level of attenuation, a filter of a higher order, whether that be a second or higher order, is required. The filter must have at least two reactive components. The impedance of the source as well as the load, along with the greatest frequency that is allowed to pass, determines the cut-off frequency. This filter, consisting of just two parts, is often referred to as a "L" filter. To account for filter resonances and ringing, designers use a metric called the damping factor to characterize the gain and the time response of the filter. In a third-order filter, there are three or more re-active components. Common names for these filters are "pi ()" and "T" filters. The greater the filter, the bulkier it becomes. For example, the third-order filter is a common filter topology. Compared to higher-order filters, first-order filters are the most cost-effective to construct because they only require a single CM choke. For example, $L = 1.99$ mH for 4 kHz cut-off into a 50 Ohm load is the value for the attenuated signal, and it is calculated by the load (in Ohms) divided by the radian frequency (2pi*f).

The attenuation level needed by the EMI filter at the desired design frequency is the primary factor in determining the values of the filter components. Typical circuit diagram of an electromagnetic interference (EMI) filter is shown in Figure 3.1. The following method is used to determine the amount of attenuation that is needed:

- The input current to the converter can be simulated or calculated.
- Using line impedance stabilization network (LISN) to determine voltage.
- Using Fourier analysis to transform the voltage from LISN to the frequency domain.
- Above-bandpass filtering around a sweep frequency Inverse Fourier analysis to convert the filtered voltage into the time domain.
- Using a quasi-peak (QP) detecting circuit to determine the noise voltage at the sweep frequency (QP).
- Time-domain voltage lowpass filtering; frequency-domain sweep repetition between 150 kHz and 30 MHz.
- Noise voltage between 150 kHz and 30 MHz was compared to regulatory thresholds.
- Attenuation must be applied at the "design frequency," which is determined by the height of the first voltage peak that appears above 150 kHz.
- In order to meet the regulations, a differential mode (DM) filter must now be implemented, which at the very least provides the necessary attenuation at the design frequency.

3.9 OPERATING PRINCIPLE

The essential elements and topology of a power line filter are:

- One- or two-stage DM (line-to-line) low pass filter.
- There must be at least one line-to-line capacitor and one set of series inductors.
- Two or more line-to-chassis capacitors and a CM inductor make up a CM filter. As a means of increasing the "outside" loop impedance, the CM inductor is positioned at the filter's input.
- One or more transient voltage suppression devices if required.
- C1 and C2 are X capacitors, which are DM capacitors with capacitances typically ranging from 0.01 to 0.47 pF.
- Since Y1 and Y2 are Y capacitors (CM capacitors), their capacitance shouldn't exceed the tens of nanofarads. Otherwise, leakage could occur.

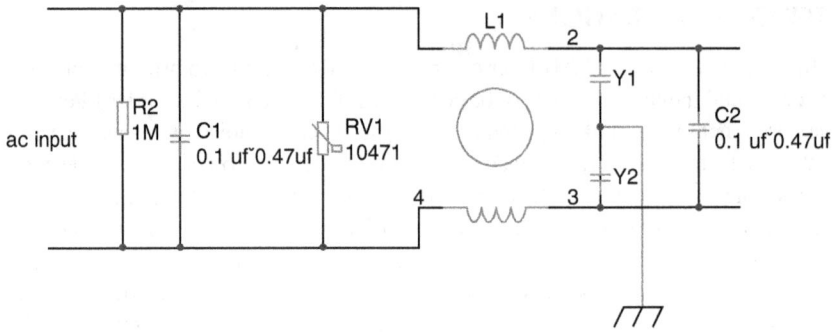

FIGURE 3.1 Typical circuit diagram of an electromagnetic interference (EMI) filter.

- A CM choke, denoted by L1, is a pair of coils twisted on the same ferrite ring in the same direction. A few millihenries are about the limit of the inductance.
- To reduce the impact of interference in common mode, the magnetic fields produced by the two coils must be aligned, and the common mode choke coil must have a high resistance.
- The mode signal is unaffected by the magnetic field generated by the two coils since the two coils cancel each other out.
- Secondary filtering is an option for enhanced performance.

One must be familiar with the filter's performance indicators in order to make an accurate assessment of its efficiency. Ratings include voltage, current, insulation resistance, voltage tolerance, temperature range, insertion loss, and more. The insertion loss is the utmost critical one.

3.10   INSERTION LOSS

The insertion loss quantifies the frequency-dependent attenuation of a filter. For every given filter, the insertion loss can be expressed numerically as the ratio of the input signal level to the filtered signal level. It is a measure of its capacity to attenuate noise in a circuit. The insertion loss of a filter varies with frequency, sometimes by several orders of magnitude across the frequency range where the filter is effective (dB). The insertion loss of a low-pass filter is small for low frequencies and large for high ones. The corner frequency is the frequency at which the filtered signal's amplitude drops to half of its unfiltered value. Filter insertion loss can be easily demonstrated with the aid of Bode graphs. You may easily compare the effectiveness of various filters by looking at their frequency-response curves, as illustrated in Figure 3.2. When comparing, it is essential that both the source and load impedances be kept constant (Figure 3.3).

Insertion loss for five different filter topologies in a 50 ohm system is depicted in the preceding figure. Each filter has a 100 nF capacitor and a 1 micro H inductor. The C-section filter offers a one-pole roll off (20 dB/decade) above its corner frequency, while the LC-section and CL-section filters offer a two-pole roll off (40 dB/decade) and the Pi-section and T-section filters offer a three-pole roll off (60 dB/decade) above their corner frequencies (60 dB/decade).

3.11   STEPS IN THE DESIGN OF A POWER LINE EMI FILTER

The filter design may start with top-level design risk mitigation. These are as follows:

- Calculate the frequency-dependent measurement of the expected DM (line-to-line) noise current's magnitude.
- Determine the frequency-dependent CM current (in phase on both or all three power lines) via calculation or measurement.

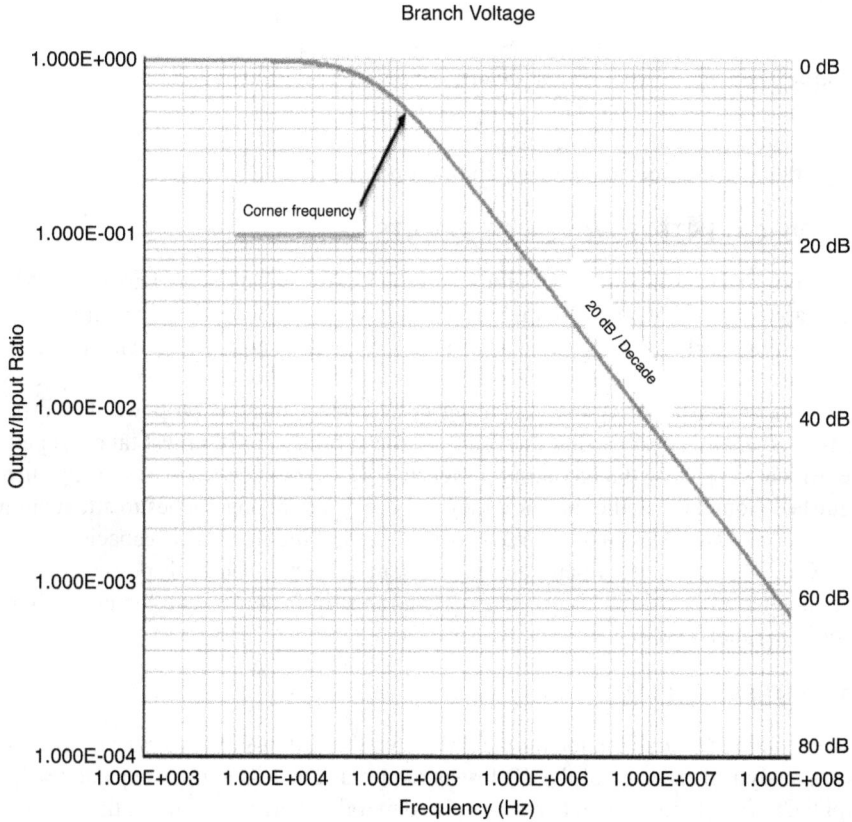

FIGURE 3.2 Insertion loss.

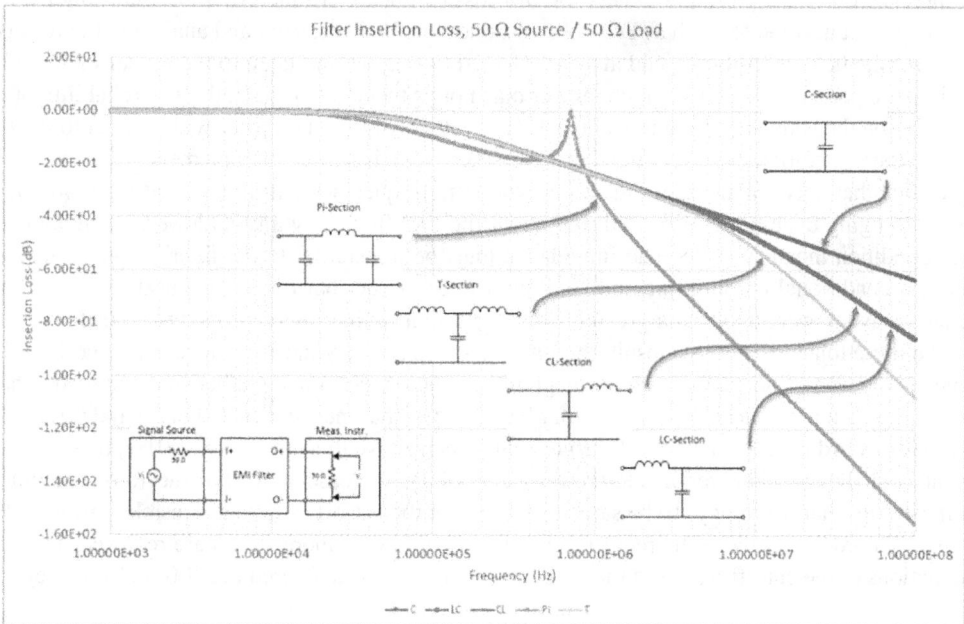

FIGURE 3.3 Graph of insertion loss of filters.

- To find out how much attenuation the filter needs, compare the emission limit with the expected noise current amplitudes as a function of frequency.
- Construct the filter's topology and determine the values for its individual parts.
- DM inductors must be designed.
- The CM inductor must be designed.
- Design the filter's actual appearance and container.

3.12 DM VS CM INTERFERENCE CURRENTS

Currents that flow in opposite directions along a circuit's two conductors are known as DM currents. CM currents are a type of AC that streams in a similar path over both wires in a circuit. That's why there's a third route back for these currents to take—finding where CM currents come from. It is common for circuits with high amplitude waveforms and very short rise times to capacitively link to the chassis of the device or system, resulting in CM noise currents on power lines. In order to return to its source, any noise current injected into the system's chassis must take the path of least resistance, which is typically the incoming power line. These two forms of interfering current must be distinguished from one another because they require different approaches to attenuation.

Inductors within an EMI filter placed between power lines and a Y capacitor can be used to dampen CM noise current, which runs in the same direction on both power conductors and returns via the ground conductor. By connecting an inductor in series with the power lines and an X-capacitor in parallel, DM currents can be dampened.

3.13 DM DESIGN GOALS

To begin achieving EMC qualification for the EMI filter, the DM loss criteria must be met. To accomplish this, an analytical estimate of the dB loss is required. The loss at this frequency will be equal to the amplitude of the fundamental harmonic plus a margin of error, etc., using this straightforward technique. The study is oversimplified and might make use of an approximate current signature for the switching converter at a certain frequency (fsw). Fast Fourier transform (FFT) sweeping can detect harmonic content if the peak current amplitude definition and the shape of the current are both precise. Filter design requirements for DM loss can be determined by superimposing the dB/frequency limit asymptote, such as DO-160, etc., onto the FFT spectrum and analyzing the results.

The harmonic composition could also be affected by factors unrelated to the model's underlying circuitry and parasitic elements, which may or may not be included in the study. The reliability of the data relies on the standard of the model used. Transparency of the DM filter with respect to the line is a necessary condition. Perfect for DC and AC setups, this is simple to implement at 50 and 60 Hz. Any device that needs significant loss at low kilohertz frequencies will find operating at 400 Hz to be a challenging endeavour. When talking about the line frequency and its lowest harmonics, the stated condition means that the load impedance must be transferred to the filter's input. Harmonic content is conditional on both line and load quality. Odd-order harmonic content distorts the form of a sine wave power supply to a greater extent with higher line impedances. Odd harmonics best describe these tones. To get this result, the filter cut-off frequency must be higher than the 15th harmonic, as harmonics above the 15th are negligible even for the lowest quality line and load. That's why the regulation operates at this particular harmonic of the line frequency. For 50 and 60 Hz, this is straightforward because the cut-off frequency is much above this target; for 400 Hz, however, this presents a challenge. However, due to the resonant voltage increase at 400 Hz, the 15th harmonic is insufficient for that frequency. To be safe, it's best to set the cut-off as high as feasible, say, at 8 kHz. However, in order to achieve the required insertion loss, two or more stages are required. Previous formulations of the cut-off had used the 10th harmonic, but these ignored the 400 Hz frequency.

3.14 DM FILTER INPUT IMPEDANCE

The input impedance of the filter Z_{in} at $N_h = R_{load}$, N_h is the harmonic number and is set equal to 15, and R_{load} is the load impedance at the same frequency.

The high-current filters are notoriously challenging to design because of the high loss required at low frequencies. Higher capacitor currents, which heat the capacitors due to the equivalent series resistance (ESR), result if the cut-off frequency allows the filter to attenuate the lower harmonics (ESR). Inductors get hotter because of DCR and higher core losses when the low-frequency cut-off raises harmonic currents through them. Because of this, the filters used for filtering power lines will run hotter than normal. The circuit's Q will increase as the resonant rise frequency is decreased due to the low cut-off frequency. High operating temperatures and/or a filter that is difficult to tune are both possible outcomes of these situations.

3.15   DM FILTER OUTPUT IMPEDANCE

The same rules apply to this part as they did to the filter's input impedance. There must be no discernible effect on the load from the filter. Normally, the output impedance target will also be attained if the input impedance target is achieved. If these two conditions are satisfied, the filter will function more effectively.

Z_o at $N_h = R$s, where Z_o is the filter output impedance, Rs is the line impedance at the same frequency, and Nh is the harmonic number. As such, the DCR of the line is the line impedance. This is typically the case up to 5 kHz, after which higher line impedances experience a sharp increase. Common line impedance at 10 kHz is around 4 ohms.

3.16   INPUT AND OUTPUT IMPEDANCE FOR A DC FILTER

Without a switching converter in the load, a DC system can easily fulfil the conditions of both of the previous sections. Here, the filter's output impedance must be extremely low at and above the switcher frequency. In order to make this claim, it is assumed that no front-end correction was implemented in the design of the switching converter. This is obviously simplified if the same person or group designs both the EMI filter and the switching converter. In principle, the output impedance of the filter, $Z_0 \ll RL@F_{sw}$, where F_{sw} is the switch frequency and the rest of the terms are the same. In the same way, the tenth harmonic of the frequency of the switcher is valid. Because of this, a slightly inductive output impedance or operating the output capacitor above its self-resonant frequency (SRF) can cause the switcher to be starved at the 9th or 11th harmonic even if the fundamental is not SRF. This is a common example of what is known as "incremental negative resistance." It is the goal of a switching converter to maintain a constant output voltage regardless of fluctuations in the input voltage. When the load current remains unchanged, the power consumption at the input remains unchanged as well. To maintain the same power output regardless of the input voltage change, the input current must be decreased by the same factor:

$$R_N = -\frac{V_{IN}}{I_{IN}}$$

where R_N = negative resistance, V_{IN} = input voltage, and I_{IN} = input current.

The switching converter will be underutilized if the filter's impedance is increased. When the filter is conducting, or "on," the output impedance should be on the order of 10% of the load impedance. The pulse width is also responsible for this. It's crucial that the switcher's voltage drop isn't too severe; otherwise, its operation will be compromised.

The DC link voltage droop during full-load switching is typically well within acceptable limits because of the hold-up capacitance added to the switch's input. Keep in mind that the input impedance, ZI, of a switching converter only appears as a negative resistance at very low frequencies. The impedance is affected by the converter's internal filter elements and the narrow bandwidth of its feedback loop at higher frequencies. If we consider the connection between the source and the load, we can write an expression that represents a voltage divider where the source impedance interacts with the load (negative resistance). In addition, the dynamic behaviour of the second-order system, which is analogous to a quadratic polynomial, is determined by the poles of the source and load. The frequency-dependent terms that govern the switcher's performance can be multiplied by the relationship in the following equation if the switcher is connected to the source impedance ZS:

$$\frac{Z_I}{Z_I + Z_s} = \frac{1}{1 + Z_s/Z_1}$$

If $|Z_s| \ll |Z_1|$ for all frequencies, we conclude that the switcher's stability and performance will not be negatively impacted. The above equation is stable if and only if its poles are located on the left side of the S-plane. The source impedance is equal to input filter, will typically feature a series inductor and DCR ohms:

$$Z_S = sL + Rdc$$

When a capacitor is connected in series with a negative equivalent resistance, the resulting impedance is the load:

$$Z_L = \frac{R_N}{1 + sCR_N}$$

For a stable system,

$$sL + R_{dc} \ll \frac{R_N}{1 + sCR_N}$$

If R_{dc} is added as in Figure 3.4, to reduce Q then a constraint for a stable system is $R_{dc} \ll |R_N|$. From a practical standpoint, the filter would be stabilized by using a damping dQ RC shunt network.

3.17　CM DESIGN GOALS

CM noise can be one of the most difficult noise sources to deal with in a switch-mode power supply. CM noise is especially difficult to deal with in the FM band, due to stringent standards. While there may not be any true half-wave-length dipole antennas in a system to capacitively couple CM noise, plenty of traces are efficient enough to couple this noise and make a design fail in this region. Typically, the biggest generator of CM noise in a buck converter design is the switch node. The switched waveform, with magnitude equal to the converter's input voltage, radiates an electric field and becomes worse in applications with higher input voltages. The amplitude of the electric

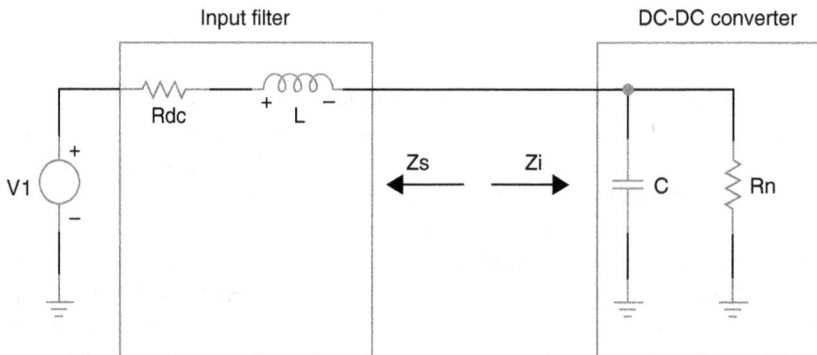

FIGURE 3.4　Source and load impedances.

field falls inversely with distance in the near field. This is a problem, as designers want to shrink implementation size but keep the switching frequency relatively fixed, which results in no change in output-inductor and switch-node sizes. One way to filter CM noise is to place mutually paired inductors, or a CM choke, in series with the input paths This method can provide sufficient CM noise rejection up into the hundreds of megahertz.

Unlike the DM, the CM does not need to fulfil any of the aforementioned conditions. If you want to cut into power harmonic frequencies, you can set the cut-off frequency as low as you like. This Zorro inductor leakage inductance must be monitored frequently (s). Leakage fluxes are the fluxes from each coil that do not go through the high-reluctance air path between the two coils but instead go around the toroid. As they travel mostly outside the toroid, the leakage fluxes that leave the coils must eventually find their way back. Part of the toroid allows these fluxes to re-enter the coil. The core's permeability drops and the path through the coil's reluctance rises as the frequency rises. When the core's reluctance grows, more of the flux generated by the coil escapes the toroid and becomes leakage.

The common mode expands to enormous proportions when the cut-off frequency is lowered, yet there is no lower-frequency limit owing to bandpass or other causes. In reality, the maximum allowed current is determined by the size and rating of the CM inductor. It is made in a way that they barely affect power factor correction circuits, switchers, or any other load. This, once more, is predicated on the assumption that the CM inductor has negligible DM inductance. The DM inductance and consequently the DM loss typically range from 1 to 2%. The typical operating area of a CM inductor is from 0.5 H to 33 mH, and even higher. This amounts to 660 H at a 2% duty cycle, which is higher than the differential inductor value in many EMI filters uses. CM will typically aid differential losses if the circuit is balanced with inductors in both the hot and return lines. Some businesses reduce coupling on purpose by physically separating windings. A distinguishing feature of the ferrite toroid is that the CM windings are equally spaced around the entire core.

3.18   ESTIMATION OF THE CM SOURCE IMPEDANCE

When selecting a CM choke, the plot for impedance versus frequency is crucial. A typical approach when determining what CM choke to use is to consider the impedance at the noise band in need of attenuation and the DC losses associated with the choke. These two values are used to determine the trade-offs for the cost and size of the CM choke. If the DC system is balanced with the hot wire and a return wire, then CM inductors can be utilized without compromising the DC load. Assuming the DM properties are small in comparison to the CM inductor, this is the case. Because of the extremely low SRF, capacitor values are typically high and the circuit is effectively open.

It is caused by high-frequency switching and parasitic capacitance in the circuit. It is now these parasitic components that are generating unwanted noise. Possible sources of the CM noise include parasitic capacitance in the transformer, diode capacitance, and stray capacitance between the switching device and ground.

3.19   METHODS OF REDUCING THE INDUCTOR
VALUE DUE TO HIGH CURRENT

There's a chance that some of these filters need a lot of power. Achieving circuit balance is one way to improve the design. For this application, the inductor is split in two, with one-half located in the hot line and the other in the return. By halving the inductance value, the inductor can be designed to produce minimal heat, tolerable flux levels, and, possibly, a manageable cost. However, other methods involving C cores employ parallel windings. It is wound so that each limb or side carries equal amounts of current. The smaller diameter makes winding the wire much simpler. Using larger capacitors helps alleviate the inductor current problem, provided the system permits it. When calculating a cut-off frequency, care must be taken to avoid setting it adjacent to the line frequency. Unless it is avoided, the line frequency will experience a resonant increase.

3.20   DESIGN CRITERIA

3.20.1   DESIGN OF THE DM INDUCTOR

- The inductor's minimum inductance must be stable up to the maximum AC current or DC current under the most extreme load conditions.
- An acceptable permeability up to the highest frequency required for usable inductance must be present in the magnetic material selected for the core.
- In order to reduce power loss, the inductor winding needs to have low resistance.
- The distributed capacitance of the winding must be low enough to enable the self-SRF of the inductor to be above the maximum frequency to be attenuated by the filter.
- Common mode inductors are multi-wound inductors optimized for low DC or power-frequency winding resistance in conjunction with high inductance and SRF.
- Common mode inductors are able to achieve these features because they are built to cancel out the magnetic flux caused by the power current by having two (or more) identical windings carrying opposite but equal currents.
- A relatively high permeability magnetic material can be used because the core of the CM inductor cancels the magnetic flux at DC or low frequencies. Because of this, a smaller number of turns can be used in each winding, which reduces the overall distributed capacity and increases the SRF.

3.20.2   DESIGN OF THE CM INDUCTOR

The CM choke is one form of the many inductor types. Its main function in a power supply is to separate the output voltage from the line voltage. Here, the impedance is the most crucial factor. For a given frequency, the inductor's impedance is typically specified as a minimum value (in OHMs). Follow this equation to determine the necessary inductance:

$$L_s = \frac{X_s}{2nF}$$

Where:

X_s = impedance in OHMs (Ω)
F = frequency in Hertz (Hz)
L_s = inductance in Henry (H)

Using the above formula, one can determine how much inductance is needed to produce a specific impedance.

1. Determine the wire size—The rule of thumb for determining wire size is 500 circular mils per amp:

$$Aw = \frac{500\,\text{cirmils}}{A} I$$

Where:

Aw = nominal area for the wire
I = current in amps
Nominal area of wire is listed on wire manufacturer's data sheets.

2. Choose a core—the best way to choose a core is through trial and error. In most cases, a ferrite material with a permeability of 5,000 or higher is chosen. The wire gauge required will determine how big a core you need. A spacer (typically 3 mm) must be able to be

placed between each winding, and the windings must all fit in a single layer on opposite sides of the core.

3. Calculate the number of turns.

Number of turns required is given by

$$L = A_L T^2$$

Where:

L = inductance in henries

T = turns

A_L = is the A_L value of the core (found on core data sheets from core manufacturer)

The following factors are to be considered during design:

- Physical layout of the windings. Bi-filar and tri-filar windings.
- Providing for adequate inter-winding insulation. Handling the high turn-to-turn potential difference.
- Keeping mechanical stress from impairing the magnetic properties of the core.
- Proper location of the common mode inductor within the filter.

3.20.3 PACKAGING AND LOCATION OF A POWER LINE FILTER

- Preventing crosstalk between the filter elements and surrounding circuits. Need for a metallic enclosure for the filter.
- Preventing the interference from bypassing the filter.
- Proper physical location of the filter in the system or device.
- Making provisions for dissipation of internally generated heat.
- Proper electrical bonding of the filter housing to the system structure.

3.20.4 IMPORTANT CONSIDERATIONS IN THE INSTALLATION OF A POWER LINE FILTER

- Making sure the filter ground bus is electrically bonded to the system structure via a very low impedance path is the single most important requirement for installing a power line filter. This path's impedance must stay low all the way up to the highest frequency that the filter is designed to reduce.
- Position the filter as close to the point of entry for the power lines as possible.
- Provision must be made to dissipate the heat generated within the filter. This is particularly important in high-power systems.

3.20.5 TESTING OF A POWER LINE EMI FILTER

- The standard setup for specifying the attenuation of a power line filter assumes a source and load impedance of 50 ohms.
- Filter attenuation must be measured with the filter carrying the maximum load current (at DC or the design power frequency) it is designed to handle.
- It is also desirable to measure the attenuation of the filter at lower load currents.

3.21 COMPONENT NON-IDEALITIES

Unfortunately, the component limitations of an LC filter pose a problem for filtering in the FM band. An inductor has a finite SRF, which can create high impedance in the FM band. For low frequencies, the non-ideal inductor impedance behaves as expected, with a positive slope related to its inductance. At some frequency, the inductor reaches its maximum impedance. At this point, the

SRF of the inductor has been reached, where the parasitic capacitances of the inductor resonates with the ideal inductance, making the device appear as a very high impedance device. Beyond this point, the impedance of the inductor often falls off rapidly. Inductor selection must factor in SRF. Often, the SRF of a particular inductor series increases with smaller package size and inductance. One way to achieve the highest level of attenuation is to select the smallest possible package size for the required inductance and current capabilities. The resonant frequency of the input capacitor also has to be acknowledged when designing a filter. Often, the equivalent series inductance (ESL) of a capacitor limits the high-frequency performance. Just as in the case of the inductor, ESL can be minimized by limiting the value and size of a given capacitor.

3.22   HIGH-FREQUENCY DM FILTERING

One of the limitations of the LC filter stage is the finite SRF of the inductor, which leads to high-frequency noise (typically greater than 85 MHz), which is not getting attenuated as greatly as the lower frequencies. This issue is extremely troublesome, as it occurs right at the start of the FM band, which often has the most stringent limit lines. Mathematically, the lowest-order harmonic that falls in the FM band is the 41st harmonic for a 2.1-MHz power converter. It may be assumed that this order of harmonic would be relatively low in energy; however, it is significant enough to make the design fail CISPR-25 Class 5 even with dithering techniques such as spread spectrum. There is some additional high-frequency energy captured in the input spectrum of a buck converter. This energy arises from high-frequency ringing on the switch node that capacitively couples back to the input power lines and/or test setup. The energy often falls in the 200- to 300-MHz range, significantly above the SRF of the low-frequency-stage inductor and the resonance frequency of the LC filter. Nevertheless, it is not a concern, as the noise will be CM and can only be attenuated by a CM choke.

3.23   FERRITE BEAD SELECTION

A ferrite bead is a device with a low inductance and small parasitics that enables a very high SRF, which provides rejection in the hundreds of megahertz. Ferrite beads filter DM noise at frequencies above the SRF of the low-frequency stage. Ferrite-bead data sheets do not provide much information; they may give only one impedance curve at a zero DC-bias current. In fact, for a buck converter, a ferrite bead could be biased in the order of hundreds of milliamps or more. To select an adequate ferrite bead, several ferrite beads rated for the buck converter's input current should be characterized with the highest impedance at frequencies close to the FM band, or at frequencies above the SRF of the low-frequency stage as shown in Figure 3.5.

3.24   FILTER DAMPING

A filter requires damping to reduce the peaking at the corner frequency of the filter. Peaking can result in significant amplification of noise at low frequencies, resulting in a non-compliant design. A poorly damped filter can also have significant effects on the stability of a power converter. The damping of the filter is directly related to the output impedance of the EMI filter. A typical approach for damping the filter is to use a parallel RC circuit. This replicates the electrical characteristics of an electrolytic capacitor, providing similar rejection at low frequencies without taking up a significant portion of the available board space.

Lacking access to a simulation tool doesn't mean the designer is out of luck; a "rule of thumb" can do in a need. The first filter element nearest to the source or load end should have the greatest possible mismatch at EMI frequencies. When the impedance of either the source or the load is low, the initial stage of filtering often consists of inductive components (100 Ohms). When the impedance of either the source or the load is large (>100 Ohms), on the other hand, a capacitive first filter element is suggested. For the designer, this means less complexity and more efficiency in the final

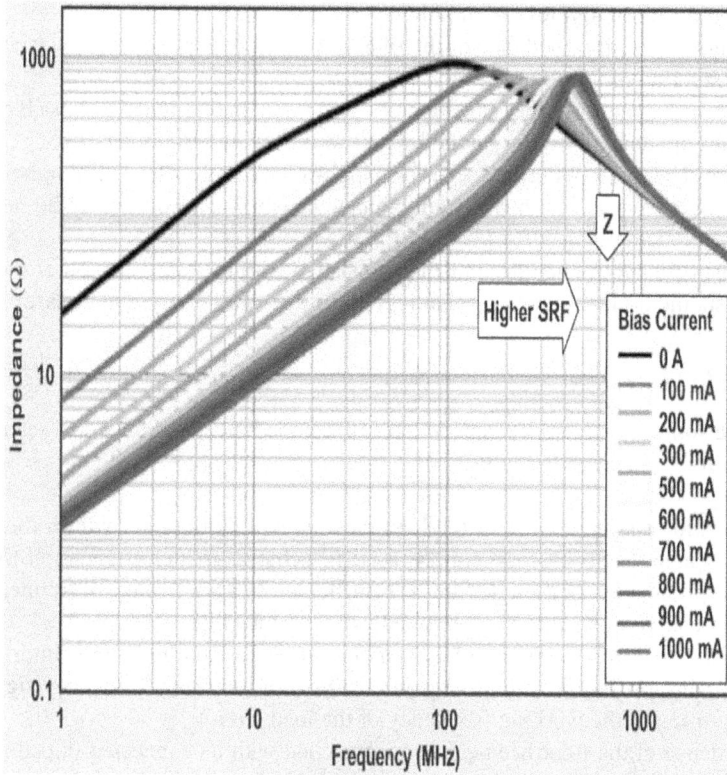

FIGURE 3.5   General derating of ferrite beads.

design. Filters can aid with a variety of immunity issues, including those caused by radiated emissions, radiated emissions from the mains power line, induced RF signals, and transients like electrical fast transients (EFT). CM and DM currents are present in every circuit. They're poles apart from one another in every respect. In the presence of transmission lines and a return path, either method may be realized. DM signals are a method of transmission that can be used to send data or a specific signal (information). Transmission in CM is an EMC nightmare, and it is an unintended consequence of DM transmission.

DM analysis is commonly used in emission prediction software that employs simulation. Radiated emissions cannot be predicted using DM (transmission line) currents alone. Interference from electromagnetic fields typically originates from CM currents. As many factors and parasitic parameters are involved in the generation of CM currents from DM voltage sources, it is possible to grossly underestimate expected radiated emissions if one only calculates DM currents. Power surges form in the power and return planes at edge switching times, and these parameters are not always easy to predict. Equal and opposite RF energy components, or DM current, are carried by the signal and return paths. By precisely establishing a 180° phase shift, RF DM currents can be cancelled. Nonetheless, CM effects may emerge as a result of ground bounce and power plane fluctuation brought on by components drawing current from a power distribution network.

3.25   FILTER REQUIREMENTS

Within the specified operating frequency range, the chosen filter must provide the attenuation characteristics demanded by the load. If not, using multiple stages in parallel to achieve the same attenuation effect would be the next best thing. Cascading multiple filters allows for effective attenuation across a broad frequency range. Criteria to be followed are:

- Performing insertion loss test on the filter.
- Comparing the output impedance of the filter circuit and the power supply, whether it will affect the stability of the filter.
- Choose a multi-stage filter that can suppress self-resonance as much as possible.
- High input impedance filter is matched with low power supply impedance.
- The filter must be able to withstand occasional high-voltage transients, such as lightning impact.
- The filter must have a certain withstand voltage capability, according to the rated voltage of the interference source.
- The current allowed by the filter must be consistent with the rated current of the circuit.
- The filter should have sufficient mechanical strength, install easily, and be safe and reliable.

3.26   FILTER SELECTION

The interference source, frequency range, voltage and impedance, and load characteristics all play a role in determining the best filter to use.

- The attenuation characteristics of the load must be met by the filter within the functional frequency range. Multi-cascade can be used to achieve greater attenuation than single-stage if the attenuation of a single filter is insufficient. In a wide range of frequencies, good attenuation characteristics can be obtained using various filter cascades.
- A filter with very steep frequency characteristics is needed to filter out the interference frequency of the suppression, only permitting the flow of important frequency signals, while still conforming to the working frequency of the load circuit.
- The impedance of the filter needs to be mismatched with its connected impedances of the interference source and the load at the necessary frequency. The filter's output impedance should be low if the load has high impedance, and vice versa if the power source or interference source has high impedance. If the impedance is low, the filter's input impedance should be high. It is challenging to obtain stable filtering characteristics if the source impedance or the interference source impedance is unknown or varies over a broad range. The filter's input and output can be connected to a fixed resistor, and it displays good filtering characteristics.
- The filter's capability to withstand voltage should be high enough, and it should be chosen in relation to the rated voltages of the power source and the interference source, so that the filter can function reliably under all foreseeable conditions of use. It should be capable of withstanding the effects of sudden, high-pressure input.
- The filter ensures that the pass is constant at the rated current for reliable circuit operation. If the rated current is high, the filter will be bulkier and heavier, and if it's low, the filter's durability will suffer.
- There needs to be enough mechanical strength, a simple design, low weight, and compact size in the filter. Simple to set up, secure, and trustworthy.

3.27   TYPICAL EMI FILTERS

These days, EMI filters come in a wide variety of shapes and sizes. The most common of these include the Cauer and RC shunts, as well as, T and L. Those filters can be doubled, tripled, or even quadrupled in some cases. There are good and bad features to every EMI filter. There are situations in which each filter type excels, and others in which it falls short.

3.27.1   THE π FILTER

According to the 220-A 50-ohm test specification, this filter appears to be excellent. At low frequencies, the capacitor on the line side will be able to interact with the 50-ohm line impedance. When taking the source into account, the filter will perform admirably. In the case of the three-phase variety, Measure one phase with the other two phases tied to ground. Each section's input and output capacitors are then increased by a factor of 2. This aids the filter designer in keeping the filter's mass and size under control while still achieving the required loss. In other words, the filter's values, package size, and weight can all be reduced while still satisfying the attenuation requirement. Figures 3.6 and 3.7 depict the single and double configurations, respectively.

When compared to the source and load capacitors, the central capacitor is significantly larger. The value of the filter's inboard capacitors is typically double that of its end capacitors when employing a multiple topology. Contrarily, this filter fails the MIL-STD-461, naval, and current-injection probe (CIP) tests. Here, a 10-F capacitor shunts the line-side capacitor to ground, increasing the current through the output current probe.

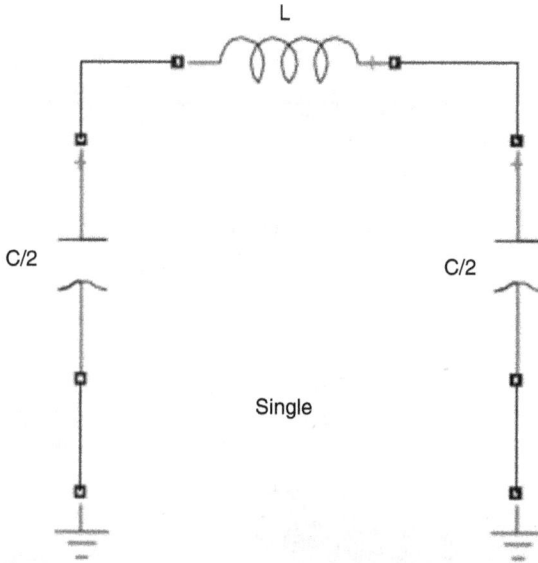

FIGURE 3.6   Single configuration.

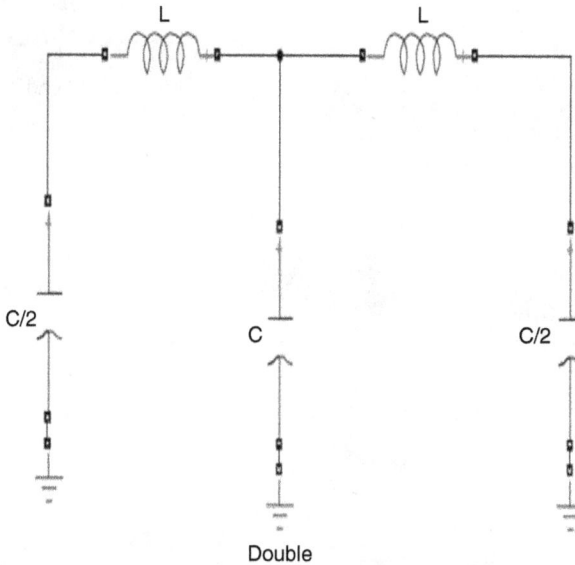

FIGURE 3.7   Double configuration.

The filter capacitor would have to be at least one order of magnitude larger than the 10-F capacitor in order to be included in the circuit. A good filter can't have a value that high. SRF is low in large capacitors, so high-quality, smaller capacitors must be shuffled in to compensate. The filter loss is measured by repeatedly switching the receiver or spectrum analyzer between the two current probes. According to the MIL-STD-461 and naval tests, the three-element appears to be a two-element L filter at lower frequencies. Each element contributes about 6 dB in loss per octave to the whole, reducing the 18-dB filter to a 12-dB loss per octave version; 18 dB is for the single and the double is 30 dB, for example. In some DC systems, the filter will also work well if the switching frequency of the power supply is high enough to keep the capacitor impedance facing the load low enough not to starve the system. And there shouldn't be any significant voltage drop due to the capacitor's impedance. One way to easily achieve

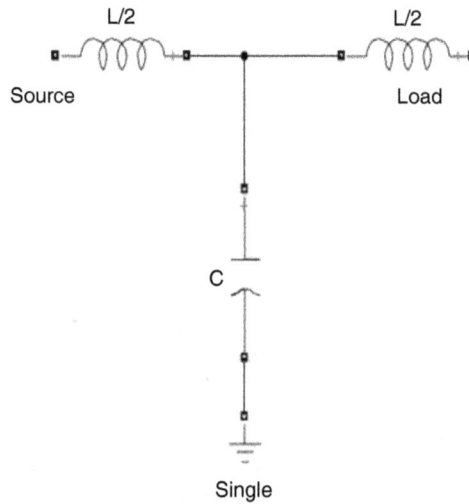

FIGURE 3.8 Single *T* configuration.

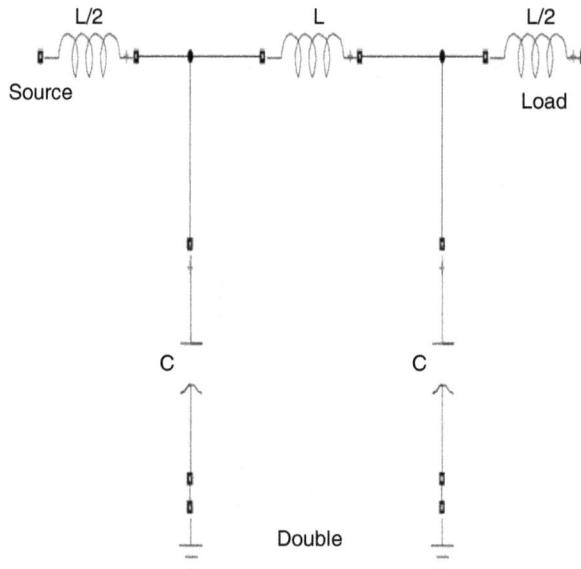

FIGURE 3.9 Double *T* configuration.

symmetry in the filter is to use only half of the required inductor in the high line and the other half in the neutral line. As a result, the filter's structure is adjusted to a more stable configuration. Since the front, or line-side, capacitor attenuation of the filter is less effective at the lower frequencies, the filter may fail testing in some cases.

3.27.2 The *T* Filter

The octave loss of a standard *T* filter is 18 dB, while that of a double *T* filter is 30 dB (6 dB per element). Low-impedance lines are ideal for *T* filters (high current requiring small inductors). Although the MIL-STD-461 loss specifications typically begin between 10 and 20 kHz, the line impedance remains very low up to at least 100 kHz. Due to the *T*'s inductive input impedance, the line impedance is further reduced. An input impedance is created for the capacitor to operate in. If the design method doesn't require excessively high values of these *T* inductors, these are also ideal for use with higher current loads. This may cause the voltage while feeding the load to spike or drop. It is not unusual for 132 V lines to be fed from 115 V AC 60 Hz sources to support a minimal load. This is because a high inductive value causes a resonant rise at a low frequency.

In a DC system, the T should never be used if the load employs pulse width modulation (PWM)-switching power converters; the high impedance of the output inductor facing the load will starve the switchers. It's possible that the switcher's designer anticipated this possibility and included a capacitor at the switch input to reduce input impedance. Since the capacitor shunts the inductor facing the load, the filter effectively becomes a maladjusted double *L*, although this is less of a problem in DC than in AC. Figures 3.8 and 3.9 depict the single-T and double-T configurations, respectively.

It's important to remember that the central inductor is normally double as big as the inductors on either end. To achieve symmetry in the *T* filter, one needs to cut the inductor in half and connect the cut-off half to the neutral leg to create an H pad. Be wary of any hidden grounds within the system that could alter the circuit configuration of these balanced filters. Dropouts in the circuit's lower half are brought about by the grounding structure. Even worse, the two capacitors are now in series with two ground inductors, making for a filter structure that is unstable and prone to frequency magnitude slope error.

3.27.3 The *L* Filter

The *L* variety is the standard. The *L* filter has a 12 dB loss per octave because it only uses two elements. These losses are all measured from a frequency above the cut-off point. DC mode is optimal for a single *L* filter if the load contains switchers. The inductor faces the DC source, and a high-quality capacitor with a high SRF would give the switcher frequency a path with low impedance. Figures 3.10 and 3.11 show the *L* filters.

For the same amount of loss, two inductors would cost less than a single inductor, and the same is true for capacitors. The following switcher, shown in Figures 3.12 and 3.13, may not be able to store enough energy with a smaller output capacitor, as shown in Figures 3.12 and 3.13.

This may result in a greater peak-to-peak voltage drop feeding the switcher, where the peak-to-peak voltage would be at the frequency of the switcher. It is imperative that the reactance of the capacitor be less than the impedance of the switcher's input. As long as the drop is not too great or the switcher frequency is high enough, the double *L* can be employed. Applications that require higher power can benefit from both a single *L* and many *L*s. Once more, to achieve symmetry in the *L* filter, divide the inductors and connect one-half of each to the neutral. However, there is a loss of 24 dB per octave when using the double *L*. Once more, validating the ground structure of the filter as well as the power system is essential in order to ensure that there are no hidden grounds that alter the circuit configuration. This is necessary in order to get a balanced filter topology.

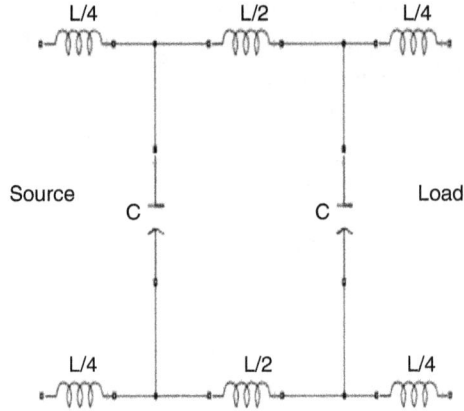

FIGURE 3.10   *L* Filter topology 1.

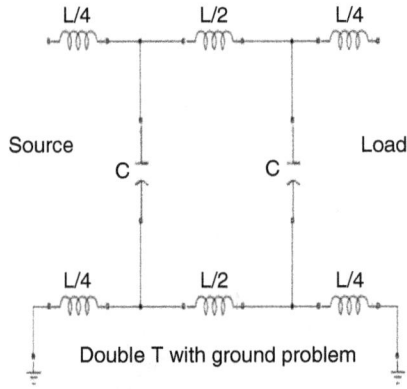

FIGURE 3.11   *L* Filter topology 2.

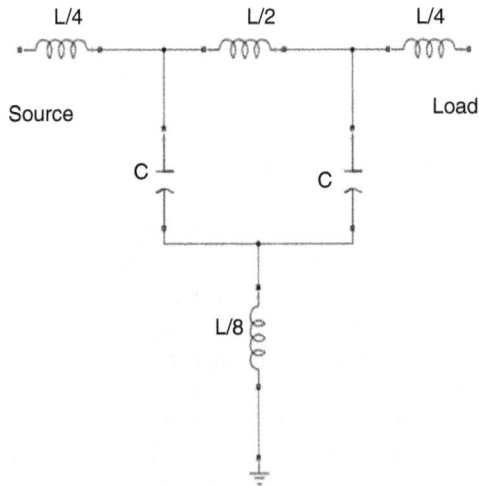

FIGURE 3.12   *L* Filter topology 3.

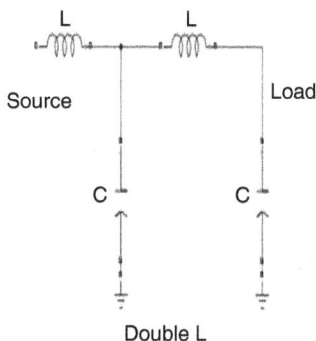

FIGURE 3.13 *L* Filter topology 4.

3.27.4 THE TYPICAL COMMERCIAL FILTER

This variety is the standard in electronic devices including computers, test equipment, and other office machinery. In this case, the maker must meet standards set by the Federal Communications Commission (FCC), Canadian Standards Association (CSA), Underwriter Laboratories (UL), Technischer Überwachungsverein (TUV), and Verband Deutscher Elektrotechniker (VDE). EMI test companies conduct the tests and provide the manufacturer with the necessary paperwork for the various regulatory bodies. These balanced type filters are typically obtained from third-party vendors and are frequently manufactured in other countries.

A CM inductor and two feed-throughs (or these might be "Y" caps to ground) give these filters a predominantly CM appearance in Figure 3.14; a capacitor across the input and output from hot to neutral and two more capacitors to ground complete the circuit. Using the two line-to-line capacitors and the CM leakage inductance, we may generate the DM. Such filters often suffer minimal loss. The two output terminals are formed by the feed-through capacitors, which must adhere to the leakage current criteria of whichever regulatory body has the strictest requirement. Washers placed in the centre of a pot core between the two windings increase the leakage inductance. To do this, a ferrite toroid core is used to wind two separate coils as far apart as feasible. The capacitors in the feed-through circuit are soldered to ground or connected to the case for a direct ground connection. Increasing the leaking inductance or adding inductors to both lines are two methods that can increase the DM loss. This results in a DM and CM, symmetrical type. Everything works out well because the FCC's loss specifications begin at 450 kHz. It is possible to use very small inductors and capacitors for these purposes.

Considering that most of the leakage inductance current is flowing through the air around the core, it is clear that this inductor cannot be saturated by the leakage inductance current. The mutual

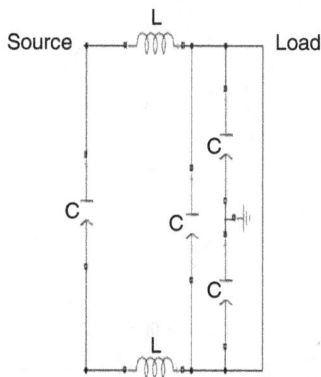

FIGURE 3.14 Commercial filter.

inductance (AL) of the ferrite or nanocrystalline cores that the CM inductor is wound on is quite high, which is why this is the case. A feed-through capacitor is used in most of these filters, but some of them employ cheaper capacitors that have longer leads, reducing the frequency at which they resonate on their own.

3.27.5 The Cauer Filter

Very low-impedance circuits are ideal for the Cauer, or elliptic, filter. Figure 3.15 shows the common configuration for these filters using numerous L and T structures. Regardless, a capacitor is often shunted across one of the core inductors. This is employed to dampen the amplitude (Q) of a resonant frequency, say 14 kHz. The network has been adjusted to operate just above the problematic frequency. While the central portion of the filter construction may be in the circuit at frequencies below the problematic resonant increase, its presence is limited there. The parallel capacitor allows the network to transmit higher frequencies. To control the amount of bypass, a resistor is typically connected in series with this capacitor. The resistor's value is normally equal to the filter's characteristic or design impedance, which is around 10 ohms. According to MIL-STD-461 standards, the design impedance is calculated by dividing the minimum RMS line voltage by the maximum RMS line current.

For a balanced filter, the inductance of the power and return lines should each be set to half the calculated value. Installing a capacitor-resistor (CR) network in parallel with each inductor will make the filter Q-limited. If you compare the values of the two CR networks, you'll notice that the capacitance is doubled and the resistor value is halved compared to an unbalanced setup. Once again, the CR networks would be connected between the inductors.

3.27.6 The RC Shunt Filter

Contrary to popular belief, the Cauer method is not the best way to restrict the filter Q; instead, another method is suggested that works better in high-impedance, low-current circuits. This filter, known as an RC shunt, requires fewer parts and is initially balanced across the line only by virtue of its design. In Figure 4.11, we see the RC shunt created from a capacitor and a series resistor.

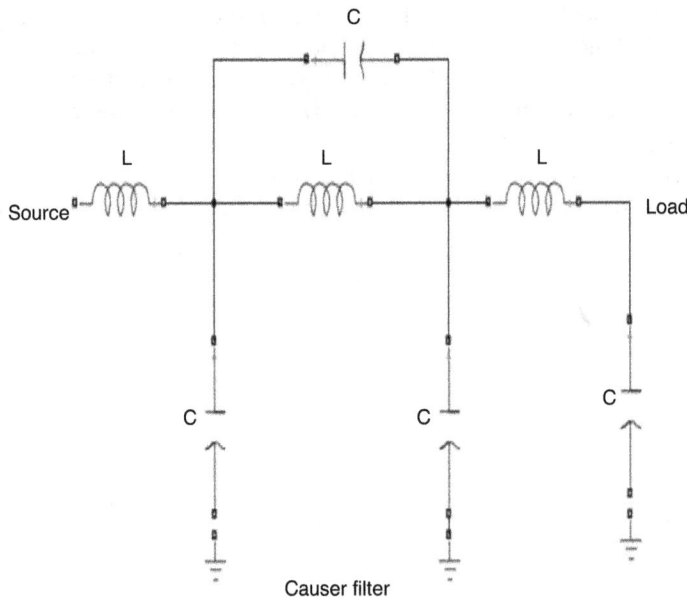

FIGURE 3.15 Cauer filter.

In reality, RC shunts provide a couple of functions. In the first, dampening is added to an LC resonant circuit to reduce its Q to below 2. When the inductive reactance and capacitance are both zero, a resonance phenomenon takes place, and the Q factor quantifies how intense that resonance is at its peak. That being the case, we have $Q=jL/RDC$. For additional protection against higher resonant frequencies caused by antistructure impedance mismatch, parasitic effects, etc., the RC shunt is used. To keep the filter loss under specified dB limits, an RC shunt can be applied to a predetermined higher-frequency resonance, attenuating the peak amplitude of the frequency of interest in the process. For PWM power supplies with incremental negative resistance, adding an RC shunt at the filter's output will alter the filter's output impedance, which may affect the supply's stability.

In most cases, the "problem" frequency will be below the filter's resonant rise or pole-Q frequency. This is especially true if the filter is a multiple filter, like a double or triple L, or T. Whereas a single L, pi, or T would not exhibit resonant rising, a quadruple would exhibit three (two less than the multiple number). For relatively small values of Q in the circuit, this is true. With the higher Q filter, every network shows a resonant peak. In the case of a problem frequency, the RC shunt must be implemented by determining the lowest resonant rising frequency and selecting the capacitor value at this frequency that equals the filter design impedance of the system. This will reduce the amplitude of any subsequent resonances at or above the offending frequency. Select the capacitor so that its impedance is equal to the design's at the frequency of greatest challenge if the design's lowest resonant rising frequencies are unimportant (above the fifth harmonic of the power line frequency and below 10 kHz).

To determine the design impedance R_d, for instance, divide the maximum load current by the minimum line voltage expected. Assume the lowest line voltage at 10 A is 100 V. At this low of a line voltage, this is the greatest current possible. The frequency of the resonant peak is 4 kHz (Figure 3.16).

$$R_d = \frac{100}{10} = 10\,\text{ohms} \quad C = \frac{1}{2\pi \times 4{,}000 \times 10} = 3.979\,\mu\text{F}$$

The capacitor has a value of 3.979 F with the 10 ohm resistor connected across the line, or 4 F when tested independently. This method can be used to lessen the amplitude of higher problem frequencies and the resonance peaking at 4 kHz by a few decibels (within a certain range). If the filter's rise is located at a frequency of negligible importance, tuning the RC shunt to match any higher issue

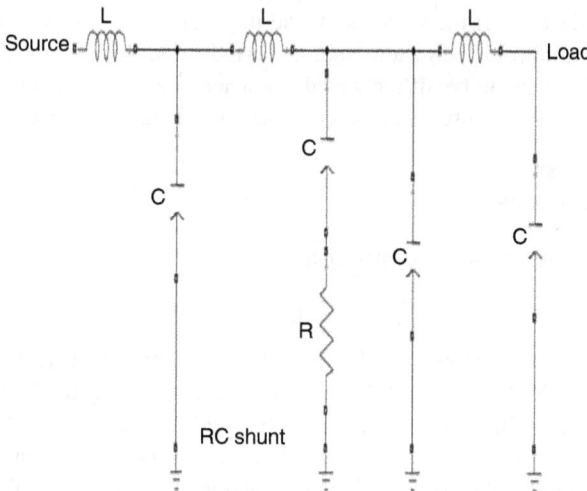

FIGURE 3.16 RC shunt filter.

frequencies may be helpful. Let's say the frequency of the issue is 14 kHz; in this scenario, we'd use this number to calculate the minimum capacitor value needed at this frequency.

$$C = \frac{1}{2\pi \times 14,000 \times 10} = 1.137\,\mu F$$

The 10-ohm resistor is linked in series with a 1.2-F capacitor that is connected across the filter. Several L filters may share a common capacitor with the former network. It reduces the load's impedance variations because of its proximity to the load. When constructing an RC shunt essential for filter dQ, the resistor would typically need to be equal to the design impedance or the filter's characteristic impedance (the necessity to control circuit Q and not higher issue frequencies). Because the shunt capacitor blocks DC current from reaching the resistor, the latter is able to retain most of the energy that would otherwise be lost to heat. At the filter's resonant (pole-Q) frequency, the capacitor's impedance must be less than the resistor's in order for the resistor to properly damp the filter. Four times the value of the DM capacitor is the norm for this.

3.27.7 CONVENTIONAL FILTERS

Traditional wave filters, including such Butterworth or Chebychev filters, are rarely designed by EMI filter manufacturers, despite the fact that some applications demand their use. Making sure the impedance between the source and the load is the same is a key difference when using this filter design to maximize power transfer. If the filters are to prevent the low-frequency losses demanded by the military, this must be the case. This is because, at these low frequencies, EMI's line and load impedances cannot support it.

Passive lossless filters, like conventional filters, have similar L and C structures; however, conventional filters are often designed to function with unknown source and load impedances, which is not the case for EMI filters. Their responsibility is to offer a fixed level of attenuation in terms of frequency and amplitude. They would typically be deployed in an LP, HP, BP, BR, or AP topology rather than a CM setup. In accordance with the frequency-magnitude relationship, they should attenuate higher-order frequencies while providing loss at the prescribed 3-dB pole-Q frequency.

The construction of a filter for a regular wave can be done in a number of different ways. The coefficient-matching approach is one example of such a method. With this method, the transfer function H(s) can be defined in terms of a specific amplitude response, such as Butterworth, even in the presence of source and load impedances. After that, the transfer function of the filter is modified until it satisfies the generic quadratic formula defining a system of order 2.

In addition, it is evident that if the source and load impedances were to vary such that RS RL, the filter's Q and passband gain would be varied, leading to a reduction in performance. Because we are not working with signals that need to be differentiated, passband attenuation is not really an issue when it comes to EMI filters. This is because we are not dealing with signals that need to be filtered out.

3.28 FILTER COMPONENTS

3.28.1 FILTER COMPONENTS—THE CAPACITOR

3.28.2 CAPACITOR SPECIFICATIONS

Capacitors are primarily governed by MIL-STD-15573. In the case of AC capacitors, the voltage rating must be 4.2 times the RMS voltage of the system, among other requirements. For instance, if the RMS voltage of a system is 220 V, the capacitor must be rated for at least 924 V (usually rounded up to 1,000 V). The multiplier for the DC capacitor is 2.5 times the total system voltage. For instance, a 50 V DC rated capacitor should be able to take 125 V DC, and ideally 150 V DC. The peak RMS voltage and the maximum applied DC voltage are employed rather than the nominal and average voltages to reach this result. To test the capacitor, multiply the peak value of 132 V by 4.2 to achieve a final result of 554 V if the system is 120 V AC, which is a realistic assumption. They are put through their paces during production, with the first test likely beginning at 1,200 V after the soldering (swedging) step. Specifications for creepage distances and corona may be required if the voltage is sufficiently high. Capacitor ranges can be limited to between a minimum and maximum value, say 10%, in a build-to-print filter specification. If these parameters are not strictly adhered to, the filter will be reported as being out of specification, regardless of how well it meets the insertion loss. No matter how well the filter meets the insertion loss, this will occur. Filters function better and generate more insertion loss if the value of the capacitor is increased to a suitable level, but this is not sufficient if the value is limited.

3.28.3 Capacitor Construction and Self-Resonant Frequency

The SRF of polyester (Mylar), ceramic, and other dielectrics used in capacitors has increased as a result of improvements in both the materials and techniques used to produce them. When compared to the SRF values that were previously available, feed-through capacitors that make use of dielectric materials such as polyester and Mylar have higher values. The tried-and-true maxim of keeping lead times to a minimum is still valid. If a long lead is required, the inductor should be used as the long lead, and the capacitor should be connected directly to it. This technique is referred to as "veeing the cap."

The lack of a ground connection in the feed-through design helps to keep the ESR and equivalent series inductance (ESL) low. Because of this, the capacitor will have a considerably greater SRF. Capacitor manufacturers often provide charts showing SRFs for all capacitor types up to 1 GHz for capacitor values below 100 pF. The difficulty is that a larger margin is needed to prevent creepage and corona when RMS voltage on the power line rises. This calls for a more substantial capacitor, which in turn drives up the price of both the capacitor and the housing it resides in. The proportion of higher-frequency harmonic current in a line grows in proportion to its frequency as line frequency rises. As a result, the ESR losses rise together with the loss from dv/dt effects across the capacitors. A higher line frequency may need an increase in the initial margin if it was close to the minimum required. Many capacitors have currents that are too high compared to their intended use. So, unless the capacitor is very close to the margin limit, the margin size is independent of frequency.

Large non-polarized can filter capacitors are used by some filter manufacturers. There are oil-infused varieties of this. Be wary of their self-resonant frequencies, although these capacitors could be excellent for use in power supplies and other applications that call for non-polarized capacitors. The SRF of these sorts is typically under 50 kHz. Given the bigger capacitor placed above the SRF, we must be wary of the resonant increase caused by this inductively dominating configuration. This is a potential failure mode at higher frequencies since it is the equivalent of connecting an inductor in parallel with the smaller capacitor. At the resonant frequency, the circuit will exhibit the characteristics of a parallel tank circuit with high impedance. Only 6 dB of gain is achieved by connecting these two capacitors in parallel, according to a study published in the IEEE magnetic guides. According to this notion, lead times should be relatively constant. Because of this, the second, smaller capacitor almost has the same ESR and ESL as the first. Due to the strong SRF and extremely low ESR and ESL of a feed-through type capacitor, using two of them together ensures a functional system. It has been observed that when capacitors are connected in parallel, significant peaks result from the feed-through capacitor oscillating with the original capacitor's effective series resonance (ESR). These non-polarized capacitors are designed to meet the EMI filter's low-frequency needs; their use is discouraged above the SRF frequency. Capacitors of 10, 15, 20, and 30 F at 480 V AC are readily available and typically cost less than their capacity and working value would indicate.

3.28.4 Veeing the Capacitor

A lot of people overlook four inductance components. A splice or joining point is represented by the intersection of the three inductance terms at the top, with a fourth inductance term appearing at the bottom. The capacitor's SRF will be reduced due to the sum of the ESL and ESR introduced by the capacitor's two lead lengths. Improving the SRF of the capacitor requires connecting the inductor self leads directly to the capacitor. The inductance is slightly increased due to the two inductor lead lengths. This raises the values of both inductors by a small amount, but they are still many orders of magnitude smaller than the values before. This is analogous to the idea of minimizing the length of the leads. Because these wires are floating freely in the air and connected to the capacitor, they add less capacitance to the circuit overall. Since all traces on a PCB are relatively close to the ground plane, capacitance prevents this approach from being used.

This method has been used for a while, but the term "veeing the capacitor" is relatively recent. The inductor's leads only increase the total inductance by a negligible amount. However, the capacitor's ESR and ESL are increased by the vertical leads that face the capacitor, reducing its SRF.

3.28.5   Margins, Creepage, and Corona—Split Foil for High Voltage

Margin, creepage, and corona need to be grasped completely. The margin calls for a separation between the capacitor's plates. Minimum voltage of 16 volts per 0.001 inch is used in the space calculation (16 volts per mil). This results in a gap between the dielectrics where no plate may be placed. Use 3/32" or 0.093 of an inch, if the operating voltage is 220 VAC. Following the steps in Figures 3.17 and 3.18, the two dielectrics and two foils or plates are wound on an arbour for the necessary turns, and the foils on the left are swedged (soldered) along with the foils on the right. These components will serve as the capacitor's two plates. If the margins are large enough, then it should be fine. Figure 3.19 depicts the simultaneous winding of two aluminium plates, two dielectrics, and two margins. The voltage demands may necessitate a total of three dielectrics to be used simultaneously. This requires simultaneous winding of eight layers, including six dielectrics and two aluminium foil.

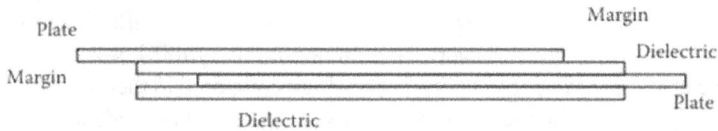

FIGURE 3.17   Design of dielectric foil Pattern 1.

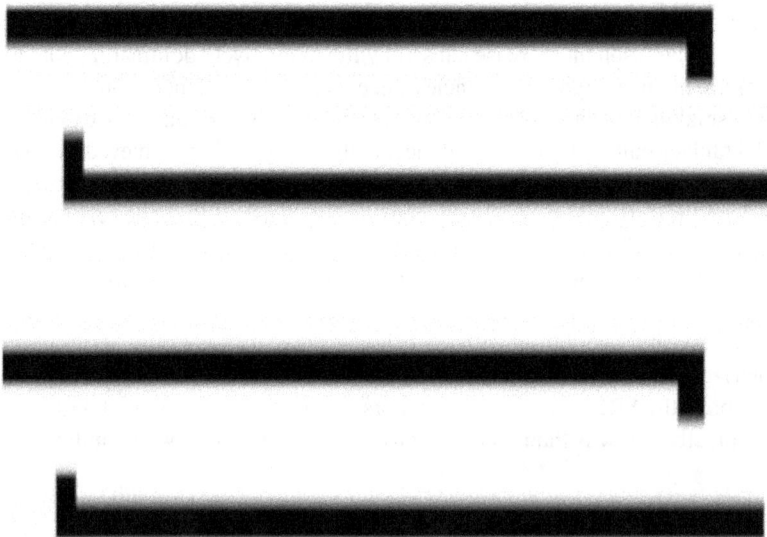

FIGURE 3.18   Design of dielectric foil Pattern 2.

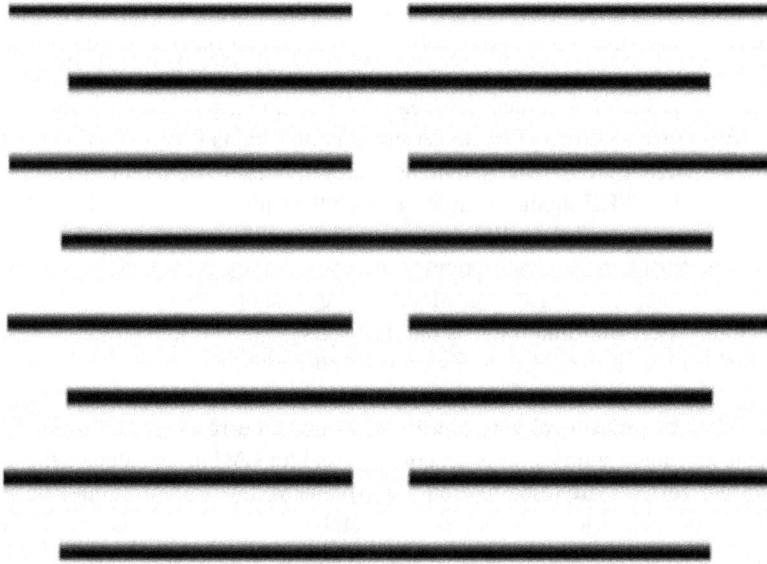

FIGURE 3.19   Design of dielectric foil Pattern 3.

If the capacitor is operated at a voltage higher than its rated capacity, the dielectric will slowly carbonize a path to the swedged plates, causing the arcing that will kill the capacitor. Although the carbonized channel will gradually expand, this will take time. However, high voltage is linked to corona. It is the sharp points, not the smooth surfaces, that are needed to prevent corona. When winding a capacitor, a splitter is used to cut the dielectric material and foil to the desired width. The capacitor maker can then split the dielectric material and foil to the required width, as they only need to purchase the bigger widths. If the process of splitting results in points appearing in a region of high voltage, then corona will occur in that region. To avoid this, you can fold over the edge along the margin to create a rounded, smooth end that meets the margin. By increasing the capacitor's diameter in this way, the capacitor's thickness grows; however, all foil's sharp edges are smoothed out. To disperse the voltage, the plates or foils in high-voltage capacitors are coiled in series.

The dielectric must be in the space between the two plates, although it is not depicted beside the edges. As a result, the voltage is split between the two primary parts and an additional section to meet the increased voltage needs. However, the amount of usable space has shrunk. Because there are now two visible halves, the voltage must be halved and the capacitor value should be multiplied by 2. Separating the two foils requires a space in the middle larger than the margin value. Add up the distance using 32 millivolts per millimetre. Both the right and left sides are squeezed together. The central foil or plate is disconnected. These capacitors are usually quite sizable, and in certain cases, you may find multiple parallel capacitors with more than one break or gap in between them.

3.28.6   Capacitor Design—Wrap-and-Fill Type

These capacitors are assembled using one of three ways, which can be further broken down into two categories. An early technique involved first affixing leads to the foil before winding the capacitor. These capacitors are now commonly referred to as "chicklets" or "inductive capacitors" since electricity must flow through the foil to charge or discharge the entire capacitor. These chicklets have an extremely high ESR and ESL and a very low SRF. For a piece of foil that is 0.5 inches broad and 0.00023 inches thick, the cross-sectional area is 0.000115 square inches. For reference, 146.42 circle miles is the result of dividing this number by 7.854 10E-7 (the typical square inch to circular mils conversion factor). The conductivity of aluminium foil is 17, so if you divide that by 146.42, you get 0.1161 ohms per foot. Since there are two plates, the average length of electron flow is half this value if the 0.23-mil aluminium foil is 100 feet in length (a typical capacitor size). Thus, 11.61 ohms is the resistance. Based on a distance of roughly 30 m, the inductance would be around 45 H. It's highly likely that production of these has ceased. In addition, you should never use these as the primary EMI filter capacitors, but you can use them as RC shunts if the frequency is low enough. Depending on the frequency, they can also be used as Cauer-type shunts for low-frequency tuning capacitors. The chicklet's SRF should be at least twice the tuning frequency. If the filter is experiencing high-frequency issues that an RC shunt could help with, a high-quality extended-foil type capacitor (described further on in this section) should be used instead of a chicklet capacitor.

Metallic and foil categories exist here. Metallized capacitors have an aluminium plate that is sprayed onto a dielectric. Aluminium foil is the other category. The standard thickness of aluminium foil is 0.00023 inches (or 0.23 mils). It's important to note that this foil is millimetres heavier than the film. The increased current-carrying capacity of foil is especially useful in pulse applications and EMI filters that must deal with powerful harmonic currents from off-line regulators and other comparable harmonic sources. This design is typical for EMI filters, albeit a thicker metal foil is used when larger currents are to be filtered. For greater voltages, the foil must be either thicker or stacked, just as the dielectric. Nonetheless, the metalized film offers a number of benefits. This type of capacitor is self-healing and can be made considerably smaller without sacrificing capacitance. Small pinholes along the length of every dielectric. A common result of the applied voltage stressing the film is a short circuit through a pinhole, resulting in the film melting. The metallized film's aluminium will then reform, allowing the capacitor to repair itself. Another advantage of this category is that aluminium can be sprayed onto both sides of the metallized dielectric. This helps with self-healing, is beneficial to longevity, and adds to the reduced size.

3.29   FILTER COMPONENTS—THE INDUCTOR

3.29.1   INDUCTOR TYPES

The following are the different types of the inductor used for filter design.

Air Core Inductor
Iron Core Inductor
Toroidal Inductor
Laminated Core Inductor
Powdered Iron Core Inductor
Axial Inductor
Shielded Surface Mount Inductor
Coupled Inductor

3.29.2   CORE TYPES

3.29.2.1   Power Cores

EMI filters typically use molybdenum permalloy toroids, but HF and Kool Mu toroids are also common. One significant benefit is that extremely high Q values are possible for all three types of cores. EMI filters typically employ MPP cores. These magnet cores are effective up to 3,500 gauss but reach saturation at 7,000 gauss. The inductors in DM filters are their principal target use. Gapping is unnecessary because there are already gaps in the structure's core at various locations. Using the provided AL values, their design is straightforward. Above 400 Hz, permeabilities higher than 125 are not advised.

Specific types of these cores reach saturation at 15,000 gauss and can be operated at up to 7,000 gauss; HF cores are frequently utilized in the production of EMI filters. Their principal function is to supply greater currents to DM filter inductors, which is why they have such a high rating.

To reiterate, gapping is not necessary for these cores because they already contain gaps dispersed over the entire core. It is strongly advised that permeabilities more than 125 not be utilized at frequencies of 400 Hz and above. In addition, the flux density should not exceed 3,500 gausses when it comes to DC filters.

Kool Mu cores are the old Sen dust cores that are made out of aluminium, and these can also be found in applications involving EMI filters. These cores reach their saturation point at 10,000 gauss but can be used up to approximately 5,000 gauss. Their principal function is to fulfil the requirements of DM filter inductors, which call for toroids that are both lighter and more cost effective. In addition, they do not need to have gaps filled because they already contain gaps dispersed throughout the core. It is strongly advised that permeabilities more than 125 not be utilized at frequencies of 400 Hz and above. In addition, the flux density should not exceed 3,500 gausses when it comes to DC filters.

Toroids made of powdered iron are occasionally used, but they are the most affordable option. However, around 400 Hz, they frequently cause issues owing to overheating. They also have a permeability level that is not too high. The primary drawback is that the reading of the inductance is far more dependent on the level of bridge drive than the readings of the other toroids. As a consequence of this, the value that is provided by the LCR bridge on another bridge will be different due to the changes in the drive levels.

To put it another way, because the permeability shifts depending on the impressed current, this core is the one with the lowest degree of stability. In any case, these components are frequently discovered in EMI filters, particularly low-current filter designs. Magnetics., Inc. and Arnold Magnetics Technologies are the two primary companies that supply the cores that are the subject of this section's discussion.

3.29.2.2 Ferrite Cores

CM EMI filters use ferrite toroids. They saturate too easily for inductors. Low H or current saturates these cores. Exception: very low current. Inductors require gapping, which is impractical. Since the tracking generator has a relatively modest driving current, the filter may function as an inductor and be allowed to pass. These inductors would pass light load testing. At 10 A, they may saturate and fail. This core's high AL values provide considerable CM inductance. CM cores cancel flux, thus high line current won't saturate them. The core is not saturated by EMI because of its low intensity.

3.29.2.3 Tape-Wound Toroids

While tape-wound toroids have some value as EMI filters, transformers are their primary function. They have a high permeability, with a workable flux range of 15,500–16,000 gauss. The biggest drawback is how cumbersome and heavy they are. These transformers exhibit both CM and normal-mode losses at low frequencies, with additional skin-effect losses. The core needs to be cut-gapped before it can be used as an inductor in an EMI setting (the energy is stored in the gap). Because of the separation, the inductance equation must be rewritten. Since the magnetic resistance (reluctance) can be calculated by dividing the magnetic route length by the permeability, the total reluctance must be utilized. Powder cores' first equation is written as (9.1). The powder cores have distributed flux density as given by

$$L = \frac{0.4\pi N^2 A_c 10^{-2}}{M_{PL}} \mu H$$

Gapping the core raises the magnetic resistance (REL). The original REL was calculated by dividing MPL by μ, thus eliminating μ from the numerator. To calculate the final result, the permeability must be multiplied by a numerator that already includes the REL gap.

$$\frac{g}{\mu_{air}} + \frac{M_{PL}}{\mu_{steel}}$$

Steel's permeability can be 2,000 times that of air's (1). This means that the gap (g) in millimetres is the most fundamental measure of the overall REL. So, the inductance L is

$$L = \frac{0.4\pi N^2 A_c 10^{-2}}{g}$$

Space (g) is denoted in centimetres. The BH curve is skewed because of the gap, therefore the permeability is less S shaped and more constant (see Figure 9.1). The unit's slightly reduced efficiency is due to its low Q. These, however, can handle currents of 300 A or more, and their inductance can reach 400 µH.

3.29.2.4   C-Core Inductors

Higher currents in EMI filters necessitate the usage of C cores since they are too much for powder cores to manage on their own. In contrast to the highest flux density of around 18,000 gausses, they can be built utilizing only 12,000 gausses. Since the inductor lacks a distributed gap, it does require that the cores be gapped, which is a simple process given that they come in two pieces to begin with. The spacing must be at least 1 mil (0.0004). Different batches of these cores have different inductances. One leg at a time, or both at once, you can wind these. Each of the two windings can be configured as either a series winding or a parallel winding carrying one-half of the current. Both "Left to Right" and "Right to Left" refer to the direction in which each leg of the latter is wound. The very low Q is the main disadvantage of the C core. We show that it is useful for DC operations and core softening to gap the C core using shims with 1/2 the gap value in each leg. They can be easily wound on a bobbin or coil shape, as opposed to the time-consuming process required for toroids. In contrast to winding a single toroid at a time, many C-core bobbins or coil shapes can be wound simultaneously, depending on wire size, to produce multiple inductors.

3.29.2.5   Slug Type

The slug is typically iron powder. It is typically between 1 and 2 inches in diameter and 1 inch in thickness. The magnetic field couples between the ends through air. As you can see, this is a major strike against them. To prevent them from acting as an inductive heater, they must be suspended in the air. The magnetic flux will flow through any nearby magnetic material, including the filter housing. In the previous example, the case would get hotter and the part's inductance would rise due to the low impedance of the new flux path if it were mounted adjacent to or close to the case. Micro metals Inc. has constructed slugs around this core, and they have achieved inductance values of 150 H. These were suspended from the ceiling by their cables. After that, pots were placed over each one. Powdered iron from Micro Metals, product number P6464-140, has an effective permeability of 5. These types of inductors have the following corresponding equations:

$$L = \frac{0.8U_{\text{EEF}}(RN)2}{(6R+9T+10B)} \rightarrow N = \sqrt{\frac{1}{R}\frac{L(6R+9T+10B)}{0.8U_{\text{EEF}}}}$$

3.29.2.6   Nanocrystalline CM Cores

In the past, these cost an arm and a leg, but today's prices are far more reasonable. Due to their high permeability (μ), these cores only need a fraction. The most common values are 20,000, 30,000, and 80,000. Metglas, MK Magnetics, and Vacuumschmelze are the major distributors (or Vaccorp). It is crucial to develop EMI filters with a high SRF. This inductance's primary purpose is to shield the system from high-frequency noise generated internally. Some strategies for improving the SRF are provided below:

- Lower the capacitance between the wire and the core. Get cores with coating and tape the ones you have, including the coated ones. A decrease in wire-to-core capacitance results in a higher SRF.

- Wind the cores in a progressive fashion (also known as the pilgrim step) by going six turns forward and four turns back, and so on. Turn-to-turn voltage difference is low. As the voltage difference between the turns increases, the capacitance does too. The SRF is improved by a small C value. It's important to ensure that the gap that forms between the beginning and ending leads stays open. The distance between the starting and finishing lines must be proportional to the difference in the number of forward and reverse rotations taken. It could be a while before we have an accurate front-to-back ratio. This approach has the potential to produce unattractive results. Uninformed clients could be suspicious of the inductor if they saw it, thus it's common practice to wrap it up using tape.
- The capacitance is lowered and the SRF is increased when (1) and (2) are combined.
- Wind the toroid in sections, making 15 turns all at once in one location, then skipping a space and winding the same amount of turns in many clusters around the core. This results in a really unusual-looking inductor. There is room between the segments and the capacitance in each segment is low, therefore the overall capacitance is minimized. Again, the SRF increases as C decreases. Fill up all the spaces with tape to prevent the different balls from slipping into one another.
- Connect a series of miniature inductors that you wound yourself. Due to an increase in SRF across the board, the number of cores required to achieve a given SRF level decreases. Placement of these inductors in quadrature will lessen magnetic coupling.
- Adding more insulation is another strategy for increasing SRF. Because of this, the turn-to-turn spacing is increased, which helps lift the copper away.

3.29.2.7 High-Current Inductors

For high currents, bigger wire diameters are necessary. The high number of strands in welding cable—833 in No. 30—makes it a popular choice for this purpose. This pliability of the wire makes it possible to shape it into the coil. To employ Litz wire and high-frequency cores, as many engineers would like to do, would drive up the price by several orders of magnitude. EMI filters, on the other hand, require either Litz or welding cable due to the simplicity of winding the coil shape. In order for the coil to be tuned, it is necessary that Q is high at the desired frequency. To achieve a low AC resistance at the tuned frequency and a bit greater at other frequencies, the core must be constructed from slimmer material and incorporate stranded wire. This reduces the thickness of the material from 12 to 4 mil. By using the square root of the frequency, we can calculate the radius of conduction in centimetres as 6.62. Up the frequency to 30,000 Hz and take the square root of it, this provides 173.2. Divide this with 6.62, which will yield the radius of conduction in centimetres. Subtract 2.54 (the number of centimetres in an inch) to obtain the radial distance in inches: 0.015. Then 0.015 times 2 yields the wire diameter in inches, therefore 0.030. To get a good approximation of 104.2 circular mils, we go with AWG No. 20 wire (0.032 in diameter). If this inductor must handle 10 A, utilizing 500 circular amps per amp, about 5,000 circular mils is needed, which would need 48 strands of No. 20 wire. Litz seems like a good option, at least on paper. The downside, other than expense, is that fewer loops of this wire can be twisted in the allotted space where 48 individual strands could fit. However, winding such a lengthy inductor requires significant effort.

3.30 SUMMARY

In this chapter, various topics related to EMI filter and their construction of filter, performance of filter, design of a filter, and types of filters, components used in a filter are discussed. It is clear that EMI filter is necessary to eliminate the noise in the case of high frequency and power conversion areas due to the switching action. EMI filters meet the requirement for mitigating the noise levels.

BIBLIOGRAPHY

1. Donald R.J. White, *A Handbook on Electrical Filters*. Germantown, MD: Third Printing, 1970.
2. Donald R. J. White, *The EMC Desk Reference Encyclopaedia*. Gainesville, VA: EMF-EMI Control Inc, 1997.
3. Mark I. Montrose and Edward M. Nakauchi, *Testing for EMC Compliance*. John Wiley & Sons, Inc., 2004. DOI:10.1002/047164465X
4. Interference Reduction Guide For Design Engineers – Volume II, National Technical Information Services, US Department of Commerce, Flushing, NY: Filtron Co Inc., August 1964.
5. Mark I. Montrose, *EMC Made Simple*, Montrose Compliance Services, Inc., US, 2014.
6. Richard Lee Ozenbaugh and Timothy M. Pullen, *EMI Filter Design*, 3rd ed. CRC Press, US, 2012.

4 EMI/EMC Design for Printed Circuit Boards

4.1 INTRODUCTION

Electromagnetic interference (EMI) is the product's unintentional electromagnetic emission. The evaluation of EMI is known as electromagnetic compatibility (EMC). The EMI can cause malfunction, performance degradation, and total failure of another system. It is vital to discover EMI in a product since it reduces the product's efficiency and damages it. In addition, EMI has consequences for the human nervous system, visual system, immunological system, etc. Common PCB signals affect the operation of the circuit and the entire system. Details on the EMI/EMC design for boards with printed circuits are provided in this chapter.

4.2 CONTROLLING EMI SOURCES—UNINTENTIONAL SOURCES

Unintended signals are frequently overlooked during the design phase. Besides, we don't want them on the board, therefore we don't frequently consider them. Unfortunately, these unintended signals account for more than 90% of a PC board's EMI emissions. No matter how meticulously the board is built, a certain proportion of these unintended signals will be there. It is necessary to evaluate these unintended signals and take the necessary precautions to ensure they do not result in excessive emissions.

4.2.1 UNINTENTIONAL SOURCES

Intentional signals that accidentally appear where they were never intended to do so are called unintentional signals. All accidental signals are the consequence of a deliberate signal coupling onto a wire, trace, or conductor that was never meant to carry that signal. Unintentional signals share the same frequency spectrum as the intended signal, but because they are present in areas where they were never intended, it can be challenging to identify and manage them. This is true even when the proper design techniques are used to minimize their effects.

4.2.2 UNINTENTIONAL SIGNALS—COMMON MODE

The return currents spread across the ground-reference plane as a result of the inherent inductance of the planes or path disruptions. The same currents are what cause inadvertent common-mode signals. The signal current on the ground-reference plane binds to the I/O connector's "ground" pin in this possible source of EMI emissions, and then "escapes" the shielded enclosure through the connected I/O cable (Figure 4.1).

The "ground" pin on the I/O connection is directly connected to the ground-reference plane, but a critical signal trace is not directly connected to an I/O connector. Existing return currents on the ground-reference plane are evenly distributed over the whole plane and are easily coupled to the I/O connector's "ground" pin. This energy is immediately connected to the exterior of the shielded enclosure once it is on the "ground" wire of the I/O connector. The majority of radiation emissions are caused by unwanted currents on cables and wires outside the device. An unshielded JJO cable's radiated emissions would increase if any unwanted or unplanned currents were present on any of the conductors, including the "ground" conductor. Even shielded cables are likely to have increased

DOI: 10.1201/9781003362951-4

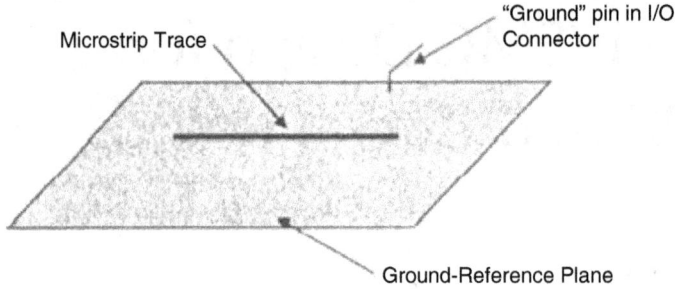

FIGURE 4.1   Example of I/O ground pin connected to the ground-reference plane.

emissions owing to noise coupling onto the cable shield unless care is taken to create a strong (low impedance) electrical connection between the cable shield and the shielded enclosure chassis.

4.2.3   CONTROLLING EMISSION FROM UNINTENTIONAL SIGNALS—COMMON MODE

Using an intentional separation between the high-speed digital circuit area and the I/O connector area is one technique to prevent the return current spread from extending into the I/O connector area (Figure 4.2).

The preceding diagram illustrates how a similar partition can be utilized to separate the I/O ground-reference area from the digital ground-reference area. This approach is capable of effectively separating "ground" I/O pins from the return current distribution. Here, extra vigilance is essential. The essential reality is that splits are neither "good" nor "bad" in and of themselves; the circumstances in which they are required must be clearly identified, and splits must be used only when necessary. Allowing a high-speed trace to traverse a split reference plane mandates that the return current find an alternative return channel, which is likely to increase emissions. Reference plane splits should never be permitted adjacent to high-speed tracks. When a low-speed I/O area is close to high-speed circuits, the "ground" pins of low-speed I/O connections can be successfully separated from high-speed return currents using splits in the reference planes.

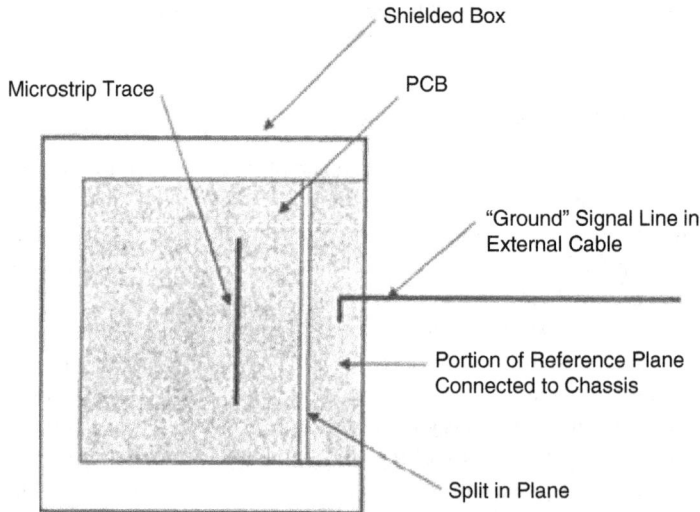

FIGURE 4.2   Example with split reference plane.

A typical computer motherboard is suited for reference plane split implementation. Numerous low-speed I/O connectors, such as keyboard, mouse, serial port, parallel port, etc., are typically found on the motherboard of a personal computer. Frequently, the "ground" pin on each of these 110 ports is directly linked to the "ground" plane on the motherboard. Due to these factors, any I/O port with an intentional data rate of less than 5 MHz is referred to as a low speed 110. High-speed traces (clock traces, high-speed bus traces, etc.) are not required to run close to low-speed I/O connectors. Splitting the ground-reference plane in the low-speed I/O region can be an effective and economical EMC design solution for isolating return currents that have spread on the reference plane from the "ground" pins of the I/O connector. To ensure the effectiveness of the split I/O ground-reference technique, a number of other major design considerations must be taken into account.

First, a low inductance (low impedance) contact must be used to attach the segment of the ground-reference plane that is separated from the digital ground-reference of the main printed circuits (PC) board to the shielded enclosure chassis. If this contact is absent, intermittent, or has suitably low impedance, the emissions can increase at specific frequencies. Keep in mind that the chassis shield is the ultimate reference point for all external wires and cables. The internal I/O signal reference voltage must match the external chassis voltage. The intended low-speed I/O data return current path is the second essential design factor. Otherwise, data errors or other system failures may arise. Inserting ferrite beads over the I/O split provides return. The return current runs at low frequencies and prevents high-frequency dispersion of the digital ground-reference return current. Never add capacitors across these splits, as they will permit high-frequency noise while preventing low-frequency return currents (Figure 4.3).

The reference plane on a PCB is shown in the diagram above with a split and ferrite beads spanning the gap. Only a few beads are needed everywhere, distributed close to the low-frequency I/O connectors because these beads are made for low-frequency return currents. High-frequency signals on the reference plane have high impedance due to the split in the ground-reference plane. The split should be at least 50 mils broad; however, it is not anticipated to be that big (Figure 4.4).

The impedance across a split plane for a $10'' \times 12''$ PC board is shown in the diagram above (Figure 4.4) as an illustration. The impedance of the split might not be as high as expected at higher frequencies because of board and slot resonances. On the internal PC board in this instance, a microstrip trace is present, but it is not in close proximity to the I/O ground pin. The ground-reference plane of the circuit board is connected to the shielded chassis in the area of the I/O connector. The emissions from the enclosure while there is no wire attached to the I/O connector are shown in the first graph. At about 425 and 750 MHz, enclosure resonances can be seen. Resonances related to the length of the external wire can be seen at frequencies like 80 MHz, 225 MHz, etc.

FIGURE 4.3 Example of a split reference plane with ferrite beads strung across splits.

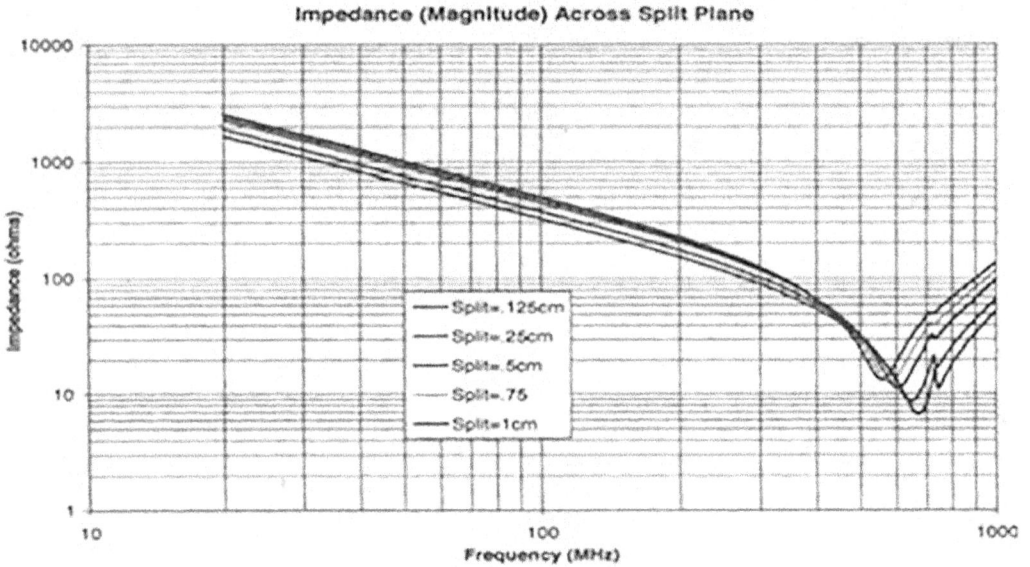

Impedance (Magnitude) Across Split Plane

Legend:
— Split=.125cm
— Split=.25cm
— Split=.5cm
— Split=.75
— Split=1cm

X-axis: Frequency (MHz)
Y-axis: Impedance (ohms)

FIGURE 4.4   Impedance across split reference plane.

4.2.4   CONTROLLING EMISSION FROM UNINTENTIONAL SIGNALS— "CROSSTALK" COUPLING TO I/O LINES

The best way to prevent crosstalk coupling to I/O lines is to sustain high traces and I/O traces apart. The two types of signals should be routed on different layers of the PCB stack with a solid plane in between them. This is the most efficient way to isolate traces. Crosstalk across layers is a possibility and a frequent occurrence when there isn't a solid plane between them (Figure 4.5).

Low-speed I/O signal traces are routed on the PC board's outer layers, whereas the inner layers route high-speed signals, as depicted in Figure 4.6.

Figure 4.6 depicts a PCB with a driven (high-speed) trace, a vulnerable (I/O) trace, and adjacent wire channels. In this design, it is recommended that the I/O trace be moved away from the high-speed trace, leaving empty wiring channels between them. The most effective way for isolating traces on the same layer is to use guard traces. A guard trace is attached to the ground-reference plane every three inches and is positioned between the high-speed trace and the vulnerable trace.

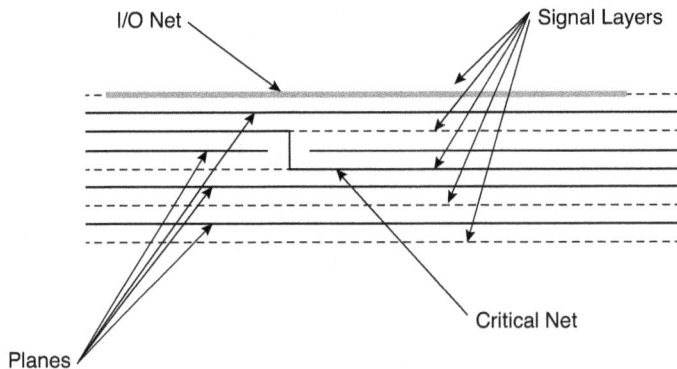

I/O Net
Signal Layers
Planes
Critical Net

FIGURE 4.5   Cross coupling example.

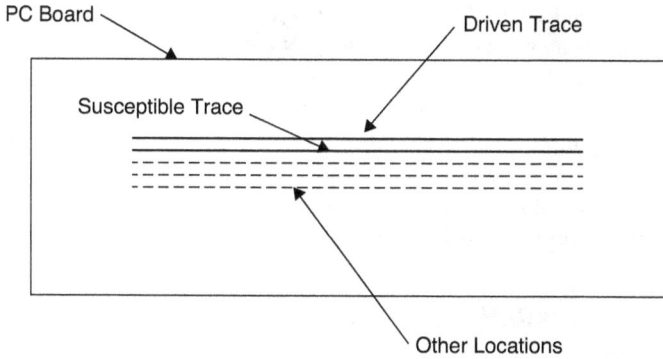

FIGURE 4.6   Cross coupling example.

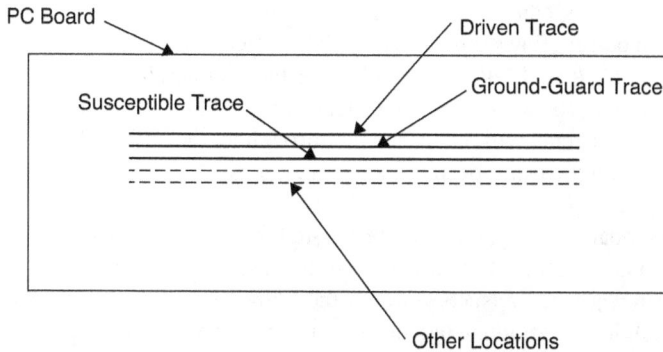

FIGURE 4.7   Trace positions with ground guard for isolation.

If there are numerous interconnected high-speed traces, such as in a bus, they can be routed along-side guard traces that are routed around the bus's exterior. The next diagram displays the identical PC board layout (Figure 4.7).

A few dB of more isolation is provided by physical separation, but the guard traces offers the greatest isolation while using less board area. There are several causes of EMC emissions. These sources are unique from one another, allowing us to independently reduce emissions from each source. All conceivable emission sources were created by the intentional signal, which is a central theme throughout the study of accidental sources. If essential signals are effectively controlled, higher-frequency EMC emissions will be less of a concern.

4.2.5   USING SIGNAL INTEGRITY TOOLS FOR EMC ANALYSIS

Engineers analyse the signal integrity of the majority of boards of PC) with high speed using commercial application analysis tools. The trace architecture of the board is routinely examined by engineers to make sure the voltage pattern at the receiver complies with operating requirements. In order for the output receiver to receive the correct voltage waveform, the values of the terminating resistors are altered, or even more drastic alterations are done. The analysis is deemed to be finished once a suitable voltage pattern has been identified. As a result, several termination resistor values are used throughout the design. As long as the termination resistor operates within an acceptable range, its value is not always maximized. The desired current of the trace might be significantly impacted by the value of the termination resistor.

The voltage pattern is not a significant issue when it comes to EMC emissions, but the current pattern is extremely important because voltage does not radiate directly but current does! Therefore, it is advantageous to analyse both the current and voltage waveforms on a trace. Sadly, only a small number of commercial software programmes permit this investigation of deliberate currents. The usefulness of undertaking more study is complicated by the time and expense involved. Some commercial instruments that provide current analysis are actually less expensive than others that do not. In a nutshell, the current cannot become a familiar current that will somehow, somewhere in the system result in emissions problems if it is never generated.

4.3　INTENTIONAL CURRENT SPECTRUM

The main source of EMI emissions is "common-mode" currents. In essence, common-mode currents happen in places where they weren't intended. Emissions may be produced if the common-mode current couples to a neighbouring I/O cable or other conductor that is leaving the shielded container.

Common-mode currents can be produced via a number of inefficient design techniques. To ensure that all return current flows directly beneath the trace in the reference plane, traces are used on printed circuit boards (usually the power plane or the ground plane). Frequently, not all return current can travel directly beneath the signal trace. In an effort to reduce the inductance in the return path, the return currents will be distributed throughout the entire plane. Not all of the return current is caught beneath the trace, despite the fact that most of it is, which leads to the occurrence of currents in unexpected places.

The layout of the board is frequently not best suited for high-speed signals. The return currents must find another route to return to their source if, for instance, a high-speed clock trace is routed via a split in the reference plane (such as when the power plane is split to accommodate several DC supply voltages). The additional inductance of the capacitor, the necessary vias, pads, etc., will prevent high-frequency return current components from being close to the signal trace even if a capacitor is placed across the split at the crossing. Another common problem appears when a high-frequency signal trace is routed through a via and the reference planes are changed. The return current must travel between two planes in order to return to its source (perhaps via a decoupling capacitor with its vias, additional inductance, etc.), and it frequently takes an erratic path.

Despite the fact that common-mode currents have many, complex, and frequently unpredictable sources, it is true that all common-mode currents originate from an intentional current. That is to say, a common-mode current was mistakenly generated in some place on the PC board by an intentional signal. It is advantageous to make sure that the critical signals are controlled so that only the essential harmonics remain and the unnecessary harmonics are deleted. When the original signal source does not have harmonic, it is more expensive to add filters to an I/O port to prevent a high-frequency harmonic from escaping a shielded enclosure. This concept was shown here using a PC motherboard. The investigation was conducted on a 133 MHz clock network. The appropriate IBIS driver and receiver models were used to characterize the driver and receiver implementing the source series resistor termination method. The termination resistor's default value for this network was 22 ohms.

The current on the trace at the receiver was also examined because the goal of this inquiry was to lower potential emissions. Clearly, compared to the other numbers in the image, the 10 resistor allows for a large increase in current flow (Figures 4.8 and 4.9).

The degree of high-frequency harmonics reduction is not adequately depicted despite the fact that this voltage and current waveform analysis is important (the prime cause of emissions problems). The frequency domain spectrum was created by Fourier transforming waveforms in the time domain. The outcome shows a significant fluctuation in the current's amplitude at each harmonic frequency. From 10 to 30, the resistor's value causes the current amplitude at each harmonic

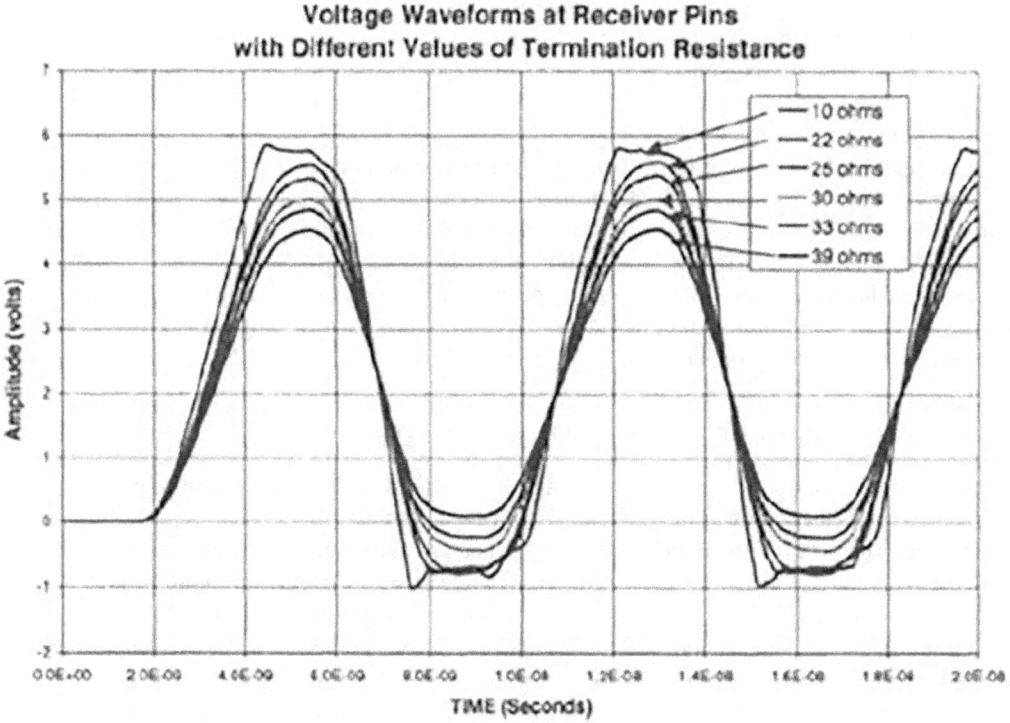

FIGURE 4.8   Voltage waveform for different values of termination resistor.

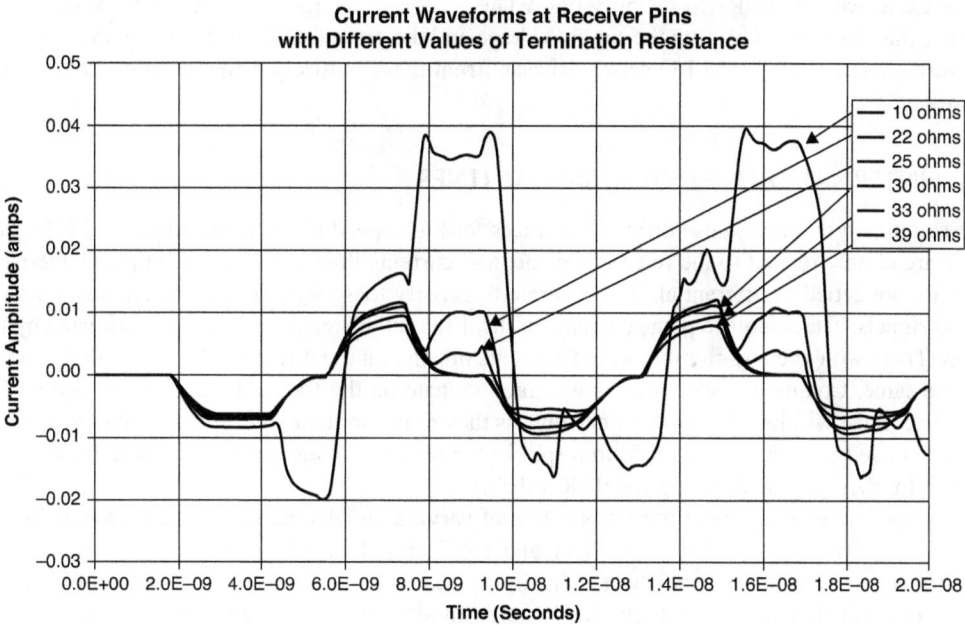

FIGURE 4.9   Current waveform for various termination resistor values.

frequency to decrease, while additional increases in resistor value cause the current amplitude at a particular harmonic frequency to change.

The graph shows a reduction in current amplitude for each harmonic frequency as the termination resistor value is changed. This graph also shows that for the majority of harmonic frequencies, the reduction in current amplitude is roughly constant, regardless of whether the termination resistor is changed from 10 to 30 or from 10 to 39. The graph shows that at some harmonic frequencies, the current reduction had an amplitude of up to 45 dB. It is essential since few product designs fall within the error margin of the EMI emission standard. It is still likely that there will be a reduction in radiated emissions even though a reduction in the harmonic current may or may not lead to a one-to-one reduction in radiated emissions (depending on the precise coupling mechanism between the current and the terminal radiation source). This is because less filtering, gasketing, etc. will be required for the finished product.

4.4   TRACE CURRENT FOR DECOUPLING ANALYSIS

The majority of the required current is utilized to power the I/O pins when an ASIC contains several output drivers, such as clock buffers, memory controllers, bus controllers, and so forth. As a result, the I/O trace's current is crucial and can be used to analyse and determine how to design the decoupling capacitor. The power pin current can be calculated using the I/O current and shoot-through current on clock buffers. The power pin of the I/O driver might be pulsed with a triangle wave. While certain devices can function with this supposed waveform, it's not necessarily as straightforward as the triangle wave.

Signal integrity tools can be used to estimate the current pulse at the power pin by measuring the current on the I/O traces. It is feasible to instantly add the current pulses in the case of clock buffers with synchronous I/O pulses to figure out the total current needed through the power pins. It is more accurate to use the average number of drivers operating concurrently in the case of a memory/bus controller because all I/O drivers are rarely driven simultaneously. The type of termination method used is also crucial. Standardly, source termination resistors are used in series exclusively on SDRAM memory lines. The latter uses a termination resistor connected to a voltage that is situated between the supply voltage and the ground-reference voltage, in contrast to the former. The DDR RAM I/O driver drives current that is entirely different from the SDRAM situation.

4.5   INTERNAL DIFFERENTIAL SIGNAL LINES

It is a common practice to use differential traces for high-speed signal applications on PC boards. They are closely related to the reference plane, and currents flow in the reference plane, therefore they are not actually differential. It is accurate to classify these signals as complementary single-ended signals. The reference plane currents of the differential signal pair will be unaffected by the splits. There won't be any common-mode currents flowing on the differential traces if their lengths are the same. Common-mode currents will predominate on the traces if their lengths differ. The reference plane will be used by these currents as they return to their source. The same problems as a single-ended trace would result from a split in the reference plane, which would stop the return current for these signals from flowing below the traces.

It is possible to analyse the impact of traces of varying lengths that are not synchronized using signal integrity tools. Differential driver and receiver IBIS models, for instance, were used. One transmission line in the pair was then slightly extended after the nominal transmission line length of ten inches was fixed. Only the common-mode current at the receiver distinguishes the two currents. The amount of common-mode current with no mismatch must be attained as a baseline.

4.6 I/O DIFFERENTIAL SIGNAL LINES

For some I/O data formats, such as Ethernet and USB data transmissions, differential signalling is the norm. These data lines can be analysed using commercial signal integrity analysis programmes. In order to manage purposeful signal output, common-mode voltage between the cables and enclosure chassis is an important factor for external I/O cables. There won't be any common-mode voltage between the cable and the chassis if the external cable delivers the differential load to the traces and their lengths match. When the cable leaves the enclosure, a common-mode voltage will be present if the trace lengths are not balanced.

The differential drivers are loaded by the cable impedance. In this case, the differential load is split in half, and a 100 ohm resistor connects the centre of the two halves to the ground reference. This figure often provides a good estimate of the common-mode impedance of an external I/O cable. There won't be any common-mode voltage across the 100 ohm resistor when the differential traces' lengths are matched. The common-mode voltage rises as the length mismatch does. As the trace length mismatch widens, the common-mode voltage's harmonic content rises.

4.7 CROSSTALK ANALYSIS

Crosstalk analysis uses methods for signal integrity. Normally, this analysis is only used to ascertain whether the level of crosstalk noise may obstruct data transmission. This examination is typically only carried out within high-speed buses, in between high-speed signal lines, etc. Crosstalk between fast signal lines and I/O lines is analysed infrequently. With the use of signal integrity tools, users can include filter elements in their circuit models, allowing for the inspection of the high-speed signal trace to the I/O trace while accounting for the effects of filters. This allows for the determination of the common-mode voltage level on the I/O connector for a number of internal sources.

4.8 PRINTED CIRCUIT BOARD LAYOUT

As other design concepts can be explored during the design processes, the printed circuit board is regarded as the most important design component in an overall EMC strategy. It is also well worth the time to focus on the PC board layout.

Mechanical restrictions and EMC design phases may require the positioning of particular connectors along one side of the board. It might be required to position an air vent next to processors with heat sinks near the edge of the circuit board, close to the CPU. To maintain signal integrity, certain ICs and ASICs may need to be placed close enough to one another. A PC board with two power/ground-reference planes may need to be designed due to cost, which is another important restriction. Without considering EMC, a less-than-ideal decision is likely to be made.

4.9 PC BOARD STACK-UP

The intended cost of the board, the manufacturing process, and the quantity of necessary wiring channels frequently determine how the PC board is stacked. There are several needs, as with the bulk of engineering designs, and the final design method is chosen after evaluating a number of trade-offs. In light of this, we move on to choose the PC board stacking utilizing the procedures listed below (Figure 4.10).

4.10 MULTILAYER BOARD

High-performance, high-speed applications commonly make use of PC boards with several layers. As ground-reference or DC power planes, several layers are used. There is often no need for varying DC voltages on a single layer because there are typically enough separate plane layers.

```
-------------- Top
================ Layer #2 Plane (ground)
-------------- Layer #3 Signal (east-west)
================ Layer #4 Plane (power)
-------------- Layer #5 Signal (north-south)
================ Layer #6 Plane
-------------- Layer #7 Signal (east-west)
================ Layer #10 Plane
-------------- Layer #9 Signal (north-south)
-------------- Layer #10 Signal (Low Speed ONLY)
================ Layer #11 Plane (multiple power)
-------------- Bottom
```

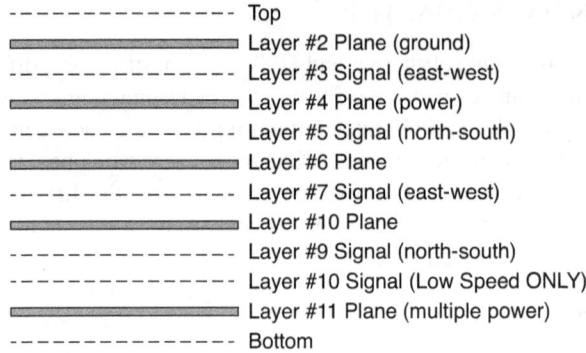

FIGURE 4.10   T-P-S-P-S-P-S-P-S-S-P-B stack-up.

For transmission lines right next to it, the aircraft will act as the return current path. The most crucial EMC duty for these plane layers is producing a good return current channel with low impedance.

Between solid plane layers, signal layers are transmitted. They can also be asymmetrical strip lines, in which two signal lines are implanted between adjacent plane layers, or strip lines, in which the distance between the signal layer and the two adjacent plane layers is the same. A number of PC board layout layouts were used in the vast majority of designs (Figure 4.11).

T-P-S-P-S-P-S-P-S-P-S-S-P-B is a popular PC board stack-up for layering a 12-layer board, where "T" denotes the top layer, "P" denotes a plane layer, "S" denotes a signal layer, and "B" denotes the bottom layer. This stack-up is used for layering a board with T-P-S-P-S-P signals. It should not be routed over significant distances on the top and bottom layers, which are used for component pads, in order to prevent direct emissions from the traces on the board. These layers are used for the pads.

Following this, it is required to prove that the plane layers must comprise a large number of power islands in order to accommodate different DC voltages. Assume that the layer 11 (the top layer) will have multiple DC voltages for the sake of this illustration. Because of this, designers are required to keep high-speed signals as far away from layers #10 and the bottom as is practically possible. This is because return current is unable to cross over the gap that exists between the planes on layer #10, hence stitching capacitors are required. As a result, the third, fifth, seventh, and ninth layers are the only ones available to serve as signal layers for high-speed transmissions.

The next step in the layout process involves planning the path that the most significant signals will take. In the majority of designs, the traces are organized in one direction as much as is practically possible in order to maximize the number of wiring channels that are available on a given layer. Levels #3 and #7 can be designated as "east-west," while layers #5 and #9 can be designated as "north-south." It is of the utmost necessity to carefully consider the shifting of layers while creating a high-speed trace as well as which separate layers will be used for each trace. Again, the most

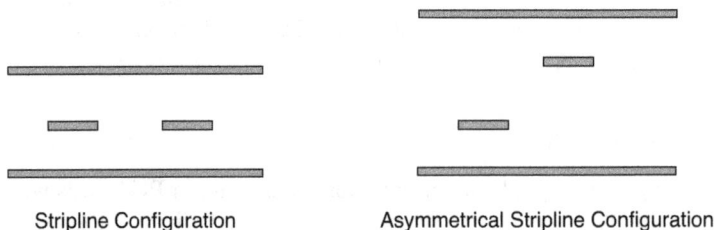

Stripline Configuration   Asymmetrical Stripline Configuration

FIGURE 4.11   Symmetric and asymmetric strip line configurations.

important thing to keep in mind is making sure that the return current may flow from one reference plane to the new reference plane. It is not necessary for the return current to switch reference planes for the optimal design; rather, it is sufficient for it to switch from one side of the plane to the other. The following combinations of signal layers can be utilized as signal layer pairings: number 3 and number 5, number 5 and number 7, and number 7 and number 9. Because of this, it is possible to route traffic both east to west and north to south in every imaginable combination.

It is not recommended to use a combination such as #3 and #9 because the return current would have to flow from the plane on layer #4 to the plane on layer #8 if that combination was used. In spite of the fact that a decoupling capacitor may be positioned close to the via on the PC board, the inductance of the leads and vias would render it ineffective at high frequencies. The cost function will increase due to the incorporation of such wiring as well as a capacitor.

It is imperative that the decoupling capacitors that are connected to the same DC voltage as the processor perform at the highest feasible level. The high-frequency performance of the on-board decoupling capacitors was substantially hindered due to the inductance that was introduced by the connecting vias, pads, and connection traces. The most straightforward method for lowering this inductance is to keep connecting traces both short and wide, while also keeping vias to a minimum in length. The portion of the through that continues to run to the bottom of the circuit board does not have any significant currents and is insufficiently long to be referred to as a "stub" antenna.

After that, it is essential to make a decision regarding the paths that high-speed signal lines will take on layers 3 and 5. It is desirable to maintain continuity between signal traces that come from an active device and are driven by the same quantity of power as the reference plane. In other words, the routing of processor signals ought to take place on layers 3 and 5, as these levels share the same power and, as a result, return currents are able to more readily return to their original source. Although only the most significant signals and ICs have been covered in this part, the same line of thinking ought to be applied to the remaining signals and ICs as well. Due to the separation of planes on layer 11, layer 10 should only be used for the routing of signals travelling at modest speeds.

4.11 SIX-LAYER BOARD

The six-layer printed circuit board is an unusual occurrence. This stack-up, which consists of four signal layers and two plane layers, is generally selected for lower-priced items since it is more cost effective. The standard layout consists of two plane layers and four signal layers. In a manner analogous to that which came before, a routing that travels east-west and north-south is generally used. Again, it is preferable to select routing layer pairings that do not require the return current to switch planes in order to get the desired results. In this particular instance, we will select layers #1 and #3 as the routing pair for one path, and layers #4 and #6 will be the routing pair for the other path. In a situation like this one, it is necessary to utilize both the top and bottom layers for signal routing. Layers 3 and 4 should never be used together as a routing pair for high-speed communications because it is more important to maintain the return current on a single plane than to bury the signal across multiple planes. Layers 3 and 4 should never be used together as a routing pair. The power reference layer will be located on layer 2, while the ground-reference layer will be located on layer 5. In the event that layer 2 is selected to serve as the ground-reference layer, the designers are obligated to ensure that all high-speed signals are routed on layers 1 and 3 in order to prevent crossing splits in the reference plane. It is permissible to route signals on layers 4 and 6 if the path of a particular signal does not traverse a divide in the power reference plane (Figure 4.12).

4.12 FOUR-LAYER BOARD

When it comes to low-cost systems, four-layer boards are used almost exclusively because of their practicality. Only two signal layers and two plane layers are typically present in a typical image.

```
------------------- Top
=================== Layer #2 Plane
- - - - - - - - - - - - - Layer #3 Signal
- - - - - - - - - - - - - Layer #4 Signal
=================== Layer #5 Plane
- - - - - - - - - - - - - Bottom
```

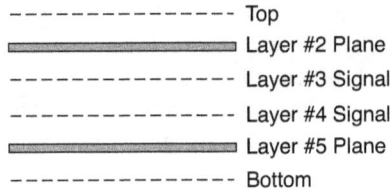

FIGURE 4.12   T-P-S-S-P-B stack-up.

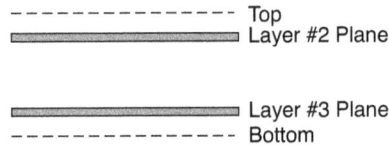

```
------------------- Top
=================== Layer #2 Plane


=================== Layer #3 Plane
- - - - - - - - - - - - - Bottom
```

FIGURE 4.13   T-P-P-B stack-up.

The east-west, north-south routing approach is utilized because it is essential to maximize the efficiency of the total number of routing channels. Concerning the flow of the return current, it is not possible to maintain the same reference plane. A decoupling capacitor needs to be placed in close proximity to the through in order to provide a return current channel. To achieve the best possible balance between inductance and impedance, the short trace that connects the capacitor pad and the via should be made as narrow as practicable.

In most cases, the planes are labelled as ground-reference and power, with the power plane being further broken into a number of different voltages. It is essential to maintain the traces over solid plane parts when the power plane is being used as the signal reference for a trace; as a result, traces do not cross these divides. When the power plane is being used as the signal reference for a trace, it is essential to do so. In the event that there is a split crossing, a capacitor needs to be placed in close proximity to the point where the trace crosses the split in order to provide a path for the current to return to its source. Again, in order to reduce inductance and impedance as much as possible, the short trace that connects the capacitor pad to the via should be maintained as short or as wide as is practically possible (Figure 4.13).

4.13   ONE- AND TWO-LAYER BOARDS

A challenge for EMC design is boards with one or two layers. Even though it is not recommended, this stack-up is frequently used for factors other than EMC. In this stack-up, there are often no solid planes and all signals, power, and power return are routed as traces. The main objective of this design approach is to keep the loop area for the signal current as small as possible and avoid allowing large loop regions for current, as seen in the figure below (Figure 4.14). These boards can nevertheless cause EMC problems even if their signal speeds are frequently slower than those of the multilayer boards covered in earlier sections. To reduce the loop area and, consequently, emissions, a signal return trace should be routed alongside the signal trace (and susceptibility to external RF interference). This design strategy is demonstrated in the diagram below (Figure 4.15). Signal traces should be kept brief to further minimize the loop's surface area. Decoupling capacitors must be positioned as close to the integrated circuits as possible and connected between the power and ground pins (Figures 4.14 and 4.15).

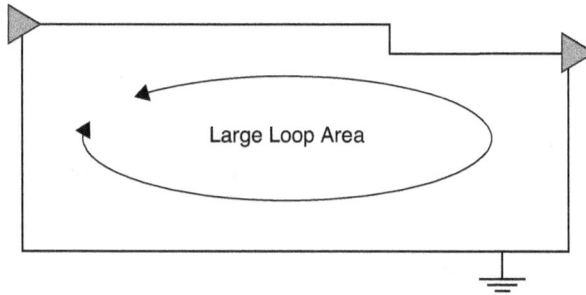

FIGURE 4.14 A single-sided PCB routing example with a sizeable loop area for signal current.

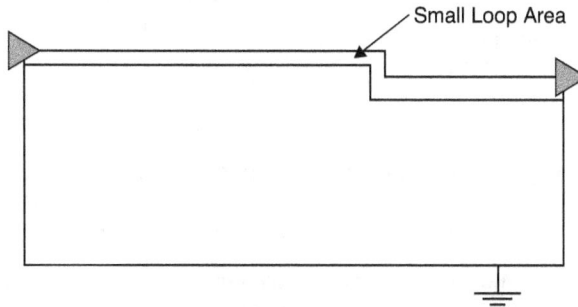

FIGURE 4.15 A single-sided PCB routing example with a minimal loop area for signal current.

4.14 COMPONENT PLACEMENT

A single-sided PCB routing example with the placement of a sizeable section of the components is typically determined by a number of other factors that improve the EMC performance of the system. One important rule is to keep the high-speed and low-speed components as far apart as is practical, and to group high-speed devices in one region of the board and low-speed devices in another. It may be even more important to take into account how the signals from these devices will be routed on the board. For instance, I/O connectors are frequently positioned along one side of the board. In order to prevent unwanted noise from high-speed devices from coupling onto the I/O traces and leaving the enclosure, it is crucial to keep the I/O driver devices near to the I/O connectors. This coupling is significantly more likely if the I/O driver is placed far from the connecting region.

High-speed gadgets typically use their pins to conduct high-frequency currents. These currents produce electric and magnetic fields that can couple directly to the I/O connector pins when the high-speed device and the I/O connector are placed close to one another. None of the filters on the board can stop signals from being coupled directly to the I/O connection pins since the coupling occurs outside the filter site. There are many ways that unfavourable coupling could develop. Not all potential combinations of devices, traces, etc., can be taken into account in this situation. However, paying a little more attention to the design can help save costs and problems, and this should be encouraged for the design.

4.15 ISOLATION

Even if the various circuits are relatively close to one another, some additional isolation is still necessary. One of the main issues is how the location and routing of high-speed devices would affect the distribution of return current in the reference planes. This return current in the reference plane can be separated from low-speed circuit sections using splits in the reference plane. If the reference plane is near a layer that high-speed signals are purposefully routed through, crossing this divide in the process, then this is literally not the preferred method. This can be a very successful technique for avoiding unwanted noise from accessing low-speed and I/O devices working at frequencies up to several hundred megahertz, provided that there are no traces routed over the area where an isolation split would be used.

Split use is demonstrated by the division of the ground-reference plane in the I/O connector region from the digital ground-reference. This design approach stops high-frequency noise from high-speed digital circuitry from spreading into the I/O area. For the intended I/O currents, a low-frequency return current channel must be provided across the reference plane split. It is possible to isolate a particular device by creating a "moat" in the power plane and ground-reference planes surrounding it. Again, extreme care must be taken to avoid routing high-speed signal traces across this moat, and a wide enough passage must be provided for the DC power current. This moat must have identical split locations for it to work and avoid accidental connection between the power and ground-reference planes.

4.16 POWER DISTRIBUTION FOR TWO-LAYER BOARD

4.16.1 SINGLE-POINT VS MULTIPOINT DISTRIBUTION

Each active component in a single-point power-distribution system has its own power and ground, which are kept separate until they converge at a single reference point. There are several 0-V reference points in multipoint systems because of the daisy-chaining of the connections. The possibility of common impedance coupling in multipoint systems exists as a result. In order to connect the regulator ground, microcomputer ground, battery negative, and chassis or shield to a single point, it is desirable.

4.16.2 STAR DISTRIBUTION

Similar to point distribution is the star distribution. It appears that all points refer to the same centrally located fixed point, as their traces are around the same length. In addition, the same reference point may be connected to its decentralized source via a big single trace. The principal distinctions are as follows:

1. A star's single reference point may have a longer trail.
2. Each of the separate traces starts near the centre of the circuit board and travels in a different direction, resulting in equal trace lengths.
3. The system clock of a high-speed computer board makes use of the star. From the edge connector, the signal travels to the circuit board's centre, where it is divided and sent to each necessary place. The signal delay from one portion of the board to the next is short since it starts at the centre of the board. A Minimal Loop Area is as shown in Figure 4.16.

4.16.3 GRIDDING TO CREATE PLANES

Gridding is the most crucial design strategy for two-layer circuit boards. Similar to the electricity grid, gridding is a system of orthogonal connections between traces that transfer to ground. The ground plane it creates provides the same level of noise reduction as four-layer boards. It does two things:

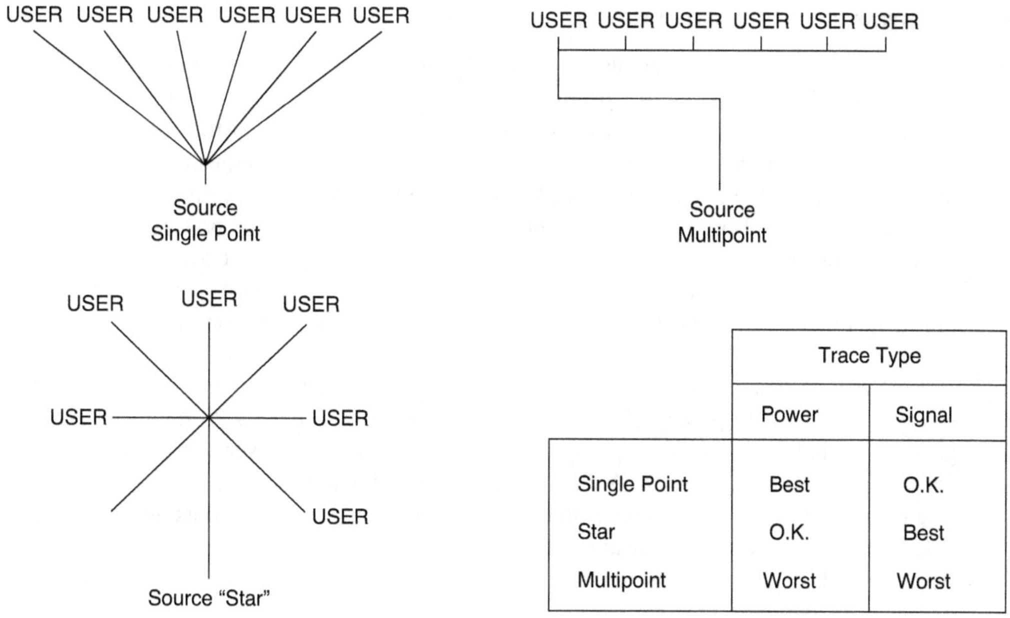
USER USER USER USER USER USER
Source
Single Point
USER USER USER USER USER USER
Source
Multipoint
USER
USER
USER
USER
USER
USER
Source "Star"

	Trace Type	
	Power	Signal
Single Point	Best	O.K.
Star	O.K.	Best
Multipoint	Worst	Worst

FIGURE 4.16 Power distribution.

1. Add a ground return line below the signal wires to extend a four-layer board's ground plane.
2. Make the impedance between the processor and voltage regulation as efficient as possible.

Gridding is the process of extending any ground traces and using ground-fill patterns to create a network of ground connections across the PCB. The bulk of the traces on a PCB, for example, flow vertically on the topside and horizontally on the bottom side. In the beginning, each ground trace is extended to fill in any empty spaces on the PCB. Ground-fill designs that have both ends anchored to the ground contribute to the grid more efficiently. When a ground-fill-pattern geometry is connected at more than one site, it transforms from a ground shield to a conductor and, as a result, prefers to grid.

- Create as many grids as possible on a double-sided board. Look for locations where minor alterations to the plan would permit additional grid link.
- Use only the through holes.
- Line widths are not required to be uniform (Figure 4.17).

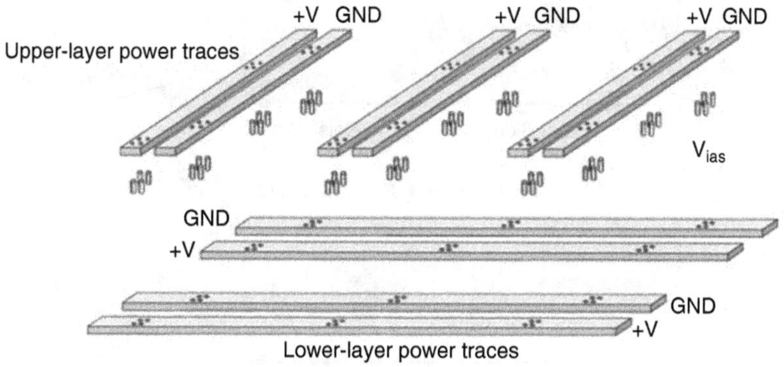
+V GND
+V GND
+V GND
Upper-layer power traces
Vias
GND
+V
GND
+V
Lower-layer power traces

FIGURE 4.17 Gridding power traces on two-layer boards.

4.16.4 BYPASSING AND FERRITE BEADS

Since the capacitance is meant to supply the switching current for the device, flowing between +V and ground at the PCB is an essential procedure. The law states that the current should flow through the leads leading to the power supply, which have the lowest impedance, if the current cannot be accessed in the bypassing loop due to excessive inductance. The higher frequencies arise from the scattered capacitance of the power route. The PCB must rely on the current provided by the ferrite bead since the ferrite bead prevents RF current from being pulled from the power line connector. It is crucial to understand and always keep in mind that the bypassing capacitor should supply all currents at or above the oscillator frequency, and that the power-routing purpose is just to recharge it. By doing the following actions, RF can be removed from power-distribution lines:

1. Use a bypassing capacitor and a ferrite bead with the capacitors housed inside the ferrite bead. Put a 1,000-pF capacitor outside the ferrite bead to make a PI filter. The micro ground should be used as the capacitor's ground connection. However, the capacitor may couple this noise back onto the +V line if there is a lot of it at this point.
2. The ferrite bead is never used on ground; it is always used on +V. When using a through-hole ferrite bead, the exposed lead must be connected to +V.
3. Apply the 3:1 length-to-width rule to the traces in the bypassing loop to reduce resistance on this high-frequency channel.

   Shorten the bypassing loops' length and circumference. When connecting the bypass capacitors for the oscillator or +V supply, try extending the microcomputer's ground instead of running a trace. Run any trace back over (or under) any other section of the loop to reduce the radiating area as seen from the top of the circuit board.
4. It will be advantageous to use ferrite beads and the same bypassing values on four-layer boards. The 1,000-pF capacitor may not be necessary on four-layer circuit boards, but it should be added in the initial design as shown in Figure 4.18.

4.16.5 BOARD ZONING

Board floor planning, which is the process of deciding the general location of components on a blank PCB before drawing any traces, is similar to board zoning. Instead of merging related tasks,

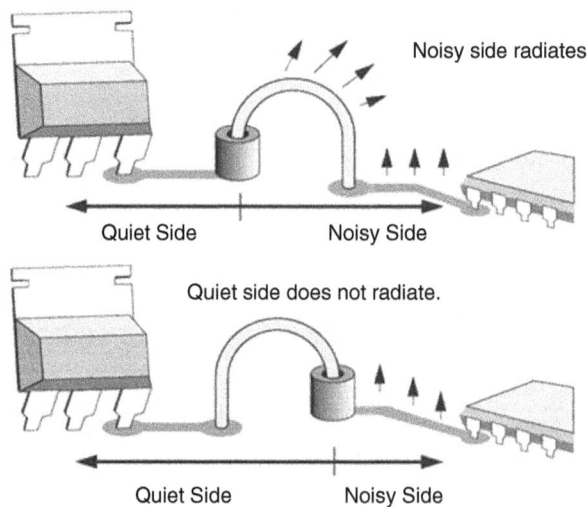

FIGURE 4.18 Ferrite bead placement closest to the noise source.

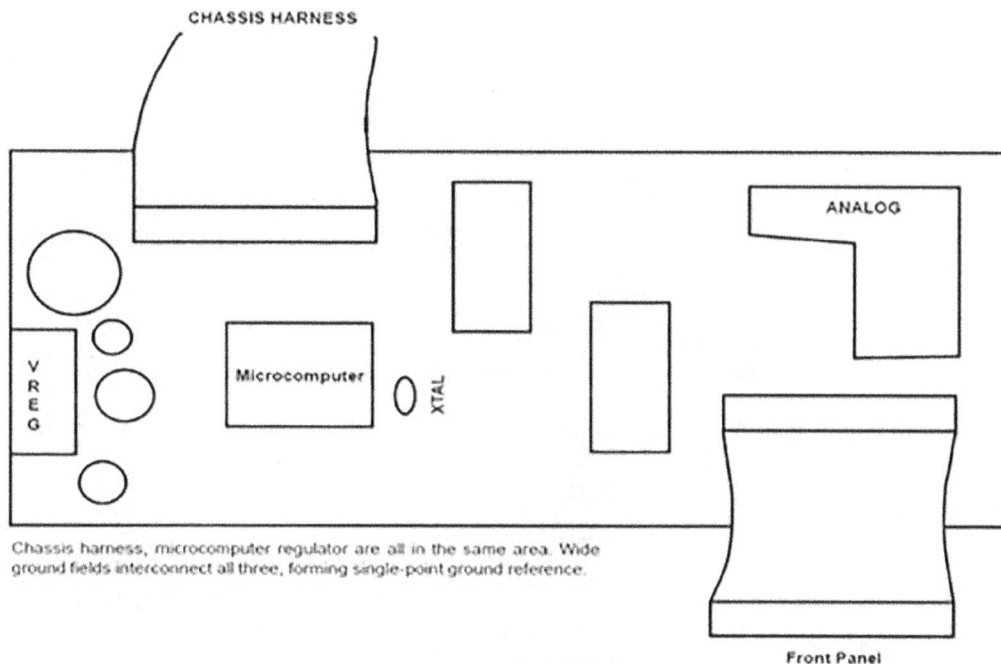

FIGURE 4.19   Board zoning.

board zoning places them on a board in the same general area. The power source is placed far from lower speed components, and analogue components are even isolated. As a result, there is a lower chance of high-speed logic contaminating other signal channels. Away from connectors, low-speed signals, and analogue circuitry is where the oscillator should be placed. This holds true for the board as well as the interior of the box that houses the board. Avoid designing cable assemblies that fold over the oscillator or the microprocessor after final assembly because they can take up noise and transmit it to the component's region.

The primary tasks for component placement in PCB design are as follows:

1. Identify the CPU, voltage regulator, and voltage regulator in that order. The voltage regulator is where the battery voltage enters the board.
2. Create a single-point ground between the three (using a gridded or solid ground) and attach the shield there (Figure 4.19).

4.16.6   Impedance Matching at Inputs

The input of a CMOS device is similar to a series inductance, producing roughly 5 pF in parallel and 5 MW to the grounded substrate. A lot of ringing and other noise will develop if the device controlling the input is not matched to the higher impedance. Emphasis is focused on the microcomputer's output because of the load's underdamped nature. In this instance, the underdamped load is a microcomputer, and the unwanted side effects are ringing and overshoot. The output impedance, as determined by the trace and the input pin, is increased by a resistance supplied at the driver, matching the high input impedance. If the input is connected to an open trace, such as the line to a switch, it is advised to use a pull-up or pull-down resistor. When the input is enabled, this increases switching current while decreasing impedance at all other times. As a result, there is less chance that the trace will be exposed to coupled sounds (Figure 4.20).

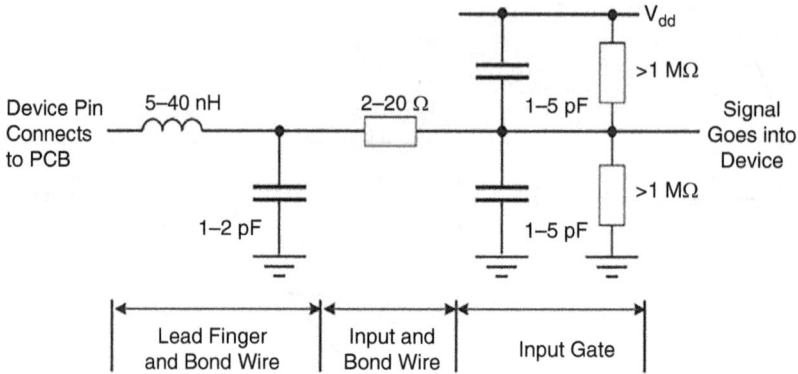

FIGURE 4.20 MOS buffer simplified schematic.

4.17 LAYOUT FOR SUSCEPTIBILITY

During the coupling of signal traces with incident electric or magnetic fields. The voltage of the trace is overlaid with the sine wave because the connected signal is an alternating current. A microcomputer's input rectifies this voltage, producing a DC offset voltage on the pin. The microprocessor won't recognize the desired switching function if this DC voltage is increased to move the input away from the switching-point voltage. The device continues in this state until the disruptive field is gone if the oscillator's input for the device has no clocks and is reset.

Large loop areas both receive more signal and produce more noise, and therefore both are amplified. Therefore, preserving a signal's radiation resistance is the same as keeping it from emitting radiation. The most important pins for immunity that affect programme control are the oscillator, reset, interrupts, and any input pins used for programme branching. The oscillator pins, crystal, crystal bypass capacitors, and the connection between the ground of the bypass capacitor and the ground of the microcomputer make up the loop that is the subject of the bulk of vulnerability problems.

Noise can also be produced through ground bounce (common impedance coupling) in circuits. The drive circuit's reference voltage may shift if the ground path has high impedance, which would cause the input to the microcomputer (RESET, OSC) to be beyond the system's switching range.

4.17.1 SIGNAL TRACES

On the board, traces are the conductors that transport current from the driver to the receiver. When these traces encounter a bend or cross, they form a complete antenna. The applicable trace design guidelines are as follows:

1. Spatial placement of traces
2. Avoid straight angles at trace corners and use a 45-degree turn
3. Guard and shunt traces must be used for clock lines (Figure 4.21)

Capacitive and inductive crosstalk can happen even when parallel traces are close together. In capacitive coupling, a rising edge on the source causes a rising edge on the victim. The victim's voltage shifts during inductive coupling in the inverse direction of the source's edge that is changing. Capacitive crosstalk predominates. Inversely proportional to separation distance, the amount of noise on the victim is connected to the parallel distance, frequency, amplitude of the voltage swing on the source, and victim impedance.

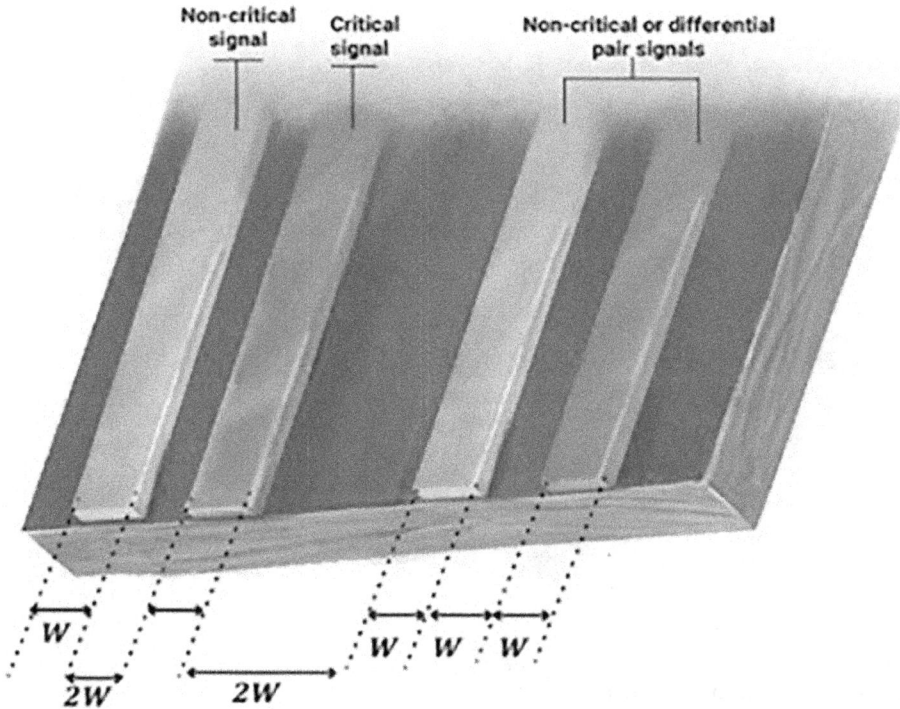

FIGURE 4.21   Adjacent traces on the same PCB layer separated by 3 W.

1. Keeping RF-noise-carrying traces connected to the microprocessor away from other signals to prevent them from taking up noise is one of the steps used to avoid crosstalk.
2. Signals that are prone to noise should have their return ground run underneath them since this decreases the noise voltage and any radiating area and tends to reduce their impedance.
3. Avoid running loud traces around the board's edge.
4. Arrange several noisy traces in a group with nearby ground traces.
5. To prevent non-noisy traces from picking up noise from their surroundings, keep them away from areas of the board like connectors, oscillator circuits, relays, and relay drivers.

The majority of crosstalk issues arise from placing the victim too close to the crystal. Do not place components closer than 1 inch from them (Figure 4.22).

Noise and ESD should exit the shield through the cable on the PCB and the bypass capacitor. The ground from the capacitor to the shield should therefore be substantial, securely connected, and fastened to the shield. The effective series resistance (ESR) should be between 50 and 500 MHz, and the bypass capacitor's value should be less than 1,000 pF. Because component lead length also affects ESR, surface-mount components are advised over through-hole technology.

4.17.2   GENERAL LAYOUT RECOMMENDATIONS

- Vias are used in multilayer PCBs to increase high-speed tracks' capacitance and inductance by 0.3–0.8 pF and 1–4 nH, respectively. Minimize vias while constructing high-speed signal tracks.
- Right-angled track turns can create inner field focus. This field might produce track noise. Turning orthogonal should be 45°.

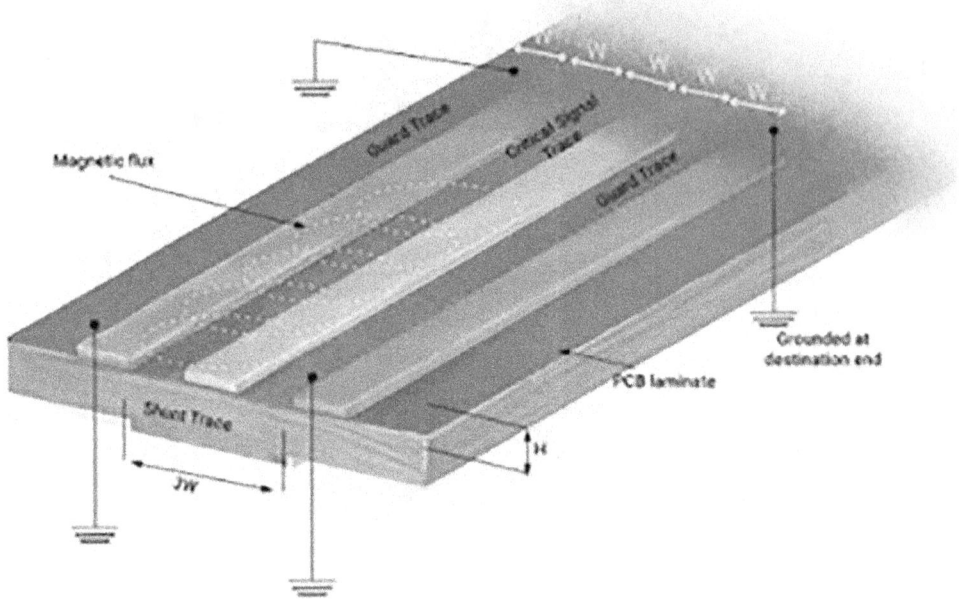

FIGURE 4.22   Guard and shunt traces.

- Stubs produce reflections. Avoid stubs with high-frequency, sensitive signals.
- All metallized patterns should be connected to the ground to prevent huge metal areas from behaving as radiating aerials.
- The digital return path and the RF return path are to be kept separate. The coupling capacitance is reduced by the wide gap between the plane surfaces. Increased coupling impedance from additional inductors prevents RF currents from flowing from one plane to another.
- The widest possible power and ground traces should be used. They should avoid creating large loops as well, as doing so will dramatically increase the self-inductance.
- Any open areas on the PCB should be filled with ground islands.
- If the IC contains multiple VCC-GND pairs, each pair requires a separate decoupling capacitor in addition to referencing the manufacturer's datasheet and application notes.
- Signals are referenced to ground; hence, it is desirable to place the capacitor close to the ground pin if the power and ground pins are separated by a large distance.
- To give high impedance to the potential noisy power and ground nets, the traces connecting the decoupling capacitor to the power and supply nets should be small and widely spaced.
- Series termination (at the driver) prevents receiver signal distortion and all reflections are removed by parallel termination (at receiver) for single-ended termination.
- Functional blocks such as power supply network, analogue, digital and RF blocks should be clearly identified, placed in separated zone and routed in their respective zone.
- To reduce the length of the interconnections, the analogue blocks should ideally be situated adjacent to the analogue signal connectors. This is to avoid crosstalk.
- Differential routing can be implemented to couple the noise.
- The 20-H rule" advises that the VCC plane be smaller than the GND plane with 20 H from the board edge in order to reduce electromagnetic radiation.
- Continuous return path has to be maintained for return current.

4.18 SUMMARY

EMI/EMC problems are considered to be inevitable in our daily life, since it is becoming much more complex daily, due to the wide usage of high-speed computers and personal communication devices. So as an EMI/EMC engineer, a suitable design should be implemented considering that every device should comply with the standards formed by each government. Experience shows that engineers will find ways to use the modelling tools that they had never before envisaged once the first reticence to develop something new is overcome.

BIBLIOGRAPHY

1. Bruce R. Archambeault, *PCB Design for Real World EMI Control.* Springer Science and Business Media, IBM, US, 2002.
2. http://www.interferencetechnology.com/
3. W. Cui, M. Li, X. Luo, J. L. Drewniak, T. H. Hubing, T. P. Van Doren, and R. E. DuBroff, "Anticipating EMI from coupling between high-speed digital and 110 lines," *IEEE Electromagnetic Compatibility Symposium Proceedings*, Seattle, WA, pp. 189–194, August 1999.
4. A design guide on "PCB Design Guidelines for Reduced EMI". Texas Instruments. https://www.ti.com/lit/an/szza009/szza009.pdf?ts=1686375490376&ref_url=https%253A%252F%252Fwww.google.com%252F

5 EMI and EMC Simulation Software

5.1 INTRODUCTION

Electromagnetic compatibility (EMC) and electromagnetic interference (EMI) have become key topics in all the devices we are using and also the devices used in industries; all of them must satisfy the EMC regulatory requirements. To do so, there is the need for simulation software to solve the real-world problems in a safe manner and make use of devices in an efficient and useful manner. The term simulation is used in different ways by different people. For the purposes of this article, "simulation" means the act of modelling an existing or prospective system. Find the driving forces behind the system and you can foresee how it will act in the future. Any system that can be represented quantitatively with rules and equations can be simulated. Problems in the actual world can be tackled in a risk-free and timely manner with the help of simulation software. It offers a valuable analysis approach that can be quickly checked, explained to others, and understood. The clear insights into complicated systems provided by simulation software are invaluable in a wide range of fields and industries. There are numerous simulation software available for EMI, The following software simulation tools are discussed in this chapter:

1. EM WORKS
   i. EMS
   ii. HF—WORKS
   iii. MOTOR WIZARD
   iv. EM WORK 2D
2. POWER SI
3. CST STUDIO SUITE
4. MOMENTUM 3D PLANER EM—SIMULATOR
5. COSMOL
6. newFASANT
7. BEM & FEM Software Design
8. EM Pro
9. Motor-CAD

5.2 EM WORKS

5.2.1 EM WORKS—EM SIMULATION SOFTWARE

EM works generate a good EM simulation software for electrical and electronics design with multiphysics capabilities. This company's products are fully and flawlessly embedded in SOLIDWORKS and Autodesk Inventor. Power and signal integrity of high-speed interconnects are just a few examples of where this technology has found use. Computer programmes that model EM fields and perform related analyses of temperature, velocity, and structure.

EMS is a software programme that is used to model EM fields, with the ability to determine various things as circuit characteristics, mechanical parameters, and losses. You can use EMS, which is a Gold Certified Add-in to SOLIDWORKS and an Add-in to Autodesk Inventor, to simulate

DOI: 10.1201/9781003362951-5

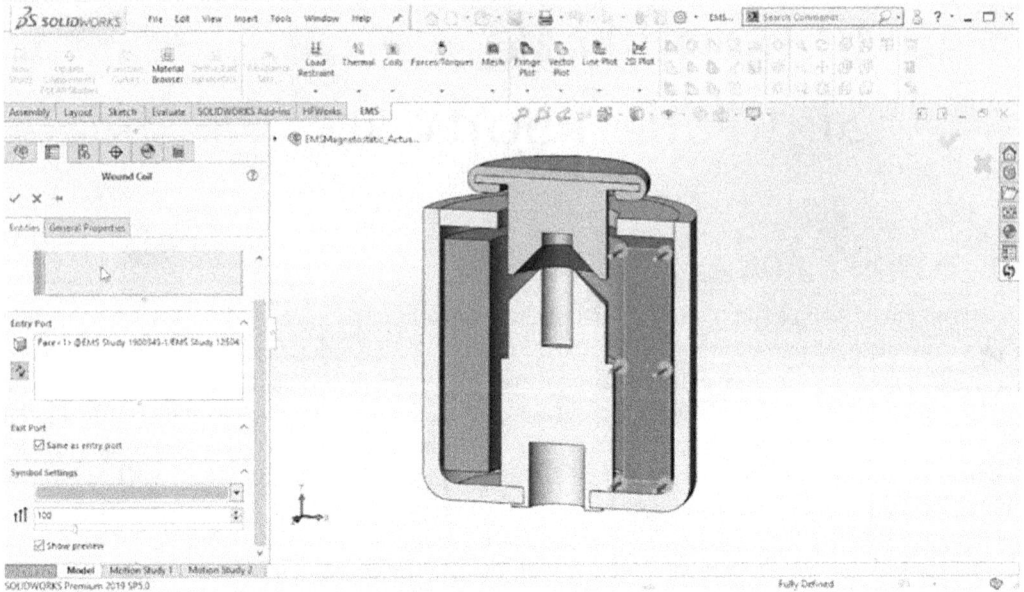

FIGURE 5.1 EMS simulation.

even the most complex electrical machines, motors, generators, sensors, transformers, high-voltage apparatus, high-power machines, electrical switches, valves, actuators, PCBs, levitation machines, loudspeakers, permanent magnet machines, NDT equipment, inverters, converters, bus bars, inductors, bushings, or biomedical equipment (Figure 5.1).

5.3 EM WORK PRODUCTS

EM works products are classified into four types:

1. EMS
2. HF—WORK
3. MOTOR WIZARD
4. EM WORK 2D

5.4 A TRANSFORMER SOFTWARE AND CALCULATION

Transformer design can be accomplished with the help of EMS. With EMS, virtual investigation of key transformer design variables can be calculated

5.4.1 ENERGY STORAGE

Energy in a transformer must be transferred instantly from input to output; it must not be stored in any way. In the real world, transformers do, unfortunately, store some energy that is not desirable. The leakage inductance, which reflects the stored energy between windings regions that are occupied by non-magnetic media, is calculated by the EMS. In a similar manner, EMS computes the mutual inductance, which reveals the quantity of unwanted energy that has been trapped in the magnetic core and the microscopic air gaps.

5.4.2 Losses and Thermal Management

The maximum "hot spot" temperature rises at the core surface inside the centre of the windings and can be calculated by EMS. To find the smallest core size that still achieves the needed power supply efficiency while staying below the maximum "hot spot" temperature, this computation is useful. The eddy loss, hysteresis loss, core loss, winding loss, and heat loss of the transformer, as well as the temperature and convection properties of the liquid in the surrounding area, are all considered by EMS to determine the aforementioned temperature increase.

5.4.3 Core Selection

To determine the optimal core material, shape, and size for any frequency and required power output, EMS analyses the magnetic flux density and saturation levels in the core. The designer of the transformer must strike a balance between meeting the required power without overloading the core, the cost of the core material, the size of the core, and the ease of manufacturing the transformer's shape.

5.4.4 Open and Short-Circuit Tests

Tests for an open circuit and a short circuit on a transformer are essential, but they are also time-consuming and expensive. EMS gives the designer the ability to practically carry out the tests in question in a manner that is both accurate and effective.

5.4.5 Insulation Coordination

Calculating the dielectric breakdown is an important step in selecting the appropriate bushings, surge arrestors, and other insulating equipment, and EMS is responsible for this calculation. The designer can more easily meet the numerous insulation coordination norms with the assistance of this form of calculation.

5.4.6 Short-Circuit Forces

Magnetic force acting on windings and core material, as well as stress and structural displacement, are all calculated by EMS. To ensure the transformer's structural soundness, these calculations are useful.

5.5 A MOTOR SOFTWARE AND A CALCULATOR

Electric motor can be designed with various design parameters as follows

5.5.1 Parameters Estimation

The inductance and resistance of the windings are crucial parameters for controlling and estimating the status of electrical motors. This is the case, for instance, when the rotor position in a Brushless DC Electric Motor (BLDC) motor is estimated without the use of sensors or when phase current is controlled in an SRM motor. If EMS is given for a range of desired values for the current and frequency, it will output a set of values for the parameters.

5.5.2 3D Modelling Problems

The EMS is a comprehensive environment for creating models in three dimensions. This allows for the simulation of crucial topologies and effects that would otherwise be impossible to study analytically:

- Skewing the slots or the rotor poles is a traditional procedure for reducing the amount of cogging force. The only way to assess its outcomes is to take into account the interaction between the stator and the rotor in all three dimensions.
- Axial flux and transverse flux machines, two types of advanced machine topologies, naturally function with 3D flux distribution and should be handled as such.
- The leakage inductance and resistance of a winding are both significantly affected by its end windings.

5.5.3 TORQUE

EMS is able to compute both transient and steady-state torque profiles for a variety of electrical machine topologies. These topologies include induction, switched reluctance, BLDC, and permanent magnet AC machines, among others. The findings of measuring torque at various rotor RPMs and winding currents are used to establish the best possible operating conditions. In addition, EMS assists in reducing the amount of cogging torque by analysing the size of this torque for a variety of air gap lengths and fractional slot pitches.

5.5.4 CORE MATERIAL

The ability to accurately depict non-linear phenomena in the core material, such as flux saturation, eddy current, and hysteresis losses, is essential to the design of a machine that will function properly. A library of predefined solid and laminated core materials is included with every instance of EMS. The designer is able to readily compare the saturation, core losses, and overall efficiency of various materials. The results of the core and winding loss can be linked with the thermal solver of the EMS to determine temperature rise and the amount of cooling that is required.

5.5.5 SHAPE AND SIZING

The torque and power rating of a machine can be determined to a large extent by its radius, length, and number of poles. On the other hand, the performance of the machine is massively affected by the more intricate geometrical details of the magnetic circuit. For instance, the form of the bars that make up a squirrel cage in an induction motor will have an effect on how torque varies with rotor slip. The influence that each of these factors has on the overall performance of the motor may be easily tested by modifying their values inside EMS.

5.5.6 PARASITIC RLC EXTRACTION

The EMS, if needed, can function as a parasitic RLC extractor. This means that it can precisely calculate the resistance, inductance, and capacitance of any arbitrary three-dimensional electric and electronic structure. When performing these calculations, the proximity effect, the skin effect, the dielectric and ohmic loss, and the frequency dependency are all taken into consideration. In other words, calculations are made for both the DC and AC versions of the parasitic RLC. When modelling a wide variety of electric and electronic devices and circuits, including the following, these parasitic values are quite helpful.

5.5.7 HIGH-SPEED ELECTRONICS

Crosstalk and distortion, connection delays and ringing, and ground bounce are all important phenomena that can only be studied with the aid of RLC models of high-speed electronic devices including ICs, PCBs, packages, and on-chip passive components.

5.5.8 Power Converters

In the realm of power electronics, RLC models find application in the simulation of bus bars, cables, inverters, and converters used in power distribution applications and hybrid and electric vehicle powertrains.

5.5.9 Touchscreen Modelling

Accurately calculating the capacitance of the screen wires is crucial for the modelling of touch-screens present in modern smart phones and laptops.

5.5.10 NDT Simulation Software

Technologies for non-destructive testing (NDT) frequently employ EM fields and waves. Eddy-current testing (ECT), magnetic flux leakage (MFL), remote field testing (RFT), magnetic particle inspection (MPI), pulsed eddy current (PEC), and the alternating current field measurement are all EM NDT procedures that can be reliably calculated using EMS (ACFM). In many cases of NDT screening, the NDT probes must be moved about. Since EMS couples to SolidWorks Motion, it is ideal for modelling such motion.

5.5.11 Solidworks and Autodesk Inventor Seamless Integration

EMS integration in the three main CAD platforms allows you to model the most sophisticated electrical machine, motor, generator, sensor, transformer, high-voltage apparatus, high-power machine, electrical switch, valve, actuator, PCB, levitation machine, loud speaker, permanent magnet machine, NDT equipment, inverter, converter, bus bar, inductor, bushings, or biomedical equipment. Individuals do not have to "reinvent the wheel." Just get a CAD model from the mechanical drafting staff and start your simulation of a magnet or a magnetic field right away without making any changes. If you want to change the CAD model you bought, you won't have to go back to the people who made it because commercial CAD programmes like Solidworks are parametric and have a hierarchy. Alter it "on the fly" as you see fit. It's likely that you can get it saved in Parasolid, ACIS, IGES, STEP, STL, CATIA, or ProE kernel from the drafting department or coworkers, even if they use a different CAD programme. From there, you can continue your EM design in SOLIDWORKS, Autodesk, Inventor, Ansys, or Space Claim by importing the data.

5.5.12 Multi-Physics Capabilities

The EMS software and simulation package is a genuine multi-physics application. It gives you the ability to combine your magnetic, and electrical design with thermal, structural, and motion analyses on the same model and mesh in a hassle-free integrated environment without the requirement to import or export any data at any point. Because this world is connected with several physics, there is no cluttering, no bouncing about, no mixing and matching, no chaos, no confusion, and no mess. In addition, it implies productiveness, correctness, and efficiency.

5.5.13 Electro-Thermal Analysis

There are electro-thermal considerations in your design. Simple and effortless! In the characteristics of your study, select "Couple to thermal" to switch between a steady-state and transient analysis. The EMS then sends that information into the thermal solver to calculate the joule, eddy, and core losses automatically. The addition of non-EM heat loadings is simple, and it can be done by adding either volume heat, heat flux, or a constant temperature. The temperature, temperature gradient, and heat flux are calculated using the EMS thermal steady-state or transient, taking into consideration environmental circumstances including convection and radiation.

5.5.14 ELECTRO-STRUCTURAL ANALYSIS

By the same token, the electro-mechanical coupling is also easy and hands-free. The "Couple to structural" option invokes the EMS structural solver, after transferring the local force distribution in relevant parts in addition to the mechanical loads and constraints, and then computes the displacements. The stress and strain are deduced subsequently and added to the "Structural Results" folder as well. Similarly, electro-mechanical coupling is simple and hands-free. After transferring the local force distribution in relevant parts, as well as the mechanical loads and constraints, the "Couple to structural" option invokes the EMS structural solver, which then computes the displacements. Following that, the stress and strain are calculated and added to the "Structural Results" folder.

5.5.15 SOLIDWORKS MOTION INTEGRATION

Moving parts and components are common in electrical machines and drives. In general, the resulting motion is either rotational, as with motors, or translational, as with linear actuators. Nonetheless, some applications, such as MagLev and Eddy-current braking, may cause all six degrees of freedom of motion. Only EMS can handle such intricate machines and equipment in this case. Why? Because EMS is compatible with SolidWorks Motion, the most versatile and powerful mechanical motion package. To learn more about this powerful package, go to https://www.solidworks.com/sw/products/simulation/motion-analysis.htm. The connection to SolidWorks Motion® is once again simple. Simply instruct EMS to couple to a SolidWorks Motion study after creating one.

5.5.16 EMS RESULTS

The designer can use EMS to compute electric, magnetic, mechanical, and thermal parameters such as:

- Electric force
- Electric torque
- Magnetic force
- Magnetic torque
- EM force
- EM torque
- Magnetic flux density
- Magnetic field
- Electric field
- Electric flux
- Current flow
- Eddy current
- Inductance
- Capacitance
- Resistance
- Hysteresis loss
- Eddy loss
- Speed
- Acceleration
- Stress

- Flux linkage
- Core loss
- Breakdown voltage
- Lorentz force
- Lorentz torque
- Skin effect
- Proximity effect
- Magnetic saturation
- Induced voltage
- Force density
- Power loss
- Temperature
- Temperature gradient
- Heat flux
- Back EMF
- Electric flux density
- Impedance
- Ohmic loss

5.6 ABOUT HF WORKS

For radio frequency (RF), microwave, millimetre wave, and high-speed digital circuits, HF Works provides EM and antenna simulation. For RF/MW frequencies and beyond, HF Works provide solutions for issues related to EM radiation, EM waves, EM propagation, EM resonance, EMI, electromagnetic compatibility (EMC), and signal integrity (SI). To calculate fields, antennas, and circuit parameters, it employs cutting-edge finite element solvers and meshing technologies. It is capable of simulating both single antenna elements and complex arrays of antennas. Time-domain calculations like Time-Domain Reflectometer (TDR) and Eye Diagram are also possible in HF Works. It can estimate a 3D structure's power-handling capacity and pinpoint weak spots in the field. It also lets you model RF microwave heating as a function of input power (Figure 5.2).

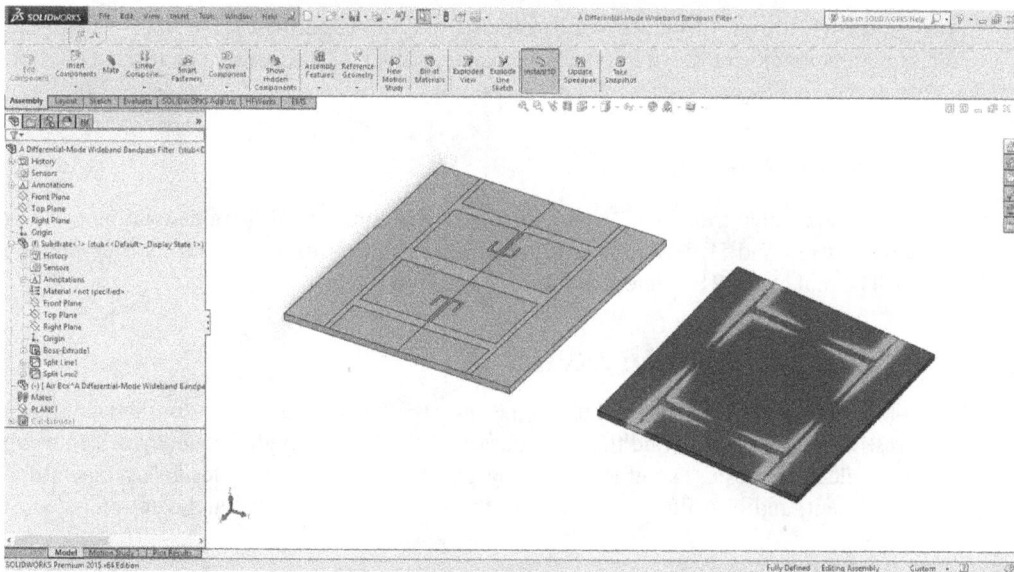

FIGURE 5.2 HF Works simulation features and capabilities.

5.7   VERSATILE HIGH-FREQUENCY AND HIGH-SPEED TOOL

If your design incorporates antennas, RF and microwave components, signal integrity, power integrity (PI), EMC/EMI, chip-packaging, PCB, connectors, cables, RF MEMS, or filters, and whether you employ planar circuit technologies, standard waveguides, or dielectric guides, HFWorks addresses your high-frequency field simulation design needs for RF frequencies and much beyond. This 3D field EM simulation tool is accurate, quick, versatile, and user-friendly; it doesn't matter whether you're interested in frequency response, resonance behaviour, EMI, matching networks, filtering characteristics; it serves multiple purposes, such as an antenna simulator, a filter, a resonator, an EMC simulator, an EMI simulator, a passive component simulator, a signal integrity simulator, and a PI simulator. Antenna, S-parameter, time domain, and resonance solvers are just some of the tools at HF Works' disposal.

5.7.1   SOLIDWORKS EMBEDDED

Gold certified by SOLID WORKS Corporation, HFWorks integrates effortlessly with Solidworks. Since no separate model for the EM simulation needs to be created, it may be used for the analysis of even the most complex high-frequency and high-speed electrical and electronic devices and circuits in record time, thanks to its seamless CAD integration. Instead of starting from scratch, you can immediately begin your 3D EM simulation by grabbing a CAD model from the drafting department or a coworker. Since it is built on top of the most common CAD software packages, you may import designs in a wide range of standard CAD and geometry kernel formats like Parasolid, ACIS, IGES, STEP, STL, CATIA, ProE, DXF, DWG, etc.

5.8   USER-FRIENDLY INTERFACE AND EMBEDDED LEARNING MATERIALS

Quite a friendly and straightforward graphical user interface. Users can rapidly build and test out simulations of sophisticated 3D models. Quickly master HFWorks and put it to use in your own work with the help of the demo viewer.

5.8.1   PARAMETRIC SIMULATION

HFWorks allows you to run a wide variety of analysis to find the optimal layout for your project. You can use parameters to analyse the impact of changing any CAD dimension or simulation variable on your design. It's a good starting point for making your designs as efficient as possible.

5.8.2   TIME-DOMAIN SOLUTION

Connectors, adapters, cable transitions, and high-speed interconnects all have discontinuities that must be accounted for, and HFWorks makes it easy for designers to do so with its time-domain analysis tool. The makers of RF connections and cables need to.

5.8.3   INTEGRATED ELECTRO-THERMAL ANALYSIS

The electro-thermal analysis module is built right into HFWorks. It provides a unified setting in which to investigate the electrical and thermal characteristics of your high-frequency design using a unified finite element mesh. Therefore, you may get the high-frequency design's temperature, temperature gradient, and heat flow in addition to the plethora of electrical and electronic design characteristics.

5.8.4 POWER HANDLING

There are two main concerns regarding the power-handling capacities of passive RF & microwave structures, and HFWorks addresses both of them. The potential for dielectric breakdown to cause a catastrophic failure is a major constraint on power handling. Safety factor maps are created in HFWorks to show where a failure has occurred or is most likely to occur in the model at a specified excitation power or stored energy. Extreme heat from RF power losses can potentially lead to a thermal runaway. The coupled thermal solver in HFWorks works in tandem with the EM solver, letting users investigate the thermal behaviour of their model in relation to the input power. Users can save time and money by not having to set up complex testing settings to determine the product's power rating and handling capabilities, thanks to these features.

5.9 HFWORKS ADD-INS

HF Works comes with two add-ins:

- Mini Atlass: is a tool for calculating the electrical and physical properties of transmission lines through analysis and synthesis. It integrates seamlessly with SolidWorks and helps you design the structure in real time.
- ECAD Importer: connects HFWorks with other EDA programmes like KeySight ADS and Cadence Allegro.

The microwave design programme has three primary problem-solving modules. These items are:

1. Antennas
2. S-parameters
3. Resonance

5.9.1 ANTENNAS

Transmitting or receiving EM waves, or converting EM radiation into electric current, is what an antenna does. The antenna is a crucial component of any radio system, as it is responsible for both the transmission and receiving of radio waves. Many different technologies rely on antennas, including television and radio transmission, wireless local area networks and mobile phones, radar, and communications between satellites.

5.9.2 S-PARAMETERS

Linear electrical networks' electrical behaviours when subjected to varying steady states by electrical signals are determined by a parameter called the scattering parameter. S-parameters are the individual components of a scattering matrix, often known as an S-matrix. The representation of circuit network in the harmonic generation, the parameters such as Z, Y, and H are used for low and medium harmonic generation. These parameters are not used for the high and very high frequency range because the open and short-circuit conditions to determine the voltage and current are difficult to achieve. On low frequency range, the impedance and the admittance matrix are widely used, but they cannot be used for the high frequency range. S-parameters can be utilized with any frequency range, although they are most commonly used to RF and microwave networks because signal power and energy are more easily quantifiable at those frequencies than they are with currents and voltages.

5.9.3 Resonance

The frequencies at which a system exhibits resonance are called its resonant frequencies. Oscillations of such high frequency and amplitude can be produced by even modest periodic driving forces due to the system's ability to store energy. Only when loss is modest, the resonant frequency corresponds roughly to the frequency of spontaneous vibrations in the system. Some systems exhibit a spectrum of resonance frequencies, each with its own unique characteristics. Mechanical resonance, acoustic resonance, EM resonance, nuclear magnetic resonance, electron spin resonance, and the resonance of quantum wave functions are only some examples of the many types of waves that exhibit resonance phenomena.

5.10 ABOUT MOTOR WIZARD

Motor Wizard is a motor design programme that operates entirely within SOLIDWORKS and relies on already templates. SOLIDWORKS users can now design and evaluate their own custom electric machines. It facilitates the study of electric machines by providing a large range of adjustable dimensions and factors that completely characterize the design of electric machines. With the use of analytical and finite element-based solvers that are seamlessly incorporated into the design process, creating an electric motor is now a simple, rapid, and precise procedure (Figure 5.3).

5.11 FEATURES AND CAPABILITIES

5.11.1 Easy Topology Designer

Motor Wizard is a high-end CAD programme with a customizable parameterization panel that provides instant access to many designs and features found in premade templates. The 2D SOLIDWORKS assembly is therefore converted into a well-defined electrical device. Altering the presets allows for the rapid production of novel motor layouts. There is a wide variety of rotor and

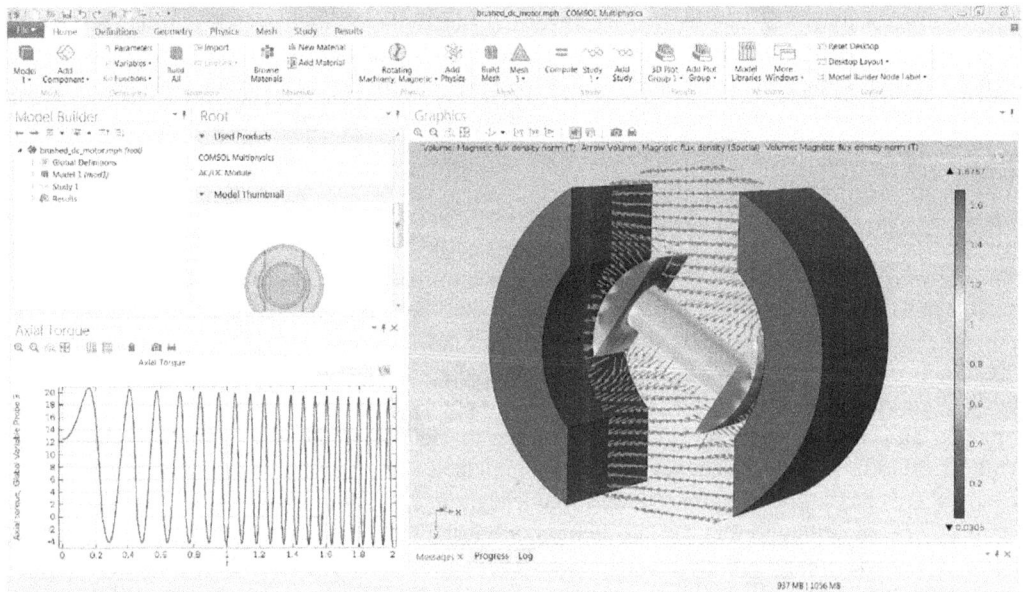

FIGURE 5.3 Motor Wizard simulation.

stator designs available for use with permanent magnet brushless DC motors. With the topology editor's auto-correction tool, which calculates sufficient dimensions depending on the specified configuration, even the most complex machine design may be completed in a matter of minutes.

5.11.2 QUICK PERFORMANCE PREDICTOR

Predicting how a motor will perform is an important first stage in the design testing process, and with the right methodology, motor designers can get rapid results after only a few mouse clicks. Torque waveform, air gap flux density, back EMF, winding flux linkage, phase inductance, co-energy, core losses, and more may all be examined and studied after the appropriate input data have been provided. Finite element analysis (FEA) is a powerful tool for properly predicting the behaviour of EM machines, and it can be used to get a variety of test-oriented outcomes. Motor Wizard allows you to run FEA simulations under varying loads, from no load to full load.

5.11.3 USER-FRIENDLY INTERFACE

Motor Wizard provides a user-friendly interface, which enables the management of the developed designs straightforward and efficient, saving users both time and effort. Using a heuristic process, one can quickly become proficient with all of the tools and their capabilities. Fast iteration through several analyses and the ability to revert back to previously used design parameters make it possible to optimize the design in a short amount of time.

5.12 MOTOR WIZARD RESULTS

When you know the main qualities of the windings and the power supply, you can easily get a wide range of machine performance characteristics, such as:

- Torque—speed curve
- Inductance profile
- Torque—Angle profile
- Winding flux linkage waveform
- Current waveform
- Co-energy

5.12.1 AIR GAP FLUX DENSITY

- The FEA results include:
- Winding flux linkage per phase
- Back EMF per phase
- Cogging torque
- Inductance per phase
- Static torque
- Magnetic field distribution
- Magnetic flux density distribution

5.13 EM WORKS 2D

Two-dimensional EM simulation software, EMWorks2D, allows you to rapidly test and refine your designs. With its user-friendly interface, EMWorks2D makes simulation a joyous and stress-free experience within the SOLIDWORKS ecosystem.

5.14   FEATURES AND CAPABILITIES

5.14.1   Simulation Speed and Seamless SOLIDWORKS Integration

- In the initial stages of product development, engineers test a great number of preliminary concepts and require a simulation software that can keep up with the fast-paced design process. EMWorks2D is a software that truly brings speed and responsiveness of 2-D EM simulation into your SOLIDWORKS project. Backed by decades of experience in CAD and simulation industry, EMWorks2D has been created to streamline and accelerate the design and simulation process. For maximum workflow efficiency, EMWorks2D automatically prepares the selected cross-section of a SOLIDWORKS assembly for a planar or axis-symmetrical EM simulation. Then, it instantly computes design parameters like magnetic and electric field distribution, forces, inductances, losses and capacitances. Relying on EMWorks2D, engineers can carry on with the product development more confidently and more rapidly.

5.15   APPLICATIONS

5.15.1   Transformer and Power Engineering Simulation Software

- High power
- Bus bars
- Insulators
- Induction heating
- Wireless power transfer

5.15.2   Multi-Physics

- Electro-thermal
- Magnetostructure
- Electrostructure

5.15.3   RF and Microwave Component

- Filter
- Transmission line

5.15.4   Electronic Design Automation (EDA) and Electronics

- Chip package board system
- Semiconductor
- Printed circuit board (PCB)
- Capacitors

5.15.5   Sensors and NDT/NDE

- Sensors
- Non-destructive testing
- Non-destructive Evaluations

5.15.6   Electrical Machines and Drives

- Motor
- Generators

5.15.7 ELECTRIC VEHICLE

- Battery charging

5.15.8 EM WORKS CUSTOMERS

- GENERAL ELECTRIC
- SAMSUNG
- FACEBOOK
- AMERICAN STANDARDS
- PFIFFNER
- AEROSPACE

5.16 POWER SI

Power SI can analyse PCBs and other PCB components in 3D in a short amount of time with high accuracy. It is possible to regulate issues with PI, signal integrity (SI), and EMC with the data from this analysis. In the analysis, numerous interpretations of EM couplings of wires are available. An interpretation of the fluctuations of supply systems depending on simultaneous switching of the signals is an important point to guarantee the stability and the function of a circuit. With Power SI frequency depending network parameter models can be extracted, which enable to visualize under-standingly complex coherences. A few years ago, the term "power-aware" was first used. Since then, the term "power-aware signal integrity" (SI) has been used to describe advanced SI simulation methods that can look at both signal noise and power noise at the same time (Figure 5.3). However, many individuals mistake the simulation technology used to assess simultaneous switching noise for the technology employed in these approaches (SSN) (Figure 5.4).

5.17 TRADITIONAL SI SIMULATION

The four components of a standard SI solution—pre-layout exploration, constraint construction, rule checking, and post-route verification—are all well recognized and understood by SI engineers.

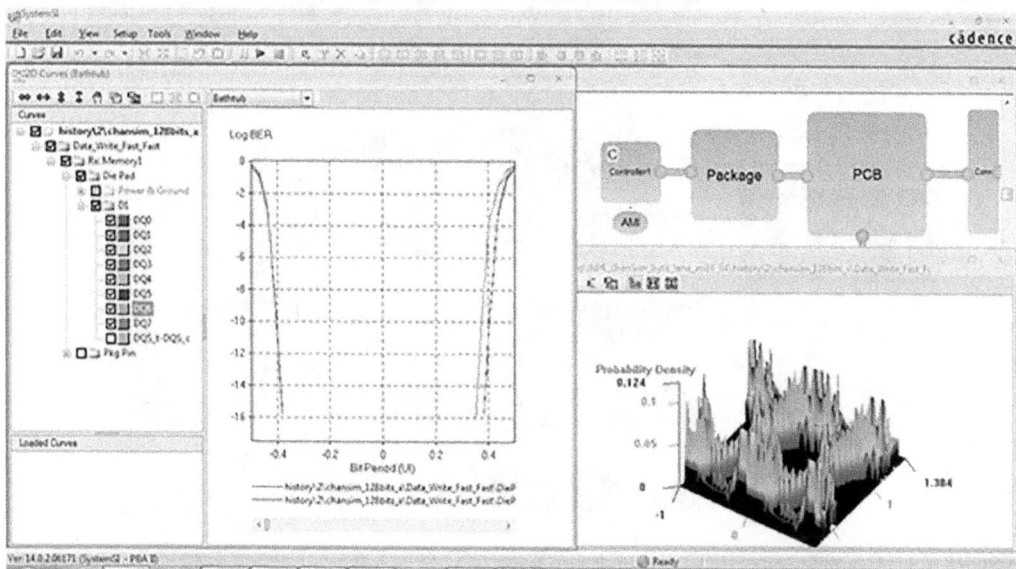

FIGURE 5.4 Power signal integrity (SI) simulation.

Time and frequency-domain simulation is used for both pre-layout exploration and post-route verification. The results help designers avoid signal degradation problems caused by things like delay, reflection, and crosstalk that happen when two things are connected. Constraints and rule checking may not use time/frequency-domain simulation at all, but they are based on the results of the field solver. The constraints (DRC) are established as quick-checking rules before or during the design, and the rule checking occurs after the route is completed to ensure that the DRC was properly applied. As the first product of its kind, the Allegro design flow uses constraints to optimize the design process. It's also widely implemented as a "design-to-signoff" solution in many modern technologies. The example flow is depicted in Figure 5.3.

The data rates of high-speed digital designs have gone from 33 MHz to over 4 Gbps on parallel buses, and there was no serial link (with a data rate of a single channel today at 28 Gbps or even higher) when the typical SI solution was designed 20 years ago. SI simulation and rule-checking methods employed in the solution do not take into account any power or ground net impacts on signals because of the lower signal speed of early high-speed devices (Figure 5.5).

Assuming ideal power/ground planes as reference, all rules or limitations and simulations of reflections and crosstalk ignore the interactions between power nets and signals. IBIS, which is the primary behavioural modelling approach for SI, does not include any specifications for how current behaves when it passes through power supply. In the early days of pre-layout tools, the device model symbols (driver and receiver) do not include any external power or ground connections. This is evidenced by the fact that these symbols no longer exist.

Since the beginning of parallel bus design, SSN has been observed, and SSN analysis is intended to capture this phenomenon. Instead of focusing on delays, reflections, and crosstalk with ideal reference planes, why didn't SI simulation tools add the ability to simulate SSN? The explanation for this is straightforward: SSN analysis consumes a large amount of computational power and time, neither of which the traditional SI tool can afford because it was designed to work on less capable systems. In addition, SSN analysis calls for the modelling of a power or ground network that is composed of planes. Traditional SI tools lacked the appropriate field solvers to complete this task successfully.

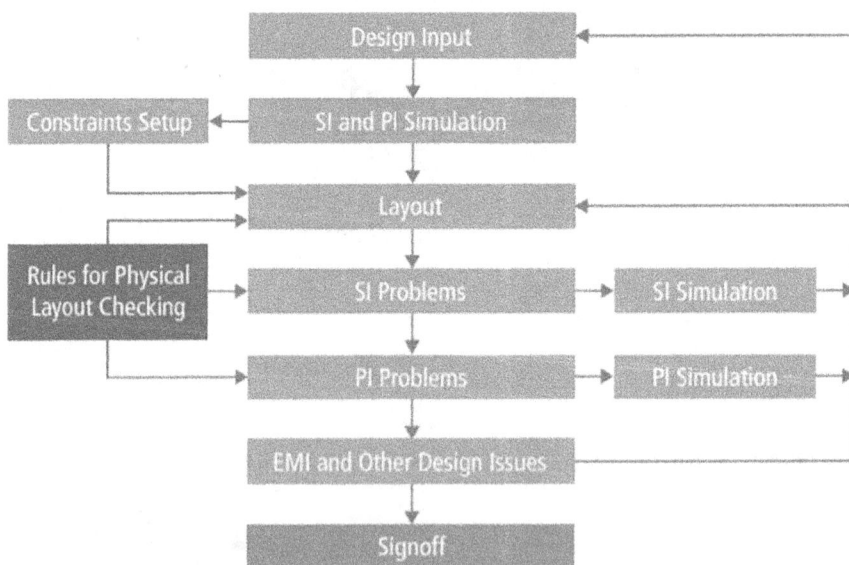

FIGURE 5.5 Example constraint—driven design flow.

5.18 POWER-AWARE SOLUTION

- Reflection: The bouncing of the plane could connect to reflections caused by an impedance mismatch in the trace.
- Crosstalk: The coupling that occurs between the power supply net and the trace results in the generation of fresh crosstalk. In addition, the crosstalk that results from the coupling of the trace is impacted by plane bouncing.
- Timing: The timing margin additionally shifts as a result of the additional reflections and crosstalk that are brought on by the power noise.
- Other: There is an effect on reflection and trace crosstalk caused by the signal's connection across the plane cavity.
- Power switching noise couples to signal nets.

A SI solution that takes power into consideration needs to be aware of all of these impacts and come equipped with the appropriate simulation and rule-checking techniques. These consequences also demonstrate that SSN only represents some of the problems that are caused by interactions between signal and power. The switching of numerous signals at the same time while maintaining the proper return current path is the primary focus of SSN. The drawing of current by other output buffers is the cause of power fluctuations in the affected signals. When the plane noise from signal via coupling through the plane cavity is injected into signals that share the same power source as the parallel bus but are not part of the multi-switching network, this creates a second power-aware situation. Voltage regulators on the PCB are another potential source of noise. This means that SSN simulation is just one component of a comprehensive approach to solving power issues.

Due to the fact that plane and signal interactions/couplings occur after routing is complete, a power-aware solution can only be developed at the rule checking and post-route analysis stages. In light of this, a comprehensive power-aware solution must provide:

- Powerful time-domain simulator that can model complex circuits (result of multiple signal and power nets) modelling using power and signal networks.
- Detailed simulation of I/O buffers.
- Fast screening for signal deterioration and power impacts is not SSN analysis.

5.19 POWER-AWARE SOLUTION AVAILABLE IN SIGRITY TECHNOLOGY

Let's examine if the Cadence toolset can deliver a power-aware solution based on the prior description. Two ways exist for post-route SI verification. Sigrity SPEED2000 contains a hybrid solver engine that does time-domain simulation directly at the layout level and accounts for all signal, power net, and through couplings. The Sigrity PowerSI tool possesses the same hybrid solver in the frequency domain and has the ability to extract S-parameter models from a large number of linked signals and power nets. The Sigrity T2B tool offers functions that can extract and generate equivalent I/O buffer models from transistor-based buffer circuitry. These models can then accommodate variations in the current flowing through power pins. In addition, the Sigritry SystemSI tool is responsible for establishing connections between all of the models (I/Os, packages, and boards) in order to carry out time-domain simulation at the system level. In the first method, T2B I/O models and SPEED2000 simulation are combined, and analysis is carried out directly at the layout level. The second method begins with the T2B I/O models and the Power SI S-parameter models, and then uses the Sigrity System SI tool to create a block-level interface that joins the two sets of models.

Since constraints are generated early in the design process, before planes and vias are in place, rule verification based on constraints cannot take into account power consumption. No power noise impacts can be recorded by doing physical rule testing at the post-route stage using the same mechanism that generates constraints. Consequently, any impedance, trace coupling, crosstalk, delay, or skew checking rules or tools based on an ideal plane assumption will overlook all power noise influence on signals. Using a rule-checking tool (that does not account for the power noise effect on signals) for design signoff is risky, as it does not and cannot discover potential design issues caused by power and signal coupling (Figure 5.6).

Figure 2's signoff process is analysed once again here. It's made abundantly obvious that the solution provides no signal-power interaction and that the derived rule set is predicated on the assumption of perfect planes.

For this reason, predicting signal and power coupling is not possible when crosstalk estimation is used. The reason for this is because the coupling between shapes and traces cannot be calculated analytically because the bulk of power nets use very big metal shapes. Having post-route validation that takes power considerations into account and provides quick results is essential for designers before they can hand off a design for simulation. Not to mention the fact that designers typically lack SI/PI expertise. Their inability to do and understand sophisticated signal/power analyses is a major limitation. An effective simulation approach for detecting troublesome signals using predefined signal compliance is the key to finding the answer.

Quickly analyse signal and power coupling via linear excitations to many signal nets with the help of the SPEED2000 technology's SI Metrics Check tool. It provides signal quality reports inclusive of all coupling noise while skipping lengthy non-linear simulations. SI Metrics Check helps designers uncover problems that standard post-route rule-checking misses because of its assumption of perfect reference planes. Using SI Metrics Check is demonstrated in Figure 4. Through-plane coupling can cause additional crosstalk, and it causes crosstalk to rise with increasing separation. There is a discrepancy between the finding and the general crosstalk rule, which states that crosstalk should decrease with increasing spacing, but in practice, this might occur due to through coupling, impedance mismatch, and plane noise. The issue is detected by SI Metrics Check's speedy screening. Nevertheless, it is undetectable by conventional methods of verifying compliance with rules (Figures 5.7 to 5.9).

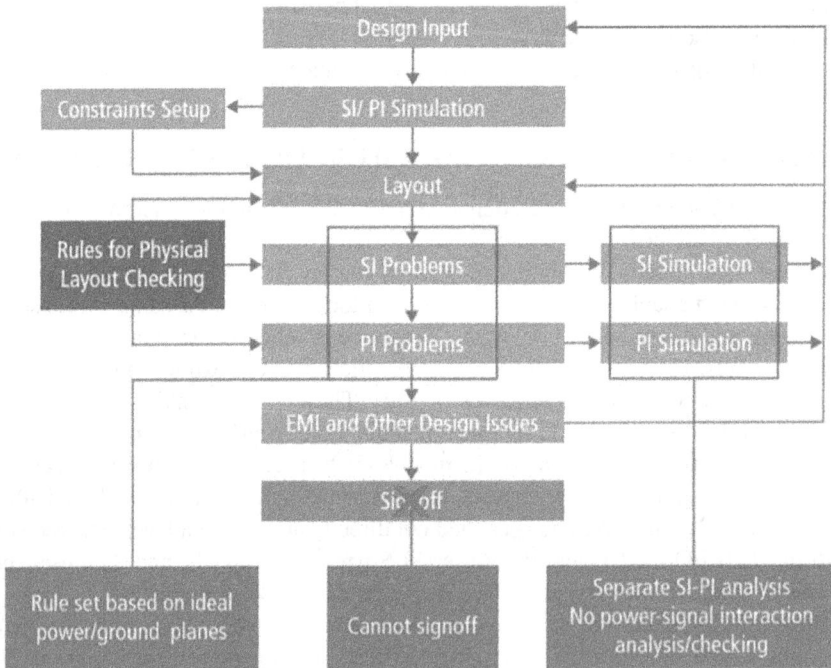

FIGURE 5.6 Flawed constraint—driven design flow.

FIGURE 5.7 Coupled nets.

Net name	NEXT Vmax (mv)	NEXT Vmin (mv)	NEXT pk-2-pk (mv)	FEXT Vmax (mv)	FEXT Vmin (mv)	FEXT pk-2-pk (mv)
SOD2_DDR3_DQ13	126	-133	259	138	-137	275
SOD2_DDR3_DQ12	124	-131	255	138	-134	271
SOD2_DDR3_DQ11	116	-118	234	122	-118	239
SOD2_DDR3_DQ15	102	-100	201	122	-118	239
SOD2_DDR3_DQ14	100	-99	199	117	-118	235
SOD2_DDR3_DQ10	100	-105	205	109	-106	216
SOD2_DDR3_DM1	95	-101	196	105	-103	208

FIGURE 5.8 Crosstalk before separating two nets with larger spacing.

Net name	NEXT Vmax (mv)	NEXT Vmin (mv)	NEXT pk-2-pk (mv)	FEXT Vmax (mv)	FEXT Vmin (mv)	FEXT 2-p (mv
SOD2_DDR3_DQ13	146	-152	298	152	-153	
SOD2_DDR3_DQ12	144	-148	291	148	-148	
SOD2_DDR3_DQ14	138	-148	286	148	-149	
SOD2_DDR3_DQ15	132	-134	265	141	-139	
SOD2_DDR3_DQ11	135	-141	276	134	-135	
SOD2_DDR3_DM1	123	-132	255	125	-126	

FIGURE 5.9 Crosstalk after separating two nets with larger spacing.

5.20 FEATURES

- Specifies requirements for integrated circuit (IC) board and package power delivery systems (PDS).
- Determines the effectiveness of EM coupling between geometries for more desirable parts.
- Uses a near-field ration display to predict energy leaks.
- Analyzes methods for decoupling capacitors and confirms effects of placement.
- Broadband modelling is made possible, and DC performance may be accurately characterized (patent pending).
- Runs what-if analyses on alternative design paths and stacking strategies.
- Works flawlessly with existing 3D IC packaging and board solutions.

5.21   BENEFITS

- Highly accurate even for designs with atypical plane structures.
- Extremely precise, especially for designs with non-standard plane structures; used in thousands of actual productions; proven through actual use.
- It is quick, and it supports several processors and distributed computing.
- Ability to accommodate elaborate layouts with multiple components and boards.
- Multiple methods of representation, including 3D fly-through.
- The ability to choose ports and manage abstractions with a high degree of freedom for the user.

5.22   CST STUDIO SUITE

The world's foremost technology and engineering firms rely on CST Studio Suite, a state-of-the-art EM and multi-physics simulation software suite. CST Studio Suite provides a wide variety of solvers for use in product design, analysis, and optimization. With the release of 3DEXPERIENCE R2018x, the electromagnetics analyst's role is now available, allowing CST Studio Suite to be incorporated into collaborative workflows on the 3DEXPERIENCE platform. When it comes to designing, analysing, and optimizing EM components and systems, CST Studio Suite is your go-to high-performance 3D EM analysis software package (Figure 5.10).

CST Studio Suite is an EM field solver that offers a unified interface for a wide variety of EM applications. Since the solvers can be combined for hybrid simulations, engineers now have more leeway in analysing complex systems with multiple parts. With the help of co-design with other SIMULIA products, EM simulation can be incorporated into the design flow and used to drive development from the outset of the process. EM analysis includes the performance and efficiency of antennas and filters, EMC/EMI, human exposure to EM fields, electro-mechanical effects in motors and generators, and thermal effects in high-power devices.

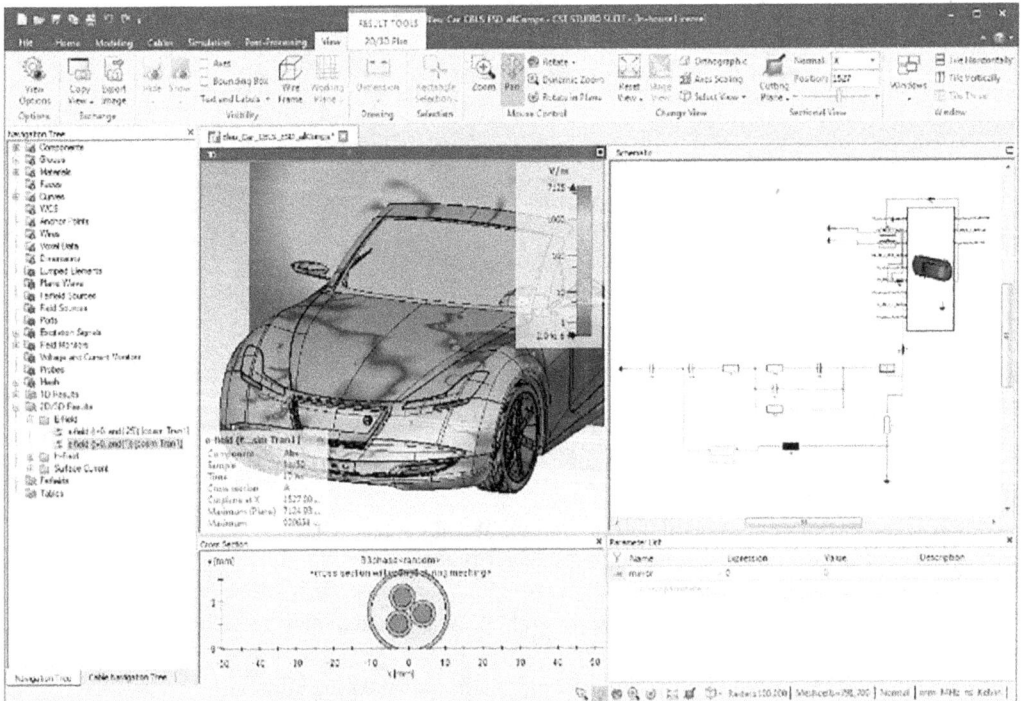

FIGURE 5.10   CST studio.

Major technology and engineering firms all over the world rely on CST Studio Suite. It allows for shorter development cycles and lower expenses, both of which are huge benefits in terms of getting products to market faster. With the help of simulation, virtual prototypes may be created and tested. Optimizing device performance, detecting and resolving potential compliance concerns early in the design phase, decreasing the need for physical prototypes, and lowering the cost involved.

5.23 DESIGN ENVIRONMENT

5.23.1 MODELLING

Creating and modifying simulation models is simple with CST Studio Suite's robust CAD interface. Because of the available import/export facilities, models can be imported from a broad variety of CAD and EDA programmes. Modifications made in CST Studio Suite can be instantly reflected in the SOLIDWORKS project, and vice versa, thanks to the parametric two-way communication.

5.23.2 MATERIALS

Complex non-linear material qualities are responsible for the signature EM effects observed in many practical contexts, including magnetics, photonics, and biological physics. Multiple material models found in CST Studio Suite make it possible to mimic a wide range of phenomena, such as plasmatic and photonic effects, ferromagnetism, secondary electron emission, and biological heating.

5.23.3 BODY MODELS

Many gadgets, especially those used in the medical and life sciences disciplines, take the human body's interaction with EM fields into account during the design process. The human body may be accounted for in CST Studio Suite since it features voxel-based and DESIGN ENVIRONMENT CAD-based body models with detailed internal structure and realistic EM and thermal properties.

5.23.4 MESHING

Integral to any successful simulation is a mesh that is both accurate and efficient. In order to improve the quality of the mesh in important areas of the model, the CST Studio Suite offers quick, automatic meshing, along with mesh refining and automatic adaption. CST Studio Suite's use of the Perfect Boundary Approximation (PBA)® keeps the speed advantages associated with a standard staircase mesh, even for models with billions of mesh cells, while allowing for the precise modelling of curved structures and complex CAD data.

5.23.5 SYNTHESIS

Several different synthesis tools are available inside CST Studio Suite, which can be used to automatically construct models of possible designs. There is the Array Wizard for antenna arrays and the Filter Designer 2D and 3D for planar filters. The software also provides access to other EM design tools from SIMULIA, such as Antenna Magus for antenna design and FEST3D for waveguide design.

5.24 SIMULATION

5.24.1 SOLVERS

The core of the CST Studio Suite is the solvers that are included with it. CST STUDIO SUITE provides the best-in-class solvers for EM simulation, including both general-purpose solvers like the time-domain and frequency-domain solvers that are applicable to a wide variety of scenarios and more specialized ones for applications like electronics, electron devices, motors, and cables. The EM solvers can be connected with the thermal and structural mechanics solvers to create a seamless workflow that allows for the simulation of multi-physics effects.

5.24.2 OPTIMIZERS

One of the best things about simulation is that it can be used to improve the performance of devices, tune them to meet strict requirements, or lower the cost of production. CST STUDIO SUITE has built-in local and global optimizers that can be used with all solvers to improve any of the model's design parameters.

5.24.3 POST-PROCESSING

Post-processing makes it possible to apply simulation results to a diverse variety of analyses, which may then be utilized to reproduce conventional measures and figures of merit. CST Studio Suite's post-processing templates provide solutions for common workflows such as eye diagrams for electronics, efficiency mapping for motors, and field analysis for MRI, in addition to versatile general-purpose templates for the creation of custom workflows. These templates can be found in the CST Studio Suite.

5.24.4 HYBRID AND SYSTEM SIMULATION

Multiple solvers can be used for a single simulation since each one excels at solving a specific type of problem. For instance, an investigation of the installation performance of a vehicle-to-vehicle (V2V) antenna on a car comprises both the time-domain solver and the efficient integral equation solver, as the former is more suited to simulating antennas and the latter to simulating big platforms like automobiles. Through the CST Studio Suite's system assembly and modelling (SAM) feature, several simulations can be merged into a single 3D model or linked automatic process. Similarly, the suite's hybrid solver task makes it possible to combine different solvers into a single simulation task.

5.25 APPLICATIONS

5.25.1 AEROSPACE AND DEFENCE

- Installed antenna performance
- Lightning strike and environmental EM effects (E3)
- Radar
- Co-site interference

5.25.2 CONSTRUCTION, CITIES, AND TERRITORIES

- Building shielding
- Cabling
- Lightning protection

5.25.3 ENERGY AND MATERIALS

- High-voltage components
- Generators and motors
- Solar panel optimization
- Transformers

5.25.4   INDUSTRIAL EQUIPMENT

- RFID
- NDT
- Motors and actuators
- Welding and lithography

5.25.5   LIFE SCIENCES

- MRI
- Implant safety
- Wearable devices
- RF diathermy
- X-ray tubes

5.25.6   HIGH TECH

- Antenna performance
- Microwave and RF components
- EMC
- SI/PI
- Touchscreens

5.25.7   TRANSPORTATION AND MOBILITY

- Antenna installed performance
- Cable harness
- Automotive radar
- Electric motors
- Wireless charging
- On-board electronics
- Sensors

5.26   BENEFITS

5.26.1   EM SIMULATION

- From statics to high frequency
- Specialized solvers for applications such as motors, circuit boards, cable harnesses, and filters
- Coupled simulation: System-level, hybrid, multi-physics, EM/circuit co-simulation.

5.26.2   MODELLING

- All-in-one fully parametric design environment
- Import/export wide variety of CAD and EDA files
- Wide range of complex material models analysis
- Powerful post-processing and visualization tools
- Built-in optimizers' high-performance computing

5.26.3 ELECTROMAGNETICS ANALYST ROLE ON 3DEXPERIENCE R2019x

- Create a shared workspace, invite coworkers, and everyone updates the same file in real time while keeping previous versions of the file under control.
- Decision-makers may get a feel of the outcomes and spend less time creating reports because of the lightweight display of model, mesh, scenario, and results.
- Direct access to geometry.
- Web-based portal to submit and monitor CST Studio Suite jobs from anywhere.
- Run CST Studio Suite in a "connected" mode, leveraging 3DEXPERIENCE.
- Capabilities for collaboration, visualization, version control, and knowledge capture.
- Supports all CST Studio Suite capabilities including continued openness to run any custom plug-ins or scripts.

5.26.4 MOMENTUM OF 3D PLANAR EM SIMULATOR

The Keysight Technologies product Momentum is the industry standard for 3D planar EM simulation of passive circuits. It uses frequency-domain method of moments (MoM) technology to correctly mimic complicated EM effects like coupling and parasitic, and it allows arbitrary design geometries (including multi-layer structures). By using reliable Momentum EM software, RF/MMIC designers, RF/High-Speed Board designers, RF Module designers, and Antenna designers can enhance design performance and increase manufacturing confidence.

In order to analyse the effects of physical layout on circuit performance, the Momentum EM programme is built into RFPro, ADS, Genesys, and Cadence Virtuoso enabling simple EM-circuit simulation setup. PCB tools like as Cadence Allegro, Mentor Expedition, and Zuken can use Momentum for RF analysis via ADS ODB++ connections.

5.27 BENEFITS OF MOMENTUM EM SOFTWARE

- Features both full-wave and quasi-static EM solvers for modelling RF passives, high-frequency (HF) interconnects, and parasitics.
- Faster simulation times thanks to efficient meshing, adaptive frequency sampling, and a threaded NlogN solver.
- Skin effect, substrate effect, thick metals, and numerous dielectrics are only some of the various EM effects that can be simulated.
- EM-Circuit co-simulation can be set up quickly and easily because of its integration with RFPro, ADS, Genesys, and Cadence Virtuoso.

5.27.1 MOMENTUM FEATURES

- With the help of Momentum, an EM-circuit simulator in three dimensions, RF and microwave engineers can greatly increase the scope and precision of their passive circuits and circuit models. Momentum is a must-have for passive circuit customization because of its capacity to analyse complex forms across multiple layers and take into account realistic design geometries when simulating coupling and parasitic effects.
- The S, Y, and Z parameters of general planar circuits can be calculated with Momentum. Momentum provides fast and precise analysis of microstrip, strip line, slot line, coplanar waveguide, and other circuit topologies. With the ability to replicate vias between layers, designers can more thoroughly and correctly model multi-layer RF/MMICs, PCBs, hybrids, and multi-chip modules (MCMs).
- The simulator uses the MoM technology, which is optimal for analysing planar conductor and resistor designs. Intended for use with ADS's harmonic balance, convolution, and

circuit envelope circuit simulators, the tool can assess multi-layer planar geometries and generate EM-accurate models.

- Circuit simulators are restricted to designs that may be built solely from existing circuit models and require explicit consideration of signal coupling. Since it takes signal coupling into implicit account, the Momentum EM simulator is able to surpass the constraints of existing models.
- Momentum RF is an alternative simulation mode that prioritizes speed over accuracy during simulation. When it comes to big geometries, which play a crucial role in the telecommunications sector, Momentum RF provides quick and easy solutions.
- Using Momentum RF is ideal for studying structures less than a half wavelength in size. By ignoring loss processes including space and substrate radiation, the quasi-static solution in Momentum RF produces precise results in a fraction of the time.

5.27.2 MOMENTUM EM SIMULATOR

- 32-bit and 64-bit solvers
- Internal ports aid in complex circuit modelling
- Side wall coupling emulates physical environment
- Box resonance calculation excludes ill-behaved S-parameters
- DC and low-frequency calculation model broad-bands accurately
- Adaptive frequency sampling enables faster frequency sweeps
- Edge mesh represents currents accurately while minimizing problem size
- Automatic arc recognition for faster simulation while preserving layout
- Optional SPICE model generator for integration with other simulation and design tools
- Integrated design flow lowers overall design cost
- Integration with ADS, Genesys, and Golden Gate Design Flows

5.27.3 MOMENTUM OPTIMIZER

- Geometry capture offers wide optimization choices
- Optimization techniques are right for the job

5.27.4 MOMENTUM VISUALIZATION

- Visualization reveals design possibilities
- Far-field plots give insight into antenna performance

5.28 APPLICATIONS OF MOMENTUM 3D PLANAR EM SIMULATOR

Momentum 3D planar EM modelling is especially valuable in the following design situations:

- **When Parasitic Coupling is Present:** Even though the circuit models are physically separated, they may nonetheless couple in unanticipated ways. Surface waves are bound to the interfaces between substrates and are excited by the presence of the correct parameters and frequencies. Similarly, stubs that appear to be suitably separated can inductively couple to each other due to a resonance condition. Parasitic coupling and radiation can be predicted using Momentum.
- **When a Circuit Model Does Not Exist:** Momentum is employed if no circuit model exists. For designers looking to do things like study tiny strip Y-junctions for which no model exists, Momentum is the way to go.

- **When There are Slots in Ground Planes:** The capacitance to ground of a spiral inductor can be decreased by having a section of the ground plane removed, or a via can be routed through the ground plane. There is an additional benefit to using Momentum because it can handle metal slots just as readily as metal patterns. For instance, coplanar waveguide circuit analysis is a breeze with Momentum.
- **When the Model Range is exceeded:** Every single model of a circuit simulator is constructed with a selection of control parameters that have a limited range (such as width, length, height, or dielectric constant,). Some models fail gradually, while others begin to produce major mistakes the moment the range boundaries are breached. Through the use of Momentum, designers are able to construct very realistic models that go beyond these built-in range constraints.

5.29   COMSOL

The simulation software COMSOL Multi-physics is being utilized in all areas of engineering, manufacturing, and scientific research because of its versatility. Starting with the definition of geometry, material qualities, and the physics that explain specific processes, this simulation platform supplies all the steps necessary in workflow modelling. Therefore, it is useful for model solving and post-processing in order to generate reliable findings. The software handles all different phenomenon of physics together, solves the mathematical and numerical model equations, and provide the model results.

The entire simulation process takes place in the same software environment only. The user needs to learn only one workflow, no matter about the physics involved or the nature of the experiments. Depending on this scenario, the user may want to model single physics, one-way multi-physics, or fully coupled multi-physics. The COMSOL software offers all the three in a single platform. The EM module provided by the software includes AC/DC, RF, wave optics, ray optics, plasma, and field of semiconductor.

5.29.1   FEATURES

- To develop, design, and to perform experiments virtually is both cost and time efficient.
- We can get accurate results, which represent the real world by studying multiple physical effects on one model.
- If the users want to add any equation in their experiments, which also includes users' own equation, the software allows and permits to work with it.
- User-friendly tool for deploying and building simulation apps.

5.30   NEWFASANT

The newFASANT is used in both academic and industrial settings as EM simulation software for research, investigation, and design of a wide variety of applications. With support for metallic, dielectric, and magnetic materials, newFASANT has been parallelized for multiprocessors/GPUs systems. The software has its own GUI, geometric editor, parallelized multi-layer meshers for surfaces and volumes, visualization, and parameter optimization tools. In the article "Altair Acquires newFASANT, Further Expanding High-Frequency Electromagnetics Portfolio©" the firm Altair announced to the public that it had acquired the high-frequency electromagnetics programme newFANSANT.

Global technology firm Altair has announced the acquisition of newFASANT, a pioneer in computational and high-frequency electromagnetics. Altair is a provider of solutions in the fields of product development, high-performance computing, and data analytics. With this move, Altair has established itself as a frontrunner in the high-frequency EM realm, a crucial technology for enabling high-speed digital communications in fields like the Internet of Things (IoT), mobile phones, cellular networks and connected devices, V2V, radar, and radio.

5.31 APPLICATIONS

1. NewFASANT's solutions address a wide range of EM problems in areas such as:
   - Antenna design and placement.
   - Automotive V2V/ADAS.
   - Infrared/thermal signatures.
2. This software can be used to study and alter such factors as RCS, ISAR, RCS maps, antennas, antenna interactions with platforms, periodic structures and FSS, microstrip circuits, Doppler effects, and radio propagation channel characteristics.

5.32 MOM MODULE FOR ANTENNA AND RCS DESIGN AND ANALYSIS

This MoM component is a computer programme that performs a 3D analysis of on-board and complex antennas, as well as analyses for EM compatibility (ECM) and RCS. The new FASANT's MoM module includes CBF as one of the helpful techniques for lowering the computational cost of rigorous computations. It achieves remarkable performance because it employs the MM in tandem with the FMLMP. To further improve effectiveness, we also incorporate macrobasis functions (CBFs) and the domain decomposition technique. This module's parametric optimizer makes it possible to design antennas and other structures with minimal wasted effort. Functionality for dynamic and parametric modelling is included as well.

5.33 BOUNDARY ELEMENT METHOD (BEM) AND FINITE ELEMENT METHOD (FEM)

The Helmholtz equation can be solved with the help of the BEM, and open BEM is a collection of open-source programmes that can do just that. This package implements anti-symmetric and half-space problems as well as codes for dealing with exterior and interior problems in two and three dimensions. The new and improved features include analytical solutions for verification, meshing for 2D and anti-symmetrical problems, working within objects, and close-surface meshing. In acoustics, numerical methods like the FEM, the BEM, and finite differences are frequently used. In order to function, the BEM needs only the domain boundary, which reduces the amount of data and processing time needed.

5.33.1 FEATURES

- Microphones, loudspeakers, calibration systems, noise barriers, object-based scattering, and so on can all be characterized with BEM.
- When the domain is unbounded, BEM offers distinct advantages over the other methods for harmonic analysis of stationary systems.
- Adaptive BEM and FEM meshing contributes to an increased level of confidence in the results of EM simulations.

5.34 FEM MAGNETICS

The FEM magnetics, sometimes known as FEMM, is an open-source software programme that uses finite element analysis to find solutions to EM issues. The programme addresses 2D planar and 3D axisymmetric linear and non-linear harmonic low-frequency magnetic and magneto static problems in addition to linear electrostatic problems.

The FEKO Software Suite was developed for the purpose of conducting analyses on a wide variety of EM issues. EMC analysis, antenna design, microstrip antennas and circuits, dielectric media, scattering analysis, cable modelling, and a great many other applications are only a few of the many possible uses. In the frequency domain, EM simulation techniques are applied through the use of XFdtd, whilst in the time domain, Maxwell's equations are solved through the use of FDTD. This indicates that the computation of the values of the EM field advances at specific points in time over the course of the calculation. A single run of the programme can produce broadband results using the time-domain approach, which is one of the many advantages of using this method.

Algorithm used: Lua

FEMM is divided into three parts:

- The interactive shell (femm.exe), a multiple document interface pre-processor and a post-processor for the various types of problems solved by FEMM.
- The triangle.exe programme, which is the segmentation of the solution region into a large number of triangles.
- The solvers (fkern.exe for magnetics and for electrostatics). Each solver takes a set of data files that describe the problem and solves Maxwell's equations to obtain values for the desired field throughout the solution domain.

The interactive shell supports the scripting language Lua through its incorporation of the language. At any given time, there is only one instance of Lua executing, building, and analysing a geometry, as well as evaluating the post-processing findings, which may all be accomplished by a single instance of Lua, which simplifies the production of a wide variety of batch runs. In addition, the Lua programming language is used to parse all of the edit boxes in the user interface. This allows equations or mathematical expressions to be typed into any edit box in place of a numerical value.

Scripting and batch processing capabilities have been added to FEMM through the utilization of the Lua extension language. Scripts written in Lua can be executed by the interactive shell by using the Open Lua Script option found on the Files menu. Alternatively, Lua instructions can be typed directly into the Lua Console Window. The scripting language Lua is available for free and in its entirety online.

5.34.1 Feko

Algorithm Used: MoM, MLFMM, FEM

FEKO is a comprehensive 3D EM simulation software suite based on state-of-the-art CEM methods that enable users to solve wide range of EM problems such as antenna design, antenna arrays, antenna placement, RFID, RCS, pattern synthesis with characteristic mode analysis, EMC/EMI, lightning and cable analysis, bio-electromagnetics, etc.

The MoM technique forms the basis of the FEKO solver. Other techniques such as the multi-level fast multi-pole method (MLFMM), the FEM, uniform theory of diffraction (UTD), geometrical optics (ray launching), and physical optics (PO) allow the solving of electrically large circuits. Several approximations and acceleration techniques are available to efficiently simulate complex EM problems.

5.34.2 XFdtd

Algorithm: FDTD Method

XFdtd, a full-featured EM simulation solver, outperforms competing approaches as the number of unknowns rises. Antenna design and placement, biomedical and SAR, EMI/EMC, microwave devices, radar and scattering, automobile radar, and many other applications may all be tackled with the help of XF's full-wave, static, bio-thermal, optimization, and circuit solvers. In addition, it's compatible with Remcom's ray-tracing solutions to provide you complete simulation power across the whole EM spectrum.

5.35 KEY FEATURES

- **2D Sketcher with constraints**: Intuitive grid/object snapping and a constraint system allow for quick creation of complex shapes.
- **Feature history for objects**: Modelling operations are chained together on each object, creating an editable history for each model in your project.
- Circuit element optimizer determines optimal values for lumped circuit elements connected directly into the EM simulation mesh.
- PrOGrid project optimized gridding simplifies grid creation by considering multiple aspects of a project to optimize the grid for both accuracy and runtime.
- XStream GPU acceleration for CPUs and GPU clusters enables calculations to finish in minutes as compared to hours.
- Unlimited memory support for problems exceeding 60 GB and billions of cells.

5.35.1 USERS

- GE General Motors
- Honda
- Honeywell
- IBM
- Texas Instruments

5.36 EM PRO SOFTWARE

Model and simulate the EM effects of high-speed and RF/microwave components in three dimensions with electromagnetic professional (EMPro). It integrates with ADS, the leading RF/microwave and high-speed design environment, and provides a state-of-the-art platform for design, simulation, and analysis.

5.37 EMPRO DELIVERS THE FOLLOWING KEY CAPABILITIES

5.37.1 MODERN, EFFICIENT 3D SOLID MODELLING ENVIRONMENT

Using EMPro, you can easily import CAD files and draw freeform 3D buildings. In EMPro, you may make 3D models, specify their material characteristics, run simulations, and examine the outcomes.

5.37.2 TIME- AND FREQUENCY-DOMAIN SIMULATION TECHNOLOGY

In EMPro, you may use the same FEM simulator found in ADS to analyse 3D structures. Oftentimes employed in RF and microwave fields, FEM is a frequency-domain technology. The finite difference time domain (FDTD) simulator is useful for large-scale electrical problems like antennas and various signal integrity assessments.

5.37.3 INTEGRATION WITH ADS PARAMETERIZED

EMPro is used to design 3D parts, which are then imported into ADS for use in the layout editor. The 2D design and 3D EM part can be simulated as a whole in ADS's 3D FEM simulator.

5.37.4 EMPro Simulation Capabilities

EM simulation can be carried out in a number of distinct ways, each of which has its own set of advantages in specific contexts. FEM and FDTD are the most well-known methods for 3D EM simulation. EMPro includes both of these technological advancements.

5.38 FINITE ELEMENT METHOD

FEM is a frequency-domain method that can deal with structures of any shape, like bond wires, conical-shaped vias, and solder balls or bumps where the structure changes in the z-dimension. Dielectric blocks and substrates of finite size can be simulated by FEM solvers.

By using volumetric meshing, in which the problem space is partitioned into thousands of smaller regions, FEM is able to represent the field in each sub-region (element) with a local function. Each tetrahedron is made up of four equilateral triangles, so the geometric model is automatically subdivided into a vast number of them. The term "finite element mesh" is used to describe this group of tetrahedra. The Keysight Technologies, Inc. FEM simulator has a wide variety of problem-solving tools at its disposal, including both direct and iterative solvers as well as linear and quadratic basis functions. Both EMPro and ADS offer access to the identical FEM simulator. EMPro enables distributed frequency sweeps in FEM and remote simulation.

5.39 FINITE DIFFERENCE TIME DOMAIN

The FDTD technique, like the FEM technique, is predicated on a systematic volumetric sampling of the electric and magnetic fields throughout the entire domain. Different from the tetrahedral cells found in FEM meshes, the rectangular (Yee) cells used in FDTD meshes are what give the latter its distinctive appearance. The FDTD technique iteratively updates the field values as time steps forward, simulating the movement of EM waves as they travel across a medium. This means that a very broad range of frequencies can be covered by a single FDTD simulation. Antenna design, microwave circuits, bio/EM impacts, EMC/EMI difficulties, and photonics are just some of the many applications that benefit from FDTD's simplicity, robustness, and ability to include a large range of linear and non-linear materials and devices. As a parallel technique, FDTD is well-suited to the processing power of today's CPUs and GPUs. As an added bonus, EMPro can run FDTD simulations with remote connections and distributed ports.

5.40 TYPICAL EMPRO APPLICATIONS

5.40.1 IC Packages

The consequences of packaging, such as wire bonding and solder balls/bumps, can have a significant impact on the performance of an RFIC, MMIC, high-speed IC, or system-in-package (SIP). In the past, designers would need to use a third-party 3D EM tool to sketch and analyse packages before importing the data back into the IC or SIP circuit-design environment. When combining 2D circuit layouts in ADS with 3D package architectures made in EMPro, the result is a highly efficient design process. This permits 3D EM modelling and circuit simulation to be used in a simplified design path for the IC, packaging, laminate, and module.

5.40.2 Multi-Layer RF Modules

Multi-layer ceramic or laminate dielectric material with RF passive components placed between layers is the common construction method for RF modules. Inaccurate solutions to such dielectric brick constructions cannot be found using planar EM simulators because they assume infinite dielectric layers and do not take edge proximity fringing into account. It would take a lot of time for a separate 3D EM tool to replicate the RF circuit layout macros that are used to draw the integrated RF components. The optimal solution for these uses is full 3D EM simulation incorporated into the circuit-design procedure.

5.40.3 RF COMPONENTS

High-frequency characterization of 3D components and connectors used in RF board designs is necessary. Resonators and other components like them can be easily damaged by their interactions with the PC board's traces and vias. Complete 3D EM simulation is possible by first creating and simulating such 3D components in EMPro, and then combining them with a board layout in ADS.

5.40.4 AEROSPACE/DEFENCE

FDTD simulation is able to manage the massive challenges typically encountered in aerospace and defence applications thanks to its enormous capacity. Such applications as radar cross-section analysis and antenna placement optimization in aeroplanes are possible because of FDTD.

5.40.5 PCB DESIGN

PCB traces are increasingly being investigated as RF transmission lines due to the rise in data speeds. High-speed signal integrity analysis, EMI, and PCB interfaces to connectors and packages all benefit from the addition of 3D EM technologies, which are not available in 3D planar simulators like Keysight Momentum.

5.40.6 RF AND HIGH-SPEED CONNECTORS

Today, it is necessary to emulate not only RF connectors but also high-speed connectors like SATA and HDMI due to the enormous data rates involved (in the terabits per second range). With EMPro, designers can construct high-frequency S-parameter models of connections and cross-verify them with FEM and FDTD simulators for increased confidence in the correctness of 3D EM modelling. The models can then be incorporated into a design kit for ADS, which can be disseminated and installed into ADS as a connector library for signal integrity analysis and design of high-speed serial channels.

5.40.7 ANTENNAS

The gain, radiation patterns, and impedance of an antenna must all be accurately simulated before it can be built. Multiple elements, a huge physical size, and a potential increase in the demand for broadband simulations all provide new difficulties when modelling phased array antennas. In order to help you model your antenna designs, EMPro offers both FDTD and FEM simulation tools.

5.40.8 EMI/EMC ANALYSIS

EMI difficulties are a primary source of product delays as more electronics are incorporated into smaller packaging. Engineers can use EMPro to confirm that their designs adhere to common EMC standards like FCC Part 15, CISPR 22, and MIL-STD-461F by simulating the radiated emissions of electronic circuits and components.

5.41   EMPRO ENVIRONMENT OVERVIEW

5.41.1   Geometry Modelling

The robust EMPro drawing environment allows for the creation of arbitrary 3D structures. There is also the option to import structures from external CAD systems. ACIS, IGES, DXF, STEP, ProE, SolidWorks, ODB++, and many others are among the supported file types. Imported or hand-drawn 3D models can have their material qualities modified by simply dragging and dropping attributes from a comprehensive material library onto the models. Once the materials have been given, they will be remembered for future CAD imports, making setup time for simulations even shorter.

5.41.2   Port/Sensor Setup

The object's source impedance can be defined when a voltage or current source is applied to it. In addition, plane wave and Gaussian beam sources from the outside world can be implemented in FDTD simulations. There are also sensors in place to measure field strengths, field directions, and field voltages and currents in both the near and far fields. Sensors for measuring SAR and HAC compliance can be added. Waveguide ports can be configured for use in finite element modelling.

5.41.3   Mesh Setup

The items' geometries are used to create a rectangular mesh for FDTD simulations. The first mesh is generated automatically, saving both time and effort. When using a fixed point meshing technique, the mesh is adjusted to fit within the borders of the object without any manual adjustment. The user has the ability to fine-tune the mesh and to further adjust the balance between precision and speed. For finite element modelling, a tetrahedral mesh is created in a similar method. Through an autonomous adaptive refining process, the mesh is tailored for speed and accuracy.

5.41.4   Simulation Setup

You can opt for either FEM or FDTD simulation. As part of FEM, a frequency strategy is developed. You can choose between direct and iterative solvers, as well as multi-core CPU support for parallel processing. Selecting a multi-core CPU or GPU card(s) with multi-threading capacity can speed up FDTD calculations.

5.41.5   Results of Post-Processing and Viewing

Data on the S-parameter can be visualized and compared to those of other projects. The fields can be seen in a variety of ways, including cut-plane views. Linking up when there are tiny spaces between conductors that would normally go unnoticed, a bird's eye view can help spot any mistakes in the design. Powerful post-processing and simulation automation can be achieved with Python programming.

5.41.6   Export 3D EM Component to ADS Library

Packages, shields, connections, and surface mount components are just a few examples of 3D structures that can be stored in an ADS library and used in both 2D and 3D designs for simulation purposes. By setting parameters in EMPro, sweeps and optimization in ADS are made possible. Both the solo component and the component combined with the ADS layout are simulated with the same FEM simulator in EMPro and ADS.

5.42 MOTOR-CAD

Motor-CAD is a patented programme created by Motor Design Ltd. that provides a comprehensive EM and thermal analysis package for electric motors and generators. Back in 1999, it was first launched.

Brushless permanent magnet (BPM) motors, outer rotor (OR) BPM motors, induction motors, permanent magnet (PM) DC machines, switching reluctance (SR), synchronous (SYN), and claw pole (CP) motor modules are all on hand.

Predictions of electrical and EM performance can be trusted because of a built-in ultra-fast finite element module (EMag).

The thermal module (Therm) optimizes the cooling system by combining lumped circuit and finite element thermal computations. Natural convection (Totally enclosed non-ventilated, TENV), forced convection (Totally enclosed fan cooled, TEFC), via ventilation, water jackets, submersible, wet rotor and wet stator, spray cooling, radiation, and conduction are all modelled as potential means of cooling. The modelling capabilities extend to many different kinds of dwellings.

To create novel layout ideas, the Lab module collaborates with the EMag and Therm modules. The duty cycle/drive cycle transient outputs and efficiency mapping are available in a matter of minutes.

In terms of model construction and attainable accuracy, thermal analysis of electric devices is considered to be more difficult than EM analysis.

Considering the current trend toward smaller and more efficient electrical machinery, thermal analysis is becoming increasingly vital. This is especially the case in the aerospace and automotive industries, where compactness, lightness, and efficiency are paramount in the development of new machinery. Motor-CAD facilitates early on in the machine design process the consideration of EM and thermal components of the design.

Computational fluid dynamics is another method that can be used for thermal modelling. It has been proven that Motor-CAD can produce just as accurate results in a fraction of the time.

Motor-CAD includes robust internal support for multi-parameter sensitivity analysis. Improve cooling by making educated design decisions based on the results of a sensitivity analysis that provide light on the most important limits on heat dissipation (Figure 5.11).

Motor-CAD uses tested formulas, some created in-house at Motor Design Ltd. and others culled from heat transfer literature, to automatically compute the resistances to conduction, convection, and radiation. The thermal transient performance takes a few seconds to calculate, whereas the steady-state performance is calculated instantly. The temperature of each node in the network, the relative power flow through thermal resistances, and the limitations on heat transfer based on these values are all analysed using the thermal schematic depicted in Figure 5.12.

When conducting a thermal transient calculation in Motor-CAD, the user is given many configuration options. In a straightforward transient analysis, the loss or load remains constant across time.

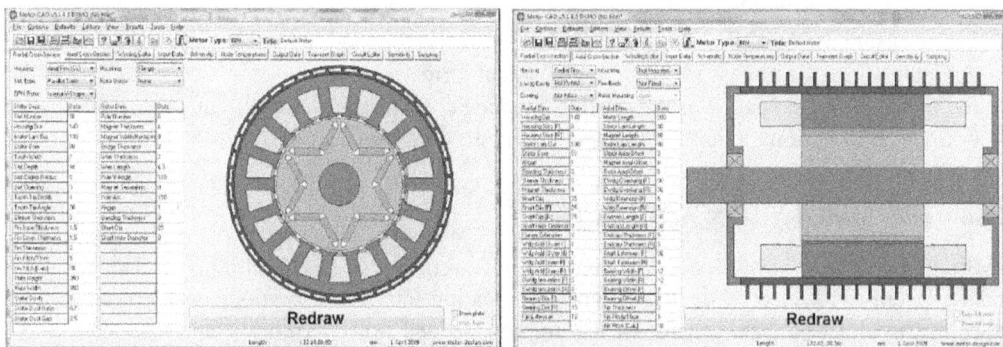

FIGURE 5.11 MOTOR-CAD radial and axial cross sectional.

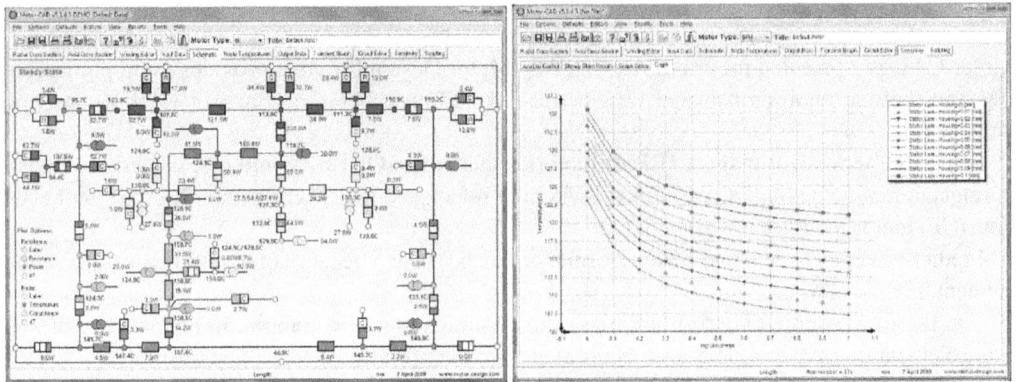

FIGURE 5.12   Motor-CAD steady-state schematic and nodal temperature plot.

Motor-built-in CAD's models can determine the frequency and temperature dependence of iron loss as a function of speed (winding resistance variation with winding temperature, and magnet loss in flux with magnet temperature). By establishing a loss and speed fluctuation over time, the user can represent any duty cycle load. The user of a permanent-magnet-driven machine can choose between two options for controlling the machine's speed, torque, and current over time. Based on the hypothesis that the torque is proportional to the current and flux, Motor-CAD provides models that allow the loss to scale with load. Motor-CAD includes a duty cycle editor where the duty cycle can be specified.

5.43   FERRITEMAGNETIC DESIGN TOOL

Application programme, Soft power—Magnetic Design Tool, for visualizing and making available material data of the DMEGC soft magnetic materials and EPCOS Ferrite cores. It's meant to be a resource for people who work on electronic circuits that incorporate inductive components. It is a reliable source of data, covering such topics as the fundamentals of winding design parameters. In addition, you can view their visual representations and other digitally stored information. A comprehensive explanation of the programme's features may be found in its user guide.

The distinct features this MD Tool are as follows:

1. Simulations based on user-defined core parameter.
2. Transmittable power adjusted for skin and proximity effects (from the wire calculation menu).
3. Calculation of the distortion factor (third harmonic) under specific circuit conditions at various temperatures.
4. Calculation of core loss as a function of signal form.
5. Display of impedance Z as the relation between core impedance and frequency.
6. Specification of wire thickness as per American wire gauge (AWG).

5.44   EMI SIMULATION ADVANCES

EMI and EMC have always been key concerns for designers of electrical devices. The RF noise radiates from every electric device, potentially creating internal disruptions or interference with other nearby devices. Next-generation RF, microwave and millimetre-wave applications like 5G wireless, the IoT and high-speed interconnects have resulted in an increasing interest in EMI simulation. As the IoT expands and embedded electronics become more ubiquitous, it will result in an

FIGURE 5.13 An example of a printed circuit board layout.

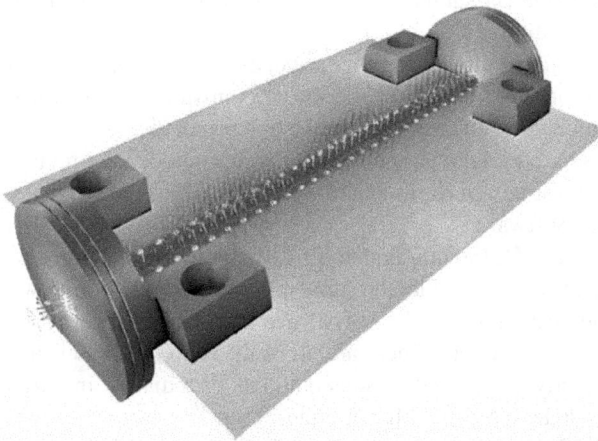

FIGURE 5.14 An example of an edge launch connector test circuit board.

unprecedented level of co-location scenario complexity. "It's expected that 5G will need to utilize higher frequency spectrums in the millimetre-wave range where the quality of the signal can be more vulnerable than the conventional lower frequency applications against the noise from outside the circuit, and the performance can be easily degraded by impedance mismatch, insertion loss and crosstalk," says Jiyoun Munn, technical product manager of RF, COMSOL, Inc. (Figure 5.13).

Every connected device can potentially increase the amount of EM radiation and RF noise, and mitigating against the problem with shielding is a costly and complex solution that most companies want to avoid. That means designers will need to engineer that interference out of the devices that they are developing earlier in the process—relying on EM simulation tools to do so (Figure 5.14).

5.45 BUZZ AROUND EMI SIMULATION

In the past, designers mostly didn't consider EMI until they presented the finished product for compliance testing. "Then they would go to the anechoic chamber and discover they don't meet the requirements for radiated emissions, or they are susceptible to internal or external inferences," says Larry Williams, director of technology at ANSYS. "EMI compliance is a process, and you have to build that process into your design activities. If you wait until the end to see if the product will radiate, you are making a mistake."

The advent of the IoT and the creation of smaller and smaller electronic components are creating greater opportunities for EMI issues, even in products where this wasn't previously even a consideration. The cost of mitigating these problems is a greater concern, particularly when it comes to lower-cost assemblies and products.

- IoT devices are smaller and relatively inexpensive, so manufacturers don't want to spend a lot of money on enclosures and shielding. Components also come from a variety of suppliers, which makes it difficult to manage coexistence or to anticipate how the system will perform in advance.
- "You need to do a lot of EM 3D simulation to understand a product well," says Minoru Ishikawa, market development manager at Mentor Graphics, a Siemens company. "You can have a virtual prototype that helps you identify these issues sooner."
- Simulation software providers have been expanding their EMI simulation capabilities through acquisition. Siemens, for example, acquired EM simulation software company Infolytica, which was folded into the Mentor Graphics mechanical analysis division. Altair acquired FEKO, while Dassault Systems acquired CST, a package of tools for EM-centric multi-physics simulation.
- Simulation aids designers by letting them realistically evaluate several design ideas, construct physical prototypes based on the most promising concepts, and examine diverse boundary conditions without risking damage to a prototype.
- The prototype is based on numerical results that accomplish the required performance in fewer design and test iterations, as stated by COMSOL's Munn, whose company specializes in simulation software.
- The difficulty with EMI, though, is that it's difficult to simulate an entire system. "You'd have to simulate the entire product the same way it would function as when you plug it into a 110v outlet and turn on the power," says Dave Kohlmeier, senior product line director of Mentor's HyperLynx product. "There are all of these digital signals and analogue signals running concurrently, producing various amounts of radiation in various directions, and it's almost impossible to recreate that stimulus."
- Kohlmeier says rules-based verification has been used to predict and address the inadvertent antennas created in these systems, an approach that can improve product performance even as the ability to run complex simulation has increased.
- "That's especially important in these IoT devices," says Swagato Chakraborty, director of engineering for HyperLynx Advanced Solvers. "You need to make sure the intended antennas perform correctly, and catch those unintended antennas."
- That activity is difficult to profile. Mentor can simulate the effect of shielding on a given system, for example, but a rules-based approach can help determine where an antenna may have been accidentally created (Figure 5.15).

5.46 SMALLER ELECTRONICS, BIGGER CHALLENGES

The fact that electronics components are getting smaller, and being incorporated into smaller products, also poses an EMI challenge. Smaller and embedded electronics are especially prone to challenges involving crosstalk, coupling, interference, and impedance mismatching—and adding shielding to these components is more complicated. "Smaller embedded devices experience the electro-thermal effects in an exaggerated manner because of the physical space limitations," Munn says. "Coupling between devices and signal lines in smaller embedded devices are prone to signal quality degradation due to the impedance mismatching caused by device miniaturization."

Engineers working on high-speed interconnects, for instance, confront a number of design challenges, including the need to match the connector's impedance to that of the remainder of the transmission line while adhering to geometry, size, and transmission limits. Keeping the impedance constant gets more difficult as the frequency increases, as imperfections in the geometry or choice of materials become more pronounced. "Analysing crosstalk is particularly challenging because in

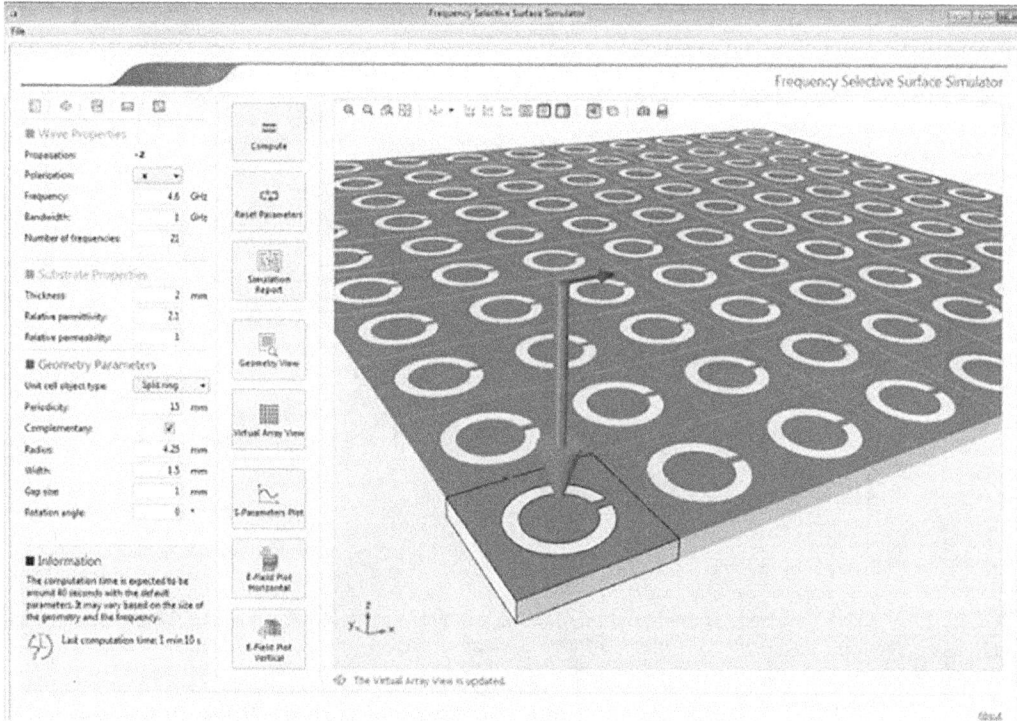

FIGURE 5.15 The frequency selective surface simulator app from COMSOL.

situations such as extreme heat or blizzard conditions, for example, you now must take into consideration not only the electromagnetic properties, but heat transfer and structural deformation to begin to get an accurate representation of the physics in the real world," Munn says. "Also keep in mind that the impact of the electro-thermal effects will most likely have a much greater impact on a smaller embedded device than it would with a larger piece of equipment."

EMI issues also often result from signal integrity or PI issues—which are also more complicated in smaller devices. "There's a lot of energy, and it has to go somewhere," Williams says. "You need good design flow that includes signal and power integrity and that can help reduce EMI problems." With smaller electronics, the need for high-speed access to memory and high performance, combined with compressing the electronics into a small space, works against EMI compliance. In the case of ANSYS, the company has built features into its software that can automate and customize design flows around signal and PI. "We've built into our tools the ability for engineers to receive a printed circuit board layout and bring those models in, and automate those parts of the process," Williams says.

5.47 MULTI-PHYSICS APPROACH

To successfully miniaturize an electronic product and reduce EMI challenges, engineers need to be able to effectively and efficiently represent the real world as closely as possible. This involves using a multi-physics simulation solution that enables modelling of components and phenomenon that you are not able to physically test in all cases.

COMSOL's AC/DC Module enables simulation of low-frequency EM and electro-mechanical components; the company's RF Module is designed to model high-frequency EM phenomena and optimize electromagnetics devices such as antennas and micro/millimetre-wave circuits.

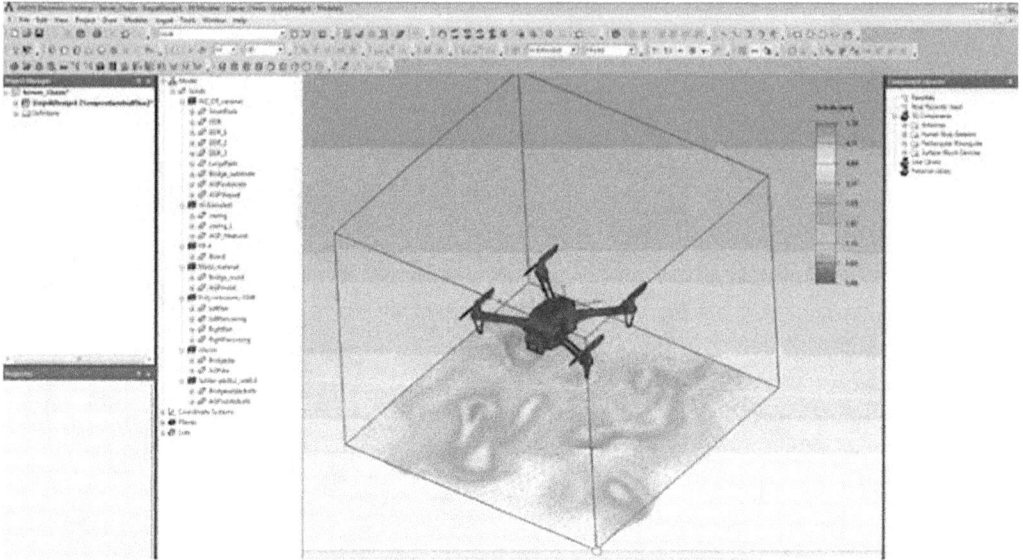

FIGURE 5.16 ANSYS Icepak is a simulation tool that can perform electrothermal and standalone thermal analyses of electronic designs.

The AC/DC Module and RF Module can be used for EM analysis. When this simulation is set up, a user can move on from a single physics electromagnetics analysis and add physics from one of COMSOL's add-on products, such as the Heat Transfer Module, CFD Module, Structural Mechanics Module, etc. "Compared to traditional electromagnetic modelling, multi-physics simulation enables the end user to include effects such as temperature fluctuations, structural deformations, and fluid flow to perform virtual prototyping that truly describes the real world," Munn says. The ANSYS HFSS simulation tools can be combined via the ANSYS Electronics Desktop to streamline workflows and reduce development time, while rapidly testing and validating EMI and EMC compliance. Engineers can use EMI simulation tools to analyse and optimize trade-offs among speed, bandwidth, signal and PI, EMI and thermal performance in these devices. "If you are putting these electrical devices inside automobiles, you want to know that they are reliable for temperature, vibration and shock," Williams says. "We can link through the ANSYS Workbench to other tools for thermal, mechanical, structural and vibration simulation and that's critical" (Figure 5.16).

5.48   EVOLVING CHALLENGES

As the number of connected devices increases, EMI-related design challenges also will grow. Collaboration across departments and specialties will be important, particularly when it comes to designers who have little or no experience running these simulations. "When designers have access to the correct resources, they can work freely with others within and outside of their company. Our ability to compete and thrive in the current highly competitive market will depend on the efforts of employees from across departments working together," Munn says. Still, the computational scale required to simulate an entire electrical system would be massive, which is what has made that approach impractical so far. A combination of rules-based verification and multi-physics simulations will be required.

New EMC and EMI models and approaches will need to be developed that take into consideration the greater vulnerability of the IoT, larger concentrations of co-located devices, greater occupation of unlicensed frequency bands, and unpredictable and highly dynamic interference scenarios. Smart devices have to juggle and balance multiple wireless signals, from the Wi-Fi coming from

FIGURE 5.17 An example of a printed circuit board layout.

a nearby access point to a GPS signal coming from space. "You have these multiple systems co-existing, and you don't want the device to destroy its own ability to operate," Williams says. "Now there are 5G systems coming and high-speed downloads. High frequency is a bigger challenge for designing devices, and having advanced field simulation is key. You can't cut corners" (Figure 5.17).

5.49 EMI SOFTWARE MEET-UP STANDARDS

5.49.1 EMC/EMI STANDARDS FOR PCB DESIGN

As early as 1979, the United States Federal Communications Commission (FCC) adopted some of the first EMC regulations. Later, the European Community set its own EMC standards, which were the foundation for the EU's later regulations. A number of other standards groups have also defined a variety of standards with a focus on certain applications or industries; these include the IEC, ISO, SAE, IEEE, and CISPR. In Europe, standards from the IEC and CISPR are mandated; however in the United States, IEEE norms are preferred. Antenna calibration tests specifically are based on the IEEE standards. MIL-STD EMC regulations are among the most demanding in the world, and are defined by the United States military.

5.49.2 COMPLYING WITH EMC STANDARDS

Companies that release devices or products that don't follow the rules could get a warning or be fined a lot of money. In addition to being unsafe, failing to follow EMC standards can hurt a company's image. Incorporating EMC considerations during design can help reduce the risk of having to pay fines after a product has been manufactured. When designing products, conformity to EMC regulations is achieved by considering EMI from two angles. Immunity from EMI design: Create a gadget that can operate in environments with high levels of EMI. Usually, this starts with the right plan for sticking up and routing. Trying to suppress both conducted and radiated EMI: Make a device that puts out the least amount of radiation possible. The layer stack, the strategy for grounding, and the placement of the components all play a role here. For PCB design, a near-field probe is used to test against EMI/EMC standards. A near-field probe was used to measure radiated EMC. Some ways to make sure you meet EMC standards are that every designer should follow some basic design rules to make sure their boards pass even simple EMC checks.

5.49.3 STICKUP, POWER, AND GROUNDING

EMC compliance starts with your layer stack. The most effective way to reduce the likelihood that your board will be affected by EMI is to design it with a ground system that has a low inductance. In order to reduce the amount of loop inductance caused by multi-layer circuit boards, the ground plane should be positioned directly below the signal layers. A decrease in the signal-to-noise ratio is the result of interference in low-level signals. Therefore, it is wise to send these signals on a lower tier of the network. These traces need to be placed between two ground planes, and your power plane needs to be placed below the ground plane that is at the bottom of your stack. The power plane's proximity to the ground plane results in a strong capacitive connection. Any noise or conducted EMI on the power plane will be radiated to the ground plane instead of affecting the communications. Maintaining tight coupling is your responsibility when passing signals from a lower layer to a higher one. To maintain the coupling to the reference plane, a neighbouring parallel via must be inserted between the ground plane and the surface layer.

5.49.4 Incorporating Shielding

The use of shielding in a strategic manner is another method to protect your board from radiated EMI. Your board's radiated EMI will be reduced as a result of this measure. In order for a wireless device to continue sending and receiving signals, simply move the antenna outside the shielding. Protecting delicate parts and wiring by enclosing them in a Faraday cage made of grounded shielding is the quickest and easiest method. Unfortunately, this isn't a fit for every layout or part. This suggests that a more sophisticated shielding strategy may be necessary. A grounded through fence around the border of the board is just as secure as using a uniform ground plane in the centre.

Shielding material for a PCB

- Shielding can for suppressing radiated EMI
- Mixed-signal layout and routing

Devices that operate wirelessly and modify digital data are by definition mixed-signal devices. Thus, the board's digital, low-frequency analogue, and RF analogue components should be physically separated. Individual spaces for these components are necessary in the ground level, but the ground level as a whole should remain uninterrupted. The grounding electrode should be positioned so that return signals from one portion do not pass under the other.

5.49.5 Bypass/Decoupling Capacitors

Lastly, bypass or decoupling capacitors can be used as high pass filters. If there is any remaining noise, connect one of these to the power pin of the active component and a grounded via. It filters in two directions thanks to this. As a result of the low loop inductance, any switching noise or ripple will have a quick path back to ground, mitigating its impact on the active components and the radiation they produce. EMI-generated noise will also benefit from a swift return to ground. Following EMI/EMC guidelines for PCB design necessitates careful consideration of the board's layout and signal integrity. Altium Designer includes all of these necessary features, plus many more, in a streamlined interface.

5.50 DOMAIN STANDARDS

5.50.1 International Special Committee on Radio Interference (CISPR)

As a branch of the International Electrotechnical Commission (IEC), CISPR developed guidelines to shield radio reception from EMI induced by the usage of electrical and electronic equipment. The CISPR specifies criteria for measuring radiated and conducted interference, as well as immunity. In order to standardize the measurement of emissions, CISPR creates these guidelines. CISPR standards include:

5.50.2 Basic Standards

- CISPR 16-1-1- Specification for radio disturbance and immunity measuring apparatus and methods—Part 1-1: Radio disturbance and immunity measuring apparatus—Measuring apparatus
- CISPR 16-1-2- Specification for radio disturbance and immunity measuring apparatus and methods—Part 1-2: Radio disturbance and immunity measuring apparatus—Coupling devices for conducted disturbance measurements
- CISPR 16-1-3- Specification for radio disturbance and immunity measuring apparatus and methods—Part 1-3: Radio disturbance and immunity measuring apparatus—Ancillary equipment—Disturbance power
- CISPR 16-1-4- Specification for radio disturbance and immunity measuring apparatus and methods—Part 1-4: Radio disturbance and immunity measuring apparatus—Antennas and test sites for radiated disturbance measurements
- CISPR 16-1-5- Specification for radio disturbance and immunity measuring apparatus and methods—Part 1-5: Radio disturbance and immunity measuring apparatus—Antenna calibration sites and reference test sites for 5 MHz to 18 GHz
- CISPR 16-1-6-Specification for radio disturbance and immunity measuring apparatus and methods—Part 1-6: Radio disturbance and immunity measuring apparatus—EMC antenna calibration
- CISPR 16-2-1- Specification for radio disturbance and immunity measuring apparatus and methods—Part 2-1: Methods of measurement of disturbances and immunity—Conducted disturbance measurements
- CISPR 16-2-2- Specification for radio disturbance and immunity measuring apparatus and methods—Part 2-2: Methods of measurement of disturbances and immunity—Measurement of disturbance power
- CISPR 16-2-3- Specification for radio disturbance and immunity measuring apparatus and methods—Part 2-3: Methods of measurement of disturbances and immunity—Radiated disturbance measurements
- CISPR 16-2-4- Specification for radio disturbance and immunity measuring apparatus and methods—Part 2-4: Methods of measurement of disturbances and immunity—Immunity measurements
- CISPR 16-4-2-Specification for radio disturbance and immunity measuring apparatus and methods—Part 4-2: Uncertainties, statistics, and limit modelling—Measurement instrumentation uncertainty
- CISPR 31—Database on the characteristics of radio services

5.50.3 International Organization for Standardization (ISO)

- ISO is an international standards organization that develops standards in all technology areas requested by industry. Members are the national standard bodies. ISO standards on EMC are most relevant for automotive EMC issues
- ISO 7637 Road vehicles—Electrical disturbances from conduction and coupling
- ISO 7637-1 Road vehicles—Electrical disturbances from conduction and coupling—Part 1: Definitions and general considerations
- ISO 7637-2 Road vehicles—Electrical disturbances from conduction and coupling—Part 2: Electrical transient conduction along supply lines only
- ISO 7637-3 Road vehicles—Electrical disturbances from conduction and coupling—Part 3: Electrical transient transmission by capacitive and inductive coupling via lines other than supply lines

- ISO/TS 7637-4 Road vehicles—Electrical disturbances from conduction and coupling—Part 4: Electrical transient conduction along shielded high-voltage supply lines only
- ISO/TR 7637-5 Road vehicles—Electrical disturbances from conduction and coupling—Part 5: Enhanced definitions and verification methods for harmonization of pulse generators according to ISO 7637
- ISO 11451 Road vehicles—Vehicle test methods
- ISO 11451-1 Road vehicles—Vehicle test methods for electrical disturbances from narrow band electromagnetic energy—Part 1: General principles and terminology
- ISO 11451-2 Road vehicles—Vehicle test methods for electrical disturbances from narrow band electromagnetic energy—Part 2: Off-vehicle radiation sources
- ISO 11451-3 Road vehicles—Vehicle test methods for electrical disturbances from narrowband radiated electromagnetic energy—Part 3: On-board transmitter simulation
- ISO 11451-4 Road vehicles—Vehicle test methods for electrical disturbances from narrowband radiated electromagnetic energy—Part 4: Bulk current injection (BCI)
- ISO 11452 Road vehicles—Component test methods
- ISO 11452-1 Road vehicles—Component test methods for electrical disturbances from narrowband radiated electromagnetic energy—Part 1: General principles and terminology
- ISO 11452-2 Road vehicles—Component test methods for electrical disturbances from narrowband radiated electromagnetic energy—Part 2: Absorber-lines shielded enclosure
- ISO 11452-3 Road vehicles—Component test methods for electrical disturbances from narrowband radiated electromagnetic energy—Part 3: Transverse electromagnetic mode (TEM) cell
- ISO 11452-4 Road vehicles—Component test methods for electrical disturbances from narrowband radiated electromagnetic energy—Part 4: Harness excitation methods
- ISO 11452-5 Road Vehicles—Component test methods for electrical disturbances from narrowband radiated electromagnetic energy—Part 5: Strapline
- ISO 11452-6 Road Vehicles—Component test methods for electrical disturbances from narrowband radiated electromagnetic energy—Part 6: Parallel plate antenna
- ISO 11452-7 Road Vehicles—Component test methods for electrical disturbances from narrowband radiated electromagnetic energy—Part 7: Direct radio frequency (RF) power injection
- 8/25/2020
- ISO 11452-8 Road Vehicles—Component test methods for electrical disturbances from narrowband radiated electromagnetic energy—Part 8: Immunity to magnetic fields
- ISO 11452-9 Road Vehicles—Component test methods for electrical disturbances from narrowband radiated electromagnetic energy—Part 9: Portable transmitters
- ISO 11452-10 Road Vehicles—Component test methods for electrical disturbances from narrowband radiated electromagnetic energy—Part 10: Immunity to conducted disturbances in the extended audio frequency range
- ISO 11452-11 Road Vehicles—Component test methods for electrical disturbances from narrowband radiated electromagnetic energy—Part 11: Reverberation chamber
- ISO 10605 Road Vehicles—Test methods for electrical disturbances from electrostatic discharge
- ISO 13766 Earthmoving Machinery—Electromagnetic Compatibility
- ISO 14982 Agricultural and forestry machinery—Electromagnetic compatibility—Test methods and acceptance criteria
- ISO 21609 Road Vehicles—EMC guidelines for installation of aftermarket RF transmitting equipment

5.51 SUMMARY

With a simulation model, you may test out your ideas on a reliable digital model of the real system. Simulation modelling is computer-based and uses algorithms and mathematics, as opposed to physical modelling, which involves producing a scale replica of a building. Whether in 2D or 3D, simulation software offers a live, interactive setting for the evaluation of computational models in action. As the number of products available in the market continues to grow, EM components will become increasingly important. New electrical and electronic gadgets are causing disruption in traditional businesses like transportation and communication, while technological advancements are creating totally new markets in sectors like medical technology, renewable energy, and building materials. Visionary design and rapid, adaptable development cycles are required to keep up with these advancements. The computational effort needed to analyse a complicated model accurately can be decreased by combining several levels of simulation.

BIBLIOGRAPHY

1. Konstantinos Baltzis, "The Finite Element Method Magnetics (FEMM) Freeware Package: May It Serve as an Educational Tool in Teaching Electromagnetics?" *Education and Information Technologies*, vol. 15, pp. 19–36, 2010, DOI: 10.1007/s10639-008-9082-8.
2. https://www.emworks.com/
3. www.cadence.com/content/cadence-www/global/en_US/home/tools/pcb-design-and-analysis/pc-design-flows/allegro-sigrity-power-aware-si.html
4. https://www.3ds.com/fileadmin/Products
5. Chang-Yu (David) Huang, "Design of IPT EV Battery Charging Systems for Variable Coupling Applications," PhD Thesis in Electrical and Computer Engineering, the University of Auckland, 2011.
6. Keysight EMPro 3D Electromagnetic Modeling and Simulation Environment Integrated with your ADS Design. https://www.keysight.com/in/en/products/software/pathwave-design-software/pathwave-em-design-software.html
7. CST Studio Suite by 2019 Dassault Systems. https://www.3ds.com/products-services/simulia/products/cst-studio-suite/

6 Instruments for EMI Measurements

6.1 INTRODUCTION

Electronics must be consistent with their electromagnetic environment, which may involve electromagnetic disturbances or sounds. The system performance should be designed such that these disruptions have a negligible effect. In addition to this, it is necessary that the sounds released by the system into the surrounding environment have as little of an effect as possible on the functioning of any other electronic systems that are located nearby. The entire category of these kinds of occurrences can be categorized as electromagnetic interference (EMI), which can be either conducted or radiated. Radiated noise travels through space in waves while conducted noise travels along electrical conductors, wires, printed-circuit traces, or any electronic component. These radiations spread via cables and wired connections, such as those found in telecommunications and electrical wiring. In terms of their nature, these emissions can either be continuous (i.e., emitting constantly at a certain frequency) or discontinuous (non-constant, occurring sporadically). Radiated noise, on the other hand, can travel through the air or through free space in the form of magnetic fields or radio waves. Radiated noise can typically be reduced by installing metal shielding that serves to contain magnetic fields or radio waves within the equipment enclosure. Radiated emissions are measured with spectrum analyser (SA), EMI receiver, and measuring antenna during electromagnetic compatibility (EMC) testing. It is routine trend in EMC testing to monitor conducted emissions using an EMI receiver connected to an impedance stabilization network (ISN) installed in the test environment. In addition, the performance of these devices should be as expected, without any vital performance being compromised inside their surroundings. Critical computer systems and licenced radio services must be protected from interference by any additional devices that may be present in that environment.

6.2 NEED OF INSTRUMENTATION FOR EMI MEASUREMENTS

The tremendous part of today's electronic applications make use of small devices that incorporate hundreds of thousands or even millions of active or passive components in order to carry out several tasks all at once. To produce such tiny devices is one of the greatest challenges every electronic hardware engineer or manufacturer encounters, and it is also one of the largest hurdles. The majority of the time, in addition to being vulnerable to those spurious sounds, an electronic equipment is also capable of creating or radiating these noise signals into the surrounding environment. It is possible to measure the EMI that is caused by the device in order to get an estimate of both its immunity and the quantity of interference it causes. The utilization of man-portable electronic equipment is required in order to carry out specific EMI measurements, which is one of the primary responsibilities.

The most important forms of EMI measurement require the following quantities:

- The occupation of radio frequency spectrum
- Path loss
- Ambient noise
- General EMC "Troubleshooting".

DOI: 10.1201/9781003362951-6

However, the man-portable equipment has several disadvantages as follows:

- It is bulky and difficult to carry and because of this, it is difficult to transfer the components, as well as build and dismantle them.
- It demands for the manual adjustment of the operational controls, as well as the human processing of a significant amount of measurement data. There is a very limited amount of automation and built-in processing that is utilized.
- It works best in a controlled laboratory or other protected setting. As such, it is not built to withstand the rigours of the field.
- Some current needs cannot be addressed because the available measuring parameter ranges are too narrow, and this situation will only worsen in the future.

Because of the aforementioned factors, determining emissions levels is a costly endeavour that necessitates the use of jet fuel. As a result, not only is the price higher, but the response time and data collection time are also affected. It also leads to a lack of information, making it hard to survey or assess EMI conditions in a specific field setting properly. This enables the opportunity for the invention of alternative instruments. The characteristics of interference waves, as depicted by EMI measurements, are essential for the engineer to solve compatibility issues. They may be classified into two main groups, and each one has two subgroups. Quantities related to the circuit (1) and quantities related to the field (2) constitute the first two categories. The next two categories are: (1) quantities in the time domain and (2) quantities in the frequency domain. The first two kinds are linked together by a transducer transfer characteristic of some kind, including the antenna factor or the coupling coefficient. Mathematical transforms, often the Fourier Transform and its inverse, express the connection between the second set of classes.

6.3 ESD GENERATOR

Different Eletro Static Discharge (ESD) testers and setups are often used for the same system-level ESD testing adding to the complexity of predicting the ESD performance of the design under different test conditions. It is widely accepted that there is a requirement for comprehensive system-level ESD simulation methodology, which can efficiently adapt to different test scenarios and provide detailed simulation results highlighting any weakness of ESD protection schemes. One of the key tests performed is the human body metal discharge test, for which ESD generators are used. The IEC-61000-4-2 has specifications for the waveforms to be used for the ESD testing of the device under test (DUT). The standard describes the characteristics of the discharge waveforms, for different levels of charge. See Table 6.1 for details on the rising time, current at 30 ns, and current at 60 ns, as well as the onset of the first peak.

To simulate ESD tests efficiently, circuit models are required so they can be easily incorporated during the design phase. Modelling an ESD generator as a circuit is a well-studied problem and

TABLE 6.1

Characteristics of Discharge Waveforms in ESD Generator

Level	Indicated Voltage	1st Peak Current of Discharge ±15%	Rise Time ±25%	Current at 30 ns	Current at 60 ns
1.	2 KV	7.5A	0.8 ns	4A	2A
2.	4 KV	15A	0.8 ns	8A	4A
3.	6 KV	22.5A	0.8 ns	12A	6A
4.	8 KV	30A	0.8 ns	16A	8A

several published works propose circuit models for ESD generators. However, since the waveforms from commercial testers can vary widely, while meeting the IEC-61000-4-2 conditions, it is very difficult to manually determine the parameters of a circuit model to match the given waveform which has a complex shape. Moreover, once the number of elements increases beyond few elements, it quickly becomes an intractable optimization problem.

However, a DNN-based methodology is a data-driven technique and requires large training datasets. To address these challenges in determining ESD generator circuit model parameters, this work proposes to automatically generate circuit models for ESD generators using a reinforcement learning (RL) technique. This process can be performed using a previously identified circuit template and the discharge current waveform from system-level ESD measurement or from 3D electromagnetic simulation. The unique contributions are as follows:

- Showing how to formulate circuit model generation as an RL problem and applying RL techniques to determine a circuit model for a given discharge waveform.
- Comparing different RL techniques and showing how to select the appropriate one.
- Demonstrating the use of domain knowledge to simplify the problem so that RL or other machine learning techniques can be efficiently applied.

The circuit template of an ESD generator is shown in the Figure 6.1. The Rd, Cd, and Cb represent the ESD generator discharging components, and Rs, Cs, and Ls represent the ground impedance. The L_{tip} is the electrode inductance and R load represents the target load impedance. Various ranges can be set for the ESD generator model parameters shown in Figure 6.1.

Table 6.2 shows the values for the ESD generator parameters.

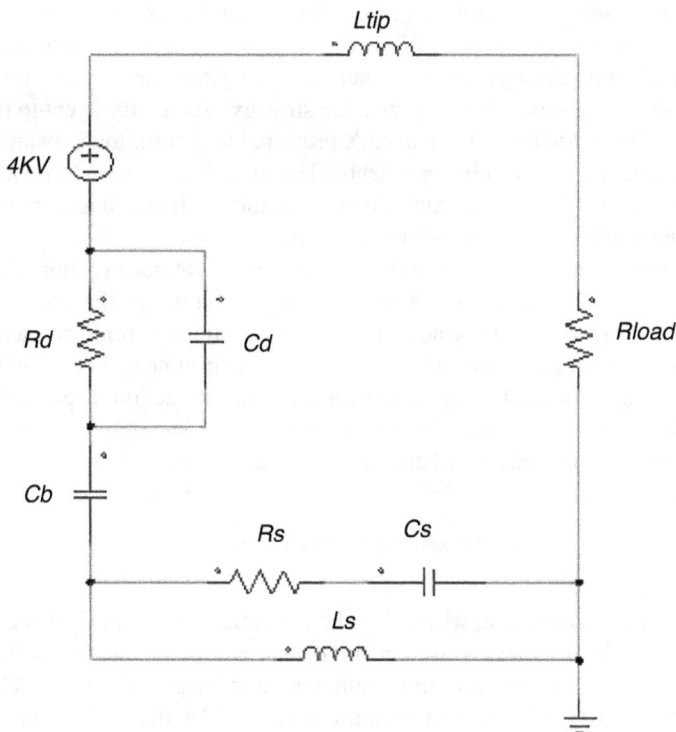

FIGURE 6.1   ESD generator model.

TABLE 6.2
ESD Generator Model Parameters

Parameter	Range
Rd	50–550 Ω
Cd	0–60 pF
Cb	0–600 pF
Rs	50–550 Ω
Cs	0–100 pF
Ls	0–10 uF

6.4   RL FOR CIRCUIT MODEL OPTIMIZATION

Most machine learning solutions are based on supervised learning, where many labelled data are used to train the model to approximate a function. RL on the other hand is based on self-learning in an environment to maximize a reward function. RL is based on the idea that an agent can learn a good strategy to solve a problem by running several experiments. RL has recently proven itself to be an effective and efficient approach to solve goal-driven problems. AlphaGO is an RL-based computer programme, which plays the board game of GO developed by DeepMind.

An agent is an entity, which takes actions to solve a given problem. Action(a) is a set of steps or moves that can be made by the agent. An action can be multidimensional. The environment(E) is the space in which the agent can move, i.e., take actions. A state(S) determines the current position of the agent in the environment. A reward(R) is the value returned to the agent by the environment when an action is taken by the agent from the current state. Note that the reward is given in return of each action taken by the agent. A policy(π) is the strategy used by the agent to take actions from a given state to maximize the reward. An agent's projected long-term total reward from the state S while acting in accordance with a policy π is denoted by the value-function ($V\pi(s)$). The action value ($Q\pi\ (s,\ a)$), which is analogous to the value-function, is the predicted long-term total benefit from being in state S and doing action a, according to a policy π.

An agent in a given state s interacts with the environment by taking an action a, receives a reward R, and potentially moves to a new state. A good action, which brings the agent closer to the goal potentially receives a higher reward. One of the key assumptions in reinforcement learning is the first-order Markov decision process for the agent and the environment. This essentially means that the next state of the agent depends only on the current state and action taken by the agent. At any time step t, when the agent takes an action, it receives a reward Rt.

The total discounted expected reward can be written as:

$$Gt = Rt + \gamma Rt + 1 + \gamma 2Rt + 2 + \ldots + \gamma kRT$$

Agent interaction with environment, where T is the last time step. In reinforcement learning, each episode is T time step long, which is usually the logical end of interaction of the agent with the environment or when a pre-determined maximum number of steps are reached. The parameter γ is the discount factor to give smaller weight to future rewards. The idea of discounted reward is used so that the agent can pay more attention to rewards, which are near term compared to longer term ones. Therefore, the objective of the agent should be to increase the profit of the discount applied to the reward at each time step.

6.5   RL-BASED METHODOLOGY FOR CIRCUIT-OPTIMIZATION

There are several different Deep Reinforcement Learning (DRL) algorithms and approaches, such as the actor-critic method, Deep Q Network (DQN), and Deep Deterministic Policy Gradient (DDPG). We looked at several RL algorithms and the two most promising algorithms for this application were DQN and DDPG. By using a deep neural network, DQN provides an equivalent of such action-value function Q(DNN). By training this DNN with several examples, the neural network becomes efficient in predicting the action to be taken so that the total reward is maximized. DDPG is an RL algorithm, which is used on continuous action-space and continuous state-space as opposed to discrete spaces. DDPG combines the actor-critic approach with DQN so that the algorithm can deal with continuous action-space.

With large action space in DQN, because of the discrete nature of the algorithm, it was difficult to optimize the circuit parameters efficiently. Conversely, the DDPG strategy has better performance than DQN because the action space is continuous, and thus the adjustment of observed state could be applied to all parameters in each time step efficiently. Based on the above comparison between DQN and DDPG, it is clear that DDPG has better performance.

The state space is defined as follows:

$$s = \begin{bmatrix} p_1 \ p_2 \dots p_i \end{bmatrix} T$$

The parameter pi indicates the ith circuit element value of an ESD generator. For DQN, the action a in action space A is selected with the ϵ-greedy policy. The ϵ-greedy policy selects an action based on either the action, which has the maximum $Q(S, a)$, or by doing a random exploration of the action space. For DDPG, the action is determined from the policy network, which outputs a single value representing the action for the given current state.

The reward function of an action is determined by the waveform using first maximum peak1, first minimum peak2, and their respective time points time1 and time2. This has been chosen because the magnitude of the first peak and its shape are the most important features of ESD discharge current waveform.

The slew1 and slew2 are determined as:

slew1 = peak1/time1
slew2 = peak1-peak2/(time2-time1)

The reward function is formulated as:

$$\sum \begin{bmatrix} 1 - (\mid x_{\text{golden}} - x \mid)/x_{\text{golden}} \end{bmatrix} \times Wx$$

where, $x \in$ [peak1, peak2, time1, time2, slew1, slew2] and $\sum Wx = 1$.

In the function, $x$golden is each of the slew values of current discharge waveform while the constants Wx indicate weights of the elements of the reward function. The above reward function considers only the first maximum peak and the subsequent minimum. To experiment with the second maximum peak, it is added to the reward function the DDPG algorithm is run to verify the performance.

The only method to increase the performance of a training RL model is to take advantage of the knowledge of the ESD testing domain to simplify the issue. Simply changing the resistors and capacitors in parallel in a synchronous manner is an additional method of leveraging knowledge from the circuit domain to simplify the issue. This will reduce the number of independent parameters to tune for the RL model and hence improve the performance. Therefore, constraints can be added to reduce the action-space of the RL algorithm based on the domain knowledge of the problem, which will reduce the training time.

In particular, DDPG RL method, which is able to optimize continuous input variables to match the given output waveform from measurement or 3D EM simulation of the ESD generator, is proven the best in getting the optimal solution efficiently.

ESD Pulse Generator 10 kV

Models	Key Features	Specifications
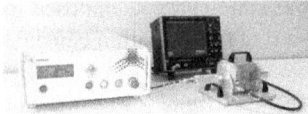 Type: **PESD-ECSS-12K** Standard: **ECSS-E-ST-20-07**, new version in preparation	1. Charging voltage: 20 V to 10 kV, positive only. 2. Voltage pulse shape: Damped oscillatory waveform. 3. Storage/working temperature: 5–50°C/20–40°C	Different configurations of the generator and its peripherals have been tested, revealing enhanced repeatability and a simplified testing approach.

ESD 300 KV Generator

Type: **PGESD300K-DP**
Standard: **MIL-STD 331C**

1. Nominal charging voltage: 300 kV ±1%
2. Voltage range: 15 to 300 kV
3. Maximum charging current: 12.5 mA.
4. Storage/working temperature: 5–50°C/15–45°C

The system can be used to perform following tests:

a. Discharge between a charged electrode and a grounded EUT.
b. Discharge between a charged EUT and a grounded electrode
c. Charged EUT for precipitation static (P-static) test.

Automotive ESD Generator 12 KV

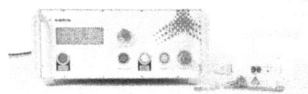

Type: **PESD-12K-EQ-IC09**

1. Output Voltage: 1–12 kV, positive and negative polarities.
2. Storage/operating temperature: 5–50°C / 10–45°C

Vehicle components' resistance to interference from high-voltage ignition systems can be tested with this generator.

6.6   LINE IMPEDANCE STABILIZATION NETWORK (LISN)

Under various test settings, the line impedance of the power supply will inevitably vary. Decoupling the device terminal from the power line is necessary for EMI noise measurements; hence, a line LISN is used.

In actual application, issues arise from the LISN's impact on the input power line's impedance (from short circuit to open circuit and also highly inductive input power line). Designers examine what happens whenever they plug in various converters. First of all, the voltage input converters have big input capacitors that might resonate with a LISN inductor. Secondly, the LISN inductor shifts the operational point of current input converters that also have an inductor at the input. The voltage drop limitations (between the source and the DUT), the fidelity test (input output (I/O) identical stable characteristics), and the repeatability for an LISN are some other standard tests that manufacturers often do.

The LISN's input impedance is stabilized due to a low-pass filter on the power line and a band-pass filter on frequencies between 0.15 and 30 MHz; both filters are compliant with the CISPR-22 standard. In a common-mode (CM) or differential-mode (DM), noise measurement is added to the LISN through an EMI separator. An LISN is made using a ferrite toroidal inductor since it is inexpensive and space-efficient.

Using an in-circuit impedance extraction technique, we look at how LISN affects EMI measurement. The CM and DM noise of a switched-mode power supply are derived using an inductive coupling technique in the proposed scheme. The CM and DM noise of a switched-mode power supply are derived using an inductive coupling technique in the proposed scheme. For the purpose of determining the LISN's conformance to the CISPR standard, a technique for impedance calibration of the network using a vector network analyser has been developed.

Again, for the objective of modelling EMI noise source impedance, a novel technique that makes use of scattering parameters and the genetic algorithm (GA) has been developed. When the circuit is shorted and loaded with noise source impedance, respectively, the transmission and reflection characteristics of a vector network analyser are used to measure the source impedance. The GA method is used to determine the LISN's ideal impedance parameters in order to optimize the network (Figure 6.2).

By isolating the power line from the DUT, an EMI measurement may be made with the help of an LISN. The quantity of the solenoid inductor, which is typically utilized in LISN, is able to be determined by the application of the Wheeler relationship.

$$L = \mu_0 n_2 a (\ln 1 + \pi a/b) + 1/2.3 + (1.6 b/a) + 0.44 (b/a)^2$$

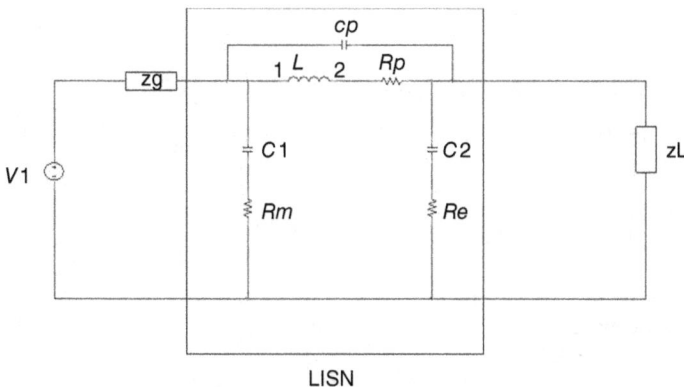

FIGURE 6.2   Line impedance stabiliszation network (LISN) with parasitic elements.

where a, b, n, and μ_0 are the radius, axial length, the number of turns of the coil, and the permeability of free space $(4 \times 10^{-7} \text{H/m})$, respectively.

The value of the capacitance is the determining factor that has the most significant impact on the deviation. Following is an example of how one may calculate the value of this capacitor:

$$Cp = (1/N - 1) \times 2e0\pi D/(1 + (1/er) * \ln(d_0/d_i) + s/d_0)^{2-1}$$

where D and N are the diameter and number of turns of the solenoid, respectively, di and do are the inner and outer diameters of the conductor, s is the air gap between conductors, $e_0 = 8.854 \times 10^{-12} \text{F/m}$ is the air permittivity, and r is the relative permittivity of the insulator.

In spite of the fact that it is utilized to decouple the power line impedance, the LISN operates in an inadequate manner over the whole frequency range. Even at low frequencies, employing a filter with a high order might result in improved decoupling performance. In addition, the application of multistage filter designs is a possibility in order to enhance the effectiveness of the EMI filter under misaligned impedance situations.

In between EMI noise generator (Vs) and the load (e.g., DC power supply) is a two-port filter. ZS and ZL symbolize the impedance of the noise source and the input DC power line, respectively. In most cases, the insertion loss (IL) is used to classify power filters. The IL is calculated by comparing the test configuration's unfiltered signal level (V20) to the filtered signal level (V2) using the filter (Figure 6.3).

$$IL = 10 \times \log(P_{20}/P_2) = 20 \times \log(V_{20}/V_2) dB$$

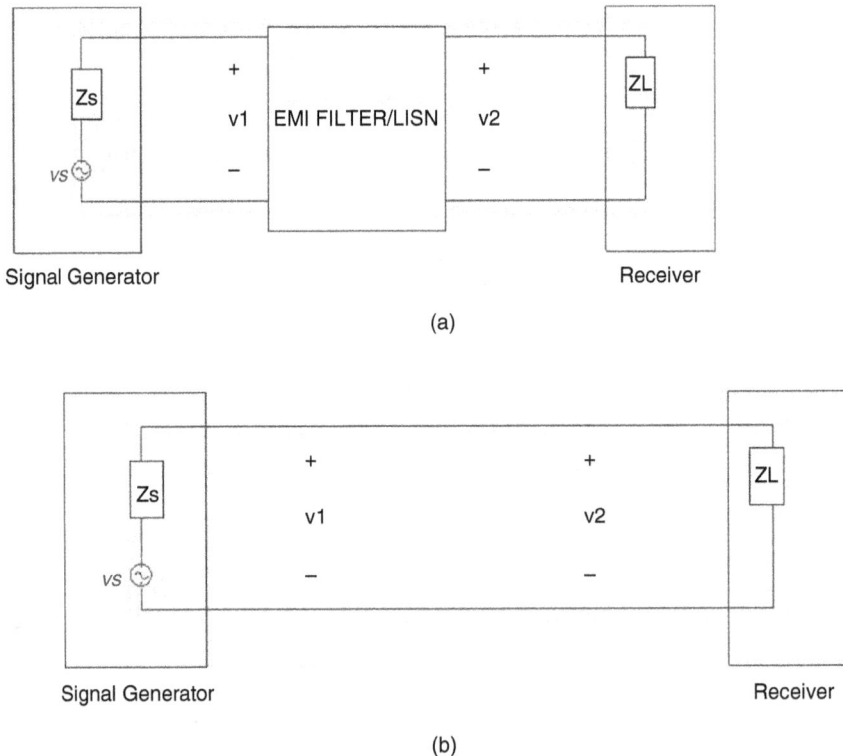

(a)

(b)

FIGURE 6.3 Two-port line impedance stabilization network (LISN)/electromagnetic interference (EMI) filter.

The following parameters can be used to determine the IL:

$$IL = 20\log\left\{\left(\left|(1-\rho SS11)(1-\rho LS22)-\rho S\rho LS11S22\right|^2\right)\Big/\left(\left|S21\right|(1-\rho S\rho L\left|^2\right)\right)\right\}$$

where ρ_S and ρ_L are the computed source and load reflection coefficients.

$$\rho S = (ZS - Z0)/(ZS + Z0)$$

$$\rho L = (ZL - Z0)/(ZL + Z0)$$

If $Z_L = Z_S = Z_0$ (Z_0 is the LISN surge impedance), then it is simplified as $I_L = -20\log |S21|$, showing that the transfer function can be analysed in terms of I_L (S21). A vector network analyser can be used to measure all S-parameters in real-world applications.

A. Parasitic Elements Optimization

For the depicted LISN, the transfer function associated with IL can be calculated as follows:

$$V_2/V_1 = \left[(1/C2S + Re)\parallel ZL\right]\Big/\left[\left((Ls+Rp)\parallel 1/CpS\right)+(1/C2S + Re\parallel ZL)\right]$$

$$= \left[a_0 + a_{1s} + a_{2s}2 + a_{3s}3\right]\big/\left[b_0 + b_{1s} + b_{2s}2 + b_{3s}3\right]$$

where ai and bi ($i=0, 1, 2, 3$) are functions of parameters shown (details are shown in Appendix A). They compare two transfer functions, with and without parasitic elements, to show the effects of parasitic elements. The transfer function only has two fictitious low-frequency poles when parasitic components Cp, Rp, and Re are ignored. Due to the parasitic elements of the inductor, this frequency has two zeros (Cp and Rp). By accounting for equivalent series resistance (ESR), third zero lowers the transfer function's peak ($C2$ and Re).

1. Deciding on Re: The damping resistor Re has an impact on IL's dynamic behaviour since the transfer function can be set up to provide the following characteristic equation:

$$1 + Re\{G1(s)/G2(s)\} = 0$$

where $G1(s)$ and $G2(s)$ are presented in Appendix A. The LISN step response and root loci are caused by varying Re values, respectively. In relation to $Re=1.8$, the damping factor $\zeta=0.707$ ($\theta=45$) has the best rise time and overshoot.

2. Deciding on Cp: The characteristic equation of the transfer function given in (6) can be easily modified as follows to illustrate the effects of parasitic capacitance Cp:

$$1 + Cp\{G3(s)/G4(s)\} = 0$$

where $G3(s)$ and $G4(s)$ are presented. The dynamic behaviour of LISN caused by the various values for Cp is demonstrated using the root locus analysis.

The third zero's ESR correction reduces the peak of the transfer function ($C2$ and Re). Although this system is stable, by carefully selecting the capacitance Cp, it is feasible to obtain a non-oscillatory response with a desired rise time. The step response illustrates that results are attainable with values ranging from 1 to 10 nF. During the process of the development of the inductors, this should be taken into consideration as a practical concern.

B. Cascading Stages

LISNs composed of a single cell are more susceptible to line impedance than those composed of two cells, which have more complex topologies and give better results. The four-stage LISN described in this article can be improved by the addition of a resistor in series with each stage's capacitor, which is the way that is recommended. (the next sections outline the procedure for calculating the proper number of phases, which is discussed in further detail later on) (i.e., the ESR of each capacitor can be raised to a higher amount as needed).

This method includes an additional zero at the end of each stage in the sequence. It offers a graphical illustration of the four phases of the projected LISN as well as the protection circuitry associated with each phase. The diode D1 and the variable resistor metal oxide varistor (MOV) are only present for the purpose of ensuring the user's safety. The D1 protects measuring instruments from electrostatic discharge, while the MOV protects circuits from power line overvoltage. The capacitance of MOV is quite low when compared to that of C7. As a result, it could be omitted during theoretical tests.

Values for components C7, L1–L4, and C1–C3 were determined in accordance with guidance provided by relevant standards. The ADS optimization subsystem has determined the values for the remaining parts. In this case, the optimization is carried out in accordance with specified limitations. Then, the **ADS environment** was used to pick component values that would result in an optimal circuit. It is important to note that the inductor's parasitic capacitance value is determined by measuring a real-world inductor, so you can rest be assured that it is correct.

1. Multistage Simulations: Consider the provided one-stage LISN with the optimized element *Re* and *Cp*, ignoring stage zeros. The graphic below illustrates the results of increasing the LISN's number of stages, where each stage has a capacitor (excluding series resistors) that adds one pole to the transfer function. The IL slope becomes more negative by −20 dB after each stage. Each stage generates a peak above the resonance frequency of its corresponding capacitor; however as the number of stages grows, the attenuation is especially noticeable between 100 kHz and 10 MHz (Figure 6.4).

2. Multistage Simulations: Adding Zeros of Stages: Our proposal to get rid of these peaks in the IL is to include zeros in the transfer function of the multistage LISN. This is what we think will work best. After removing the leading zeros, the simulation results are shown. To get rid of the peaks, just add a 0 to the stage number, although the total attenuation slopes will be reduced. After zeros are added, the attenuation slopes for all simulated multistage are roughly the same at 30 MHz, starting with a two-stage and going up to a five-stage LISN.

3. Choosing the Best Number of Stages (the AIC Index): An increase in attenuation is a trade-off for the complexity introduced by each additional stage in the circuit. Hence, it is crucial to analytically demonstrate a guideline to select the appropriate amount of phases. The Akaike

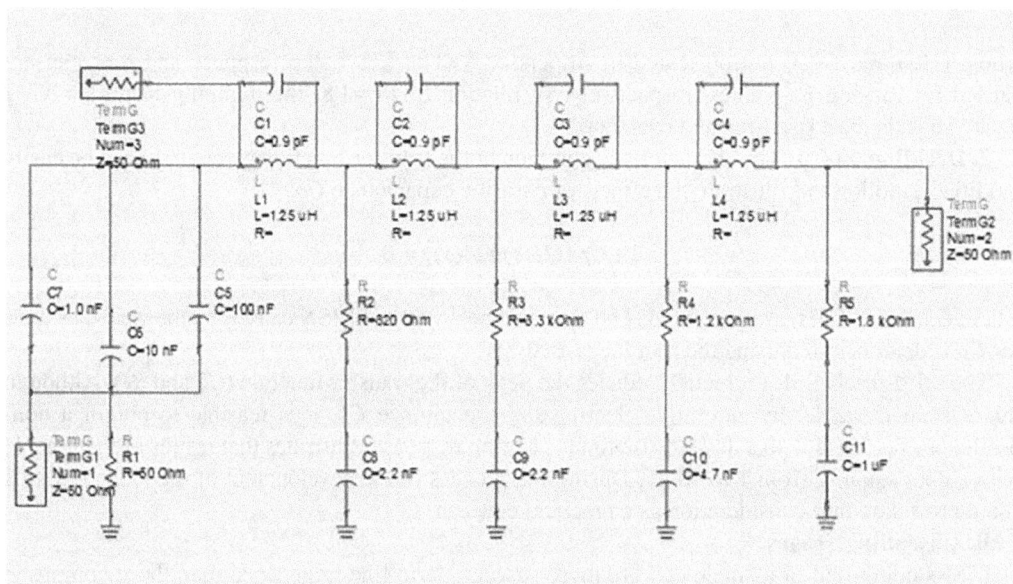

FIGURE 6.4 Modelled multistage line impedance stabiliszation network (LISN) in the ADS.

information criterion (AIC) is a straightforward method for choosing the optimum model that is both effective and objective. To evaluate distinct models, the AIC provides the following criteria:

$$AIC = -2 \times \ln(Li) + 2p$$

where p is the total number of model parameters and Li is the greatest likelihood (i.e., the number of variables). In the aforementioned AIC equation, terms 1 and 2 stand for precision and complexity, respectively. Here's how you can calculate the AIC:

$$AIC = -n * \ln(\sigma^2/n) + 2p$$

where n is the number of observations, σ^2 is the sum of squared error (SSE). The AIAA-specified noise limit is taken into account while calculating the SSE. A multistage LISN has a circuit consisting of capacitors and inductors, and the value of parameter p represents the total number of these components. We can infer that the studied circuits will be more fit if the AIC is smaller. For an LISN with N stages, where N is an integer between 1 and 5, the AIC has been calculated. The LISN grows by a factor of N3 with the addition of three components (capacitor and inductor) at each stage. The findings are shown. The four-stage LISN introduces the smallest AIC of the five that were determined.

C. Transfer Function of the LISN

The shown LISN has the following transfer function, which includes 12 poles and 12 zeros:

$$H(S) = \left\{ K(S - Z_1)(S - Z_2)\cdots(S - Z_{12}) \right\} / \left\{ (S - P_1)(S - P_2)\cdots(S - P_{12}) \right\}$$

An LISN inductor is an air-core inductor, meaning its coil winding is coiled around a non-magnetic, non-flammable, and insulating substance rather than a magnetic core to prevent the magnetic core from becoming saturated by the high current flowing through it, e.g., Bakelite. Inductors are the only exception; the rest of the components can go on the back of the printed circuit board (PCB).

The pictured setup has been used to measure several LISN features to confirm the LISN's quality. The GPS-1250 is a radio frequency (RF) signal generator that can produce a sinusoidal wave and sweep between 100 kHz and 30 MHz (the sweep time is about 1 s). A SA is used to monitor the propagation of this wave as it passes through the LISN. (KEYSIGHT MXA Signal Analyser N9020B). An EMC-box has been implemented in order to mitigate the impact of the surrounding noise.

LISN provides an adequate level of isolation between the terminals for its input and output. In addition to this, because of its impedance characteristics, it complies with the requirements of the standard AIAA (S-121A-2017). The stability study has been carried out to ensure that the LISN does not exhibit any signs of self-oscillation and that it has the appropriate rise time for the step response. In addition, the output qualities of LISN are superior to those of its competitors in certain aspects, such as the noise level.

LI-1100C

Models	Key Features	Applications
 1. 150 KHz–30 MHz 2. CISPR 16-1-2 3. ANSIC63.4 4. 50 microhenry/100 Amps	1. Conforming to ANSI C63.4 and the CISPR 16-1-2 standards. 2. This LISN avoids saturation and permeability variation with the help of air-core conductors.	1. It is possible to get phase, isolation, and insertion loss. 2. Single-conductor networks, each enclosed in its own enclosure, which are connected in series with the wires carrying electricity in a three-phase, two-phase, or direct current power supply.

LISNLISL100

LI-125C

1. Compliant with CISPR 16-1-2 and ANSI C63.4.
2. This LISN avoids saturation and permeability variation with the help of air-core conductors.

1. It is possible to get phase, isolation, and insertion loss.
2. Single-conductor networks, each enclosed in its own enclosure, which are connected in series with the wires carrying electricity in a three-phase, two-phase, or direct current power supply.

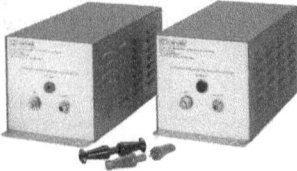

1. 150 KHz–30 MHz
2. CISPR 16-1-2
3. ANSI C63.4
4. 50 microhenry/100 Amps

LI1100C

Line impedance stabilization network (LISN) Model: LI-3100 can handle up to 100 Amps AC (70 Amps DC) current and provides the necessary measurement platform for performing power line conducted emissions compliance testing as required by DO-160 & MIL-STD 461F and CISPR 25 and CISPR 16 from 10 kHz to 400 MHz.

1. As opposed to current or voltage probe tests, where the impedance of the lines is generally uncontrolled, series connections allow a fixed impedance to be imposed on the lines being tested.

1. LISN-100 Amps
2. MIL-STD 461
3. CISPR 25 and CISPR 16

LI-3P-1X

1. Frequency range 150 KHz to 30 MHz.
2. Fully Compliant with CISPR 16-1-2/ANSI C63.4
3. Remote switching of line under test.
4. Current ratings of 16, 32, 63 and 100≈Amps.

1. Transient protection
2. Remote or local operation

1. 15 KHz to 30 MHz.
2. CISPR 16-1-2
3. ANSI C63.450

6.7   NETWORK ANALYSER

In a microwave system, we have to transfer signals with minimum distortion and maximum efficiency. Scattering parameter signal behaviour can be measured and analysed with a vector network analyser. Gain, return loss, voltage standing wave ratio (VSWR), reflection coefficient, etc., are all observable manifestations of the signal's behaviour. It evaluates a network's performance by gauging its amplitude and phase responses. Our demonstration system receives its energy from the signal generator.

The system uses a separate source in our conventional method. In that method, they use either voltage controlled oscilloscope, which are cheaper or based on more expensive synthesized sweepers which provide better performance. But the main disadvantage of the voltage controlled oscillator (VCO) is that it produces much noise and it degrades the system accuracy. Today, the signal sources have excellent frequency and resolution. The main function of the splitter is to separate the signal. It is usually resistive. They are non-directional devices and can be very broad band. Due to their inherent high pass response, it can be limited for 10 MHz. It has two main functions They are:

- It measures the portion of the incident signal to provide a reference signal for the receiver.
- It separates the incident and the reflected wave at the input of our DUT.

The signal detection in the network analyser can be done in two ways.

6.7.1 DIODE: DIODE DETECTORS CONVERT THE RF SIGNAL TO THE PROPORTIONAL DC LEVEL

Advantages

a. They provide broadband frequency
b. Inexpensive

Limitation

a. Medium sensitivity and the dynamic range
b. No phase information

Application

It is used in frequency translating devices.

6.7.2 TUNED RECEIVER

The tuned receiver uses a local oscillator to mix the RF frequency down to an intermediate frequency. The IF filter is a bandpass filter, which narrows the bandwidth and improves the sensitivity. It is displayed in the form of linear and log formats, polar plots, Smith chart, etc.

Vector network analyser has precision cables because the phase and loss of a standard cable would vary too much for a slight movement. Both ends of this cable are attached to the unit being tested (DUT). From the own sources, the variable frequency signal is generated and their output is given to test the DUT in either one direction. One signal goes to the receiver as a reference, and the other goes to the directional coupler 1 through the precision wires, so the splitter serves a dual purpose. The third port detects the reflected power and this is connected to the receiver. The same process takes place in directional coupler 2. The signal processed by the receiver is fed to the processor and displayed.

Newer VNAs can do more than merely measure S-parameters. Two ports will be available on the basic VNA, with a maximum of 24 available. As shown, VNAs are typically employed in tuned receiver mode, where the phase of the signal is measured relative to an external reference. The problems with VNAs are the bad image and the accidental injection of noise signals and their harmonics (Figure 6.5).

A. PNA Family

Model	Applications	Key Features

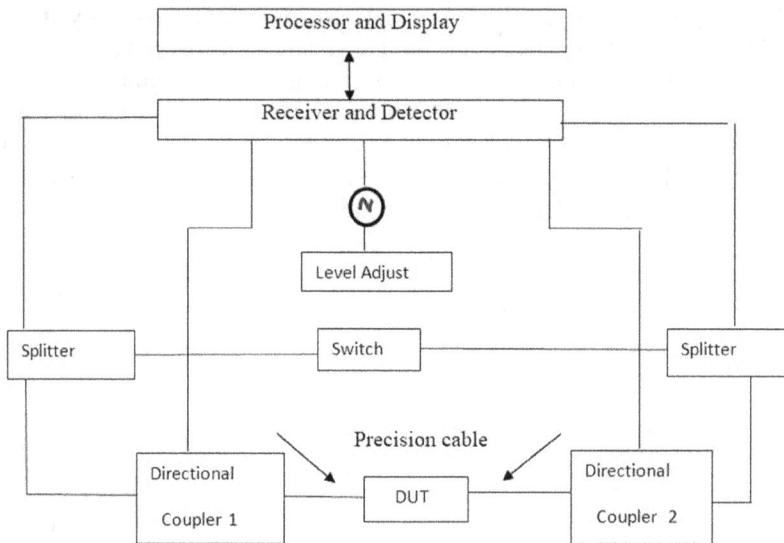

FIGURE 6.5   Block diagram of vector network analyser.

PNA-X Series N524Xb

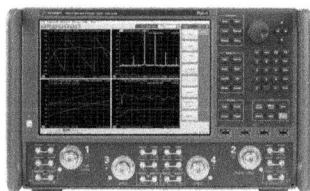

1. Most advanced integrated and flexible VNA
2. 10 MHz to 8.5/13.5/26.5 GHz
3. Up to 1.5 THz with extenders

1. Rather than using multiple pieces of equipment, only use one instrument.
2. Characterization of all active, linear, and non-linear devices.

1. Low spurious and very low phase noise.
2. Multiple measurements can be taken with just one link.
3. There is a combiner already installed.
4. Flexibility to add signal conditioning system.
5. Phase noise, AM noise, and residual noise measurements.
6. To the DUT, add testing apparatus without repositioning test cables.

PNA Series N522Xb

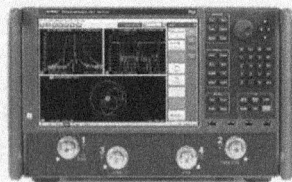

1. High-performance microwave VNA
2. 10 MHz to 13.5/26.5/43.5/50/67(70) GHz
3. Up to 1.5 THz with extenders.

1. Analysis of passive components with the highest possible performance, up to 70 GHz.
2. Identification of the active substances.
3. Metrology and call lab.

1. Superior RF performance in an integrated VNA.
2. Most accurate in the world. Choice of metrology for the best S-parameter measurements.
3. Phase noise, AM noise, and residual noise measurements.

PNA-L Series N523xB

1. Economy microwave VNA
2. 300 kHz to 8.5/13.5/20 GHz
3. 10 MHz to 43.5/50 GHz

1. The S-parameter test for microwaves.
2. Signal integrity
3. Material measurements

1. Measurements for S-parameters and materials.
2. With a configurable test set, you can be flexible.

B. ENA Series

Model	Applications	Key Features

E5080B

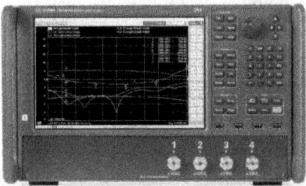

1. High-performance ENA
2. 9 kHz to 4.5/6.5/9/14/18/20 GHz
3. 100 kHz to 26.5/32/44/53 GHz

1. General-purpose RF/microwave component tests.
2. High-volume manufacturing test.
3. Uses for both active and passive measurement.
4. Better time-domain analysis for applications in signal integrity.

1. With 2 or 4 ports, RF to microwave frequency ranges.
2. Best measurement performance in its class.
3. DC sources, bias tees, and pulse generators/modulators are all built in.
4. Uses for measuring a wide range of things.
5. Four-port models have two internal sources.

E5072A

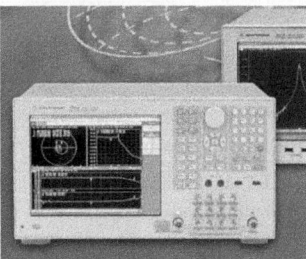

1. RF VNA with configurable test set
2. 30 kHz to 4.5/8.5 GHz

High-power RF component tests.

You can measure both passive and active (non-linear) devices with a wide source power range (−85 to +16 dBm).

C. PXI VNA Family

Model	Applications	Key Features

M980XA

1. High-performance multiport VNA
2. 9 kHz to 4.5/6.5/14/20 GHz
3. 100 kHz to 26.5/32/44/53 GHz

1. Multiport component tests.
2. Tests of both active and passive parts made in large quantities.
3. Multi-site (parallel) manufacturing tests.

1. Performance measurement at the top of its class.
2. Electronic calibration (ECal) modules make it possible to calibrate more than one port quickly and easily.
3. Uses for measuring a wide range of things.

M937XA

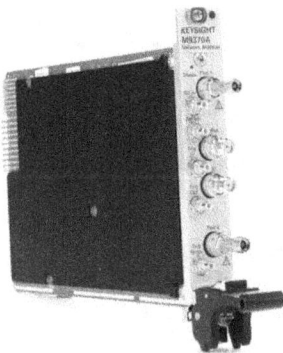

1. Full 2-port VNA that fits in one slot
2. 300 kHz to 4/6.5/9/14/20/26.5 GHz

1. Active component tests (PA, LNA, T/R modules, etc.).
2. Multiport component tests.
3. Manufacturing tests for beam former IC and front-end module (FEM) for 5G FR2.

1. Combines vector network analysis and wideband modulation distortion analysis.
2. Excellent accuracy.
3. Lower-cost solution.
4. Scalable and flexible configurations for multiport.

VNA Simulator

Model	Applications	Features

1. S94050B (VNA simulator—standard)

1. Test programme development for automated test.
2. Post data process without VNA hardware.

1. Eliminates the need for VNA hardware for the operation.
2. The standard version (S94050B) supports operations of standard S-parameter measurement class.

6.8 SPECTRUM ANALYSER

Swept frequency SA for high frequency is readily available and inexpensive, making them a good option for near-field scanning. The signal's intensity is all it can resolve. It is necessary to have extra devices and components in order to calculate phase from magnitude-based SA measurements in the event that phases and magnitudes of submitted data are both sought.

In order to gather information regarding the phase at a single frequency, it is necessary to do three measurements using a variety of setup. As a result, this phase detection method was constructed for use with a single frequency. The SA method only works if the main probe signal is about the same size.

An attenuator, whose setting can be adjusted to compensate for the magnitude difference between the two input signals, is inserted into the signal path taken by the reference probe. One of the problems with SA-based phase measurement employing a hybrid coupler is that it can't make a broadband measurement. To obtain the right phase based on magnitude alone SA measurements, the system employs a switch to select among various phase shift cable lengths, and then takes numerous measurements. The SA method is used to find out a DUT's magnitude and phase, which are then compared to the measurements from the VNA and oscilloscope. In addition, the measured data are compared both with the scope and the measured data. The advantages of each method and the performance for different measurements.

A. Basic Spectrum Analyser

Model	Applications	Key features

N9320B Spectrum Analyser

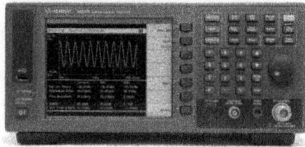

1. High measurement speed.
2. 10 Hz to 1 MHz.

The most important part of RF testing is being able to pick up weak signals and separate frequencies that are close together. An N9320B spectrum analyser will be able to handle these tasks easily because it has one of the best combinations of sensitivity and narrow resolution bandwidths (RBW).

1. Low-cost instrument for measuring RF.
2. High accuracy.
3. The frequency counter gives a more accurate reading of the frequency.

N9322C BSA

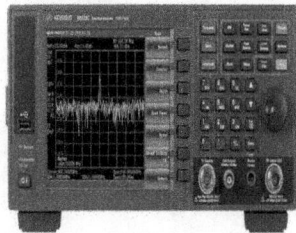

1. 9 kHz to 7 GHz frequency range
2. ±0.3 dB absolute amplitude accuracy
3. 10 Hz to 3 MHz resolution bandwidth

1. Robust measurement features for easily characterizing your product.
2. Support transmission and reflection measurements with built-in VSWR bridge support.
3. Demodulation mode allows signal analysis easily and cost effectively.

1. Simple test setup.
2. Wideband modulation distortion analysis.
3. High spectrum accuracy.
4. Automation and communication interface with industry standard.
5. USB and LAN connectivity choices.

B. MXA Family

Model	Applications	Key features

N9020B Signal Analyser

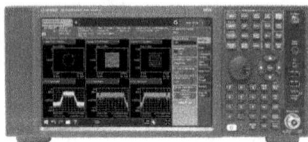

1. Frequency range: 10 Hz to 3.6, 8.4, 13.6, 26.5, 32, 44, 50 GHz, mixers to 1.1 THz
2. Analysis bandwidth: 25 (standard), 40, 85, 125, and 160 MHz
3. Display average noise level (DANL): −172 dBm at 2 GHz

1. Utilize the streamlined, multi-touch user interface to complete most tasks in just two steps.
2. Views and measurements that make troubleshooting a breeze, even for the most intricate tasks.
3. Effects on the spectrum when using high-resolution FFT-based measurements

1. Cellular and wireless application.
2. Video and general-purpose application.

N9021B Signal Analyser

1. Frequency range: 10 Hz to 32, 44, and 50 GHz.
2. Analysis bandwidth: 255 and 510 MHz
3. Display average noise level (DANL): −172 dBm at 2 GHz

1. Keep up with the ever-changing needs for testing wireless devices.
2. Updates and sweeps on the display happen quickly.
3. The X Series software can take several measurements with the push of a single button, greatly simplifying test procedures.
4. Spectrum analysis in real time, utilizing full-band transient signals.

1. Analyse phase noise in both frequency domain and time domain.
2. Cellular and wireless application.

C. GW-INSTEK Family

Model	Key Features	Applications

GSP-818-TG 1.8 GHz Spectrum Analyser

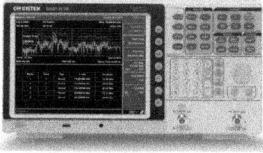

1. Built-in AM/FM demodulation.
2. Bandwidth zoom function.
3. Options: tracking generator, EMI filter, & detector.

1. Characteristics of the spectrum are being checked and analysed.
2. Examine the features of AM and FM transmissions.
3. For a compact test system.
4. RF cable, attenuator, filter, and amplifier frequency response measurements.

1. Frequency Range: 9 kHz–1.8 GHz
2. RBW: 10 Hz–3 MHz,
   10 Hz–500 kHz in 1–10 steps
3. Sensitivity: –148 dBm/Hz

GSP 730–3 GHz Spectrum Analyser

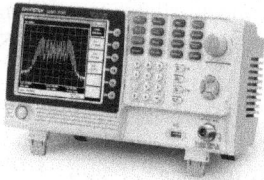

Pass/fail function: Function may run the "Pass" and "Fail" inspection with efficiency. Lower and upper limit lines provided as judgement lines, then the LCD will display "Pass" or "Fail" according to the input signal meeting the required condition or not and it is indicated by the examined outcome.

1. Examine phase noise using both frequency and time-domain analysis.
2. Frequency counter offers more accurate frequency readout.

1. Frequency Range: 150 kHz–3 GHz
2. Noise floor: 100 dBm

D. Portable Spectrum Analyser

N9963B 54 GHz Spectrum Analyser

Model	Key Features	Applications
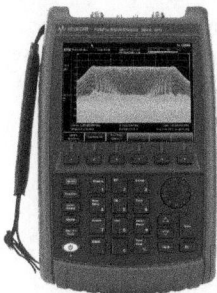	1. Options for a real-time spectrum analyser and a tracking generator will allow you to expand your skills. 2. Rugged enough to meet MIL-standards. 3. Simplify complex setups, provide quick and accurate results with confidence.	1. For devices such as mixers and converters, the scalar transmission response has been characterized. 2. It is possible to exercise remote control over it with the use of PC software and either a wireless or connected LAN connection.

1. Up to 120 MHz bandwidth
   Built-in power meter
2. Pulse measurements
3. Channel scanner
4. GPS receiver
5. Real-time spectrum analyser
6. Base model includes cable and antenna analyser
7. Measure all four S-parameters with 115 dB system dynamic range.

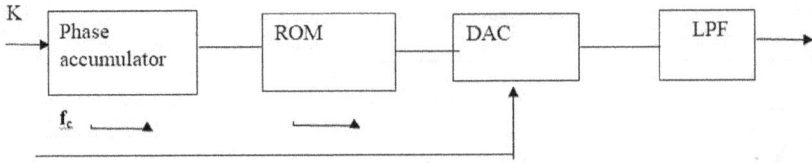

FIGURE 6.6   Block diagram of signal generator.

6.9   SIGNAL GENERATORS

The signal generator, which serves as an instrument for testing and generates the test signal with the specified parameters, has limited practical utility but a wide range of applications. This also serves as an instrument for testing and generates the test signal with the specified parameters, has limited practical utility but a wide range of applications. Due to their cumbersome size, poor quality, lack of precision, etc., analogue signal generators are being phased out in favour of their digital counterparts. Frequency synthesis, phase-locked loop, and direct digital synthesizer (DDS) technologies are the primary lenses through which the digital signal generator. Phase accumulator, waveform storage ROM, a D/A converter, and a low-pass filter are the standard components of every DDS (Figure 6.6).

The signal generator has been employed because to its ability to generate waveforms quickly, cheaply, and easily. In order to make signal waveform generation easier and more efficient, the panel displays the waveform after the user sets the signal waveform, amplitude, frequency, phase, channel, and DC offset voltage. Respectively high-frequency resolution and high precision are available in the signal that seems to be output, which are both advantages. When compared to the analogue signal generator, the system design benefits from being more compact, having a higher level of precision, being more versatile, and having an easier time being used. It could be used in a lot of different ways.

N9310A RF Signal Generator

Models	Key Features	Applications
 1. Frequency range 9 kHz to 3 GHz (RF 20 Hz to 80 kHz (LF)) 2. Switching speed < 10 ms 3. Output power –127 to +13 dBm	1. Provides reliable RF signal generation at low cost. 2. Easy, comprehensive sweep function for general-purpose testing. 3. Obtains reliable results in installation and maintenance.	1. Low-cost measurement solution. 2. Analysis for transmitter and receiver test. 3. Reliable performance.

LS1291B 12 GHz Dual Channel Signal Generator

1. 12 GHz RF analogue signal generator
2. Extremely fast switching speed of <100 us
3. AM, FM, PM, sweep & pulse modulation
4. USB and LAN interfaces

1. This instrument's graphical user interface (GUI) and full set of drivers make it possible to programme in environments as diverse as Labview, Python, C++, and Matlab, giving you full command over the instrument's capabilities.
2. The Ethernet and USB interfaces built into this series allow for PC-based remote programming.

Signal integrity and purity:
Whenever it comes to modern testing and measurement, a high-quality signal is essential, and this type of signal generator provides it.

LS1292B 12 GHz Dual Channel Signal Generator

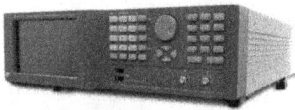

1. 3, 6, and 12 GHz multi-channel RF analogue signal generator
2. Extremely fast switching speed of <100 us
3. AM, FM, PM, sweep & pulse modulation
4. USB and LAN interfaces

1. This instrument's Graphical User Interface (GUI) and full set of drivers make it possible to programme in environments as diverse as Labview, Python, C++, and Matlab, giving you full command over the instrument's capabilities.
2. The series has an ethernet and USB interface for PC-based remote programming.
3. Easy to use.

1. Signal integrity and purity:
With one such type of signal generator, you can get high signal quality, which is one of the most important things in testing and measuring today.
2. Easy to use.

TABORLS30843GHz Four Channel Signal Generator 1URACK Module

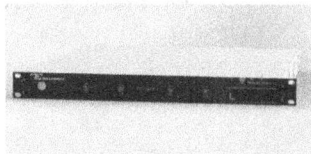

1. 3 GHz multi-channel RF analogue signal generator
2. Four-phase coherent channels in a single 1U box
3. Extremely fast switching speed of <100 us

1. This instrument's Graphical User Interface (GUI) and full set of drivers make it possible to programme in environments as diverse as Labview, Python, C++, and Matlab, giving you full command over the instrument's capabilities.
2. The series is equipped with ethernet and USB interface enabling remote programming from PC.
3. Easy to use.

Single bursts and chirps are now commonplace in aerospace and defence applications.

GW-INSTEK USG-0103 100 MHz to 300 MHz, USB RF Signal Generator

1. Frequency range: 100 MHz–300 MHz
2. Output power range: −30 dBm– 0 dBm
3. Continuous wave signal without any modulation
4. Support fixed frequency, frequency sweep, frequency hopping, & power sweep mode
5. External PC software support Different operating system

1. Frequency resolution-10 KHz.
2. PC USB interface powered and controlled.
3. This model supports applications programmes such as Windows 2000/XP/Vista/7/8, Linux ad MAC OS X through the USB interface.

1. Reference source for PLL and ADC.
2. Tracking generator for spectrum analyser.

6.10   EMI MEASUREMENT USING ANTENNA

Antennas are used to measure how much a product is radiating, but they can also be used to troubleshoot digital products by picking up their harmonic emissions. Despite the fact that most EMI antennas are optimized for E-fields, the far field exhibits no clear preference for either E- or H-field. Antenna calibration, also known as the antenna impedance method, is a recently discovered technique for measuring EMI. Theoretical research on antenna configurations and achievable accuracy is conducted. Simply placing the antenna under calibration on top of a metal ground plate at different heights and measuring the impedance of the antenna is all that is required for this calibration approach. This is because the antenna impedance, a key component of the antenna factor, fluctuates with antenna height in a quasi-periodic fashion.

ETS-AMS-8800 Antenna Measurement System

Models

1. 700 MHz–10 GHz frequency range (Upgradable to higher frequencies with optional antennas)

Key Features

1. The AMS-8800's distributed axis positioning system is made up of an azimuth rotator for turning the DUT about the phi axis and a theta arm position for raising the measurement antenna above the DUT.
2. Test laptops, desktops, small appliances, and even equipment at the customer's location with the supplied table top mount.

Applications

1. Antenna pattern measurement for design and development.
2. Over-the-air performance testing.

ETS-AMS-8600 Antenna Measurement System

1. Shield test per MIL-STD-285.
2. Need shielded control room.

1. Antenna pattern measurement for design and development.
2. Over-the-air performance testing.
3. Shielded room testing.

1. Minimum 400 MHz frequency; maximum 6 GHz.
2. Physical specifications:
Internal working area (L) 12.5 m (41.01 ft)
Internal working area (W) 3.05 m (10.01 ft)
Internal working area (H) 3.05 m (10.01 ft)

ETS-AMS-8500 Antenna Measurement System

The system can be used to measure antennas in both near- and far-field test situations to obtain data on more generalized antenna attributes.

1. Both passive antennas and active wireless base stations' frequency responses were measured (cell phones).

1. Full size rectangular anechoic chamber 7.32 m L × 3.66 m W × 3.66 m H (24 ft. × 12 ft. × 12 ft.)
2. Frequency range: 700 MHz – 6 GHz

ETS-AMS-8055 MIMO Single Cluster Environment Simulator

1. Self-contained moveable system.
2. Compact anechoic RF enclosure.

1. VNA (For chamber calibration)
2. Spectrum analyser
3. Data acquisition and analysis software.

1. Efficient MIMO over-the-air (OTA) performance
2. Measurement SISO OTA performance
3. Measurements 2D/3D antenna pattern
4. Compact, anechoic RF enclosure

ETS-AMS-7000 Wireless OTA Reverb Test System

	Shield and cable kit including:	1. MIMO antenna testing
	1. DUT positioner	2. Communication tester
	2. Z fold tuners	3. Channel emulator
	3. Measurement antennas	
	4. RF-absorber loading elements	

1. Faster test times.
2. Freestanding moveable cart configuration.

6.11 SURGE GENERATORS

Surge generators, also known as impulse generators, are electrical devices that generate brief, high-frequency spikes in either voltage or current (Table 6.3). The two main categories into which these devices fall are:

• Impulse voltage generators
• Impulse current generators

Figure 6.7 shows a simplified circuit schematic of the generator. The generator's ability to supply the specified open-circuit voltage surge and short-circuit current is determined by the values chosen for the various components Rs1, Rs2, Rm, Lr, and Cc.

TABLE 6.3
Relationship between OC Voltage and SC Current

Open-Circuit Peak Voltage ±10% Generator Output	Short-Circuit Peak Current ±10%
Generator output	
0.5 KV	0.25 KA
1.0 KV	0.5 KA
2.0 KV	1.0 KA
4.0 KV	2.0 KA

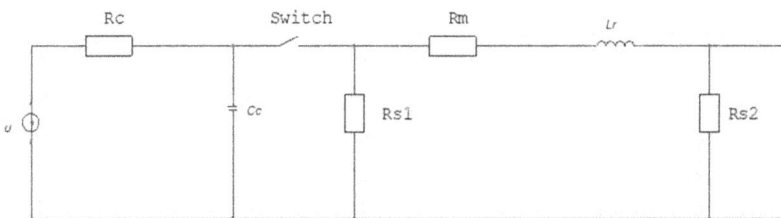

FIGURE 6.7 Surge generator model.

The roles of the components from Figure 7 are as follows:

U = high-voltage source
Rc = charging resistor
Cc = energy storage capacitor
Rm = impedance matching resistor
Lr = rise time shaping inductor

SGIEC-645

Models	Key Features	Applications
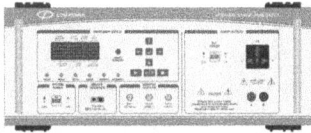 Standard: IEC 61000-4-5	1. Voltage: 100–250 VAC 2. Frequency: 50/60 Hz 3. Power: 300 VA maximum 4. Connects to a high impedance oscilloscope input for verification of surge voltage level.	Surge immunity testing

SGTEL-168

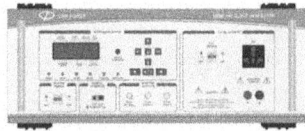 Standard: CS03	1. Open circuit voltage (1000:1) ± 10% 2. Short-circuit current (100:1) ± 10% 3. Built-in 16 Amps AC, 250 Vac single phase. 11 Amps DC, 350 Vdc. 4. Universal multi-configuration outlet for EUT and IEC 60309 industrial socket for mains.	Simulate the effect of lightning on telecom equipment test.

6.12 CURRENT PROBES

Magnetic field pickup devices like current probes can be clamped onto a bundle of wires or cables to measure the CM RF current in the bundle. A core of broadband ferrite or a similar material is often used in a toroidal shape. Using a pickup, the frequency response and sensitivity will be affected by the material and the number of wraps applied to the core. Correction, transfer impedance, and transducer factors all refer to the resistive network employed in emission-only probes to adjust impedance and smooth out the response. If these networks are not present and the windings around the core of the current probe are strong enough, it is possible to use the current probe as an injection probe, which is more commonly referred to as a bulk current injection (BCI) probe. These instruments are utilized in a variety of tests in order to get conducted emission measurements. However, as a tool for debugging, they can be quite helpful. It is possible to determine which cables are the primary sources of radiated emissions by measuring the current flowing through certain cables. A reduction in such noise (currents) on those lines can frequently result in a reduction in the radiated emissions coming from the equipment that is being tested.

ETS-94111-1 Current Probe

Models	Key Features	Applications
1. For standards testing and pre-compliance applications 2. Measures RF current without direct connection	1. Frequency minimum: 1 MHz 2. Frequency maximum: 1000 MHz 3. Load impedance: 50 +/− j0 4. Maximum primary current amps: a. 400 Hz: 200 b. DC-60 Hz: 200 c. Pulse: 50 d. RF(CW): 20 4. Transfer impedance: 1 to 10 Ω	Diagnostic tool employed primarily for the search of cable common-mode current sources of radiated emissions and for the measurement of cable common-mode coupled currents during radiated susceptibility testing.

ETS-93686-8 Current Probe

Models	Key Features	Applications
1. For standards testing and pre-compliance applications 2. Measures RF current without direct connection	1. Frequency minimum: 10 KHz 2. Frequency maximum: 200 MHz 3. Load impedance: 50 +/− j0 4. Maximum primary current amps: a. 400 Hz: 200 b. DC-60 Hz: 200 c. Pulse: 62 d. RF(CW): 62 4. Transfer impedance: 8 Ω	The measuring of high-level pulsed currents with a low duty cycle.

6.13   OSCILLOSCOPES

Oscilloscopes, which operate in the time domain rather than the frequency domain, are also useful for identifying and fixing EMC problems. For instance, oscilloscopes have the ability to measure transitory occurrences, whereas SAs can only examine continuous periodic waveforms. Make sure the oscilloscope and probe's bandwidth is larger than the measurement you're making while working with high-frequency clock signals. Maximize signal strength by shortening the probe's signal return cable. Due to significant self-inductance caused by the measuring loop area, ringing will occur if a standard probe ground wire of 4–6 inches is used. The best oscilloscopes can be used with the small solder-in probe sockets that are available from many companies. Another option is to utilize a 1/4-inch (or shorter) probe ground or signal return connection, or to solder the probe directly into the circuitry. One can use an oscilloscope to diagnose clock ringing, characterize switching power supply noise, and spot crosstalk. Pulses of noise are often synchronized in time with crosstalk or other temporally related EMI issues. It can be also used in conjunction with near-field probes (H- or E-field). In fact, you can use one channel as a reference to find correlations between a known noise source and other signals by probing with the other. Today, many oscilloscopes have a math function called fast Fourier transform (FFT) that changes signals from the time domain to the frequency domain. While this could be potentially helpful in troubleshooting, one of the issues is lack of dynamic range. Most inexpensive oscilloscopes can only record eight bits of data, which means that very small signals can get lost in the noise floor and be hard to see (Figures 6.8 and 6.9).

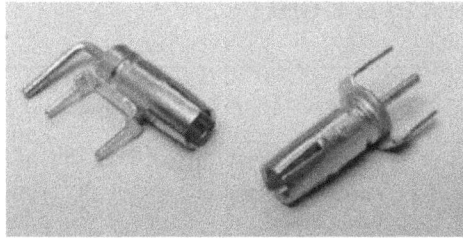

FIGURE 6.8   Miniature oscilloscope.

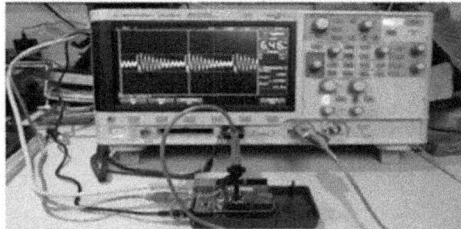

FIGURE 6.9   A high bandwidth digital oscilloscope.

6.14   EMI RECEIVERS

Rather than the very wideband SA technology, the EMI receiver is a tuned measurement instrument with a narrow frequency bandwidth exposed to the detector. This keeps out-of-band signals from overloading the receiver and keeps gain compression and other types of distortion from happening. As specified by MIL-STD 461 and DO-160, the resolution bandwidth roll-offs of modern EMI receivers are typically 6 dB. Nonetheless, some lack CISPR bandwidths and non-peak detectors (without add-on equipment). Outside its passband, the EMI receiver is far more tolerant of high-powered broadcast and two-way radio transmissions, making it perfect for use in outdoor test ranges.

GW-INSTEK GSP-818-EMI GSP-818:

Models	Key Features	Applications
 Frequency range: 9 kHz–1.8 GHz	1. Built-in AM/FM demodulation 2. Bandwidth zoom function 3. Measurement function: Bandwidth, freq. counter, noise marker, limit line 4. Built-in 20 dB pre-amplifier standard 5. EMI Filter & detector 6. Interface: Lan, USB 7. Screen: 10.4″ SVGA Output (800*600) 8. Options: Tracking generator	Serves a frequency range of 1,8 GHz and specifies testing criteria for RF products during development and production.

N9048B PXE EMI Receiver, 2 Hz to 44 GHz

Standard: **CISPR 16-1-1: 2010 and MIL-STD-461F** EMI receiver requirements

1. Frequency: 3 Hz to 3.6, 8.4, 26.5, or 44 GHz,
2. CISPR, MIL-STD 6 dB, and standard 3 dB resolution bandwidths.
3. Quasi-peak, EMI-average, RMS-average, and diagnostic detectors.
4. Extensive diagnostic capability.
5. Intuitive multi-touch user interface.

1. Functionality for EMC measurement are simultaneous detectors, signal lists, scan tables, limit lines, and correction factors.
2. Real-time scan provides uninterrupted signal capture and analysis in a bandwidth of up to 350 MHz, displaying simultaneously in frequency domain, time domain, and spectrogram view.

N9038A MXE EMI Receiver

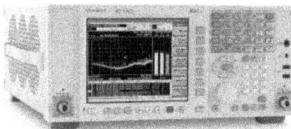

CISPR, MIL-STD 6 dB, and standard 3 dB resolution bandwidths

1. Frequency 3 Hz to 3.6, 8.4, 26.5, or 44 GHz.
2. Quasi-peak, EMI-average, RMS-average, and diagnostic detectors.
3. Extensive diagnostic capability

1. Functionality for EMC measurement are simultaneous detectors, signal lists, scan tables, limit lines, and correction factors.
2. It is capable of both time domain and frequency domain.

6.15 NEAR-FIELD PROBES

Sniffer probes, also known as near-field probes, measure electric or magnetic fields nearby to pinpoint the source of emitted signals. Electric field versions are essentially stub antennas, with the end of the coaxial wire put into a resistive load (e.g., 50 ohms). The magnetic field version is a small loop that sometimes actually ended in a load (e.g., 50 ohms). The effective frequency range and the capacity to pinpoint the source are both constrained by the size of the stub or loop, which in turn defines the probe's sensitivity. Users can easily make these near-field probes out of regular or semi-rigid coaxial cables.

It's important to keep in mind that near-field probes may be extremely reliable or completely false. More sensitive, larger-sized sensors can take up ambient readings from powerful broadcast radio and television. Sensitivity to environmental signals can be measured, for example, by monitoring the FM broadcast radio band's frequency range of 88 to 108 MHz with a specific probe. To avoid missing your favourite show because of background noise, adopt precaution when scanning the oscilloscope or SA for your favourite station. In order to accomplish this, pull the probe away from the device and, if at all feasible, power down the unit. If the signal does not disappear, you can consider that frequency to be part of the background noise and disregard it. When positioned in the same plane as the wire, cable, or circuit trace, H-field probes couple together most effectively. This is because it enables the greatest number of H-field lines of flux to pass through the loop.

Most H-field loop probes are insulated for E-fields; however, the capacitance between the shielding and circuit can generate a high-frequency resonance (about 700 to 1,000 MHz, depending on the probe design). It is possible to prevent this resonance by building unshielded loops; however, doing so will result in less effective rejection of E-fields. Typically, H fields are increased in proximity to

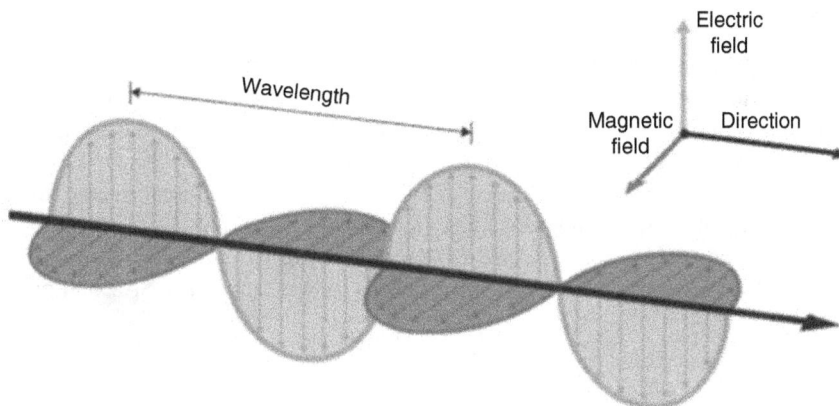

FIGURE 6.10   EM wave composed of both E and H fields.

circuit traces due to their low impedance and, by extension, high current. Hot signal sources, wires, or circuit traces are typically located with the aid of H-field probes. H-field probes are commonly used to track down the origin of a high-frequency signal, as well as its accompanying cable or circuit trace. However, E-field sensors excel at finding leakage in chassis seams or holes where E-fields may be very concentrated (Figure 6.10).

U1818A 100 KHz to 7 GHz active differential probe

Models	Key Features	Applications
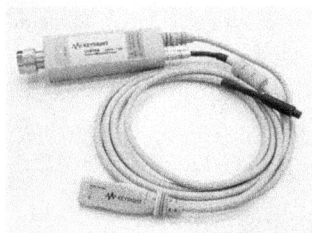	1. Wide frequency range with +/−1.5 db of flat frequency response. 2. At 10 MHz to 7 GHz, the noise floor is less than −130 dBm/Hz.	1. Makes sure that measurements are accurate and helps you develop the most effective product specs. 2. Allows us to take readings with a relatively weak signal.

ETS-7405 E & H Near-Field Probe Set

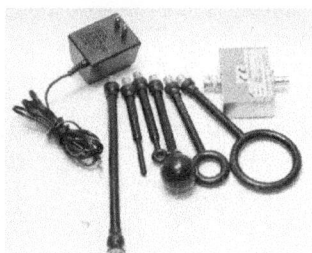	1. Frequency range: 100 KHZ to 1 MHz. 2. Noise Figure: 3.5 dB. 3. Saturated output Power: 12.0 dBm.	1. An additional pre-amplifier is now available to boost signal strength. 2. Linear response. 3. Unique forms and measurements for the sensitive. 4. Discovers the origins of electromagnetic and radio waves. 5. Broadband frequency range.

6.16   SUMMARY

During the manufacturing process, EMC testing is essential for risk management. It is important that devices can function in close quarters without causing interference to one another or being hindered by ambient noise. Products like microwaves, cellphones, laptops, and satellite TV dishes need to be subjected to EMI and eEMC testing to guarantee they will not generate or pick up any interference that could be harmful, as well as to ensure they will function normally when exposed to interference. Specific EMC testing procedures are defined by the nature of the equipment being evaluated, its intended application, and applicable regulatory requirements. A typical EMC testing lab may utilize ESD generator, LISN, Network Analyser, power amplifiers, SAs, and more.

BIBLIOGRAPHY

1. Leopoldo Angrisami and Antondo Pietrosanto, *A Technique for Electromagnetic Interference Measurements on Instruments*. IEEE, DOI:0018-9456, 1998.
2. Shubhankar Marathe, Zongyi Chen, Kaustav Ghosh, Hamed Kajbaf, Stephan Frei, Morteon Sorenson, David Pommerenke, and Jin Min, *Spectrum Analyzer-Based Phase Measurement for Near-Field EMI Scanning*. IEEE, DOI:0018-9375, 2019.
3. Zhenyu Zhao, Lianming Wang, Jufang Chen, Yang Lv, Zengyu Cai, and Yuan Feng, *The Design and Implementation of Signal Generator Based on Direct Digital Synthesiser*. IEEE, pp. 920–923, DOI:978-1-5090-3822, 2017.
4. Shinya Kaketa, Katsumi Fujji, Akira Sugiura, Yasushi Matsumoto, and Yukio Yamanaka, *EMI Antenna Calibration using Antenna Impedance Measurement*. IEEE, pp. 792–795, DOI:0-7803-7831, 2003.
5. Jungkuy Park, Guseon Mun, Dachoon Yu, Boweon Lee, and Woo Nyun Kim, *Proposal of Simple Reference Antenna Method for EMI Antenna Calibration*. IEEE, pp. 90–95, DOI:978-1-4577-0811, 2011. https://www.keysight.com
6. Proposing an Improved DC LISN for Measuring Conducted EMI Noise IEEE Transactions on Electromagnetic Compatibility Manuscript Received April 8, 2020; Revised August 11, 2020; accepted September 6, 2020. (Corresponding author: Reza Amjadifard) Digital Object Identifier 10.1109/TEMC.2020.3025459. DOI: 10.1109/TEMC.2020.3025459
7. Electromagnetic Compatibility Requirements for Space Equipment and Systems, AIAA Standard S-121A–2017, 2017. https://arc.aiaa.org/doi/book/10.2514/4.105203
8. D. L. Sengupta and V. V. Liepa, *Applied Electromagnetics and Electromagnetic Compatibility*, 2nd ed. Hoboken, NJ: Wiley, 2006.
9. Andrei-Marius Silaghi, Aldo De Sabata, Adrian Graur, and Radu Fechet, *Simulation of Surge Pulse Generator and Applications in Automotive Immunity Testing*. IEEE, pp. 117–120, DOI:978-1-7281-6870, 2020.

7 EMI Using MATLAB

7.1 ABOUT THE SOFTWARE

The MathWorks company created the multi-paradigm numerical computing environment and proprietary programming software known as MATLAB (Matrix laboratory). It is a fourth-generation high-level programming language and interactive environment for numerical calculation, visualization, and programming. It is also possible for matrix manipulation, function and data plotting, algorithm implementation, user interface creation, and interface with other language programmes such as C, C++, C#, Java, Fortran and Python. The MuPAD symbolic engine is used by an optional toolbox in MATLAB, which is primarily designed for numerical computing but also has access to symbolic computing capabilities. Graphical multi-domain simulation and model-based design for embedded and dynamic systems are added via an additional programme called Simulink.

7.1.1 FEATURES OF MATLAB

- It offers an interactive environment for iterative exploration, design, and problem-solving. It is a high-level language for numerical computing, visualization, and application development.
- It offers a significant library of mathematical operations for solving ordinary differential equations and performing linear algebra, statistics, Fourier analysis, filtering, optimization, and numerical integration.
- It offers tools for making personalized plots as well as built-in graphics for visualizing data.
- The programming interface for MATLAB provides tools for development that enhance code quality, maintainability, and performance.
- It offers resources for creating programmes with unique graphical user interfaces.
- It offers integration tools for MATLAB-based algorithms with third-party software and programming languages like C, Java, .NET, and Microsoft Excel.

7.1.2 USES OF MATLAB

With applications in physics, chemistry, math, and many engineering courses, MATLAB is a widely used computational programme in science and engineering. It is utilized in a variety of applications, such as control systems, test and measurement, signal processing and communications, image and video processing, computational finance, and computational biology.

7.1.3 SIMSCAPE

Within the Simulink® platform, SimscapeTM enables you to quickly develop models of physical systems. Simscape allows you to create models of physical parts based on their actual connections, which are then easily integrated into block diagrams and other modelling paradigms. By arranging basic parts into a schematic, you may mimic systems like electric motors, bridge rectifiers, hydraulic actuators, and refrigeration systems. Simscape add-on products give users access to more sophisticated parts and analysis tools.

Simscape aids in the development of control systems and the testing of system performance. The Simscape language, which is based on MATLAB® and allows text-based writing of physical modelling components, domains, and libraries, allows you to develop bespoke component models. MATLAB

DOI: 10.1201/9781003362951-7

variables and expressions can be used to parameterize your models, and Simulink can be used to develop the control schemes for your physical system. Simscape offers C-code creation for deploying the models to different simulation environments, such as hardware-in-the-loop (HIL) systems.

7.1.4 SIMPOWER SYSTEM TOOLS—SIMULATING MOTOR DRIVES

IGBTs, MOSFETs, and GTOs are examples of forced-commutated electronic switches used in the variable speed control of AC electrical machinery. Today, asynchronous machines fed by voltage-sourced converters (VSC) with pulse width modulation (PWM) are gradually replacing DC motors and thyristor bridges. You may achieve the same flexibility in speed and torque control as with DC machines by using PWM in conjunction with contemporary control approaches like direct torque control or field-oriented control.

7.1.5 MATLAB SIMULINK MODEL

Model-based and simulation-based design Simulink is a block diagram environment for model-based design and multi-domain simulation. It provides continuous testing and verification of embedded systems, simulation, automatic code creation, and system-level design. For modelling and simulating dynamic systems, Simulink offers a graphical editor, adaptable block libraries, and solvers. Because of its integration with MATLAB®, you may export simulation results for additional analysis and include MATLAB algorithms into models.

7.1.6 KEY FEATURES

- Visual editor for creating and maintaining hierarchical block diagrams.
- Predefined block libraries for both continuous-time and discrete-time modelling.
- Fixed-step and variable-step ODE solvers in the simulation engine.
- Data and scope displays for simulation results.
- Tools for handling model files and data, including project management software.
- Tools for model analysis that can improve model design and speed up simulation.
- The MATLAB function block for adding MATLAB algorithms to models.
- Models can import C and C++ code with the legacy code tool.

7.1.7 CREATING NEW MODEL

Simulink Editor is used to build new models. The below process can be explained for creating a new model.

1. Open MATLAB®. From the MATLAB toolstrip click the **Simulink** button shown in the Figure 7.1

2. Click on the **Blank Model** template as shown in Figure 8.1. The Simulink Editor opens as shown in the Figure 7.2.

3. Choose Save as from the File menu, as displayed in Figure 8.2. Enter a name for your model in the File name text box, such as simple model and press Save. The file extension is used to save the model. slx.

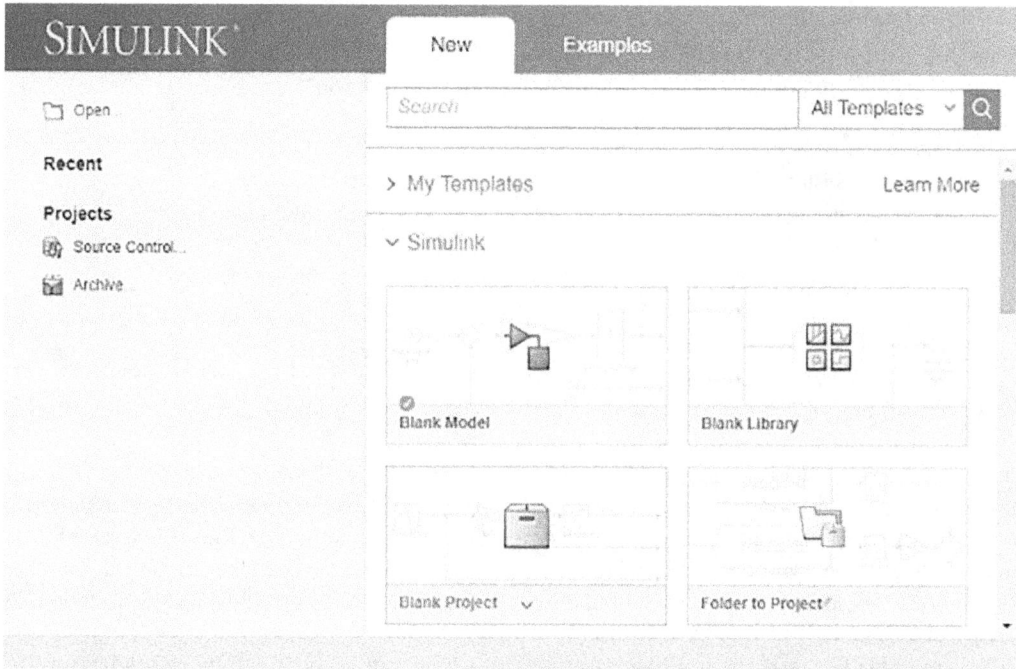

FIGURE 7.1 Simulink editor.

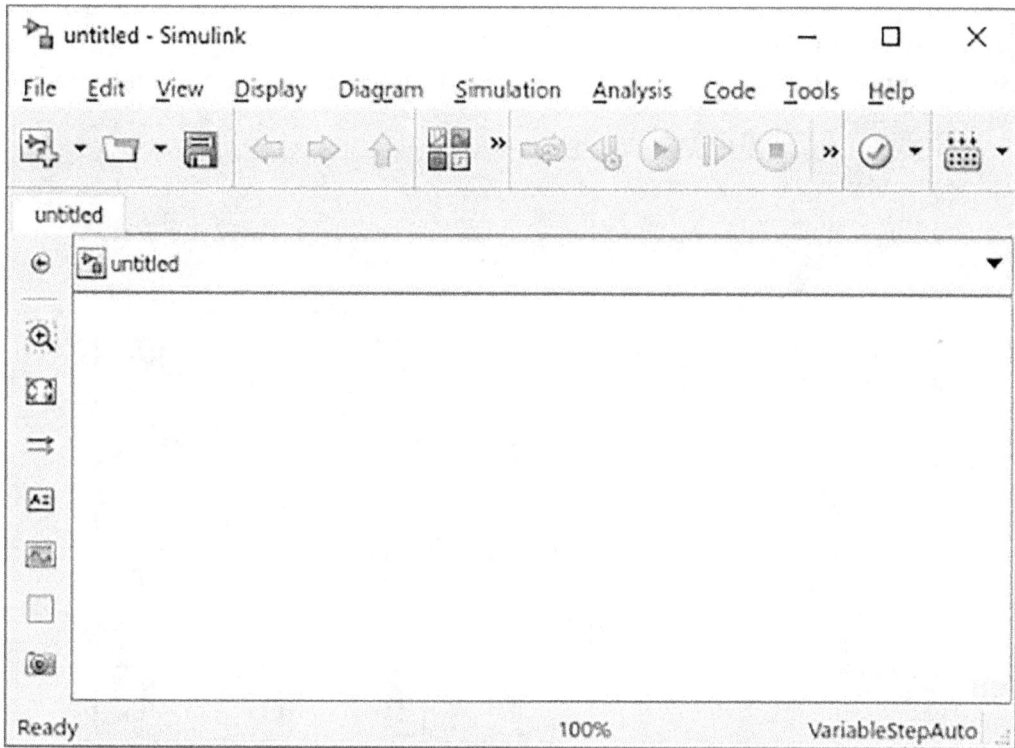

FIGURE 7.2 Blank model.

7.1.7.1 Simulink Library

Simulink offers a collection of block libraries that are arranged in the library browser according to functionality. The majority of workflows share the following libraries:

1. Continuous—Blocks for continuous state systems.
2. Discrete—System building blocks for discrete state systems.
3. Math Operations—Building blocks for logical and algebraic equations.
4. Blocks that store and display the signals that connect to them are known as sinks.
5. Sources—Components that produce the signal values used to power the model.

Click the Library Browser button on the Simulink Editor toolbar 🖵 shown in the Figure 7.3

As seen in Figure 7.3, configure the Library Browser so that it always appears on top of other desktop windows. Choose the Stay on top option from the Library Browser toolbar ⬛. Choose a category, then a functional area, in the left pane to browse the block libraries. Enter a search item to search across all of the accessible block libraries. For example, to find the Pulse Generator block, enter "pulse" in the search box on the browser toolbar, then press the enter key. Simulink searches the libraries for blocks with the word "pulse" in either the block name or the block description, as illustrated in Figure 7.4

It is possible to learn about a block using help, which can be obtained by right-clicking the Pulse Generator block. The block's reference page appears when the Help browser first loads. Typically, blocks have a number of parameters. By double-clicking the block, you may access all parameters.

7.1.7.2 Adding Blocks to a Model

Go through the collection in the search list and add the blocks to begin creating the model as instructed below.

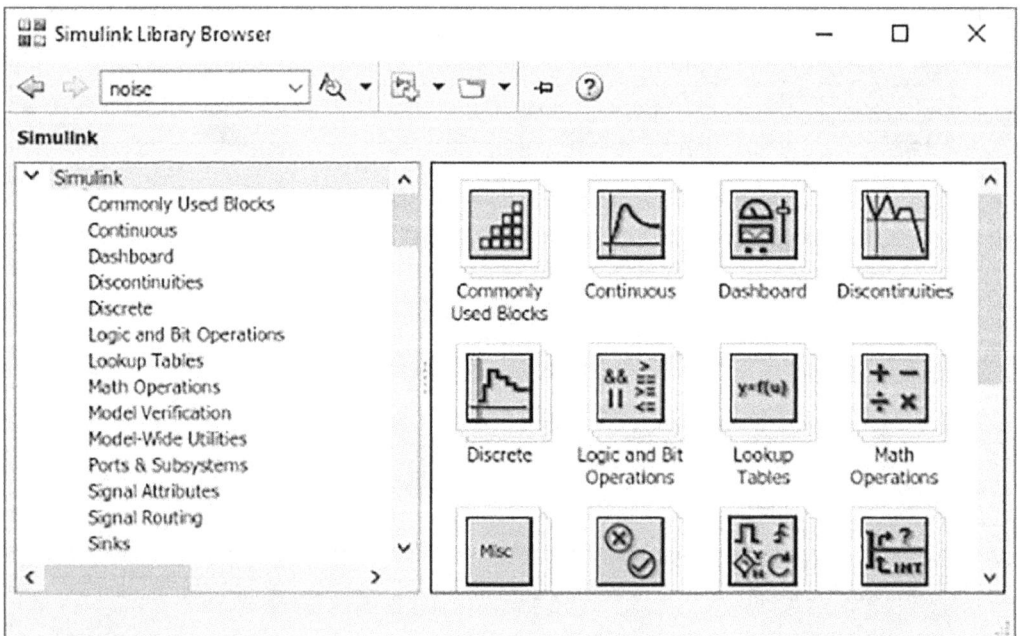

FIGURE 7.3   Library browser.

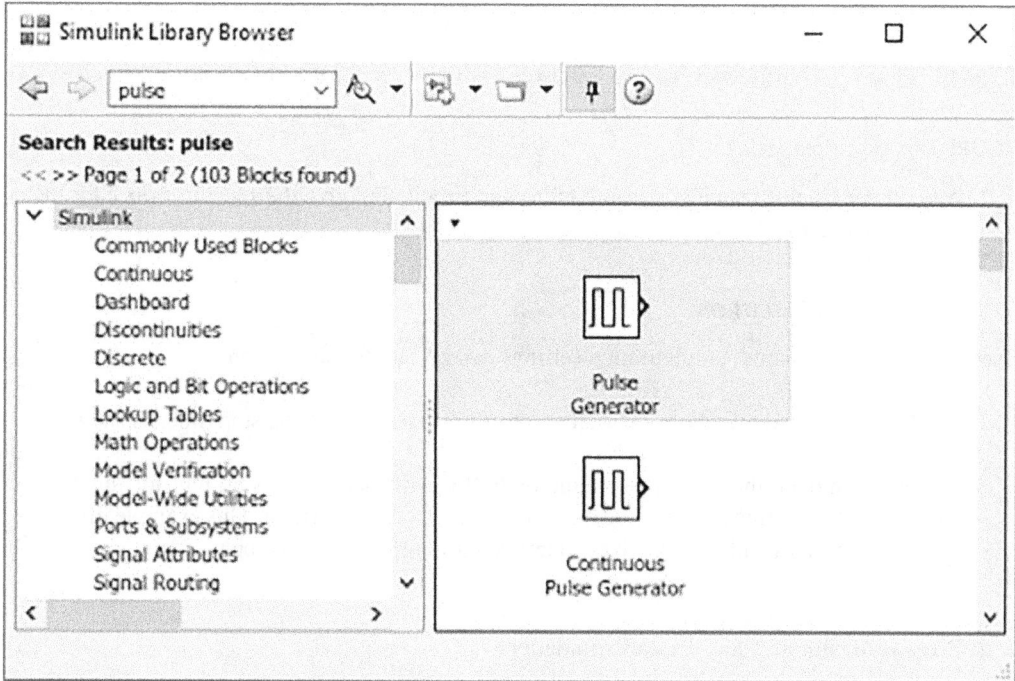

FIGURE 7.4   Library browser for selecting pulse generator.

1. Drag the Pulse Generator block to the Simulink Editor from the Sources library. A copy of the Pulse Generator block appears in your model with a text box for the amplitude parameter's value, as shown in Figure 7.5

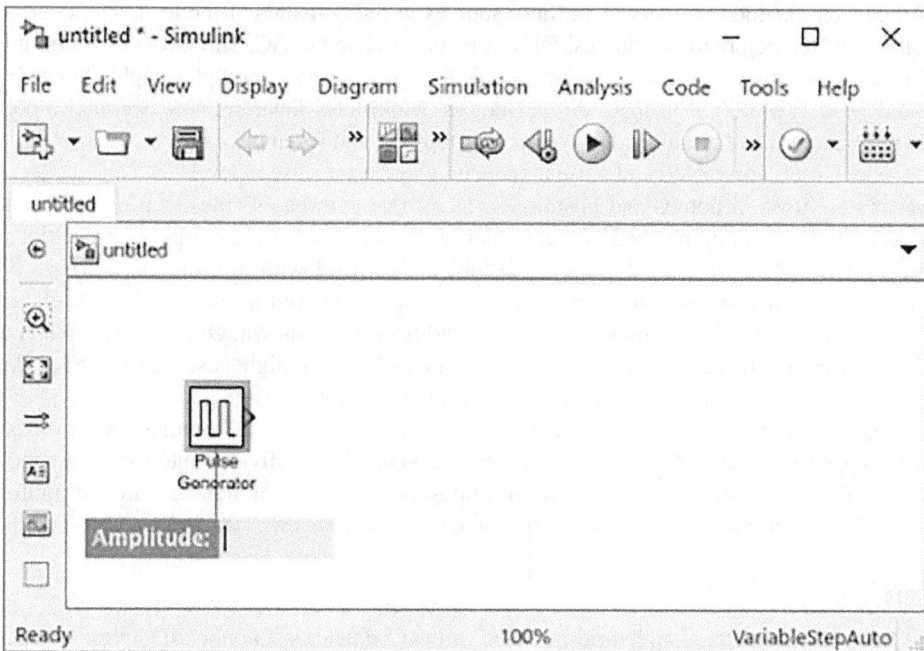

FIGURE 7.5   Library browser for adding amplitude parameter's value.

FIGURE 7.6 Stop time setting.

2. Add all the blocks, which you wanted for your circuit, and give the specific values for the respective blocks

7.1.8 Simulation Execution

Once the model is prepared, configuration settings can be sent for simulation.

1. On the model window, adjust the toolbar value to set the simulation stop time, as shown in Figure 7.6.

    For this model, the default stop time of 10.0 is suitable. There is no unit for this time value. Simulink's time unit is based on how the equations are written. This example mimics a motion for 10 seconds; alternative models can use milliseconds as the measure of time.

2. Press **Run** button . To start simulation.

7.2 VOLTAGE SOURCE THREE-LEVEL INVERTER

7.2.1 Introduction

In this section, a setup with a motor drive system is used to suggest an energy management approach while taking into account the real switching in MATLAB software.

Non-conventional energy sources are typically renewable ones like solar, wind, tidal, geothermal, etc. The concept behind this is to adapt these sources to the demands of the load. The majority of loads should be fed before these sources' DC output is converted to AC. Simple two-level traditional inverters create a square wave, which is unsuitable for the majority of complex applications. A pure sinusoidal wave is preferred in these circumstances. Multi-level inverters now become important. The output waveforms of a multilayer inverter are step-like and resemble a near sine wave. The sine wave is more closely resembled and noise is reduced with an increase in level.

Inverters are used in homes and businesses. Due to contaminated sinusoidal output, standard basic-level inverters cannot be used in many microgrid, industrial, and aerospace applications that use photovoltaic (PV), wind, fuel cells, etc. Problem is solved with multi-level inverters. Multi-level inverters produce sinusoidal waveforms. More levels improve the sinusoidal wave. A normal three-level inverter MATLAB simulation is done and results are shown. One voltage input is used. Switched capacitor cells determine output levels. Repeated blocks might cascade to different levels. This layout avoids concerns with several active switches and complicated gating circuits.

A group of power semiconductor switches and capacitors known as a multi-level inverter can output a stepped sine wave. Multi-level inverters can generally be divided into three categories—cascaded multi-level inverters, flying capacitor multi-level inverters, and diode clamped multi-level inverters. Multi-level inverters have low distortion output voltages, low dv/dt stress, and other benefits.

7.2.2 Design of Converter

Studies have been conducted on hybrid-powered railway vehicles with onboard energy storage systems. In a hybrid diesel motive unit (DMU), the diesel engine output power is regulated to reduce fuel consumption. In this case, engine output power is sufficient to control LiBs of SOC. EDLCs are connected in parallel to the overhead line voltage by a two-quadrant chopper, and energy storage device (ESD) management achieves peak power cut of line voltage. EDLCs have no load power constraints in this case. A hybrid 3Lv VSI uses an ESD voltage increase to improve acceleration and deceleration to save energy. This circuit's PWM limits the ESD's power. PWM links the inverter's output voltage, power, and ESD SOC in this circuit.

VSI is anticipated to be developed through these investigations in order to conserve energy. Inverters are most commonly used in the following applications:

- Variable speed induction motor drives
- AC drives with adjustable speeds
- Induction heating
- Uninterruptible power supply (UPS)
- Standby power supply
- HVDC power transmission variable voltage and variable frequency power supply
- Battery-operated vehicle drives

7.2.2.1 Circuit Topology

Lithium-ion batteries (LiBs), an example of an ESD, are linked in series with the line voltage on the DC input side. This connection serves as the three-level inverter's input voltage source. The inverter is able to increase its AC output voltage by adding the ESD voltage to the line voltage. With the two-quadrant DC chopper system for ESDs, this circuit architecture has the advantage of removing the large inductor and the loss of the power semiconductor devices. Comparing the hybrid 3-level VSI to the two-level conventional VSI, the hybrid three-level VSI increases its output AC voltage by using ESD. The ESD maintains the increased DC voltage. As a result, the ESDs' energy must stay within their lower and upper bounds. A six-step bridge inverter circuit topology is shown in Figure 7.7.

A six-step bridge inverter is a basic three-phase inverter. A minimum of six thyristors are used. Three-phase inverters are more prevalent than single-phase inverters for delivering power with adjustable frequency to industrial applications. A step in an inverter topology is described as a change in the firing from one thyristor to the next in the correct order. For a six-step inverter, each step would have a 60-degree interval for a 360-degree cycle.

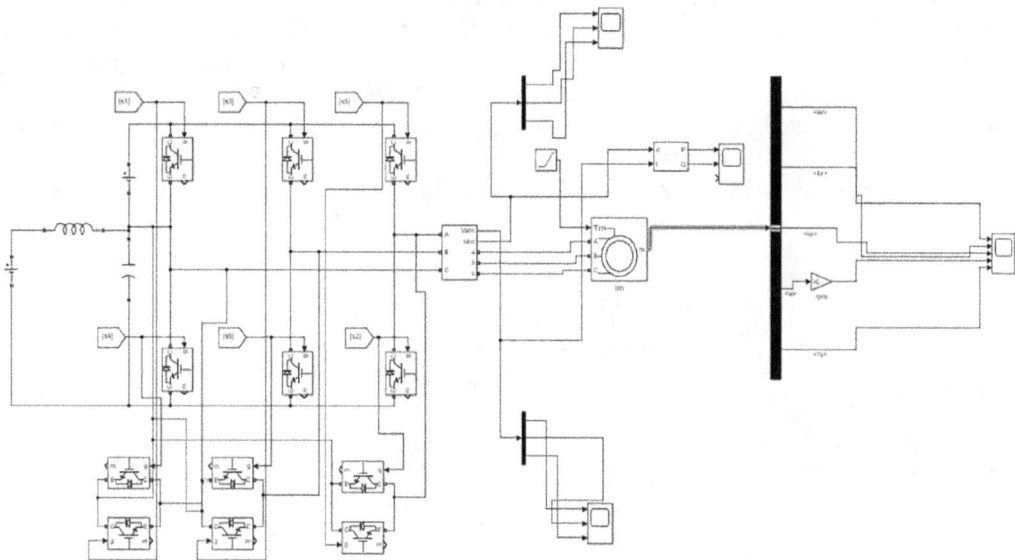

FIGURE 7.7 Circuit topology.

An inverter's job is to convert a dc input voltage to an AC output voltage with the correct magnitude and frequency. At a fixed or variable frequency, the output voltage could be either fixed or variable. By adjusting the input DC voltage and keeping the inverter gain constant, a variable output voltage can be achieved. However, if the DC input voltage is fixed and uncontrollable, it is still possible to achieve a variable output voltage by changing the inverter's gain, which is typically done through PWM control. The ratio of the AC output voltage to the DC input voltage can be used to define the inverter gain. Inverters should have sinusoidal output voltage waveforms. The waveforms of actual inverters, however, are not sinusoidal and include certain harmonics. Square-wave or quasi-square-wave voltages may be acceptable for low- and medium-power applications, whereas low-distorted sinusoidal waveforms are needed for high-power applications. With the advent of high-speed power semiconductor devices, switching strategies can considerably reduce or minimize the harmonic contents of output voltage. Table 7.1 shows the switching pulse table.

In this mode, each switch conducts for 180° duration in each cycle of the output voltage. Each leg of three-phase inverter consists of two switches, one is a part of positive group switches and other is a part of negative group switches. When a positive group switch of a leg conducts for 180° duration, its corresponding negative group switch of same leg conducts for next 180° duration as one complete cycle is equal to 360° duration. For example, if the switch S1 of leg 1 is ON for 180° duration during 0° to180°, then the switch S4 of leg 1 is ON for 180° duration during 180° to 360°. In this switching scheme, three switches from three different legs are conducted at a time. Hence, two switches from the same leg are not switched on simultaneously. The one complete cycle of switching can be operated into six modes and each mode operates only for 60° duration.

7.2.3 Simulation of Circuit in MATLAB

The complete and simplified synchronous machines, the asynchronous machine, and the permanent magnet synchronous machine are the most popular three-phase machines found in the machines collection. Each machine has two operating modes: generator mode and motor mode. They can be used to model electromechanical transients in a network of electrical devices by combining them with linear and non-linear components like transformers, lines, loads, breakers, etc. To simulate drives, they can also be used in conjunction with power electrical equipment. You may mimic devices such as diodes, thyristors, GTO thyristors, MOSFETs, and IGBTs using the blocks in the Power Electronics library. A three-phase bridge could be constructed by connecting many blocks together. For instance, six IGBTs and six antiparallel diodes would be needed for an IGBT inverter bridge. Firing circuit blocks have to be added as shown in Figure 7.8.

And inverter circuit blocks are also added in a same Simulink model as shown in Figure 7.9.

TABLE 7.1
Switching Pulse Table

FIRING ANGLE (°)	TRIGGERING SCR
0–60	T1, T6, T5
60–120	T1, T6, T2
120–180	T1, T3, T2
180–240	T4, T3, T2
240–300	T4, T3, T5
300–360	T4, T6, T5

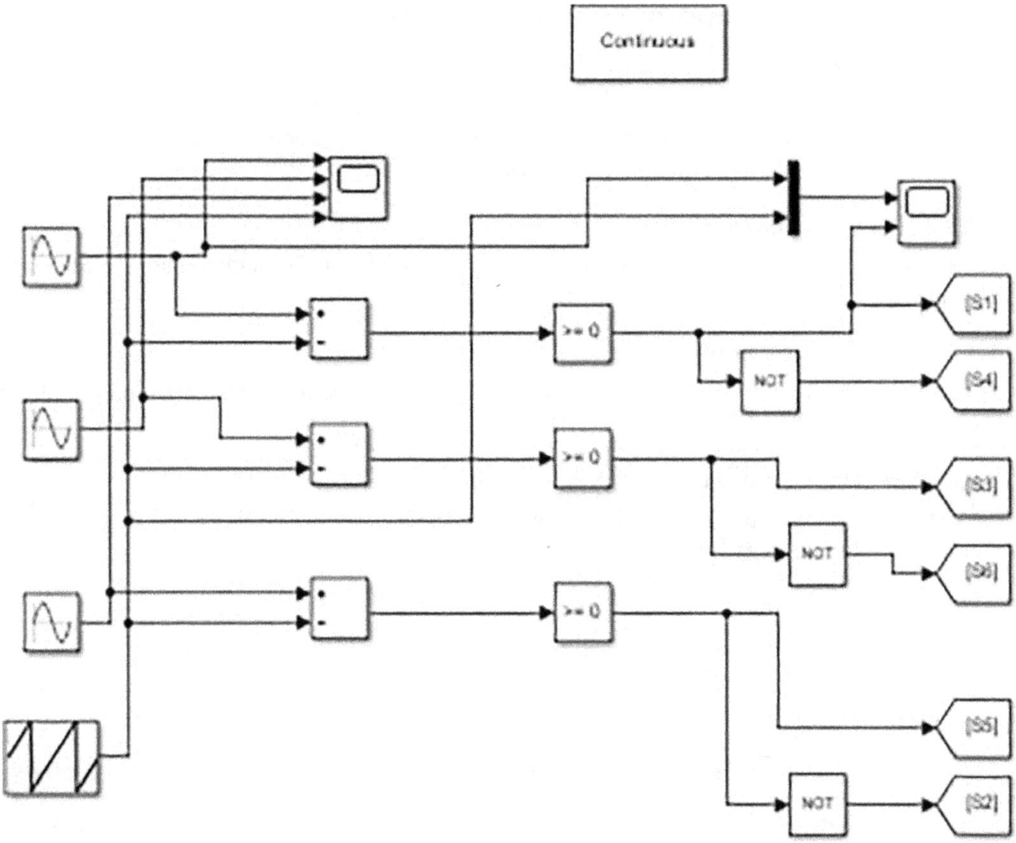

FIGURE 7.8 Firing circuit.

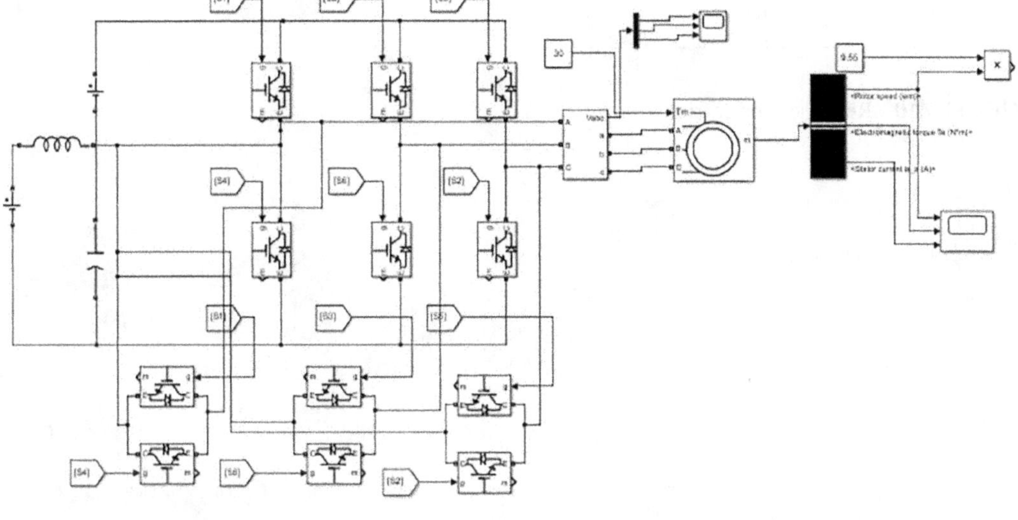

FIGURE 7.9 Inverter circuit.

7.2.4 Output of Each Block of a Circuit in MATLAB

7.2.4.1 Sinusoidal Pulse Width Modulation Technique

A sinusoidal PWM is used for the gate pulses for the switches used in the inverter. SPWM technique is obtained by sine wave generator and repeating sawtooth wave generator with the help of logic gates SPWM technique can be obtained as shown in Figure 7.10.

7.2.4.2 SPWM Technique Output

With the help of the logic gates and the sine wave and sawtooth wave generator, the SPWM output can be achieved, as shown in Figure 7.11.

7.2.4.3 Phase Voltage of Inverter

The phase voltage of nearly 150 V is scoped with the help of scope in the respective phases of the circuit, as shown in Figure 7.12.

7.2.4.4 Line Voltages of Inverter

Line voltages are the same as the supply voltage of 200 V that is three-level in a 180-degree mode, as shown in Figure 7.13.

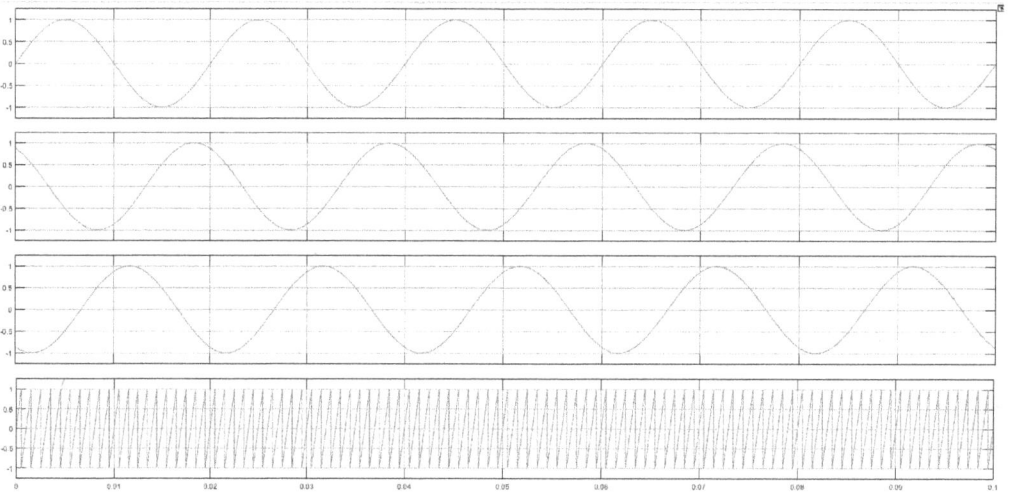

FIGURE 7.10 Reference and carrier wave.

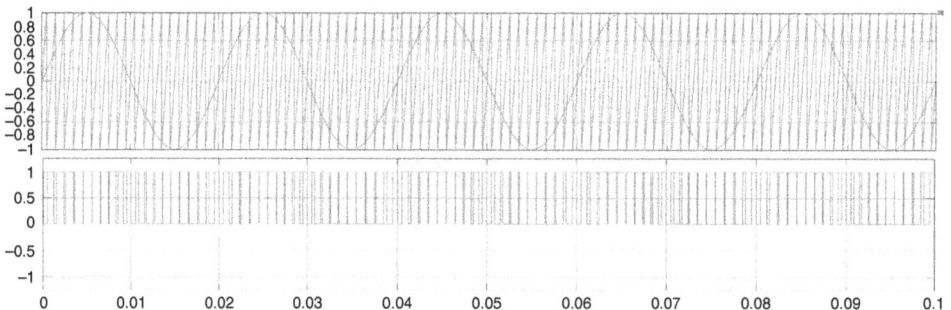

FIGURE 7.11 SPWM technique output.

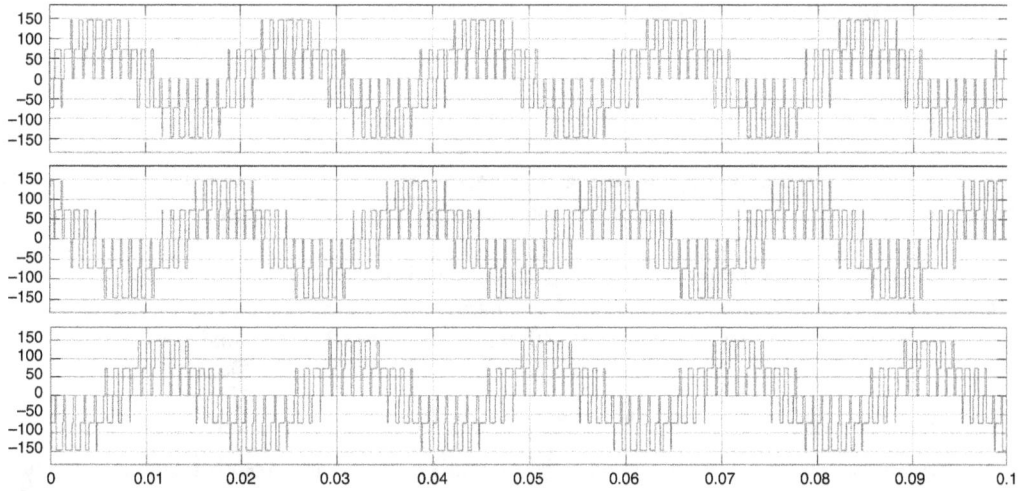

FIGURE 7.12 Phase voltage of inverter.

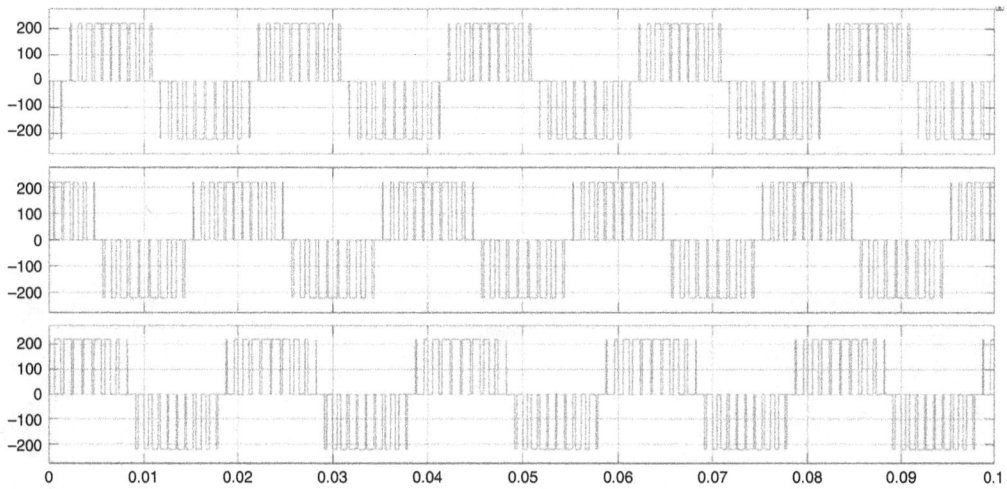

FIGURE 7.13 Line voltages of inverter.

7.2.4.5 Stator Voltage of Induction Motor

Stator voltage of the induction motor is 100 V scoped by using de-mux from the IM, as shown in Figure 7.14.

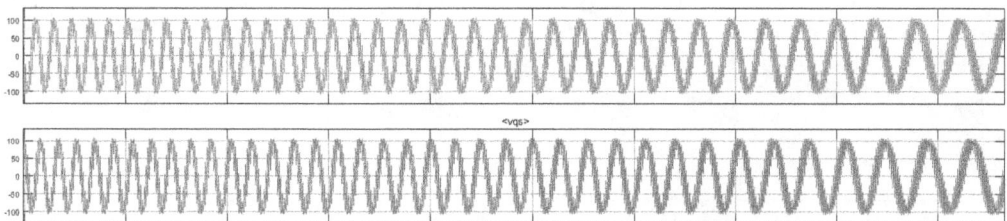

FIGURE 7.14 Stator voltage of induction motor.

7.2.4.6   Rotor Current of Induction Motor

Three-phase rotor current of 18 A is obtained from the induction machine, as shown in Figure 7.15.

7.2.5   S-Parameter Calculation of a Circuit

S-parameters can be obtained by the using rf budget analyser; the touchstone file or s2p file is important to get the S-parameter plots in the analyser. The MATLAB programme is used to get the s2p file, as shown in Figure 7.16.

RF budget analyser with the s-parameter block window is shown Figure 7.17.

Once this execution is over, the corresponding graphs S11 S12 S13 S14 are plotted as shown in Figures 7.18 and 7.19, and they are obtained with the help of the RF budget analyser.

7.2.6   Conduction Emission of a Circuit

Conducted emissions impacts' power quality brought on by electrical and magnetic coupling, semiconductor device electronic switching and other electromagnetic compatibility-related problems in electrical engineering. By preventing or reducing their deliberate generation, propagation,

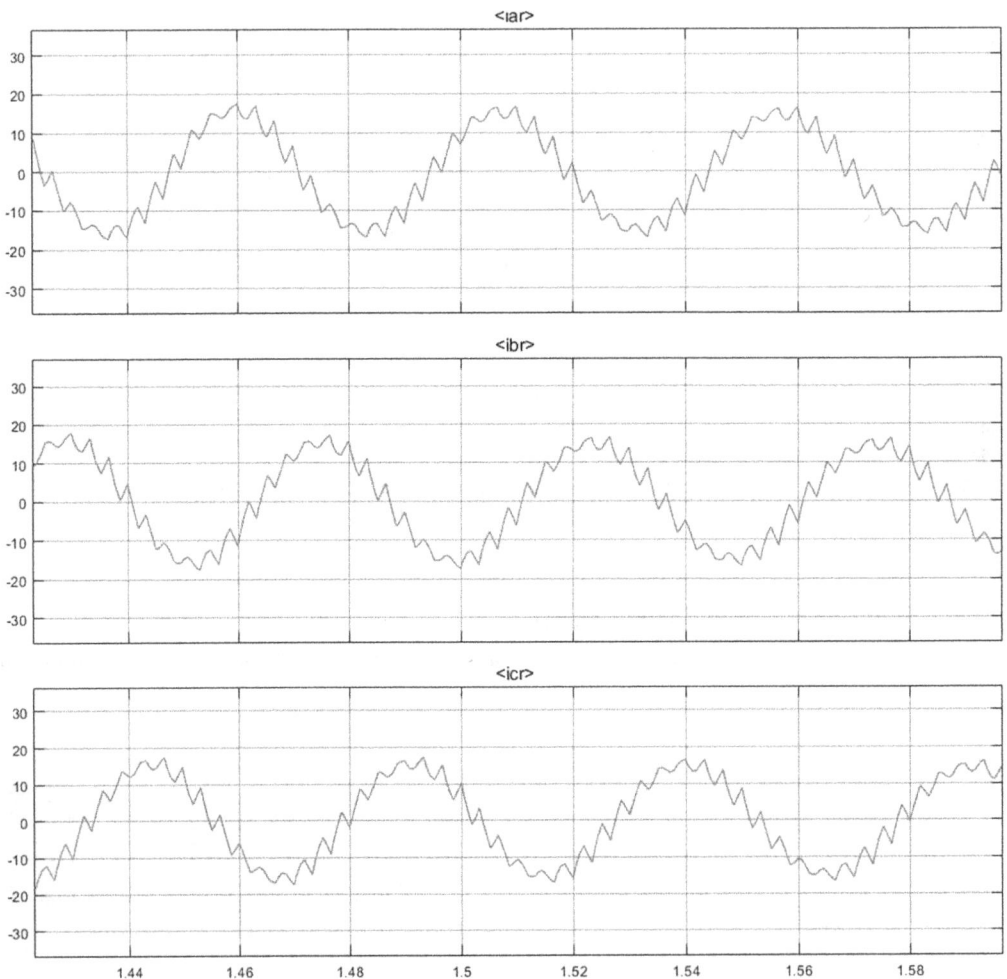

FIGURE 7.15 Rotor current of induction motor.

```
1 -   syms F
2 -   syms I [3 2 1]
3 -   syms V [6 12]
4 -   syms Z0 La Lb M s
5
6 -   nI=6; % number of branch currents
7 -   nV=4; % number of node voltages
8
9     % F = [Fconstitutive; Fconservative]
10 -  F = [
11        V1 - Z0*I1
12        V1 - V2 - (La-M)*I3*s
13        V2 - M*I4*s
14        V2 - V3 + (Lb-M)*I5*s
15        V3 - I2*Z0
16        I1 + I3
17        I4 - I5 - I3
18        I2 + I5
19        ]
20 -  J = jacobian(F,[I; V]);
21 -  syms rhs [nI+nV 2]
22 -  syms x v S t
23
```

(a)

```
% Compute S-parameters of cascade
rhs(:,:) = 0;
rhs(nI+1,1) = 1/Z0;  % rhs for driving input port
rhs(nI+nV,2) = 1/Z0  % rhs for driving output port
x = J \ rhs;
v = x(nI+[1 nV],:);
S = (2*v - eye(2));
matlabFunction(S,'file','mutualInductorS.m','Optimize',false);

La = 0.000001;
Lb = 0.000001;
Z0 = 50;
k = 0.763;
M = k*((La*Lb)^(1/2));

freq = linspace(1e9,2e9,10);
s = 2i*pi*freq;
s_param = zeros(2,2,10);
for index = 1:numel(freq)
    s_param(:,:,index) = mutualInductorS(Lb,Lb,M,Z0,s(index));
end

Sobj = sparameters(s_param,freq);
rfwrite(Sobj,'mutualtalinductance.s2p');
n = nport('mutual inductance.s2p');
b = rfbudget(n,20e9,-30,10e3);
show(b)
```

(b)

FIGURE 7.16   Programme of S-parameter.

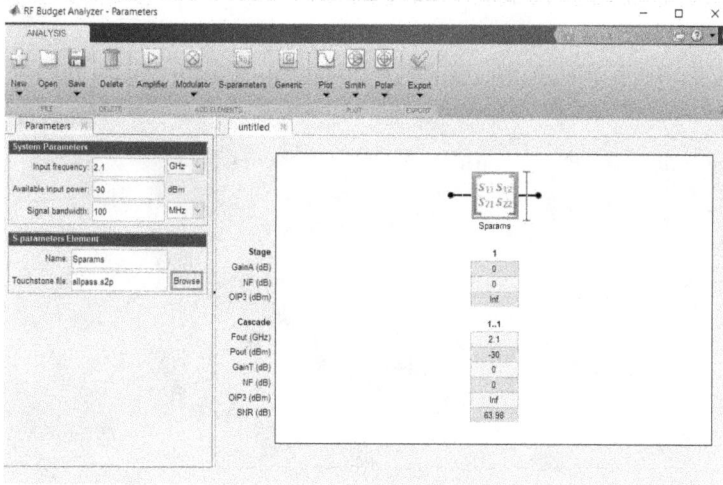

FIGURE 7.17   RF budget analyser with the S-parameter block.

and receipt of electromagnetic energy, these have an impact on the capability of all interconnected system devices in the electromagnetic environment. Conducted emissions are a component
of electromagnetic interference in circuits that primarily cause problems with the quality of the
power that is delivered. This interference is brought on by harmonics that are produced by linear
and non-linear loads that are present in the electric system, primarily as a result of the growing
use of switched mode power supplies and other consumer electronics. The provided electric power
quality from the main electricity system impacts how well electrical home equipment function as a
result of these combined interferences. These can include dwindling light output from bulbs, flick-
ering and inadequate heating of the induction coil in kettles, as well as heating elements in other

(a)

(b)

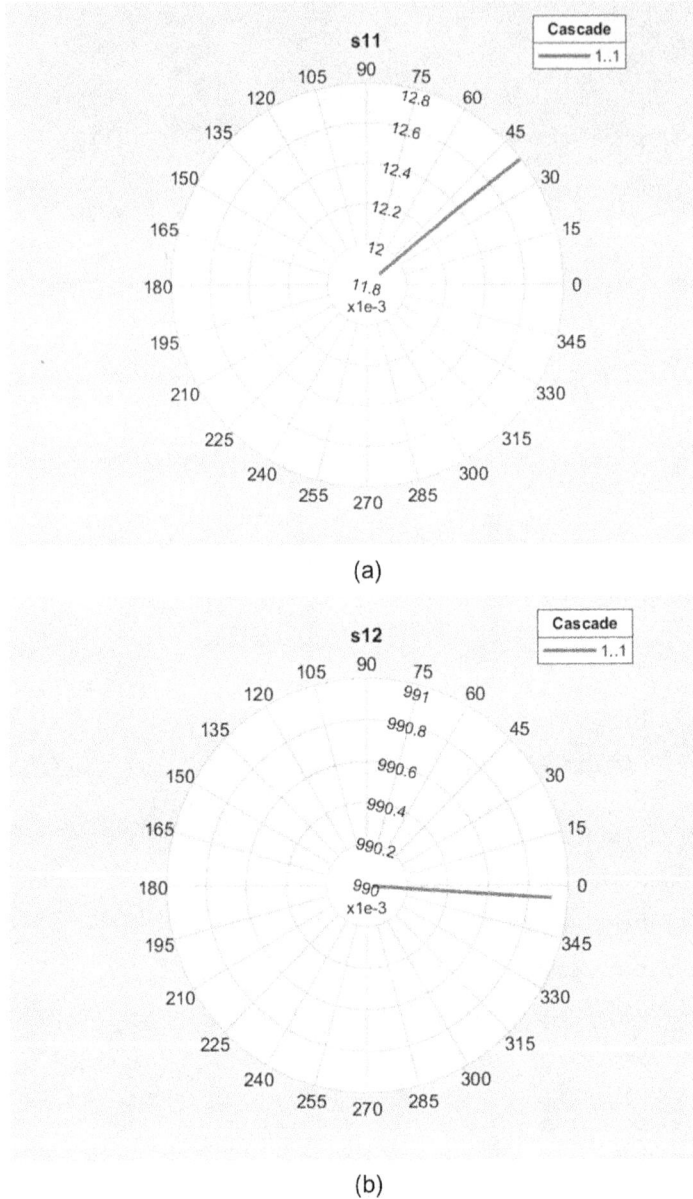

FIGURE 7.18　Plots of S11, S12.

commonly used home appliances. The circuit diagram of Line impedance stabilization network (LISN) is shown in Figure 7.20.

Conductive emissions on power lines are measured using LISNs. The LISN delivers power to the test-related equipment from a standard wall outlet (EUT). The LISN separates any RF noise produced by the EUT and feeds it to the spectrum analyser for recording or measurement.

A LISN's primary job is to give the power line stability, normalized impedance. This function is crucial because, depending on how and where the wiring is connected, the power line impedance through a typical wall outlet might vary significantly. The difference between the impedances of the power line and the EUT has a direct impact on the quantity of EUT noise that is present at the LISN measurement port. Only a portion of the noise voltage may actually reach the measurement

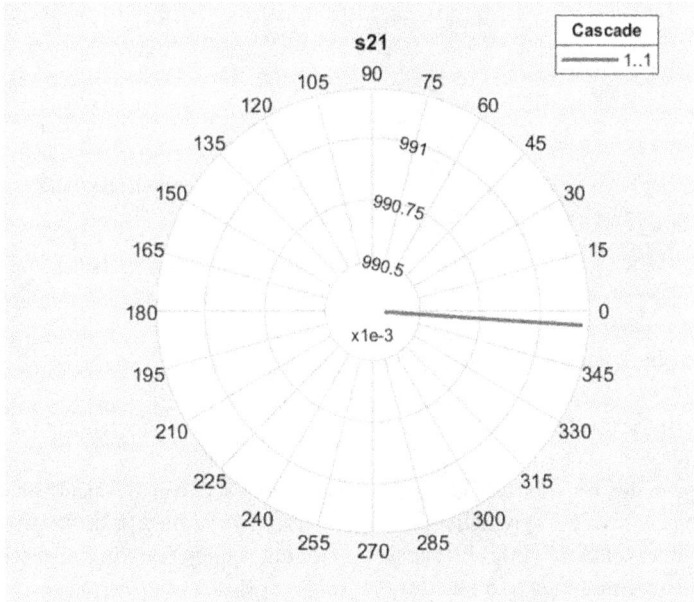

(a)

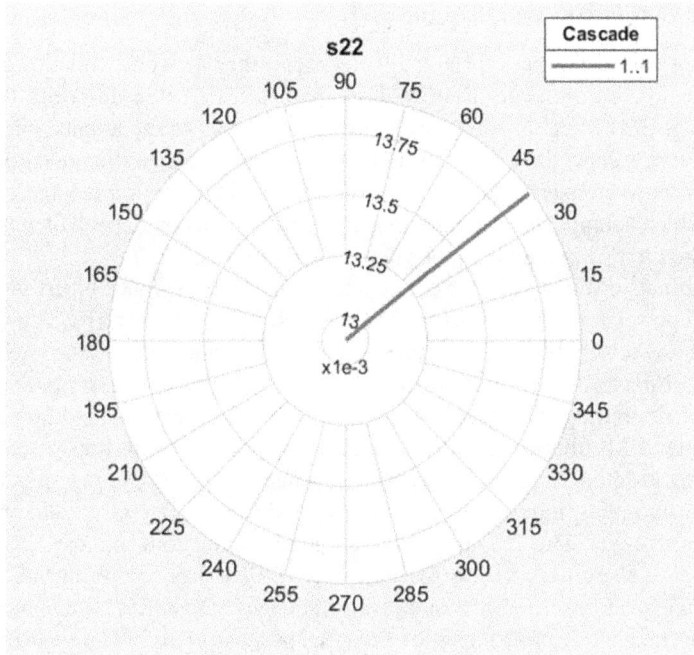

(b)

FIGURE 7.19 Plots of S21, S22.

port due to these two impedances, which effectively construct a voltage divider network for the EUT noise. As a result, the LISN impedance during the test must be dependable and consistent in order for measurement findings to be reliable and consistent.

Isolating any outside noise that may be present on the power line is the LISN's second crucial job. A LISN with a 50 micro-Henry inductance, as an illustration, provides high resistance to the external RF noise while enabling the lower frequency power to pass through to the EUT.

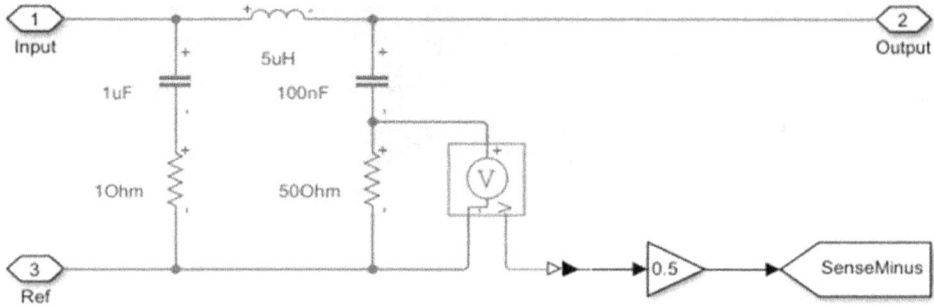

FIGURE 7.20   Line impedance stabilization network (LISN) block.

Together with the inductor, the first 1.0 microfarad line-to-ground capacitor creates the first stage of a two-stage filter.

The measuring meter for this test is commonly an EMI receiver or spectrum analyser. These gadgets' input stages are extremely delicate and vulnerable to damage. The resistor lowers the low frequency power line voltage, while the capacitor facilitates easy coupling of the EUT's low-level RF noise to the measuring meter's input.

7.2.7   SUITABLE EMI FILTER FOR A CIRCUIT

Specific frequency bands will pass through an electrical filter network while lower or higher frequency bands are filtered out. The insertion-loss characteristic often describes the fundamental characteristic of an EMI filter. This characteristic, which refers to the attenuation of the EMI filter, is often frequency dependant. Numerous factors make measuring of insertion loss difficult. Different types of measurement settings can vary how an EMI filter's input and output terminals are configured, which complicates the measurement process on its own. Lack of defined impedance terminations at the filter's input and output sides is another issue.

The input terminals of the EMI filter are coupled with the impedance of the power supply network. The type of power network, the current load, as well as the operating frequency of the test signal, all affect the power supply network's current impedance value. Impedance, which is typically unknowable and unstable in the time domain, is routinely injected into the filter's output. While conventional wave filter designers consider poles, zeros, group delay, pre-distortion, attenuation and the order of the filter, EMI filter design experts focus on attenuation, insertion loss, and filter impedance. EMI filtering is not an exact science, unlike, for instance, an active low-pass filter that might be employed as an anti-alias filter in a data-gathering application, even if the mathematical concepts are the same in both cases. The filter in this situation must safeguard the ADC against HF-folded spectral components; therefore, precise placement of the 3-dB corner frequencies is crucial. In order to achieve adequate insertion loss at a given frequency range, EMI filters must provide high impedance to that frequency range. This shows that the purpose of an EMI filter is to pass unwanted frequencies through the filter intact while producing the highest amount of mismatch impedance at those frequencies.

The EMI filter can also be identified in the RF budget analyser with the help of rf elements and the corresponding values should enter in the RF element block, as shown below in Figure 7.21.

The plots of the rf elements are signal to noise ratio vs input frequency and noise figure vs input frequency. These plots can be obtained in the rf budget analyser application, as shown in Figure 7.21. The S-parameter will change after the EMI filter implementation in **Voltage Source Three-Level Inverter**; the changed S-parameter plots will have the lesser noise level. If the noise is not in the allowable range against the standard, the design has to be redone and the same process is to be followed.

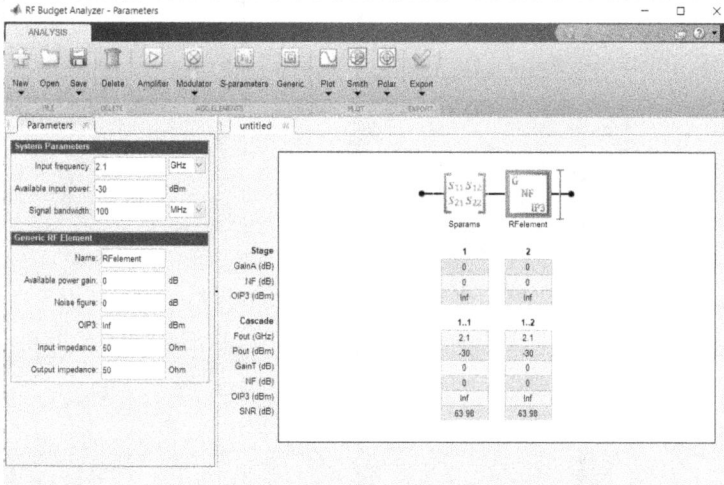

FIGURE 7.21 RF budget analyser R elements.

7.3 HIGH GAIN DC–DC CONVERTER USING VOLTAGE MULTIPLIER

This section describes a DC–DC multilayer buck-boost converter with a single switch and a very high voltage gain ratio. Multi-level buck-boost converter (MLBBC) is based on PWM with 2N-1 diodes and 2N-1 capacitors for N-level boost converter. This topology combines the buck-boost converter with the Cockcroft–Walton voltage multiplier. The primary benefit of this topology is the ability to raise output voltage by increasing the number of capacitors and diodes on the output side without affecting the main circuit. There are various innovative power electronic converters for integrating low-voltage DC input sources, such as PV solar panels, to a high-voltage DC bus in a 200–960 V DC distribution system. This converter operates in continuous conduction mode (CCM) and offers desirable characteristics such as low component voltage stresses, continuous input currents, and the ability to combine many independent DC input sources. Initially, a family of scalable interleaved boost converters with voltage multiplier cells (VMC) are shown. Several feasible Dickson and Cockcroft–Walton VMC combinations are demonstrated and evaluated in terms of voltage gain, number of components, and input current sharing. In this instance, the effective frequency observed by the magnetic element is several times the switching frequency, allowing for the employment of smaller magnetic devices. Each suggested converter includes a theory of operations, steady-state analysis, component selections, simulation, and efficiency analysis.

7.3.1 DESIGN TOPOLOGY

The topology of the converter is shown in Figure 7.22.

It functions in CCM. As typical buck-boost converters work in buck or boost mode depending on duty cycle, the suggested design only operates in boost mode due to the voltage multiplier stage. The circuit design for an N-level DC–DC boost converter consists of a single inductor, 2N-1 diodes, 2N-1 capacitors, and a single switch. This operates similarly to a standard buck-boost converter. The circuit's operation is divided into two modes.

7.3.1.1 Mode 1 (Switch S1 Closed)

When the switch S1 shuts, the inductor is charged from the input voltage source, while capacitor C1 discharges capacitor C2 is charged from both the input voltage and capacitor C1 via diode D2. After capacitor C2 is fully charged, capacitor C4 begins charging with voltage V_{in}+C1+C3 to throw diode D4. Capacitors C6 charge with V_{in}+C1+C3+C5 throw diode D6 and capacitor C8 charge

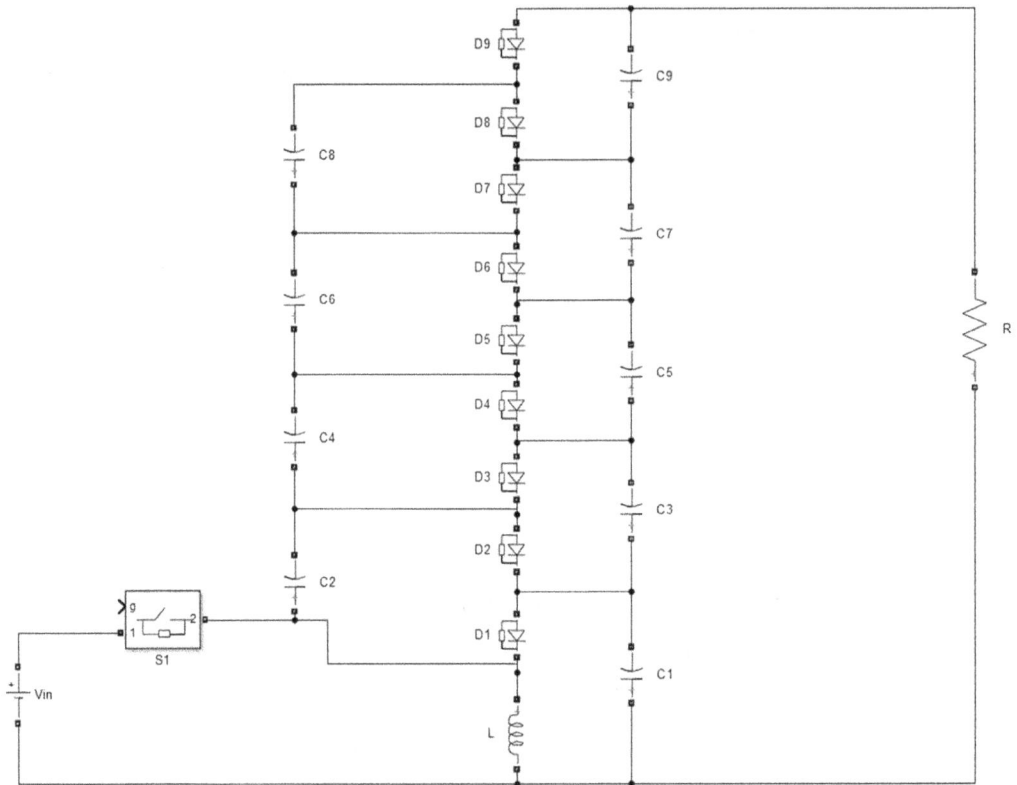

FIGURE 7.22　Circuit topology.

with $V_{in}+C1+C3+C5+C7$. This operation continues up to level N in a very brief amount of time. During the ON phase, an even number of capacitors are charged by both sources of input voltage, while an odd number of capacitors discharge an even number of diodes.

7.3.1.2　Mode 2 (Switch S1 Open)

When switch S1 is open, inductor L transfers its stored energy to the output capacitor (odd number capacitors). There is an isolation of input voltage sources from the circuit while the switch is in the OFF position. Simultaneously, capacitor C1 is charged by diode voltage throw inductor D1. After capacitor C1 is fully charged, capacitor C3 begins charging from the inductor voltage and capacitor C2 via diode D3. Capacitor C5 charges with VL+C2+C4 throw diode D5. Capacitor C7 charges with VL+C2+C4+C6 throw diode D7. As depicted in Figure 8.22, capacitor C9 is charged by VL+C2+C4+C6+C8. In OFF mode, an odd number of capacitors are charged by an even number of capacitors, which then discharges an odd number of diodes. Calculation of parameters is done using standard formulas. Using Simulink, these circuits are drawn. Buck-Boost converter Simulink model is shown in Figure 7.23.

7.3.2　S-Parameter

Scattering parameters or S-parameters (the elements of a scattering matrix or S-matrix) characterize the electrical behaviour of linear electrical networks subjected to various steady-state electrical signal stimuli. The characteristics are useful for numerous fields of electrical engineering, including electronics, the design of communication systems, and microwave engineering in particular.

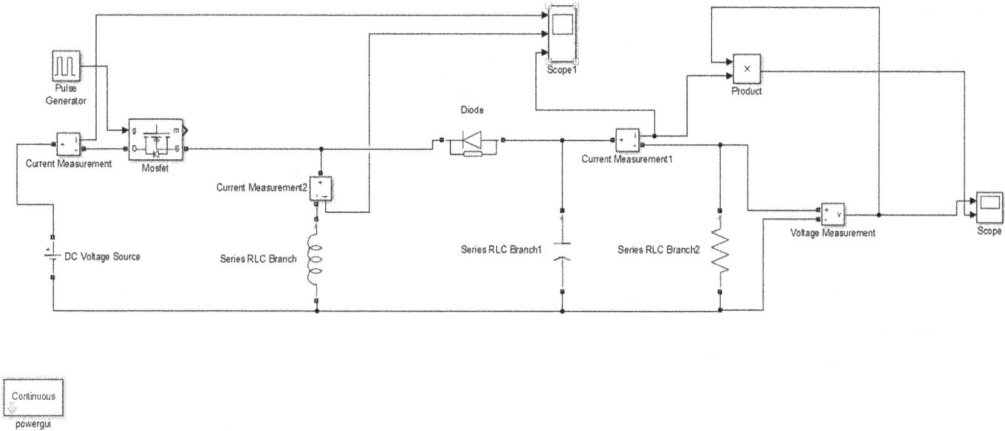

FIGURE 7.23 Simulink model—Buck-Boost converter.

S-parameters are related to Y-parameters, Z-parameters, H-parameters, T-parameters, and ABCD-parameters, among others. S-parameters differ in that they do not use open or short-circuit circumstances to characterize a linear electrical network; instead, they use matched loads. These terminations are simpler to utilize than open-circuit and short-circuit terminations at high signal frequencies. Contrary to common assumption, quantities are not measured by their power (except in now-obsolete six-port network analysers). Modern vector network analysers evaluate the amplitude and phase of voltage travelling wave phasors utilizing virtually the same circuit as digitally modulated wireless signal demodulators.

Using matched impedances, S (scattering) parameters are utilized to characterize electrical networks. Scattering refers to the manner in which travelling currents or voltages are impacted by a transmission line discontinuity. S-parameters in linear (dB) units always refer to power, whereas S-parameters in linear (dB) units always refer to amplitude (voltage or current). When it is crucial to define a network in terms of amplitude and phase against frequencies, as opposed to voltages and currents, S-parameters are employed to do so.

VSWR (Voltage Standing Wave Ratio) is defined as the ratio of the maximum voltage to the minimum voltage in a transmission line structure's standing wave pattern. At a reference plane, a reflection coefficient is defined as the ratio of reflected wave to incident wave.

7.3.2.1 To Extract S-Parameter from Circuit

Consider a two-port network, which is defined using S-parameters, as shown in Figure 7.24.

The RF Toolbox ends each port with the reference impedance Z_0 in order to extract the S-parameters from a circuit into an S-parameters object. The RF Toolbox then drives each port j separately with Z_0 and solves for the port voltages V_{ij}. The Norton counterpart of driving with current sources is driving with a 1 V source and a series resistance of Z_0.

Measure the port voltage V_{ij} at node i when node j is driven.

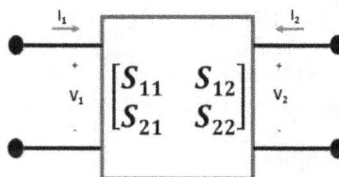

FIGURE 7.24 Two-port network.

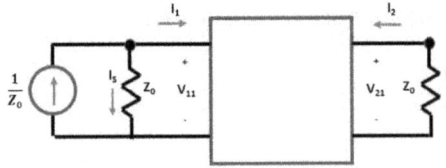

FIGURE 7.25   Circuit topology of two-port network.

- If $i \neq j$, the S-parameter entry S_{ij} is simply twice the port voltage V_{ij}, and this is given using the equation $S_{ij} = 2 \times V_{ij}$.
- The diagonal entries of S-parameters when $i = j$ are given using the equation $S_{ij} = 2 \times V_{ij} - 1$.

The circuit topology of two-port networks is shown in Figure 7.25.

Constitutive and conservative equations of circuit:

In the RF Toolbox, circuits are represented in the node-branch format. In the circuit, there are four branches, one for the input port, two for the two-port n port object, and one for the output port. This indicates that the circuit includes four unknown branch currents IS, I1, I2, and IL as well as two unknown node voltages V11 and V21. To express the circuit in node-branch form, four constitutive equations are required to represent the branch currents, and two conservative equations are required to represent the node voltages.

Source Code:

```
symsF IS I1 I2 IL V1 V2 Z0
symsS11 S12 S21 S22
nI = 4; % number of branch currents
nV = 2; % number of node voltages
% F = [Fconstitutive; Fconservative]
F = [
    V1 - Z0*IS
    V1 - Z0*I1 - S11*(V1+Z0*I1) - S12*(V2+Z0*I2)
    V2 - Z0*I2 - S21*(V1+Z0*I1) - S22*(V2+Z0*I2)
    V2 - Z0*IL
    IS+I1
    I2+IL
    ]
```

To solve s-parameters of circuit

Create a right-hand vector with two columns to represent the driving of each port.

Source Code

```
symsrhs [nI+nV 2]
symsx v S
% Compute S-parameters of cascade
rhs(:,:) = 0;
rhs(nI+1,1) = 1/Z0;   % rhs for driving input port
rhs(nI+nV,2) = 1/Z0   % rhs for driving output port
```

7.3.2.2   Create Object for RF Toolbox

In order to create S-parameters object, the parameters must be determined at a set of frequencies. To do so, define the variables for mutual inductor. If you would like to test multiple values for your variables and automatically update your S-parameters object, use numeric sliders in the control drop-down under the Live Editor tab. Then, use the Symbolic Math Toolbox's MATLAB function to automatically generate a function, mutual Inductor to compute the analytic S-parameters at a set of frequencies. Finally, use the S-parameters object to create S-parameter.

Consider a mutual inductor as shown in Figure 1 with the inductors L_a and L_b. This example uses the Symbolic Math Toolbox to extract the analytical S-parameters of the mutual inductor and write them an RF Toolbox™ object.

One way to model a mutual inductor in the RF Toolbox is to draw the mutual inductor as an equivalent of a two-port network of inductors in a T configuration. Such a mutual inductor has the mutual inductance M and the coupling coefficient k. Inductors in a T configuration can have negative values when there is a strong coupling between the inductors or if the M is greater than L_a or L_b.

Use constitutive and conservative equations to represent the circuit in node-branch form. There are eight unknowns, five branch currents, and three node voltages. Therefore, there are eight equations in the node-form, five constitutive equations for the branches, and three conservative equations obtained from Kirchoff's Current Law for the nodes. The constitutive equation for a resistor is derived from Ohm's Law, $V = IR$, and the constitutive equation for an inductor is given by $V = sLR$, where s is a complex frequency. Results of RF Budget Analyser is shown in Figure 7.26

7.3.2.3 S-Parameter Measurement for Circuit Frequency

Implementation in MATLAB

S-Parameter Measurement for Circuit Frequency code is shown in Figure 7.27.

Plot measured on Z-Smith Chart

S11 Parameter

Plot measured on Z-Smith Chart is shown in Figure 7.28.

7.3.3 CONDUCTED EMISSION

The conducted emission test examines the portion of the electromagnetic energy emitted by your device that is conducted onto the power cord. The objective is to limit the amount of interference a device may couple back onto a power source. In addition, it isolates RF signals from the power source.

Conducted emissions are a subset of electromagnetic interference in circuits that primarily cause problems with delivered power quality, due to interference caused by harmonics arising from linear and non-linear loads present in the electric system, primarily as a result of the increasing prevalence of switched mode power supplies.

Conducted emissions are the noise currents produced by the device-under-test (DUT) that travel via the power cord or harness to other components/systems or the power grid. These noise currents can be measured with either the voltage or current technique. When we speak of conducted EMI,

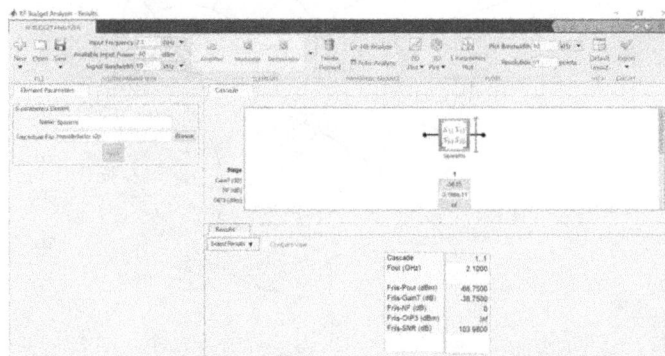

FIGURE 7.26 RF budget analyser.

```
Command Window
>> S_measuredBJT = sparameters('samplebjt2.s2p');
freq = S_measuredBJT.Frequencies;
>> leftpad = circuit('left');
add(leftpad,[1 2],inductor(1e-9));
add(leftpad,[2 3],capacitor(100e-15));
setports(leftpad,[1 3],[2 3]);
S_leftpad = sparameters(leftpad,freq);
>> rightpad = circuit('right');
add(rightpad,[1 3],capacitor(100e-15));
add(rightpad,[1 2],inductor(1e-9));
setports(rightpad,[1 3],[2 3]);
S_rightpad = sparameters(rightpad,freq);
>> S_DUT = deembedsparams(S_measuredBJT,S_leftpad,S_rightpad);
>> figure
hs = smithplot(S_measuredBJT,1,1);
hold on;
smithplot(S_DUT,1,1)
hs.ColorOrder = [1 0 0; 0 0 1];
hs.LegendLabels = {'Measured S11','De-Embedded S11'};
```

FIGURE 7.27 S-parameter measurement for circuit frequency code.

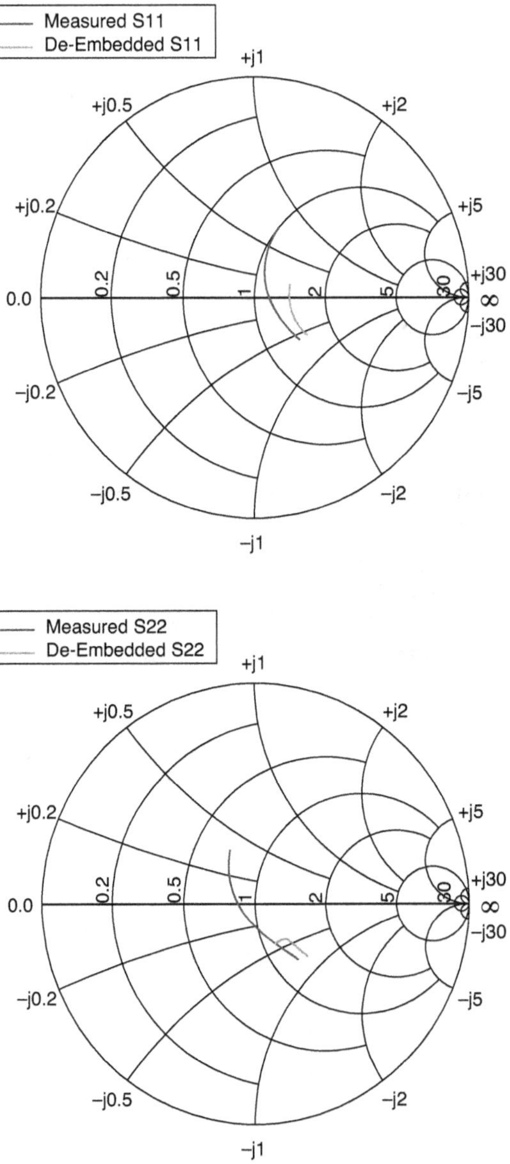

FIGURE 7.28 S11 and S22 parameters.

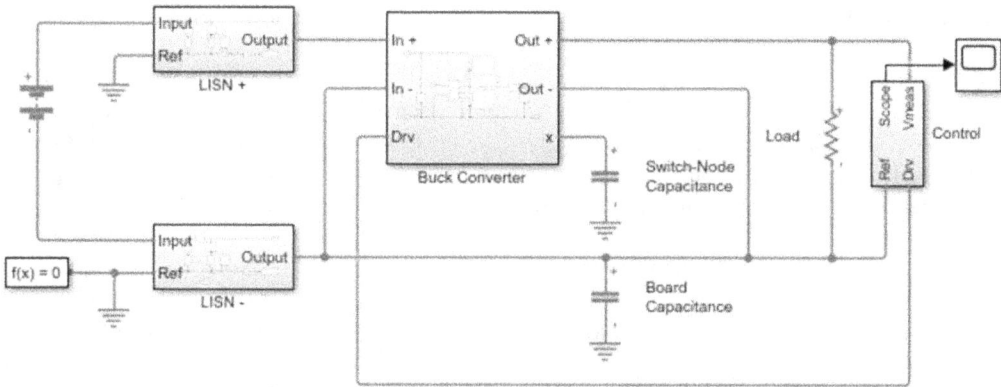

FIGURE 7.29 Conducted measurement setup.

we're referring to noise that is generated by one device or subcircuit and conveyed to another via cabling, PCB traces, power/ground planes, or parasitic capacitance.

7.3.3.1 Measurement of Conducted Emission

Observe a buck converter configured for a source measurement of common- and differential-mode noise. The model of conducted measurement setup is shown in Figure 7.29.

7.3.3.1.1 Model

Capacitance between the switching node (between the high and low side transistors) and the reference plane is also included in this circuit. The capacitive connection between the circuit and a reference plane must be modelled in order to approximate common-mode noise. Between each terminal of the power supply and the inputs of the buck converter, ideal LISNs are positioned to provide standard impedance and measuring port for common- and differential-mode noise. A steady-state

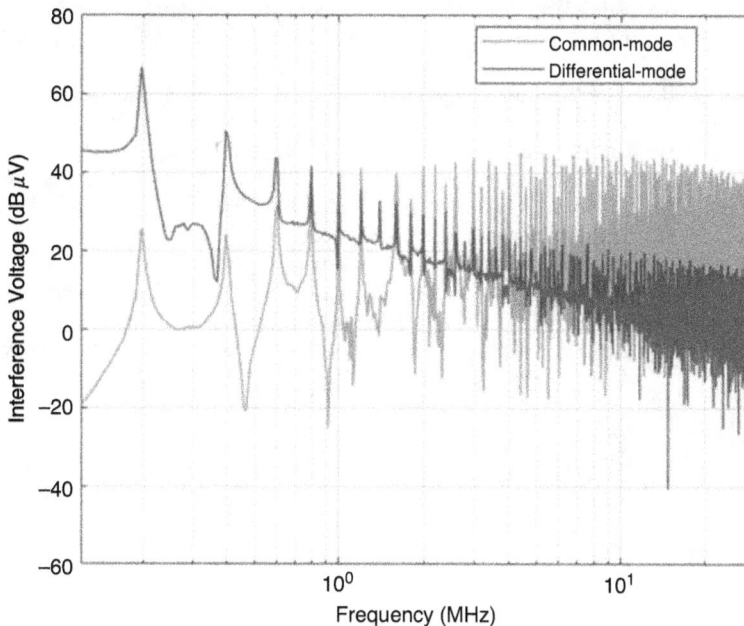

FIGURE 7.30 Common-mode and differential-mode noise.

operating point has been saved and serves as the starting point for the noise simulation in order to obtain the noise measurement under steady-state conditions.

Common-mode and differential-mode noise can be plotted and Figure 8.28 depicts the common- and differential-mode noise for a 200 kHz switching frequency buck converter. The noise level with the required frequency level can be observed. If the noise level is above the standard, EMI filter need to be designed and the conducted emission has to be measured again. The common-mode and differential-mode noise is shown in Figure 7.30.

7.4   CONCLUSION

The converter circuit is designed using MATLAB and simulation outcome supports the suggested topology. Conducted emission can be measured using LISN block with filter and without filter. Characteristics of the transmission line can be analysed using the S-parameter graph. Noise level will be improved upon implementing the EMI filter.

BIBLIOGRAPHY

1. Yosuke Dairaku, Tadashi Mizobuchi, Kouhei Aiso, Keiichiro Kondo, Takeshi Shinomiya, and Katsumi Ishikawa, Power Flow Control of a Hybrid Voltage Source Three Level Inverter for Energy Saving of Railway Vehicle Traction System. 978-1-7281-4878-6/19 ©2019. IEEE.
2. Naser Abdel-Rahim and John E. Quaicoe, Three-phase Voltage-Source UPS Inverter with Voltage-Controlled Current-Regulated Feedback Control Scheme. 0-7803-1328-3/94 1994. IEEE.
3. Justin John and Jenson Jose, A New Three Phase Step Up Multilevel Inverter Topology for Renewable Energy Applications. 978-1-5090-1277-0/16 ©2016. IEEE.
4. Yuya Kojima, Toshikazu Sekine and Yasuhiro Takahashi, Generalized Indirect S-parameter Measurement Method of n-ports Circuit using T-parameter of (m, n)-ports Fixture. 978-1-5386-3974-0/ 2017. IEEE.
5. Muhammad H. Rashid, *Power Electronics Devices, Circuits, and Applications*, 4th ed. Pearson Education Limited. 2014.
6. M. D. Singh and K. B. Khanchandani, *Power Electronics*. New Delhi: Tata McGraw-Hill Publishing Company Limited. 2008.
7. Rong-Jong Wai, Wen-Hung Wang, and Chung-You Lin, "High Performance Stand-Alone Photovoltaic Generation System", *IEEE Transactions on Industrial Electronics*, vol. 55, no. 1, pp. 240–250, January 2016.
8. J.C. Rosas-Caro, J.M. Ramirez, and F.Z. Peng, "A DC-DC Multilevel Boost Converter", *IET Power Electronics*, vol. 3, no. 1, pp. 129–137, 2012.
9. Mojtaba Forouzesh, Yam P. Siwakoti, Saman A. Gorji, Frede Blaabjerg, and Brad Lehman, "Step-up DC–DC Converters: A Comprehensive Review of Voltage-Boosting Techniques, Topologies, and Applications", *IEEE Transactions on Power Electronics*, vol. 32, no. 12, pp. 9143–9178, 2017.
10. J.C. Rosas-Caro, J.C. Mayo-Maldonaldo, A. Gonzalez-odriguez, E. N. Salas-Cabrera, M. Gomez-Garcia, O. Ruiz-Martinez, R. Castillo-Ibarra, and R. Salas-Cabrera, "Topological Derivation of DC-DC Multiplier Converter". Proceedings of the World Congress on Engineering and Computer Science 2010 Vol II.
11. https://www.mathworks.com/help/rf/ug/extract-s-parameters-from-circit.html.
12. https://www.mathworks.com/help//physmod/sps/ug/conducted-emission-of-a-buck-converter.html.

8 EMI Using PSPICE

8.1 PSPICE INTRODUCTION

MicroSim PSpice A/D is a simulation programme that models the behaviour of any combination of analogue and digital components in a circuit. PSpice A/D models mixed-signal circuits without performance deterioration due to tightly connected feedback loops between the analogue and digital parts since the analogue and digital simulation algorithms are incorporated into the same software. In addition to models for resistors, inductors, capacitors, and bipolar transistors, PSpice A/D includes models for the following components.

1. Models of transmission lines incorporating delay, reflection, loss, dispersion, and crosstalk
2. Non-linear models of the magnetic core, including saturation and hysteresis
3. MOSFET models, including version 3 of BSIM
4. GaAsFET models, such as Parker–Skellern and the TOM2 model of TriQuint
5. IGBTs
6. Digital elements with analogue input/output models
7. Utilize components from MicroSim's wide library: The model libraries include more than 10,200 analogue and 1,600 digital models of North American, Japanese, and European components.
8. PSpice A/D allows analogue and digital modelling of behaviour, allowing you to describe functional blocks of circuitry using mathematical expressions and functions.

PSpice is available in three versions:

- PSpice A/D
- PSpice A/D Basics
- PSpice

This chapter is meant to provide a comprehensive overview of how to utilize all PSpice A/D capabilities. Given that PSpice A/D Basics and PSpice are limited versions of the complete PSpice A/D software, their functionality is also explained in this manual. Wherever in this manual particular PSpice A/D capabilities or functions are described that are unavailable in one of the other two versions, this limitation is denoted with one of the following special icons.

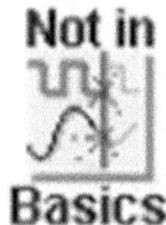

DOI: 10.1201/9781003362951-8

- The feature or function discussed in the chapter is accessible in PSpice A/D but not PSpice A/D Basics.

Not in

PSpice

- The feature or function described in the related section is only accessible in PSpice A/D.

If you are currently utilizing PSpice A/D Basics or PSpice and find that you require the increased capabilities given by PSpice A/D, you can upgrade without losing any data or needing conversion utilities. Any circuit file developed for PSpice A/D Basics or PSpice will simulate PSpice A/D.

PSpice A/D models analogue-only, analogue/digital hybrid, and digital-only circuits. The analogue and digital algorithms of PSpice A/D are embedded into the same software so that mixed analogue/digital circuits with tightly connected feedback loops between the analogue and digital portions can be simulated without performance deterioration. After you construct a design for simulation, Orcad Capture generates a circuit file set. PSpice A/D reads the circuit file set including the circuit net list and analysis commands for simulation. PSpice A/D converts these into useful graphical plots, which can be displayed immediately from the schematic page using markers.

8.1.1 Comparison

Table 8.1 shows the comparison of PSpice features in various versions.

8.1.2 User-Configurable Data Files in PSpice A/D

The user-configurable data files in PSpice A/D are shown in Figure 8.1.

These files can be generated using Orcad applications such as the PSpice Stimulus Editor and the PSpice Model Editor. These tools automate the creation of files and give graphical methods for validating the data. You may also manually insert the data using the Model Text view in the PSpice Model Editor (or any text editor like Notepad).

Performance analysis in simple and complex circuits:

- Direct current (DC), alternating current (AC) and transient analyses allow to examine the response of your circuit to a variety of inputs.

TABLE 8.1
Comparison of PSpice Versions

Features	PSpice A/D	PSpice A/D Basics	PSpice
Digital primitives	All	Most	None
Digital model library	1,600+	1,600+	0
PSpice Optimizer	Yes	No	Yes
Network Licensing	Yes	No	Yes
Unlimited circuit size	Yes	Yes	Yes

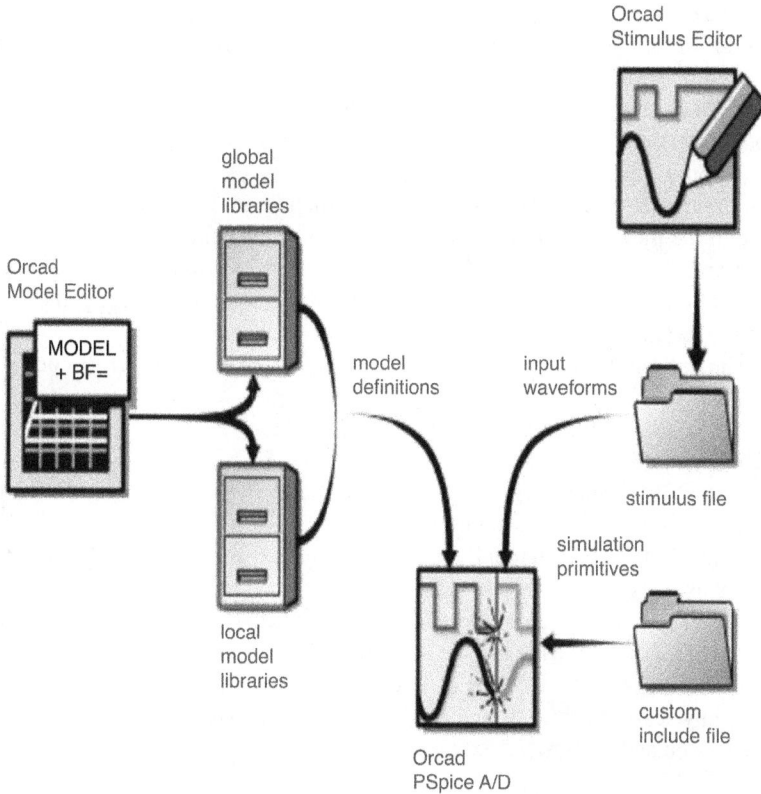

FIGURE 8.1   User-configurable data files in PSpice A/D.

• Parametric, Monte Carlo, and sensitivity/worst-case studies—it can be observed how the behaviour of the circuit evolves as component values change.
• Digital worst-case timing analysis to assist in identifying timing issues that only occur with specific combinations of slow and fast signal transfers.

Schematics as preparation for simulation schematics is a design entry programme, which is necessary for preparing a circuit for simulation. Schematics are also the control point for running other simulation design flow applications.

In schematic windows, one can do the things listed below:

• Placing and connecting component symbols;
• Defining component values and other attributes;
• Defining input waveforms and enabling one or more analyses; and
• Marking places in the circuit

8.1.3   ANALYSING SETUP IN PSPICE

Analyses that can simulate analogue-only, mixed-signal, and digital-only circuits are supported by PSPICE A/D. It fully supports digital analysis by replicating the timing behaviour of digital devices, including worst-case timing, within a typical transient analysis. All of the aforementioned analyses can be performed on analogue/digital hybrid circuits. If the circuit is entirely digital, only transient analysis can be performed.

a. **MODEL LIBRARY**

A model library is a file containing the electrical definition of several components. PSpice uses these data to determine how a component will react to various electrical inputs. These definitions are presented as follows:

- Model parameter set, which defines the behaviour of a component by fine-tuning the PSpice-built underlying model.
- Sub-circuit net list, which explains the structure and function of the component through the interconnection of other components and primitives. Orcad model libraries included with your programmes contain the most frequently used models. The model library names are suffixed with.LIB. However, if necessary, you can construct own models and libraries by:
  - Manually utilizing the Model Text view of the PSpice Model Editor (or any text editor like Notepad)
  - Automatically utilizing the PSpice Model Editor

b. **STIMULUS FILE**

A stimulus file contains definitions of analogue or digital input waveforms based on time. You can build a stimulus file by:

- Creating the definition manually using a regular text editor such as Notepad (a typical file extension is.STM).
- Utilizing the Stimulus Editor automatically (which generates a.STL file extension). Include file. An include file is a file defined by the user that contains PSpice commands.
- Additional text comments to be included in the PSpice output file.
- Using any text editor, such as Notepad, you can create an include file. Include file names often end with *.INC extension.

8.1.4 Configuring Model Library

Stimulus include PSpice examines model libraries, stimulus files, and inclusion files for any information required to define a component or conduct a simulation. The files that PSpice searches for depend on how your model libraries and other files are configured. A significant portion of the setting is set up automatically; however, you can manually configure the following:

- Modify the configuration by adding and removing files.
- Modify the file's scope, i.e., whether it applies to a single design (local) or to all designs (global).
- Modify the search order.

8.1.5 Sample Circuit Simulation

The sample simulation circuit of diode clipper is shown in Figure 8.2

8.1.5.1 To Create a New PSpice Project

- Select the Orcad application folder and the Capture shortcut from the Windows Start menu to launch Capture.
- From the Project Manager File menu, select New and then Project.
- Choose the Analogue or Mixed-Signal Circuit Wizard option.
- Enter the project's name in the Name text box (suitable name).
- Select the location for the project files using the Browse option, then click OK.
- Choose Create a blank project in the Create PSpice Project dialogue box.
- Select "OK".

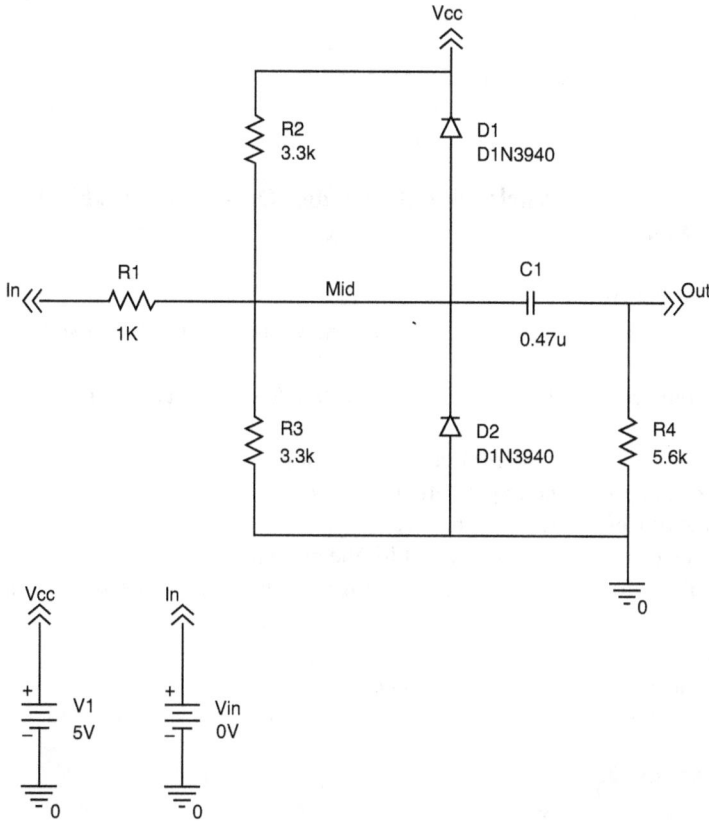

FIGURE 8.2 Diode clipper circuit.

At this time, no special libraries must be configured. A new page will be displayed in Capture, and the Project Manager will be configured for the new project.

8.1.5.2 To Place the Voltage Sources

- Navigate to the schematic page editor in Capture.
- Add the library for the required components.
- To add a library, click the Add Library button and choose SOURCEOLB (from the PSpice library) and press the Open button.
- Type VDC in the Part text box.
- Select "OK".
- Place the first component by moving the pointer to the desired location on the schematic page and clicking.
- To position the second section, move the pointer and click again. Select End Mode with a right-click to stop placing components.

8.1.5.3 To Place the Diodes

- From the Place menu, select Part to open the Place Part dialogue box.
- Include the library for the necessary components.
- Click the button labelled "Add Library" and select DIODE.
- Click Open followed by OLB (from the PSpice library).

- In the Part text box, enter D1N39 to get a list of diodes.
- Choose D1N3940 from the Part List, then click OK.
- Press r to rotate the diode into the proper position.
- Click to position the first diode (D1), then click again to position the second diode (D2).
- To stop placing pieces, right-click and select End Mode.

8.1.5.4   To Transfer the Text Related to the Diodes (Or Any Other Object)

- Click the text to select it, then move it to a new spot.

8.1.5.5   To Connect the Parts

- Select Wire from the Place menu to begin wiring components. The cursor transforms into a crosshair.
- Click the connection point (extreme tip) of the pin on the off-page connector at the circuit's input.
- Click the closest input resistor R1 connection point.
- Connect R1's opposite end to the output capacitor.
- Connect the diodes and the wire between them:
  - Click the cathode connection point for the bottom diode.
  - Click the wire between the diodes after moving the cursor straight up. The wire terminates, and the intersection of its segments becomes visible.
  - Click the connector again to continue wiring.
  - Click the anode pin of the higher diode.
  - Continue connecting components until the circuit is completely wired

8.1.5.6   To Save the Design

- In the File menu, select Save.

8.1.6   CLASSES OF PSPICE A/D ANALYSES

8.1.6.1   Standard Analyses

Table 8.2 shows the classes of PSpice in A/D analyses.

TABLE 8.2
Classes of PSPICE A/D Analyses

Analysis	Setup/Dialogue Box	Swept Variable
DC sweep	DC sweep	Source
		Parameter
		Temperature
Bias point	Load bias point	–
Frequency response	AC sweep	Frequency
Noise (requires a frequency response analysis)	AC sweep	Frequency
Transient response	Time domain transient	Time
Fourier (requires a transient response analysis)	Time domain transient	Time
DC sensitivity	Sensitivity	–
Small signal DC transfer	Transfer function	–

8.1.6.2   Simple Multirun Analyses

Parametric	Parametric Sweep	–
1 Temperature	Temperature (sweep)	–

8.1.6.3   Statistical Analyses

Monte Carlo	Monte Carlo/Worst Case	–
2 Sensitivity/Worst case	Monte Carlo/Worst case	–

8.1.6.4   AC Sweep Analysis

A frequency response analysis is AC sweep. PSpice A/D computes the linearized, around the bias point, small-signal response of the circuit to a combination of inputs. Before PSpice A/D executes the small signal analysis, non-linear devices, such as voltage- or current-driven switches, are linearized about their bias point value. Digital devices retain the states determined by PSpice A/D when solving for the bias point. Since AC sweep analysis is a linear analysis, it just evaluates the gain and phase response of the circuit and does not restrict voltages or currents. The most effective technique is to set the source magnitude to one using AC sweep analysis. In this manner, the measured output equals the gain relative to the source of input at that output.

a. **Setting up an AC stimulus:**
   To conduct an AC sweep analysis, it is necessary to position and connect one or more independent sources and then configure the magnitude and phase of each source's AC output.
   To set up an AC stimulus:

STEP 1:
   • Insert and connect any of the following symbols in your diagram:
   • For voltage input—volts alternating current (VAC) (only AC sweep analysis), voltage source (VSRC) is employed (Multiple analysis types including AC sweep)
   • For current input—IAC (just an AC sweep analysis), the current source (ISRC) value is employed (Multiple analysis types including AC sweep)

STEP 2:
   • Double-click the instance of the symbol. A dialogue window displaying the symbol instance's attribute values is displayed.

b. **Setting up an AC analysis:**
   To set up the AC sweep analysis

STEP 1: Select Setup in the Analysis menu, which is shown in Figure 8.3
STEP 2: Select AC sweep, which is shown in Figure 8.4
STEP 3: Choose the AC sweep type and specify the number of points in the AC sweep dialogue box in Figure 8.5, and its explanation in Table 8.3

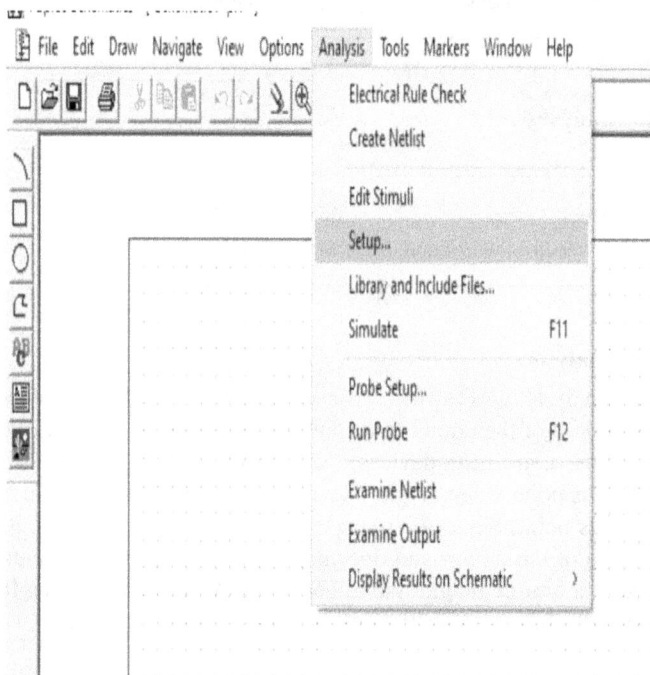

FIGURE 8.3   Analysis setup.

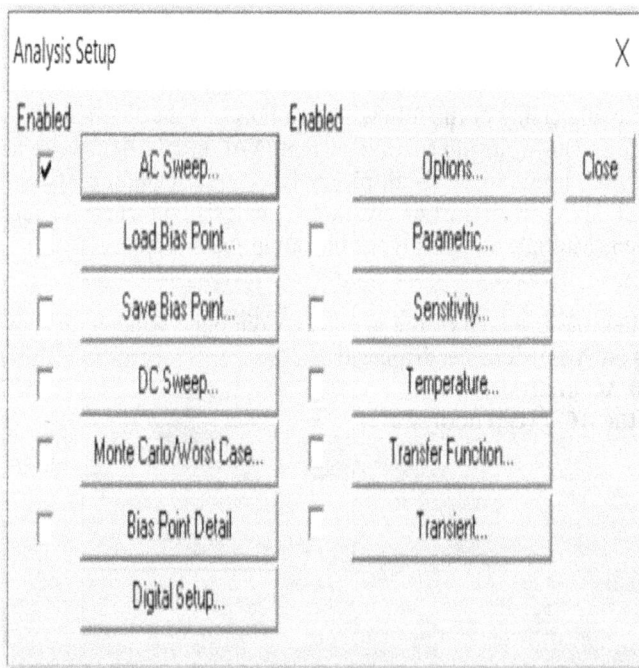

FIGURE 8.4   Analysis setup—(alternating current) AC sweep.

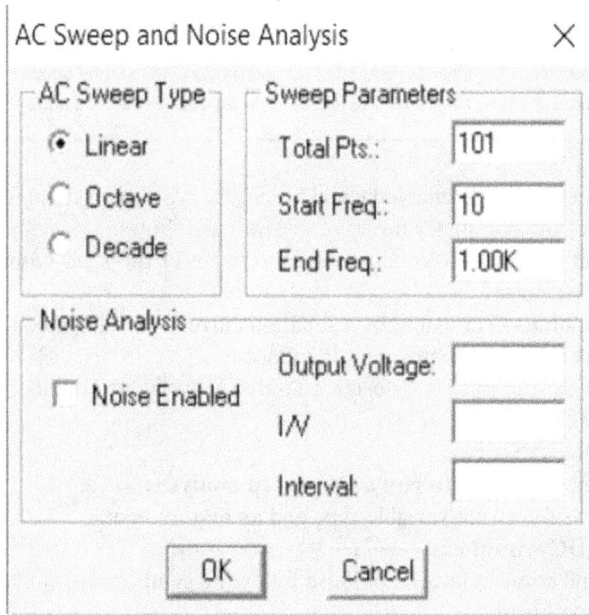

FIGURE 8.5 Alternating current (AC) sweep and noise analysis.

TABLE 8.3

AC Sweep Type Analysis

Linearly	Choose linear and set total points to the total number of points in the sweep.
Logarithmically by octaves	Choose octave and set points/octave to the total number of points per octave.
Logarithmically by decades	Choose octave and set points/decade to the total number of points per decade.

STEP 4: Enter the starting and ending frequencies for the sweep in the Start Freq and End Freq text boxes.

STEP 5: Select OK.

c. **NOISE ANALYSIS:**

PSpice A/D calculates and presents the following for each frequency provided for an AC sweep analysis when performing a noise study:

1. Device noise, which is the noise contribution propagated to the specified output from every resistor and semiconductor device in the circuit; for semiconductor devices, the device noise is also decomposed into constituent noise contributions where applicable.

2. Total output and comparable input noise.

Output noise is the RMS total of all device contributions to an output.

Input noise is the comparable noise required at the circuit's input source to generate the computed output noise

Setting up noise analysis:

To set up the noise analysis:

STEP 1: Select Setup from the analysis menu.

STEP 2: Select AC sweep

STEP 3: Choose the AC sweep type and set the number of points in the AC sweep dialogue box

Linearly	Choose linear and set total points to the total number of points in the sweep.
3. Logarithmically by octaves	Choose octave and set points/octave to the total number of points per octave.
Logarithmically by decades	Choose octave and set points/decade to the total number of points per decade.

STEP 4: Select the noise enabled checkbox in the AC Sweep dialogue box.

STEP 5: Enter the parameter for noise analysis as follows:

Output voltage—A voltage output variable of the kind required to determine total output noise.

I/V Source—The name of a separate current or voltage source for which you wish to calculate the equivalent input noise.

Interval—An integer n designating that at every nth frequency.

STEP 6: Click OK.

d. **DC SWEEP ANALYSIS:**

Minimum requirements to run a DC sweep analysis:

Table 8.4 shows the swept variable type and its requirement.

Setting up a DC stimulus:

STEP 1: Place and connect anyone of these following symbols in your schematic:

For voltage input

VDC, a DC sweep analysis only.

VSRC, multiple analysis types including DC sweep.

For current input

IDC, a DC sweep analysis only.

ISRC Multiple analysis types including DC sweep.

STEP 2: Double click the symbol instance. A dialog box appears listing the attribute values for the symbol instance.

To set up the DC sweep analysis:

STEP 1: From the analysis menu, select Setup

STEP 2: Click DC sweep.

STEP 3: In the DC sweep dialog box, choose the DC sweep type and set the number of points as follows; also shown in Figure 8.6.

Linearly	Choose linear and set total points to the total number of points in the sweep.
4. Logarithmically by octaves	Choose octave and set points/octave to the total number of points per octave.
Logarithmically by decades	Choose octave and set points/decade to the total number of points per decade.

STEP 4: In the Start value and End value text boxes, enter the starting and ending frequencies, respectively, for the sweep.

TABLE 8.4
Requirement for DC Sweep Analysis

Swept Variable Type	Requirement
Voltage source	Voltage source with a direct current (DC) specification.
Temperature	None
Current source	Current source with a DC specification
Model parameter	PSpice A/D model
Global parameter	Global parameter defined with a parameter

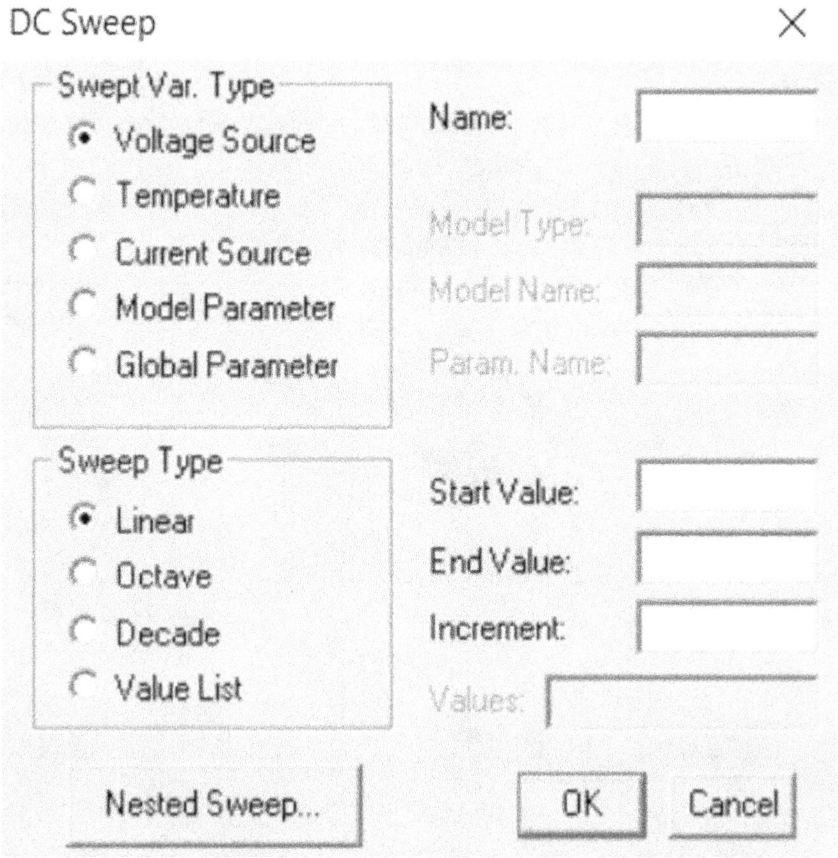

FIGURE 8.6   Direct current (DC) sweep setup.

8.1.6.5   Bias Point Detail

Bias point detail analysis can be enabled in the Analysis Setup dialogue box and the bias point detail analysis is calculated for all analyses. When bias point detail analysis is disabled, the output file contains simply analogue node voltages and digital node states.

When the Bias Point Detail analysis is enabled, the output file contains the following information:

- A listing of all voltages at analogue nodes.
- A listing of all node statuses.
- Currents and total power of all voltage sources.
- A listing of all small-signal device parameters.

Below are the steps to enable the bias point analysis.
    If Bias point is enabled:

STEP 1: Choose Setup from the Analysis menu.
STEP 2: In the Analysis Setup dialog box, click Options and click ok which is shown in Figure 8.7
STEP 3: In the Yes/No options, select Yes to confirm.

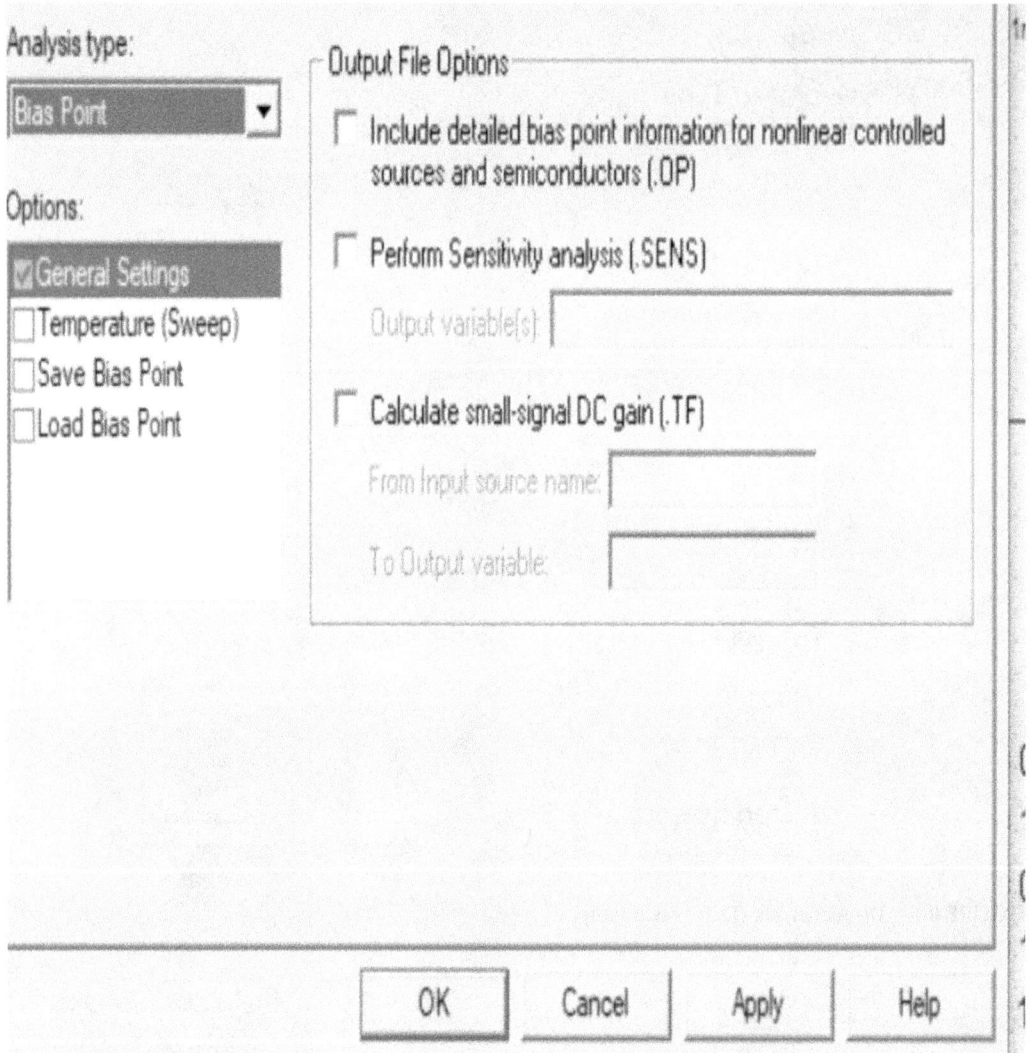

Analysis type:

[Bias Point ▼]

Options:

- ☑ General Settings
- ☐ Temperature (Sweep)
- ☐ Save Bias Point
- ☐ Load Bias Point

Output File Options

☐ Include detailed bias point information for nonlinear controlled sources and semiconductors (.OP)

☐ Perform Sensitivity analysis (.SENS)

Output variable(s): [   ]

☐ Calculate small-signal DC gain (.TF)

From Input source name: [   ]

To Output variable: [   ]

[OK] [Cancel] [Apply] [Help]

FIGURE 8.7 Analysis type.

8.1.6.6 Small-Signal DC Transfer

The small-signal DC transfer analysis linearizes the circuit around the bias point in order to determine the small-signal transfer function. Calculations are performed on the small-signal gain, input resistance, and output resistance. In the Transfer function dialogue box, you must enter an output voltage or current through a voltage source to compute the small-signal gain, input resistance, and output resistance. Alongside the input and output resistances, the gain from the input source to the output variable is also output. Entering $V_{(OUT2)}$ as the output variable and V_1 as the input source, results in the calculation of the input resistance for V1, the output resistance for $V_{(OUT2)}$, and the gain from V_1 to V_{OUT2}.

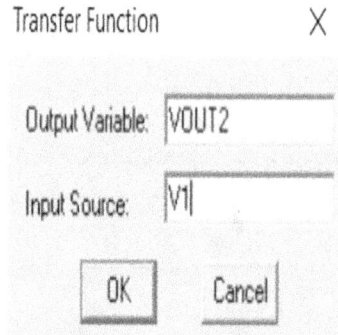

FIGURE 8.8   Transfer function tab.

Minimum requirements to do a Small-signal DC Transfer analysis:

STEP 1: The circuit should include an input source like VSRC.

STEP 2: Click the Transfer function button within the Analysis Setup Dialog. In the Transfer Function dialogue box, indicate the desired input source.

STEP 3: If necessary, check the transfer function box in the Analysis Setup dialogue box to enable it, which is shown in Figure 8.8.

STEP 4: Finally, start the simulation.

8.1.6.7   DC Sensitivity

DC sensitivity analysis computes and provides the sensitivity of one node voltage to each device parameter for the following sorts of devices:

- Resistors
- Voltage and current source independently
- Voltage- and current-controlled switches
- Diodes
- Bipolar transistors

Calculating the sensitivity involves linearizing all devices around the bias point.

Minimum programme setup requirements:

STEP 1: Click the Sensitivity button in the Analysis Setup dialogue box. In the Sensitivity analysis dialogue box, specify the desired output variable.

STEP 2: If necessary, select the Sensitivity check box in the Analysis setup dialogue box to enable it, which is shown in Figure 8.9.

STEP 3: Finally start the simulation.

8.1.6.8   Transient Analysis

The transient response analysis calculates the circuit's reaction between TIME = 0 and a specified time. The analysis should span the time interval from 0 to 1,000 nanoseconds, and values should be provided every 20 nanoseconds to the simulation output file. Either anyone or all of the independent sources may have time-varying values during a transient analysis. The sole source whose value varies over time is V1 (VSIN portion) with the following attributes: VOFF = 0 v VAMPL = 0.1 v FREQ = 5 Meg. The value of V1 fluctuates as a 5 MHz sine wave with a 0 v offset voltage and 0.1 v peak amplitude. Multiple sources typically have time-varying values, such as two or more clocks in

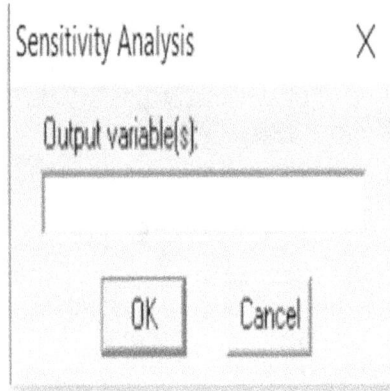

FIGURE 8.9   Sensitivity analysis.

a digital circuit. The transient analysis calculates a bias point independently, employing the same method as explained for the DC sweep. This is important due to the fact that the starting values of the sources may differ from their DC values.

8.1.6.9   Fourier Components

Fourier Components Fourier analysis is enabled using the dialogue box for configuring transient analysis. Fourier analysis computes the DC and Fourier components of the outcome of a transient analysis. The first through ninth components are computed by default, but more can be provided. Before conducting a Fourier analysis, you must conduct a transient analysis. The Fourier transform uses the same sample interval as the print step set for the transient analysis. When Fourier is selected to do a harmonic decomposition analysis on a transitory waveform, just a piece of the waveform is utilized. Probe can calculate a Fast Fourier Transform (FFT) of the entire waveform and its spectrum.

Minimum circuit design requirements:
Circuit should contain one of the following:

1. An independent source with a transient specification, which is shown in Table 8.5
2. An initial condition on a reactive element
3. A controlled source that is a function of time
   Minimum programme setup requirements:
   STEP 1: Click the Transient button in the Analysis Setup dialogue box. As appropriate, complete the Transient dialogue box.
   STEP 2: If necessary, select the Transient check box in the Analysis Setup dialogue box to enable it, which is shown in Figure 8.10.
   STEP 3: Start the simulation.

TABLE 8.5
Stimulus Symbols for Time-Based Input Signals

Specified By	Symbol Name	Description
Using the stimulus editor	VSTIM	Voltage source
	ISTIM	Current source
	IF_IN INTERFACE	Interface ports
	DIGSTIM	Digital stimulus

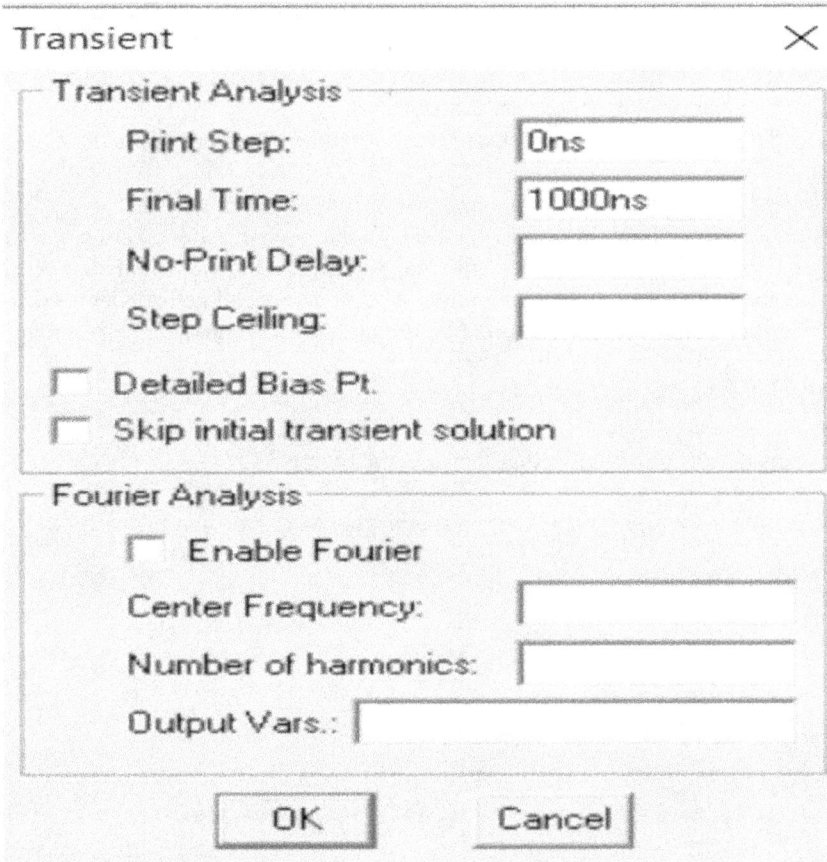

FIGURE 8.10 Transient analysis.

8.1.7 Schematic Window

Schematic window in PSpice is important, where we create a new project pr circuit and connect it and run the simulation of our circuits. Let us have a look at all the parts in schematic windows in ORCAD PSpice, which is shown in Figure 8.11.

The first-row toolbar in schematic window has file, edit, place, view, macro, PSpice, accessories, options, window, and options. Second row toolbar has save file, print file, and zoom options, and third row toolbar has run option, voltage probe, current probe, voltage differential markers, voltage, current, and power display keys.

- From the file option, we can create new project or open a saved project. It also has print and print setup, import design and export design options.

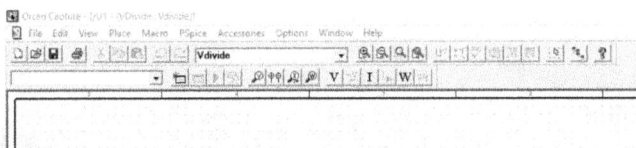

FIGURE 8.11 Schematic window in PSpice (Top toolbar).

- In edit, we can do, undo, redo, repeat, cut, copy, paste, and select all. Properties, link database part, device database part, rotate and find options are available.
- In view option, ascend and descend hierarchy, go to and zoom option, grid and grid reference, and toolbar display options are available.
- Place is the main and major part in schematic windows, the functions in the place part is shown in Figure 8.12.
- In macro option, we have configure, play, and record options.
- In the PSpice menu, we can create a new simulation profile. After creating, we have the option to edit the simulation profile. After that, we can run and view simulation results. All of these are available in the PSpice option. Also, it can create netlist, view netlist, place optimizer, run optimizer, and markers like voltage, current probes, and bias points.
- In accessories we have rotate aliases option.
- In the window button, we have new window, cascade, tile horizontally and vertically, and session log options.
- Help menu has the following option shown in the Figure 8.13.

The toolbar present at the right side of the schematic window is shown in Figure 8.14.

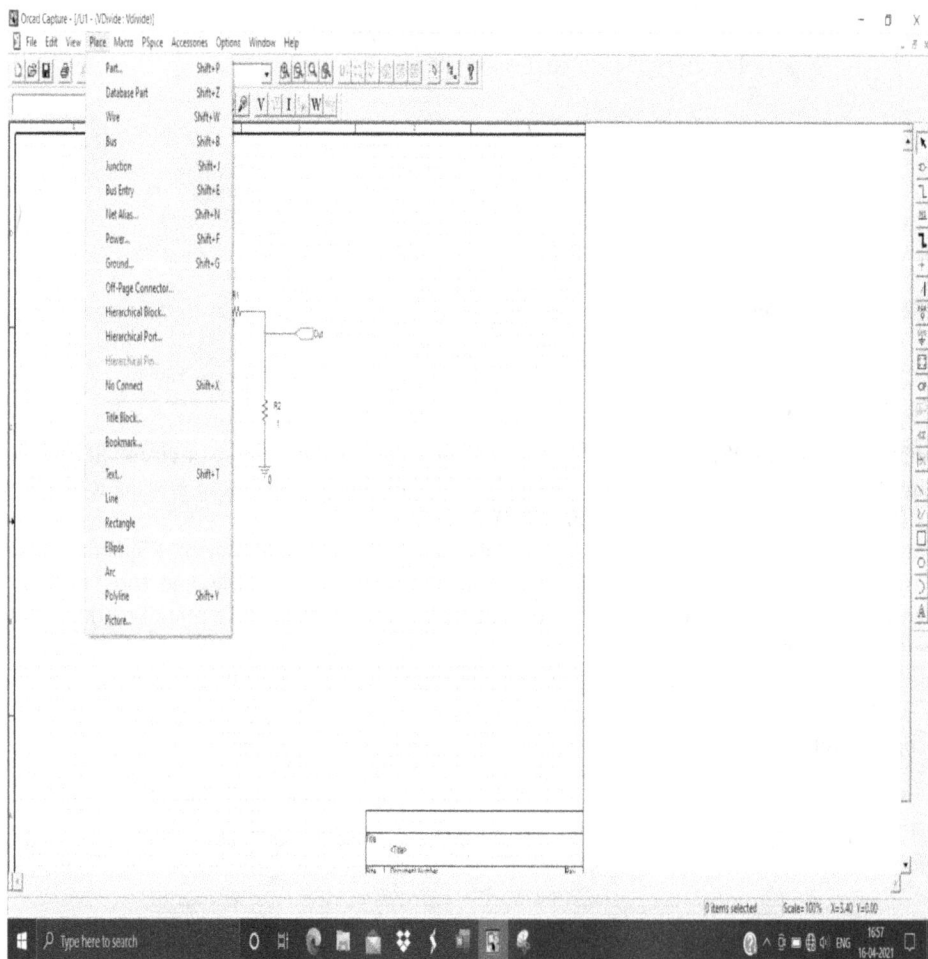

FIGURE 8.12   Place option schematic window toolbar in PSpice.

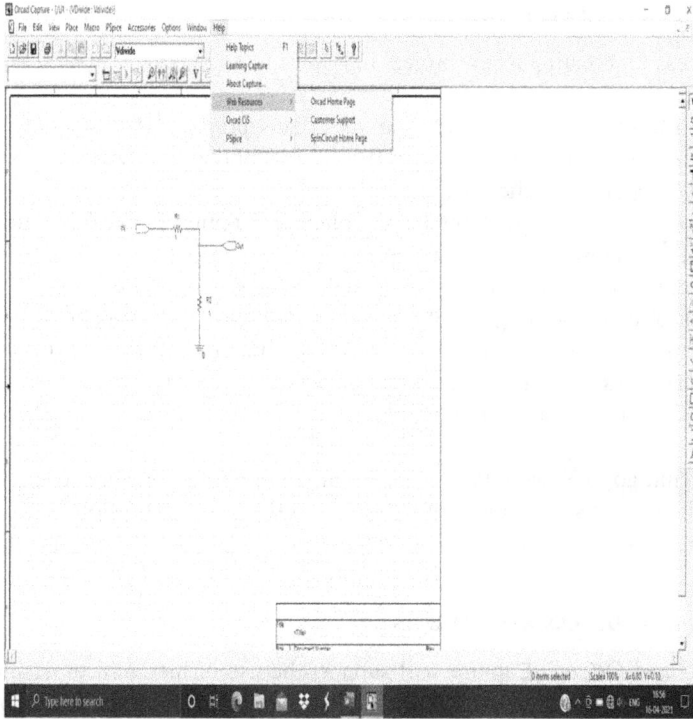

FIGURE 8.13 Help option schematic window toolbar in PSpice.

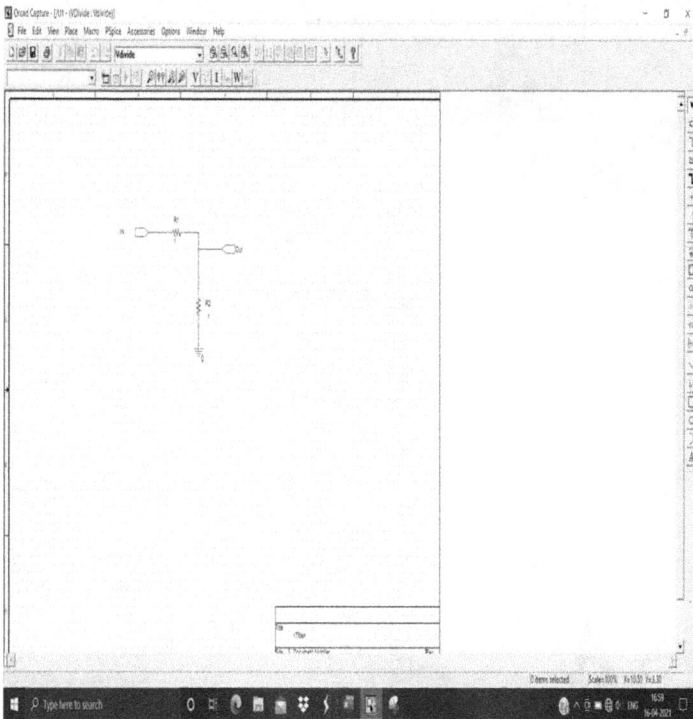

FIGURE 8.14 Options available in schematic window right side toolbar in PSpice.

8.1.8 Creating New Project and Run the Simulation

Step by step process for creating a new project in PSpice is as follows.

Step 1: Go to file and select new and then select new project. After selecting new project property window will appear. In that, give the name of the project and select the analogue or mixed A/D option and click ok.
Step 2: Place the components with the help of place part option available on the right side of the schematic window.
Step 3: Connect the basic RLC circuit and give the values as in Figure 9.17.
Step 4: Save the circuit. For this go to PSpice option and select new stimulation profile. In that, give the name of the project, following which stimulation property window will appear. Now give the necessary setup options and give ok.
Step 5. Press F5 to run the simulation

To simulate a circuit, go to New → Project and then create a new Analogue or Mixed A/D project to design a circuit up to its simulation. This is shown in Figures 8.15 to 8.19
From Figure 8.19, we can cross-check the values of current and voltage by manual calculations.

8.1.9 Hierarchical Blocks and Symbols

Both PSpice A/D and OrCAD Capture will support the use of hierarchical design with the help of hierarchical symbol and hierarchical blocks. In the schematic, both the hierarchical blocks and symbols serve as graphical representation; usage and application of both hierarchical symbols and blocks are also different.

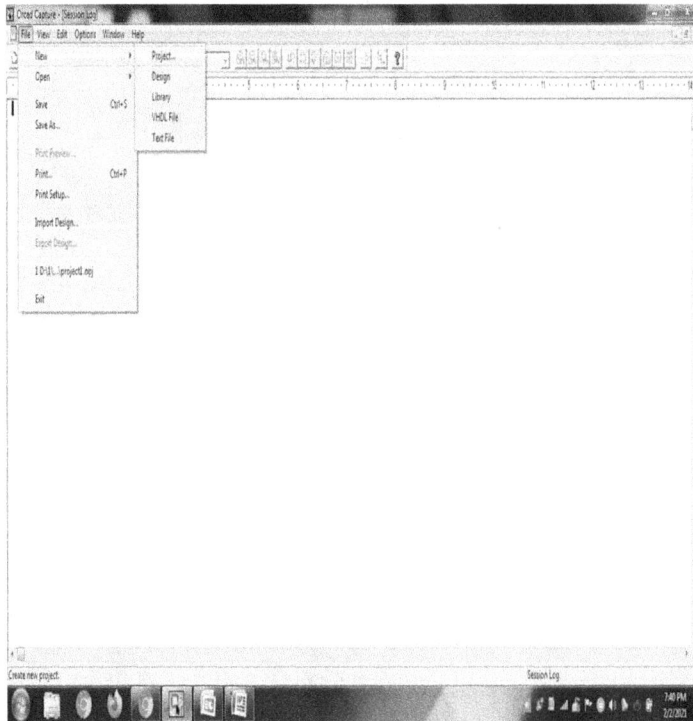

FIGURE 8.15 New project property setup.

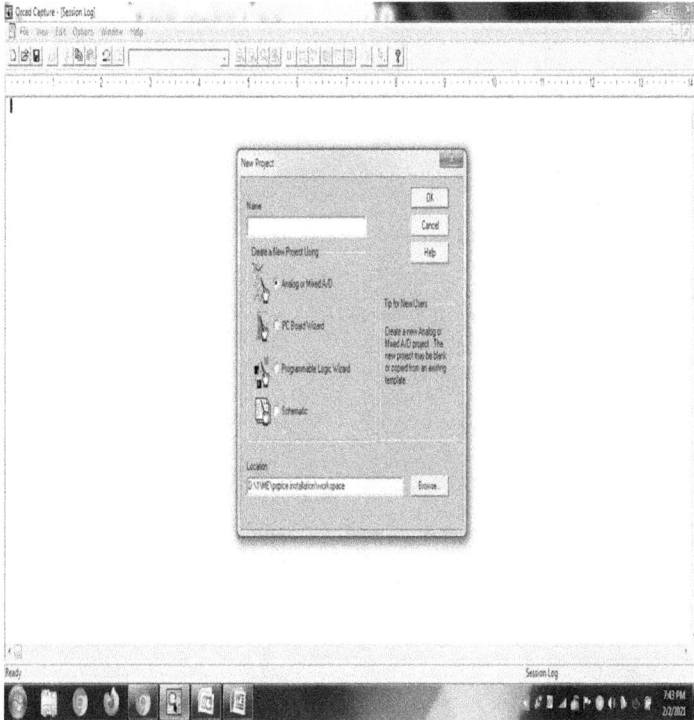

FIGURE 8.15 (*Continued*)   New project property setup.

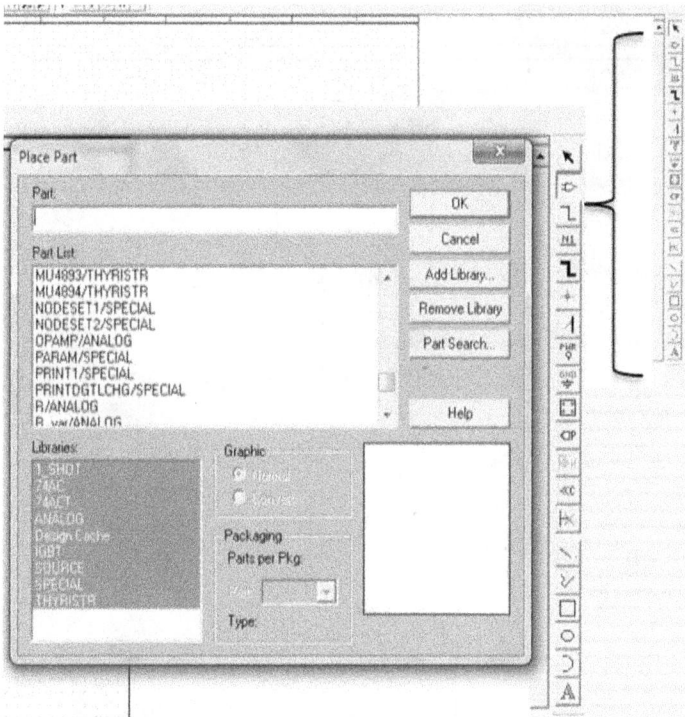

FIGURE 8.16   Place part window.

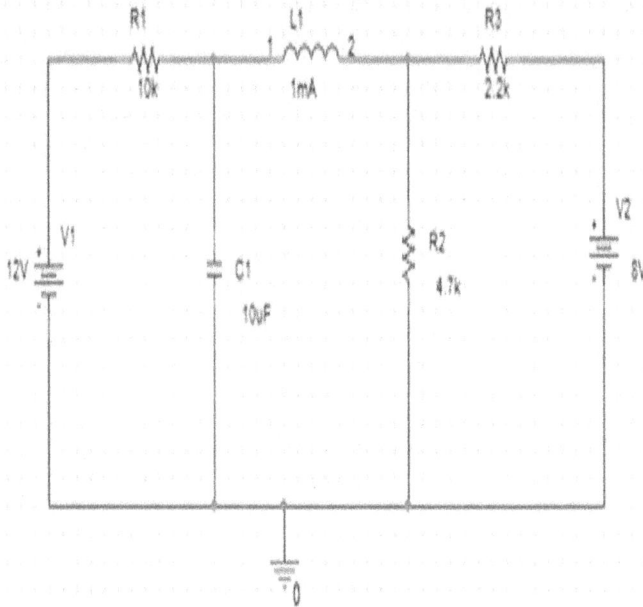

FIGURE 8.17 RLC circuit.

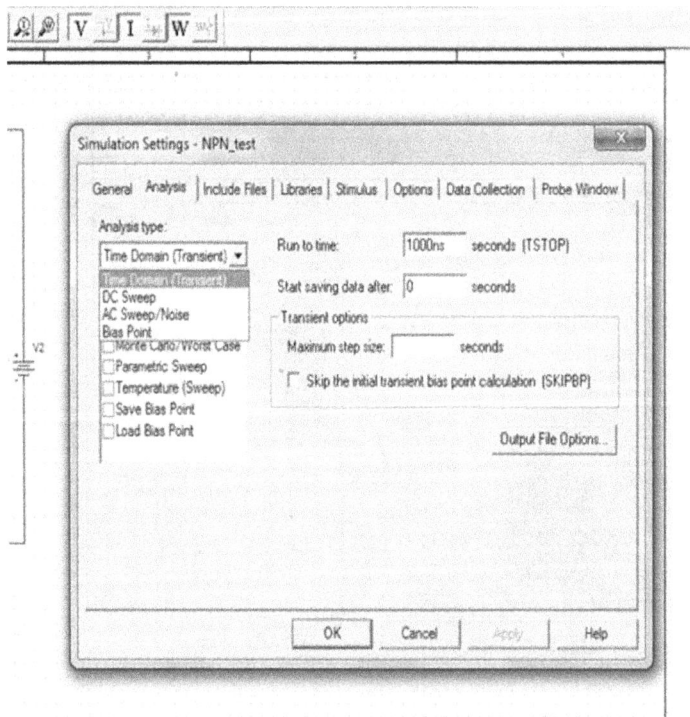

FIGURE 8.18 New simulation property and simulation setting.

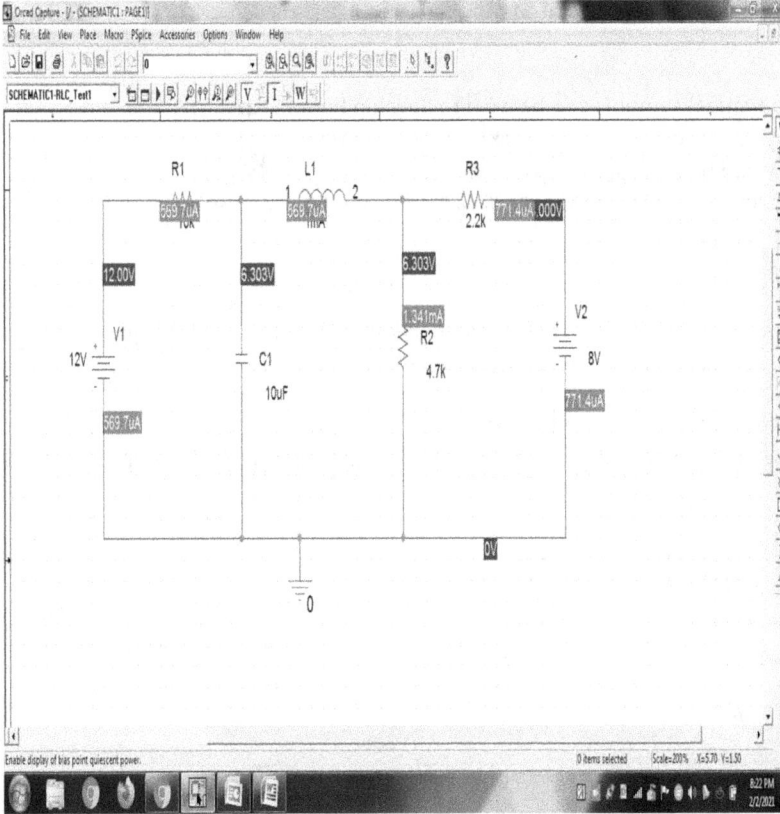

FIGURE 8.19 RLC circuit after simulation.

8.1.9.1 Hierarchical Blocks

Generally, these hierarchical blocks are utilized for the implementation of top-down designs. In a top-level design, the hierarchical blocks are placed. The hierarchical block pushes from the top-level design to the lower-level design; they are also local to a schematic design and blocks are said to be a part of the design database. As blocks are not stored in a library file, they are not available for other designs

Pros and Cons of usage of Hierarchical Blocks

Pros

- Hierarchical blocks can be utilized to develop top-down designs.
- They can be created only in the schematic page and stored as schematic file.

Cons

- It cannot be a reference file for another design window.
- To implement the blocks, we have to create this block for each design.

8.1.9.2 Hierarchical Symbols

Generally, hierarchical symbols are utilized in bottom-up designs. First, the lower-level schematic is drawn. The symbol is then positioned to symbolize a connection to a design at a higher level. As with regular/primitive symbols, hierarchical symbols are kept in symbol libraries and are available for use in other designs.

Pros and Cons of Hierarchical Symbols

Pros

- Hierarchical symbols can be used to develop bottom-up designs.
- They can be created from an existing schematic.
- They can be used in multiple designs.
- They can be stored in symbol library file, and they are also reusable.

Cons

- Main disadvantage in using hierarchical symbols is that they require schematic reference for tracking the file always while using.

8.1.9.3 Creating Hierarchical Blocks

Below are the steps to create and run the hierarchical block with the help of VDivide circuit as example,

Step 1: In the schematic window, go to place option; in that click on the Hierarchical Block as shown in Figure 8.20.

FIGURE 8.20 Hierarchical block option.

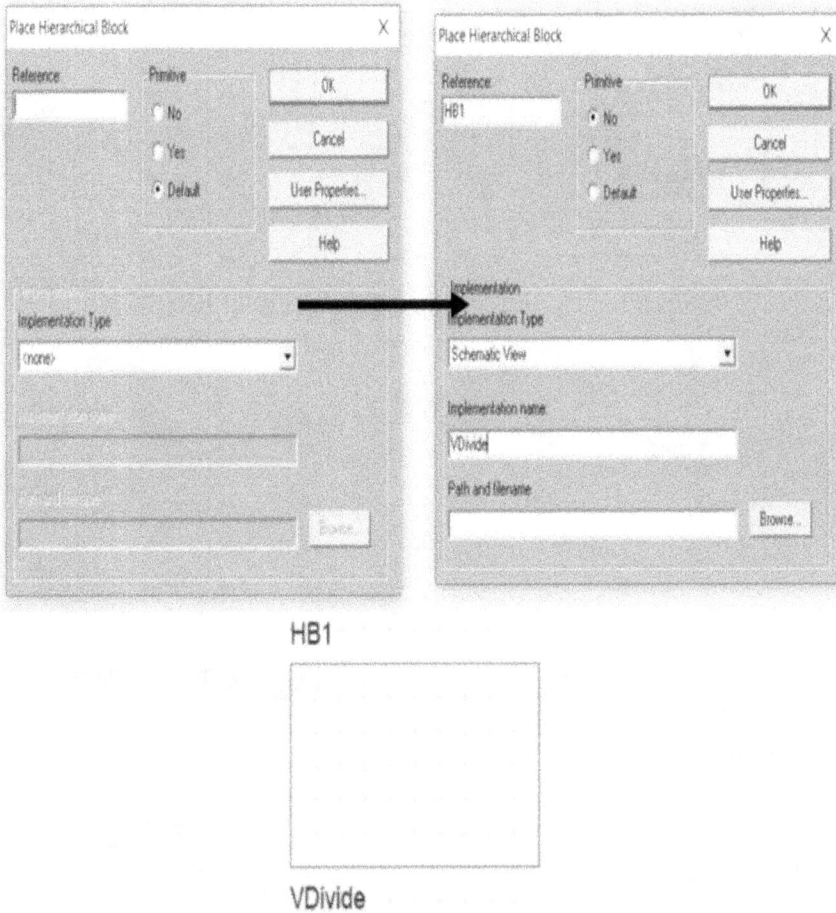

HB1

VDivide

FIGURE 8.21 Placing the hierarchical block in window.

Step 2: Place hierarchical blocks, in that reference as HB1, in primitive option select No and give implementation type as schematic page. After giving implementation, name will enable. In that give VDivide (name of circuit), then press ok.Now Hierarchical block will be created as shown in Figure 8.21.

Step 3: Select the hierarchical block that has been created and go to place option and click the hierarchical pin option, which is shown in Figure 8.22.

Step 4: Place Hierarchical Pin window will open. In that, give name as IN, width as Scalar, and type as input for input pin and press OK. Now have the input pin as viewed in the figure. In a similar way, create another pin for output. Give name as Out and type as output and press okay, which is shown in Figure 8.23.

After creating both the pins, VDivide hierarchical block will appear, as shown in Figure 8.24.

Step 5: Select the created hierarchical block and right click and select descend Hierarchy option as shown below. Now new schematic page for VDivide block will be created, name it as VDivide, and press OK, which is shown in Figure 8.25.

Step 6: In new schematic page, the pins are placed as shown in the left figure. Now place the required components for VDivide circuit from the place part option and connect it with the help of wire, as shown in Figure 8.26.

FIGURE 8.22 Hierarchical pin option.

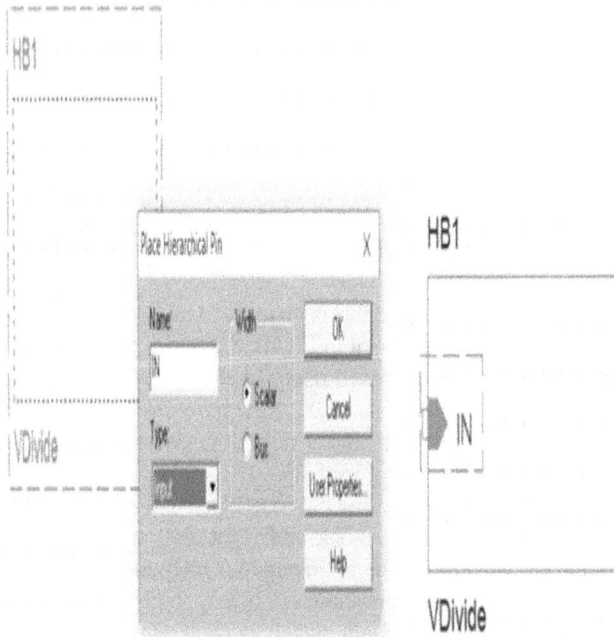

FIGURE 8.23 Placing hierarchical pin in hierarchical block.

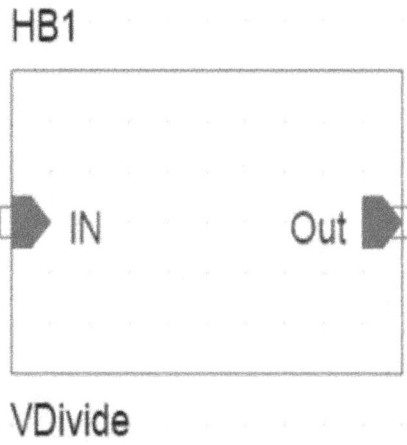

FIGURE 8.24 VDivide hierarchical block.

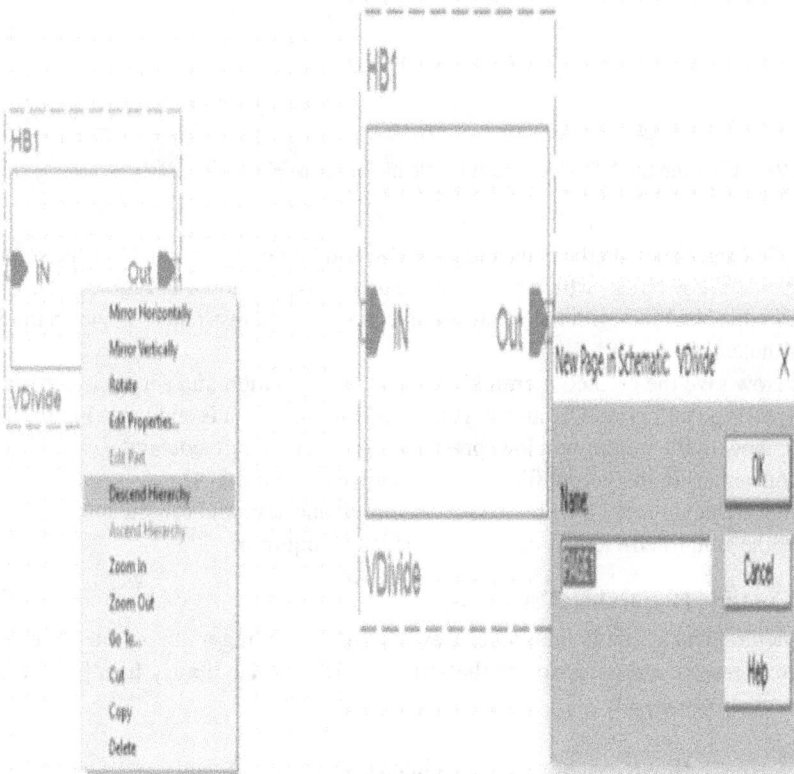

FIGURE 8.25 Descend hierarchical block.

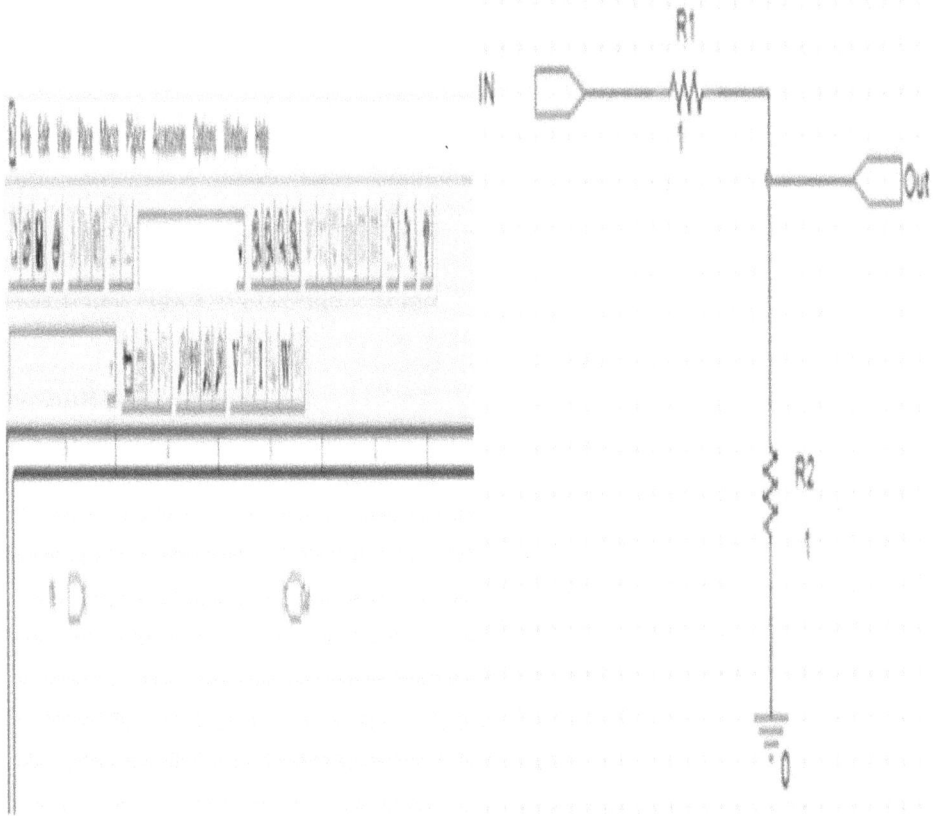

FIGURE 8.26 Creating the VDivide circuit inside hierarchical block.

Step 7: Now right click on the page and press Ascend hierarchy option. Now the screen will go to the hierarchy block, which has been created early, as shown in Figure 8.27.

Step 8 : In that block, connect the voltage source, ground, and resistance, as shown in Figure 8.28 for simulation.

Step 9: Now save the created hierarchical block in simulation and set up select Bias point as analysis type and press OK and start the simulation, which is shown in Figure 8.29.

Step 10: Now in the output window, press file option at the left side and we will see the node voltage values in the output file, which is shown in Figure 8.30.

Hierarchical blocks are created for the circuit and save and run the blocks. This helps to show the simplified view of our circuit, which is required.

8.1.9.4 Creating Hierarchical Symbols

Creating hierarchical symbols in another way is akin to creating a library file for easy and convenient usage. Here, we are going to see the steps for creating the library files for that same voltage divider circuit as discussed.

Step 1: Create a Hierarchical block for voltage divider circuit name as VDivide, as shown in Figure 8.31.

Step 2: Now go to file and go to New option and click library as shown. Now the library file will be created for our project, which is shown in Figure 8.31.

FIGURE 8.27   Ascend hierarchy circuit.

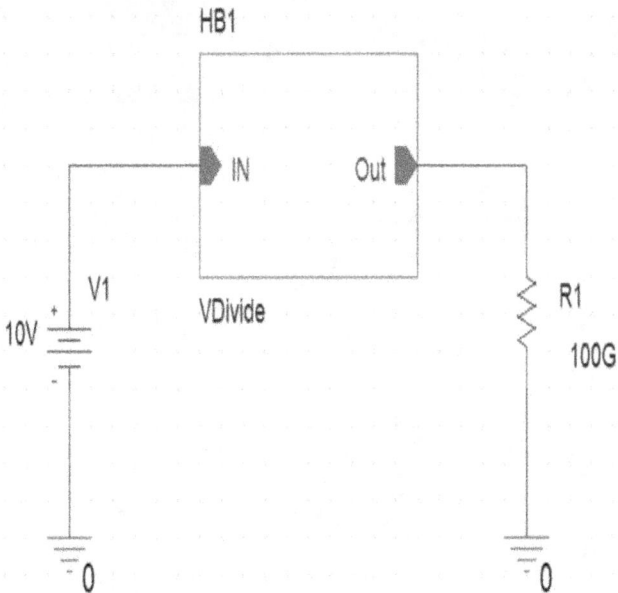

FIGURE 8.28   VDivide circuit.

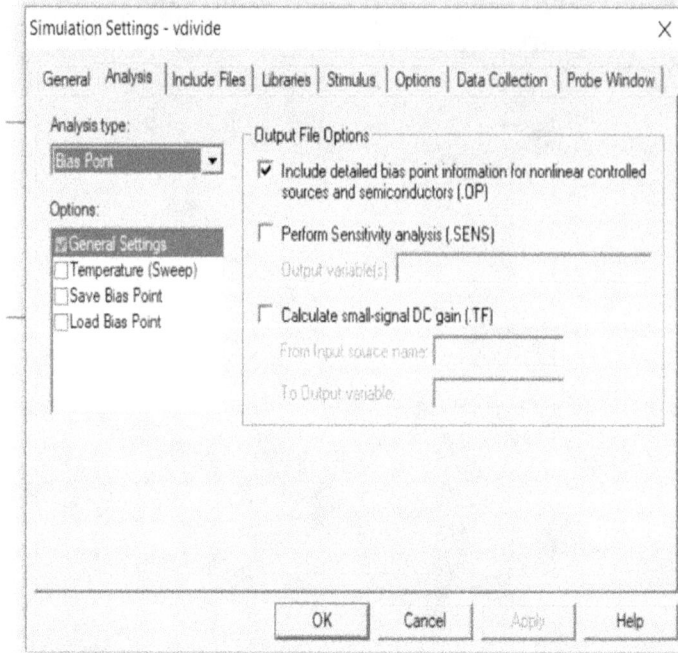

FIGURE 8.29 Simulation setting for VDivide circuit.

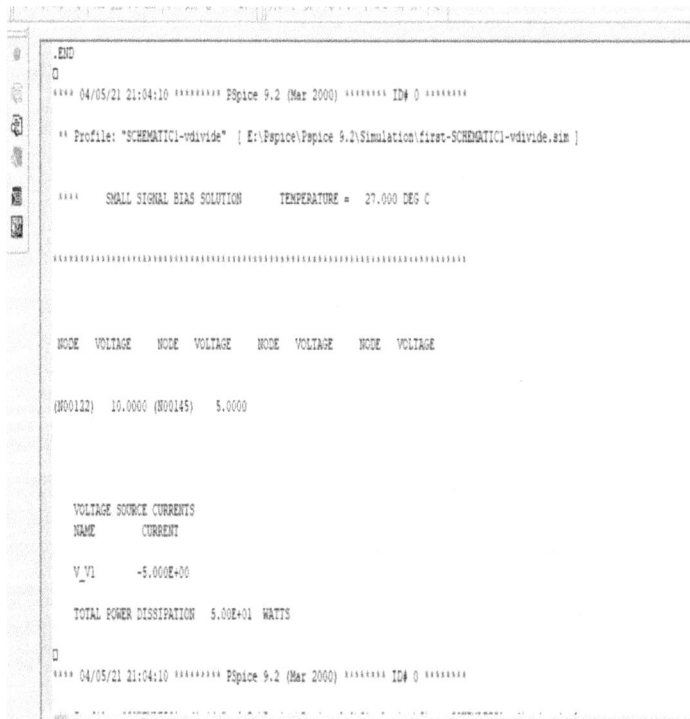

FIGURE 8.30 Output file for VDivide circuit.

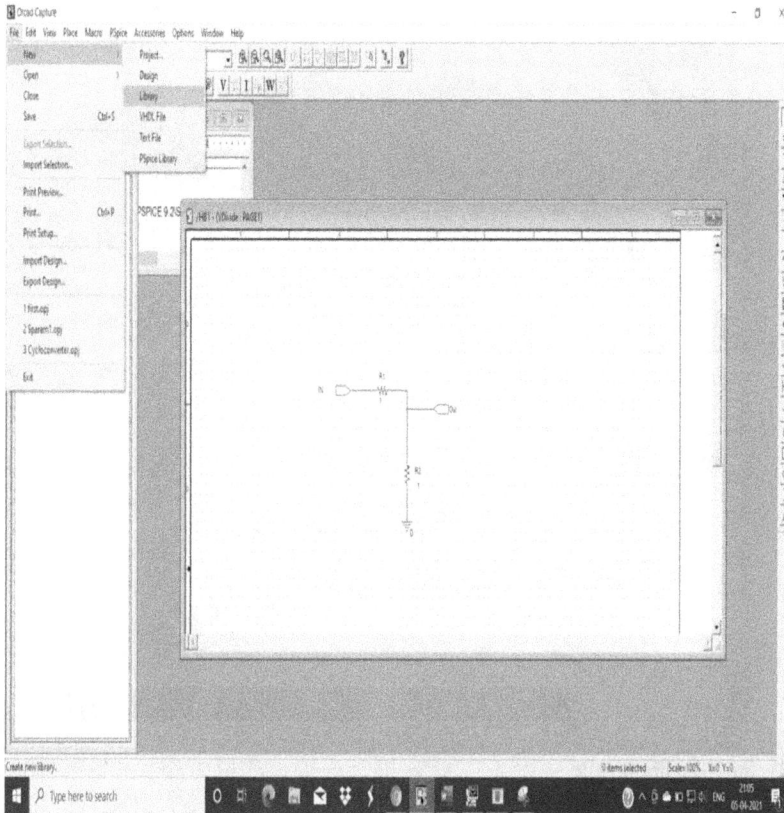

FIGURE 8.31 Library file option in toolbar.

Step 3: In that library file, right click the library1.olb and press save as. Now save the libray file as VDivide, as shown in Figure 8.32.

Step 4: Now right click the created VDivide library, and select new part option, as shown in Figure 8.33.

Step 5: Now new part properties windows will appear. In that, give name as VDivide and in select attach implementationtype as schematic view and implementation as VDivide, which we have created already, as shown in **Figure 8.34**, now press OK in both windows, as shown in Figure 8.34.

Step 6: Now the U box will appear. After that go to place option and select rectangle and draw it inside the box, as shown in Figure 8.35.

Step 7: Now go to place and select pin option, or we can also select the pin in the left side toolbar. Now place pin window will appear, wherein we are going to create pins as input and output pins. For input pin give name as IN and number as 1 and type as input and for output pin give name as OUT and number as 2 and type as output shape as short for both the pins and press OK, as shown in Figure 8.36.

Step 8: After creating each pin, place the pin in the box as shown below and right click the Value and edit as VDivide (circuit name). Now our final library block will appear as shown in Figure 8.37.

Step 9: Now go to.dsn file, which you have created. Right click it and open a new schematic and right click on the new schematic page and select a new page. Now save the entire project done so for. After that right click the created schematic and select make root option, as shown in Figure 8.38.

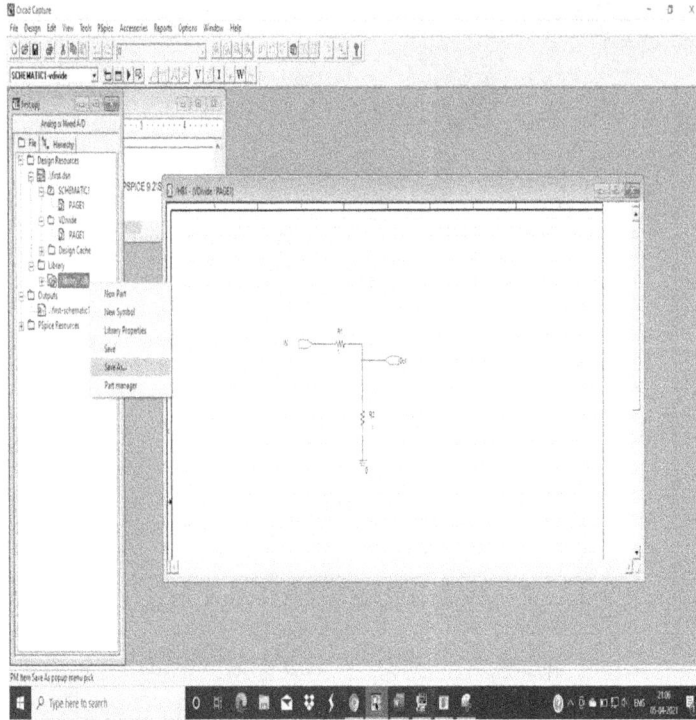

FIGURE 8.32   Save the library file.

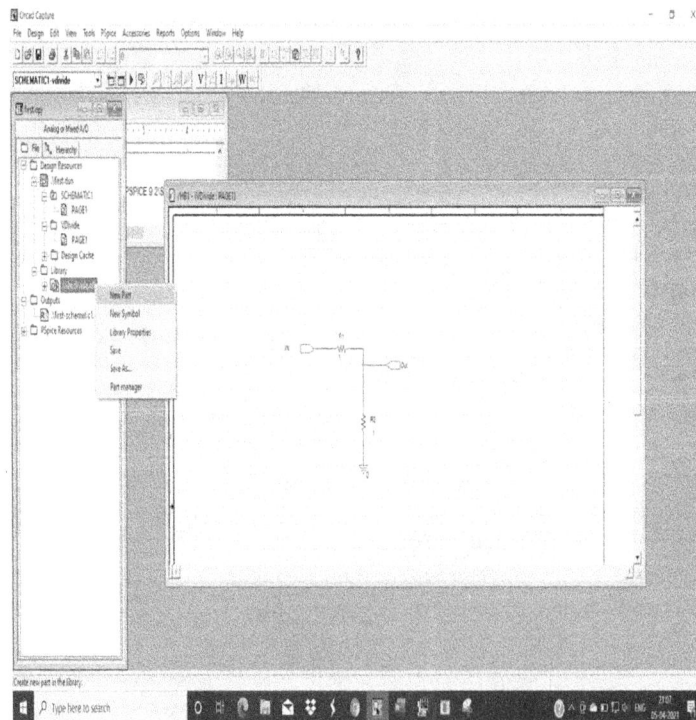

FIGURE 8.33   Creating new part for library file.

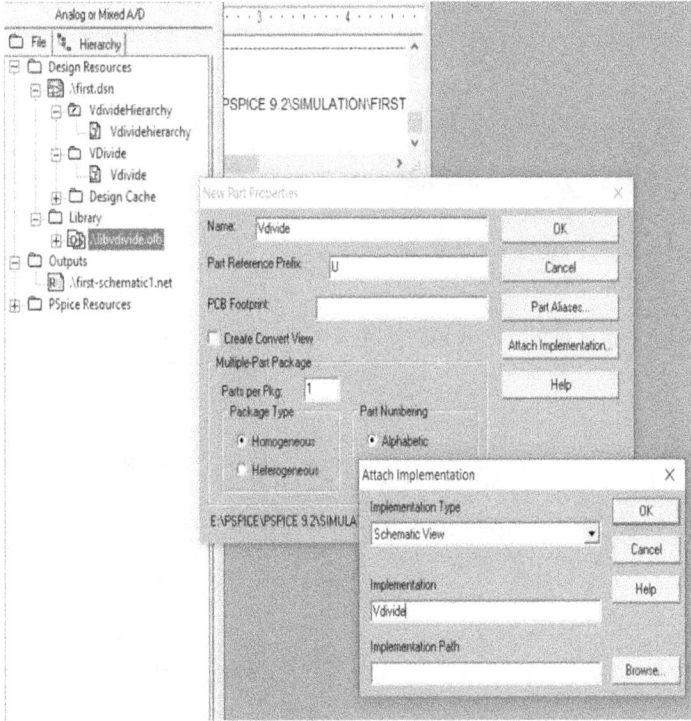

FIGURE 8.34 Creating path for library file.

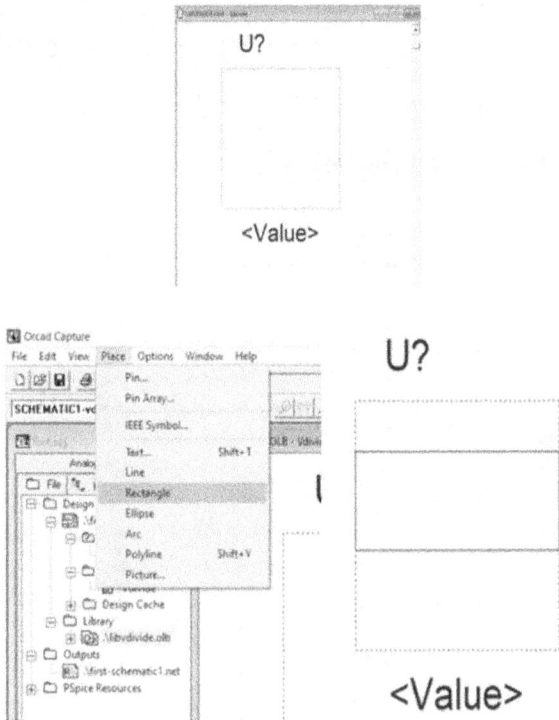

FIGURE 8.35 Draw the hierarchical symbol.

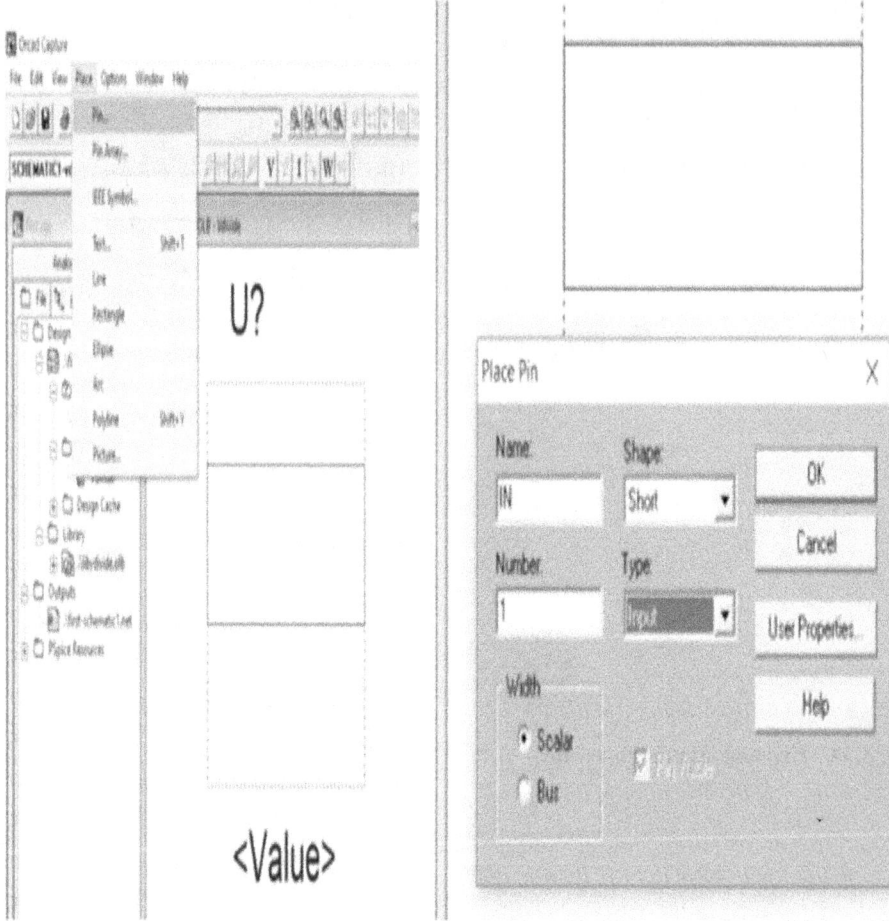

FIGURE 8.36   Place pin setup for hierarchical symbol.

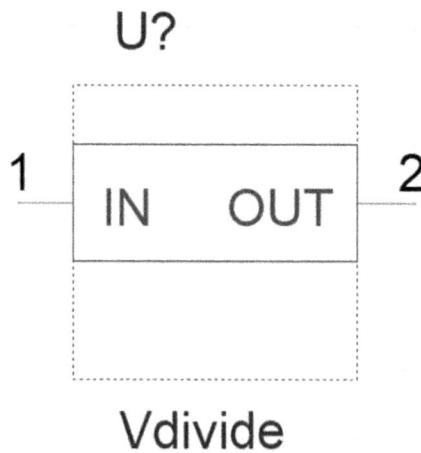

FIGURE 8.37   VDivide hierarchical symbol.

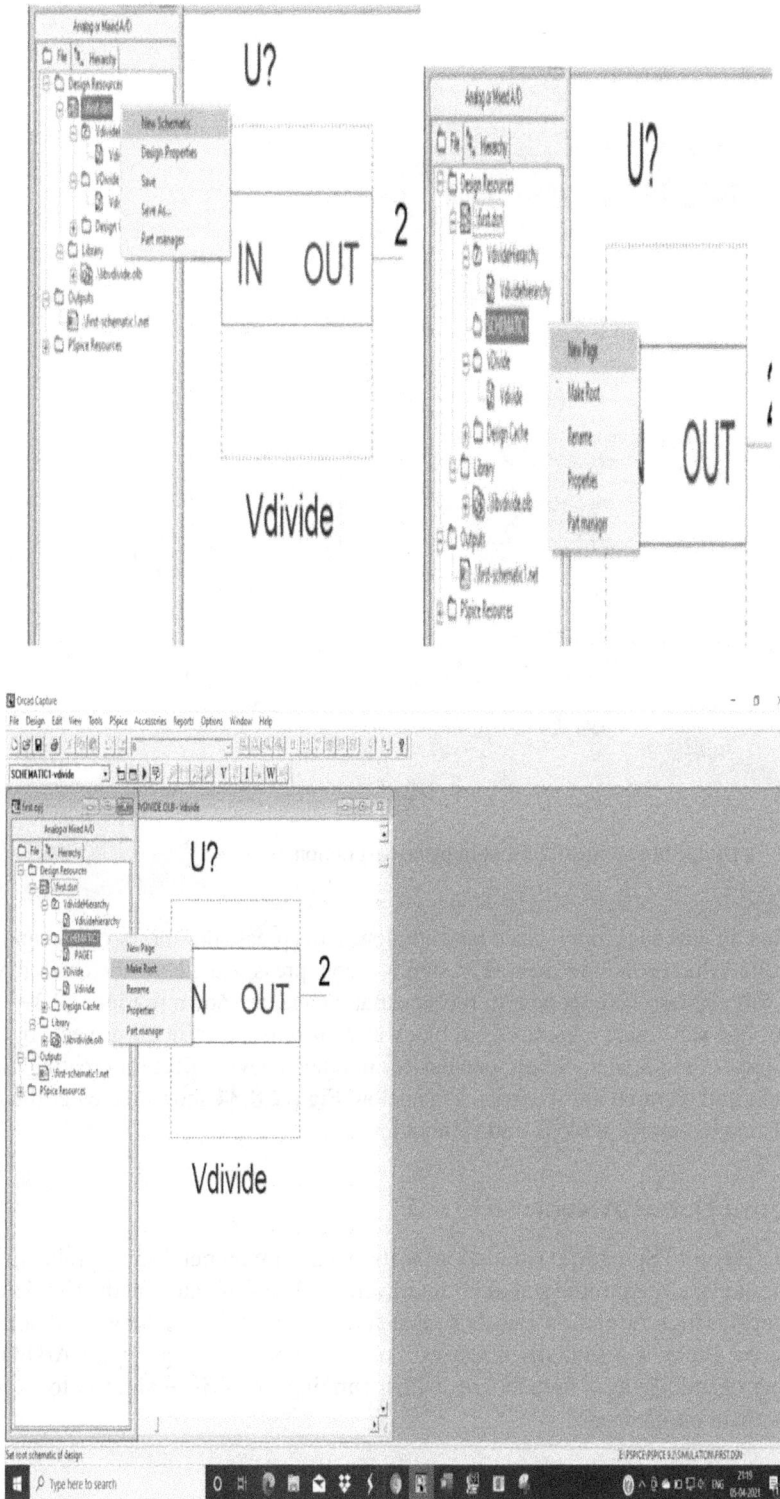

FIGURE 8.38 Steps for root file creation of VDivide hierarchical symbol.

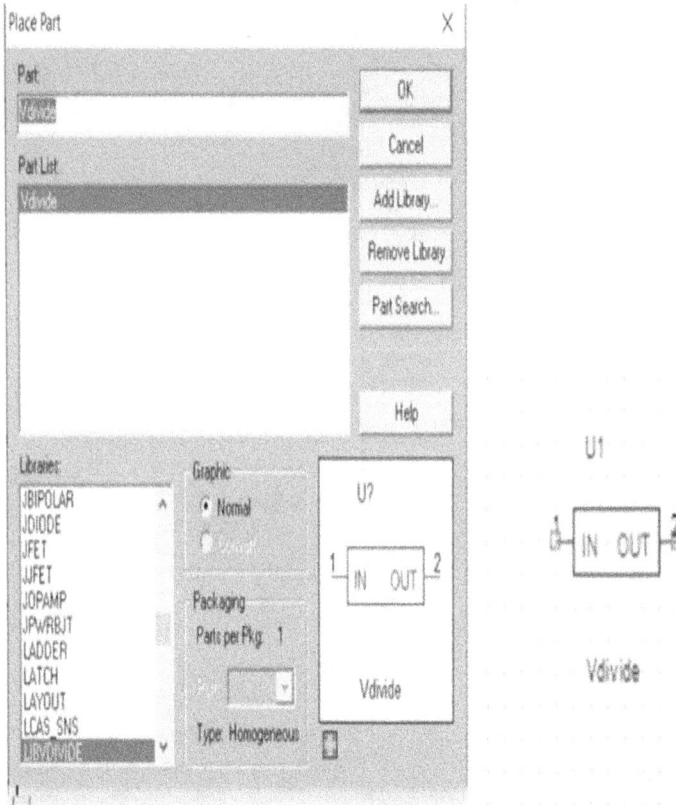

FIGURE 8.39 VDivide hierarchical symbol in place part option.

Step 10: Now in new page, open place part window. Now press add library option and add the library file, which you have created in step 3. Then, press okay. Now the created VDivide circuit will appear as component in the schematic page, as shown in Figure 8.39.

Thus, one can create a hierarchical block and make use of it in all projects by creating library files. And for each circuit, one desires to open a new project for creating hierarchical blocks and hierarchical symbols. This below **Figure 8.34** shows the enter file creation for VDivide hierarchical block and symbol circuit.

8.1.10   OUTPUT DISPLAY WINDOW

Waveform analyser for PSpice A/D simulations is the MicroSim output display probe. In the display window, you can visually examine and alter the waveform data generated by the simulation of a circuit. Probe employs high-resolution visuals so that simulation results can be viewed both on-screen and in hard copy. Probe is essentially a software oscilloscope. Running PSpice A/D is analogous to constructing or modifying a breadboard, whilst running Probe is analogous to examining the breadboard with an oscilloscope.

8.1.10.1   Output Waveform analysis

Example circuit: Unipolar Junction Transistor (UJT) Simulation circuit as in Figure 8.40

Step 1. Run and simulate the circuit.
Step 2. Open the output schematic window.

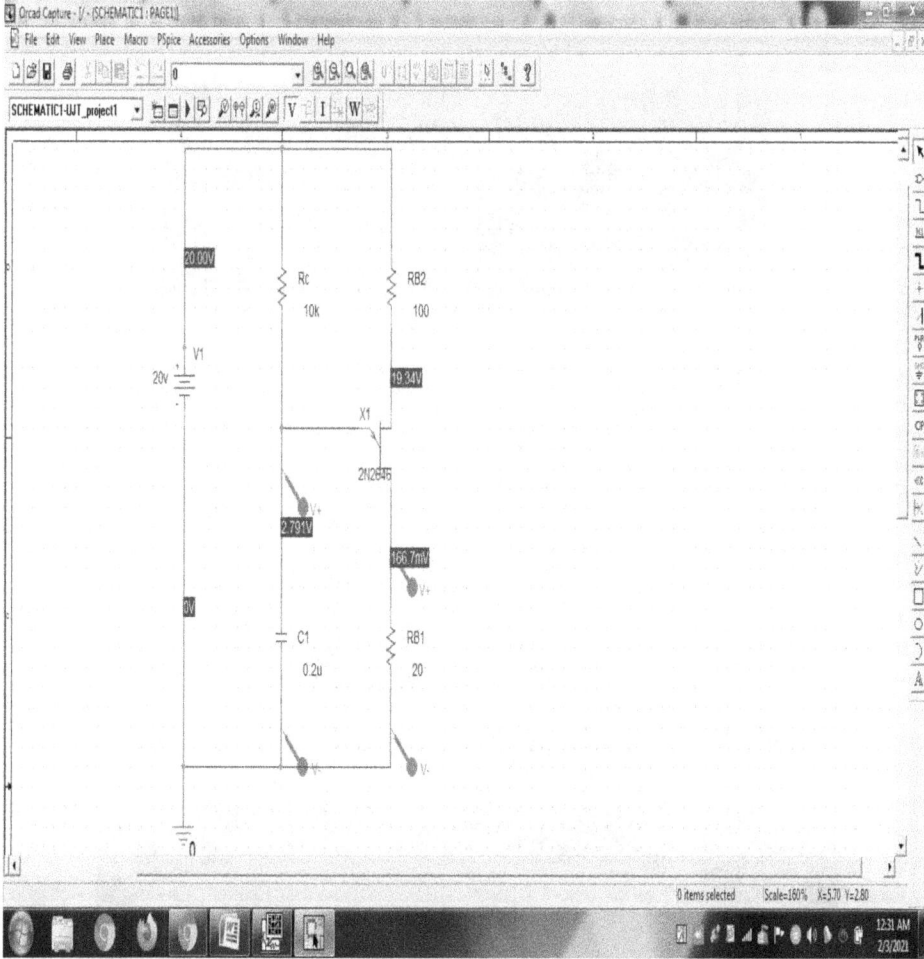

FIGURE 8.40   UJT simulation circuit.

Step 3. Right click the wave form. We have options like information, properties, cursor 1, and cursor 2, as shown in Figure 8.41.

Step 4. By clicking Properties option, trace properties window will appear, in that it is possible to change the colour of the waveform, width and type..

Step 5. By right clicking any area in the wave, it is possible to set the property option with the help of the setting option.

Step 6. By selecting the setting option, Axis setting will open. In that, we can change the X-axis and Y-axis properties, as shown in Figure 8.42.

8.1.10.2   Toolbar in Output Schematic

In PSpice, output window has more options. The toolbar in Figure 8.43 is an important toolbar one wants to know in output schematic window.

From left to right, the details of symbol in the toolbar are as follows:

1. Log in x option; by selecting this the X-axis waveform will change for log value.
2. FFT option; by selecting this fourier option, the wave form will change for FFT.
3. Performance analysis option.

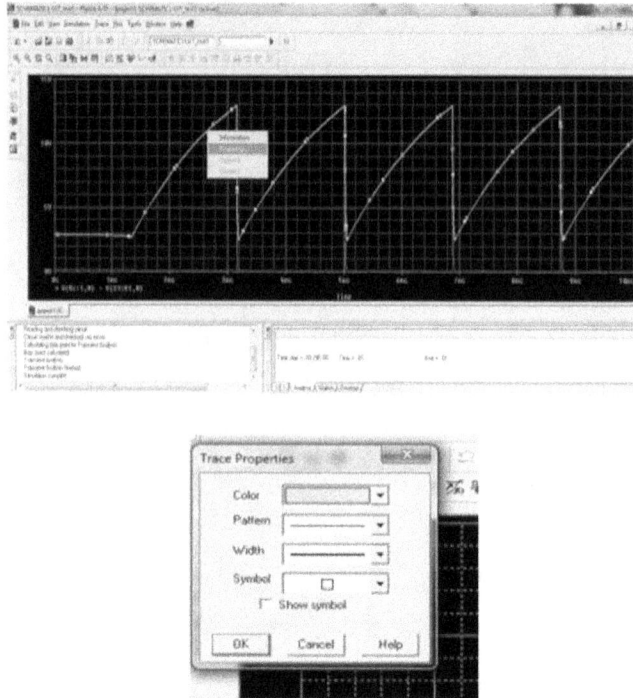

FIGURE 8.41   Waveform colour change property in output window.

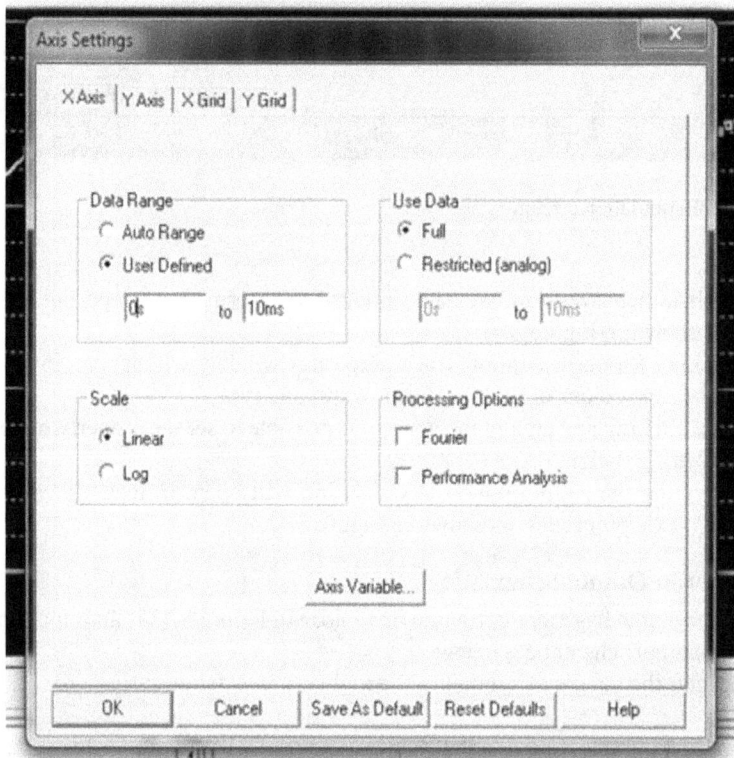

FIGURE 8.42   Axis setting property in output window.

FIGURE 8.43   Toolbar in output window.

4. Log Y-axis option; by selecting this, the Y-axis waveform will change for log value.
5. Add trace option; by selecting this add trace window will appear. From that one can choose the values to display the waveform.
6. Eval goal function option.
7. Text label; from this we can name our output waveform.
8. Mark data point option.
9. Toggle cursor option; with the help of this one can move the cursor and note the values in the toggle window.

8.1.11   PARAMETRIC AND TEMPERATURE ANALYSIS

8.1.11.1   Parametric Analysis

Multiple iterations of a defined standard analysis are conducted while modifying a global parameter, model parameter, component value, or operating temperature during parametric analysis. The effect is the same to running the circuit multiple times, once for each value of the variable being swept.

a. **Minimum requirements to run a parametric analysis:**
   1. Configure the circuit based on the sweeping variable type, as shown in Table 8.6
   2. Set up a DC sweep, AC sweep, or transient analysis.
b. **Minimum programme setup requirements:**
   STEP 1: Select the Parametric button in the Analysis Setup dialogue box. Complete the required fields in the Parametric dialogue.
   STEP 2: If necessary, select the Parametric check box in the Analysis Setup dialogue box to enable it, as shown in Figure 8.44.
   STEP 3: Run the simulation.

8.1.11.2   Temperature Analysis

PSpice A/D reruns typical analyses enabled in the Analysis Setup dialogue box at different temperatures when a temperature analysis is executed. Temperature analysis permits the specification of zero or more temperatures. If no temperature is given, the circuit will operate at 27 degrees Celsius. If many temperatures are specified, the impact is the same as running the simulation multiple times, one for each temperature setting in the list.

TABLE 8.6
Parametric Analysis Circuit Design Requirements

Swept Variable Type	Requirement
Voltage source	Voltage source with a direct current (DC) specification (VDC, for example)
Temperature	None
Current source	Current source with a DC specification (IDC, for example)
Model parameter	PSpice A/D model
Global parameter	Global parameter defined with a parameter block (PARAM)

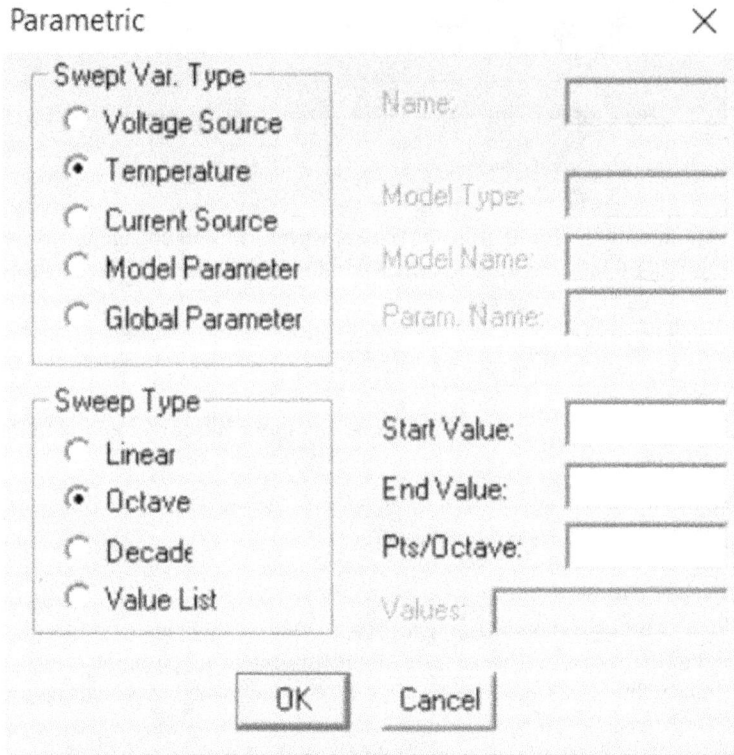

FIGURE 8.44 Parametric setting.

a. **Minimum circuit design requirements:**
   None
b. **Minimum programme setup requirements:**
   STEP 1: Click the Temperature button in the Analysis Setup dialogue box. In the Temperature Analysis dialogue box, specify the temperature or select from the list of temperatures.
   STEP 2: If necessary, select the Temperature check box in the Analysis Setup dialogue box to enable it, as shown in Figure 8.45.
   STEP 3: Run the simulation.

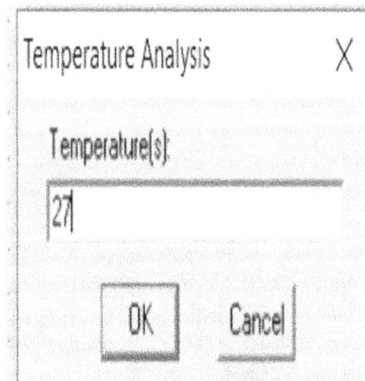

FIGURE 8.45 Temerature setting tab.

8.1.12   ANALYSIS OF WAVEFORM IN PROBE WINDOW

The waveform analyser for PSpice A/D simulation is MicroSim Probe. Probe allows you to visually examine and alter the waveform data generated by circuit simulation.

Probe employs high-resolution visuals so that simulation results can be viewed both on-screen and in hard copy. Probe is essentially a software oscilloscope. PSpice A/D is analogous to constructing or modifying a breadboard, while Probe is analogous to examining the breadboard with an oscilloscope.

With Probe, the following can be possible:

*   View simulation result plots in various windows
*   Display simple voltages, currents, and noise data
*   Display sophisticated arithmetic expressions that use the basic measurements
*   Show Fourier transformations of voltages and currents or arithmetic equations using voltages and currents

PSpice A/D produces two types of output. They are simulation output file and Probe data file. The calculations and outcomes given in the simulation output file serve as a simulation audit trail. Nonetheless, the most instructive and adaptable approach of reviewing simulation results is the graphical analysis of information in the Probe data file.

8.1.12.1   Probe Elements

A single Probe plot comprises the analogue (lower) and digital (upper) regions. Multiple plots can be displayed on the screen. If only analogue waveforms are displayed, the entire plot will be an analogue area. Similarly, if only digital waveforms are displayed, the entire plot will be digital (Figure 8.46).

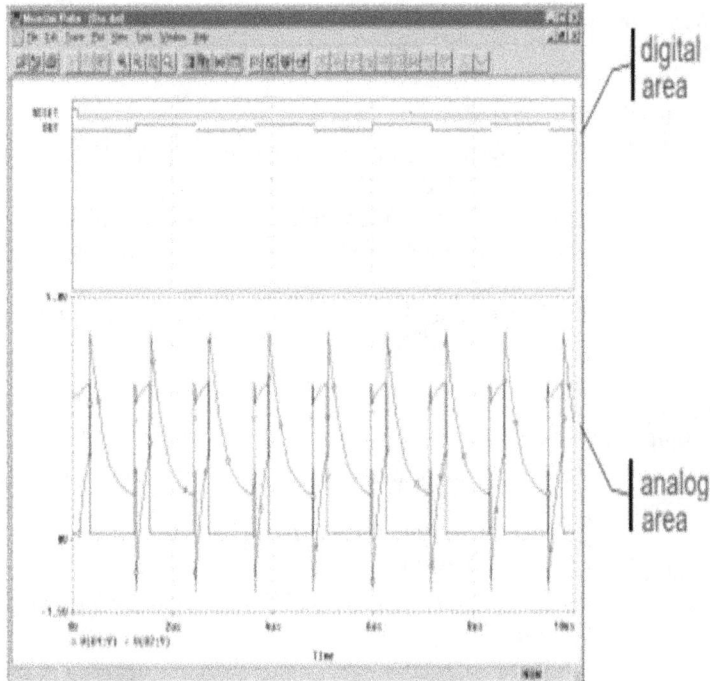

FIGURE 8.46   Probe plot-analogue and digital areas.

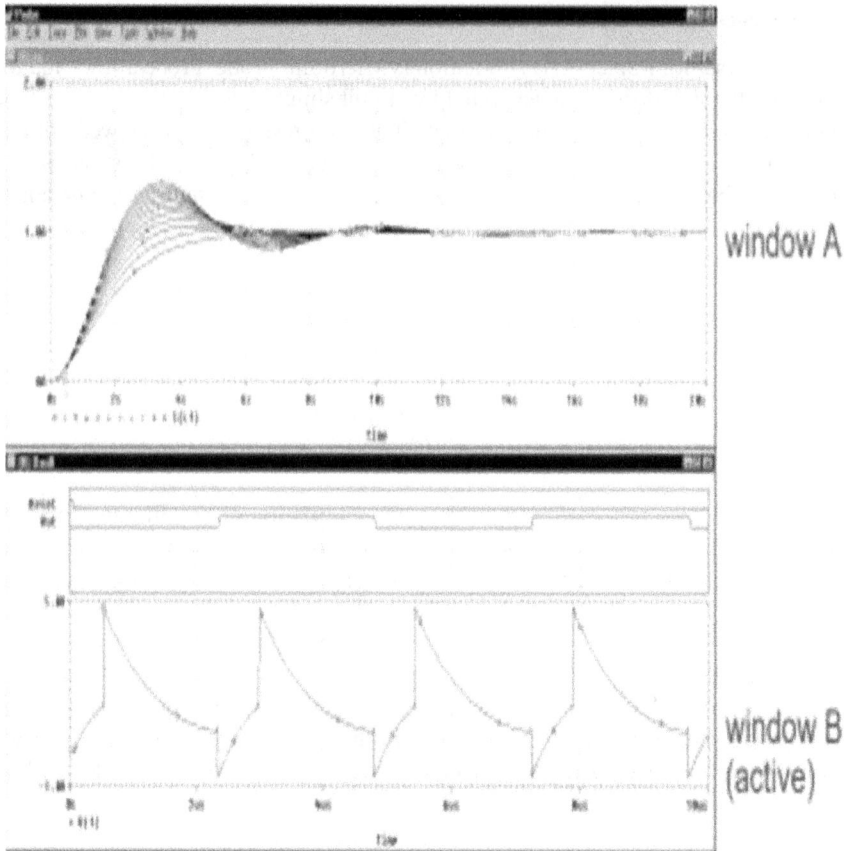

FIGURE 8.47   Probe window with two plot windows.

8.1.12.2   Elements of a Probe Window

A plot window is a waveform display area that is maintained independently. Multiple analogue and digital plots can be displayed in a plot window (Figure 8.47).

Due to the fact that a plot window is a window object, you can minimize and maximize it, as well as move and resize it inside the Probe window area. The Probe window can display a toolbar that applies to the active plot window.

It is possible to open many Probe data files on a single plot window by performing one of the below:

- Using the Design Journal function in Schematics, configure Probe to load open working schematics and checkpoint files automatically.
- After the initial file has been loaded, manually append additional files to the same plot window in Probe.

8.1.12.3   Managing Multiple Plot Windows

Plot windows can be opened in any quantity. Each plot window operates independently. It is possible to display the same Probe data file in many plot windows. At any given time, only one plot window can be activated, as indicated by its highlighted title bar. Menu, keyboard, and cursor operations only influence the currently selected plot window. You can activate another plot window by clicking anywhere within the window.

All or selected plot windows can be printed, with up to nine windows per page. A list of all open plot windows is presented when you pick Print from the File menu. In the title bar of each plot window is a unique identification enclosed in parentheses.

The Page Setup dialogue box can be used to modify the layout of plot windows on the page. You can print in either portrait or landscape orientation. Using Print Preview, you may also examine all plot windows as how they will appear when printed.

8.1.12.4   Setting Up a Probe
  a. **CONFIGURING PROBE COLOUR:**
       Probe display and print colours can be configured in:
  - The configuration file, msim.ini, and
  - The Probe Options dialog box.
  b. **EDITING DISPLAY AND PRINT COLOURS IN THE msim.ini FILE:**
       In the msim.ini file, the following print and display colour options can be adjusted:
  - The colours of the display traces.
  - The colours Probe employs for the plot window's foreground and background.
  - The sequence in which Probe employs colours to display traces.
  - The number of colours Probe employs to display traces.

8.1.12.5   Edit Display and Print in the msim.ini File
1. Open the msim.ini file using MicroSim Text Editor or any other text editor.
2. Navigate to the file's [PROBE DISPLAY COLORS] or [PROBE PRINTER COLORS] section.
3. Add or change a colour entry. Refer to Table 9.6 for colour entry defaults. Valid item names include:
  - ACKGROUND
  - OREGROUND
  - RACE_1 through TRACE_12
4. Set NUMTRACECOLOURS = n to the updated number of traces (1n12) if you have added or deleted trace number entries. This item indicates the number of trace colours displayed or printed before the colour order repeats.
5. The file should be saved.
6. In case of copying the Probe plots to the clipboard and pasting them into a black-and-white document, consider the following colour settings: BACKGROUND = BRIGHTWHITE FOREGROUND = BLACK.

Table 8.7 shows default probe item colours.

TABLE 8.7
Default Probe Item Colours

Item Name	Description	Default
BACKGROUND	Specifies the colour of window background	BLACK
FOREGROUND	Specifies the default colour for items not explicitly specified	WHITE
TRACE_1	Specifies the first colour used for trace display	BRIGHTGREEN
TRACE_2	Specifies the second colour used for trace display	BRIGHTRED
TRACE_3	Specifies the third colour used for trace display	BRIGHTBLUE
TRACE_4	Specifies the fourth colour used for trace display	BRIGHTYELLOW
TRACE_5	Specifies the fifth colour used for trace display	BRIGHTMAGENTA
TRACE_6	Specifies the sixth colour used for trace display	BRIGHTCYAN

In the Probe Options dialogue box, the colours and colour order supply configured in the msim. ini file are utilized to display traces in a Probe plot window with the following features:

- Unique colour for each trace.
- Same colour for all traces belonging to the same Y-axis.
- Available colours in order for each Y-axis.
- Same colour for all traces belonging to the same data file, including data files for check-point schematics.

8.1.12.6 Customizing the Probe Command Line

It is possible to specify command files, *.prb files, and arguments in the Probe command line. Probe recognizes these choices whether launched automatically following simulation or from Schematics by selecting Run Probe from the Analysis menu.

To modify the probe command line, the following can be done:

1. In Schematics, pick Editor Configuration from the Option menu.
2. Select App Settings inside the Editor Configuration dialogue box.
3. Edit the Command text field in the Simulate Command frame.

This command line is stored in the msim.ini file.

a. **CONFIGURING UPDATE INTERVALS:**
   You can specify the refresh rate of Probe's waveform display as follows:
   - At user-defined time intervals (every n seconds)
   - Based on the percentage of simulation completion (every nano seconds)
     The default configuration (Auto) permits Probe to update traces whenever it receives fresh simulation data.
b. **TO CHANGE UPDATE INTERVAL:**
   1. From the menu Tools, select Options.
   2. In the Auto-Update Interval frame, select the interval type (seconds or percentage) and then enter the interval in the text box.

8.1.12.7 Running a Probe

a. **STARTING A PROBE:**
   If you are utilizing Schematics, Probe can be launched automatically after a simulation or independently from Windows 95 or NT. When starting Probe, you can use either the default.prb file or a custom.prb file.
b. **AUTOMATICALLY BEGIN A PROBE FOLLOWING SIMULATION:**
   - In Schematics, pick Probe Setup from the Analysis menu.
   - Select the Probe Startup tab within the Probe Setup Options dialogue box.
   - Select Automatically Run Probe after Simulation from the Auto-Run Options window.
   - Select any additional settings you wish to utilize.
   - Select OK.
c. **TO BEGIN INVESTIGATING AND MONITORING RESULTS DURING SIMULATION:**
   - In Schematics, pick Probe Setup from the Analysis menu.
   - Select the Probe Startup tab within the Probe Setup Options dialogue box.
   - Select Monitor Waveforms in the Auto-Run Options window (Auto-Update). If this option is not available, proceed as follows:

    a. Select the Tab for Data Collection.

    b. Clear the checkbox for Text Data File Format (CSDF). The option to Monitor Waveforms (Auto-Update) should now be accessible via the Probe Startup tab.

- Click the ok button.
- Select Simulate from the Analysis menu to begin the simulation. Automatically launches and shows a single window in monitor mode.
- Select the waveforms to be watched by using one of the following:
  - Within Probe, pick Add from the Trace menu and enter one or more trace phrases.
  - In Schematics, choose and place one or more markers from the Markers menu (and marker colour, as needed).

d. **TO START PROBE FROM SCHEMATICS:**

    From the Analysis menu, select Run Probe.

  1. TO START PROBE IN WINDOWS 95:

    Select the MicroSim application folder and then the Probe shortcut from the Windows Start menu.

  2. TO START PROBE MANUALLY USING WINDOWS RUN COMMAND:

    1. Select Run from the Windows Start menu.

    2. Enter the location of the Probe command file and command line arguments or a data file (.dat). Using the Windows Run command, type probe options* data file to launch Probe.

e. **TO ADD TRACES:**

    Markers can be placed on a schematic to indicate where the waveform findings should be presented in Probe. Markers can be placed as required with the following consideration:

- Before simulation limit results recorded to the Probe data file and automatically display those traces in Probe.
- During or after simulation display traces in the active plot window.

    The colour of each individual marking can be determined. In Probe, the colour selected for a marker will also be the colour of its trace.

    The Markers menu gives extra options for influencing the display of marked results in Probe after the initial placement of markers and throughout simulation.

f. **TO PLACE MARKERS:**

  1. In Schematics, from the Markers menu, select the marker type you want to place, as shown in Table 8.8.

8.1.12.8   A nalogue Example

An example of an analogue circuit is shown in Figure 8.48.

TABLE 8.8
Marker Selection

Waveform	Markers Menu Selection	Symbol Selection
Voltage	Mark Voltage/Level	not required
Voltage differential	Mark Voltage Differential	not required
Current	Mark Current into Pin	not required
Digital signal	Mark Voltage/Level	not required
dB*	Mark Advanced	VDB (voltage)
		IDB (current)
Phase	Mark Advanced	VPHASE (voltage) IPHASE (current)

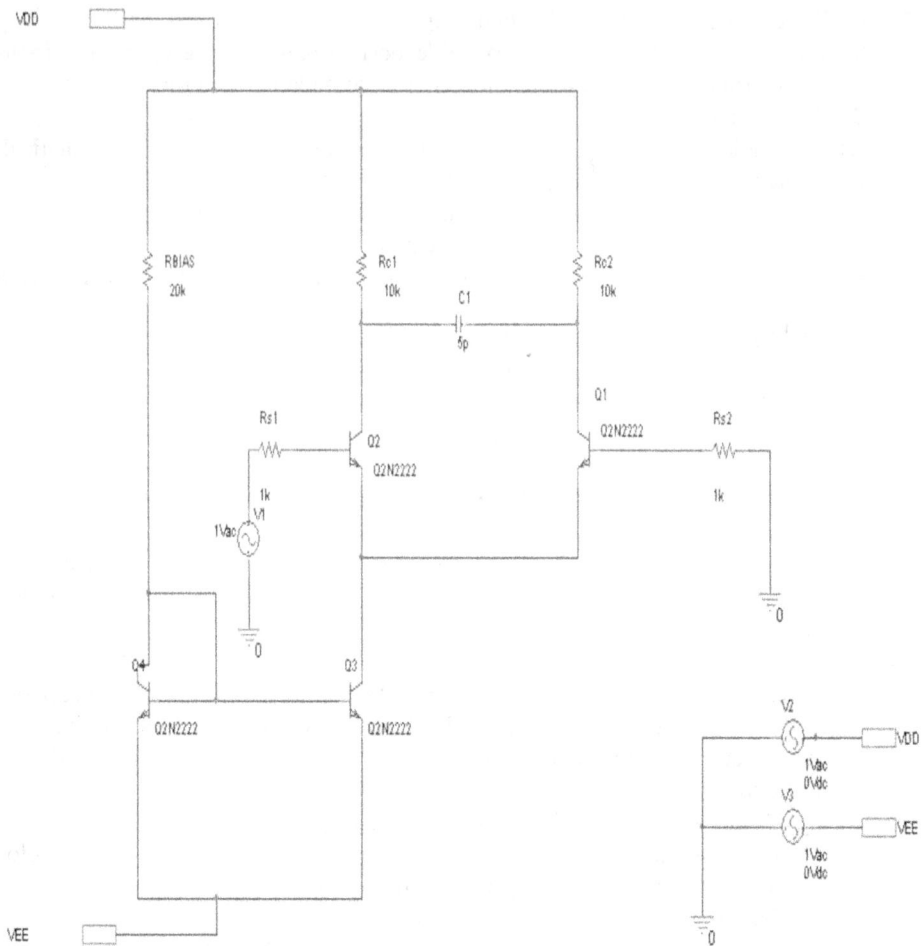

FIGURE 8.48   Analogue example.

8.1.13   SIMULATION

With the Bias Point Detail, temperature and transient analysis enabled, the simulation is run. The analytical temperature is set to 35 degrees. The setup for the transient analysis is as follows:

- Print Step 20 ns
- Final Time 1,000 ns
- Enable Fourier selected
- Centre Frequency 1 Meg
- Output Vars V(OUT2)

8.1.13.1   To Start the Simulation
1. Begin with Schematics. If Schematics is already running, ensure that the schematic editor is active.
2. Open the following file from the location where your MicroSim programmes were installed.
   Examples\Schemat\Example\Example.sch
3. Select Simulate from the Analysis option to begin the simulation.

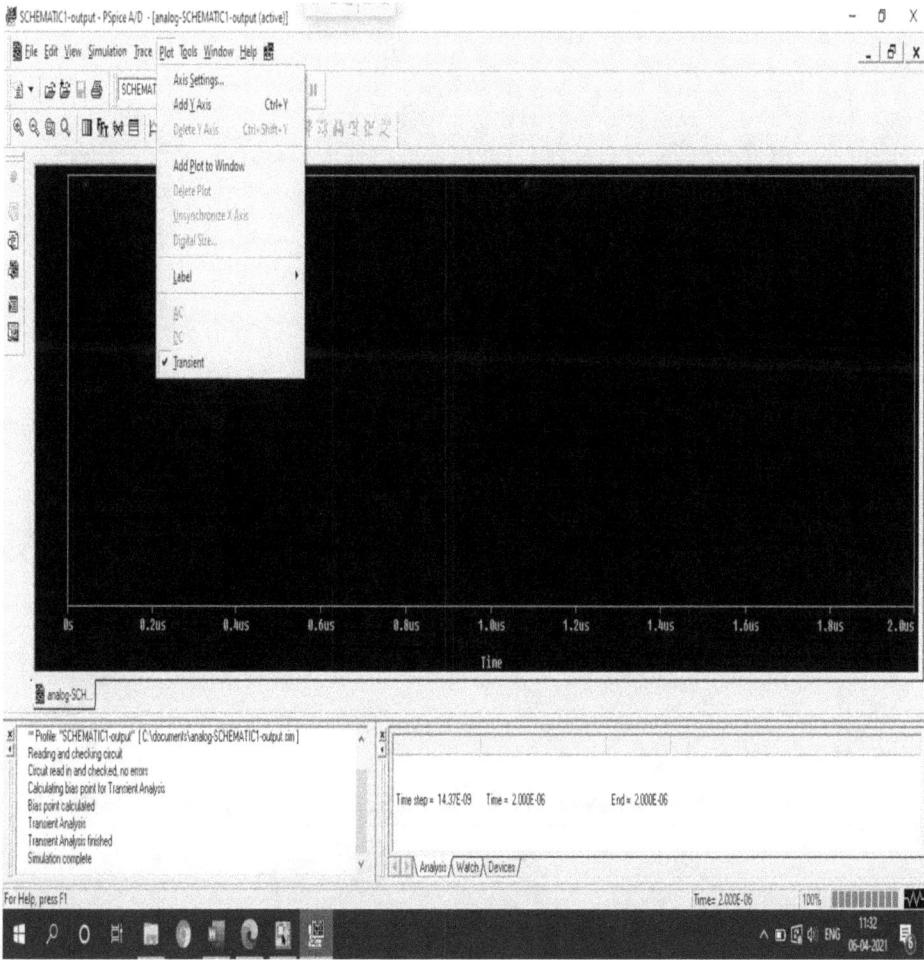

FIGURE 8.49   Simulation window.

PSpice A/D outputs a binary Probe data file that contains the simulation results, as shown in Figures 8.49 and 8.50.

In the title bar, the name of the data file, Example.dat, is displayed. All Probe commands are accessible via menu options. Notice that the transitory command is selected on the Plot menu, suggesting that the data now loaded are the findings of a transient analysis.

8.1.13.2   To Display Output

1. In the Trace menu, choose Add. Probe populates the Simulation Output Variables frame with a set of valid output variables.
2. Click V(OUT1) and V(OUT2), followed by OK.

8.1.14   Hierarchical Block and Symbols

Both PSpice A/D and OrCAD Capture will support the use of hierarchical design with the help of hierarchical symbol and hierarchical blocks. In the schematic, both the hierarchical blocks and symbols serve as graphical representation, usage and application of both hierarchical symbols and blocks are also different.

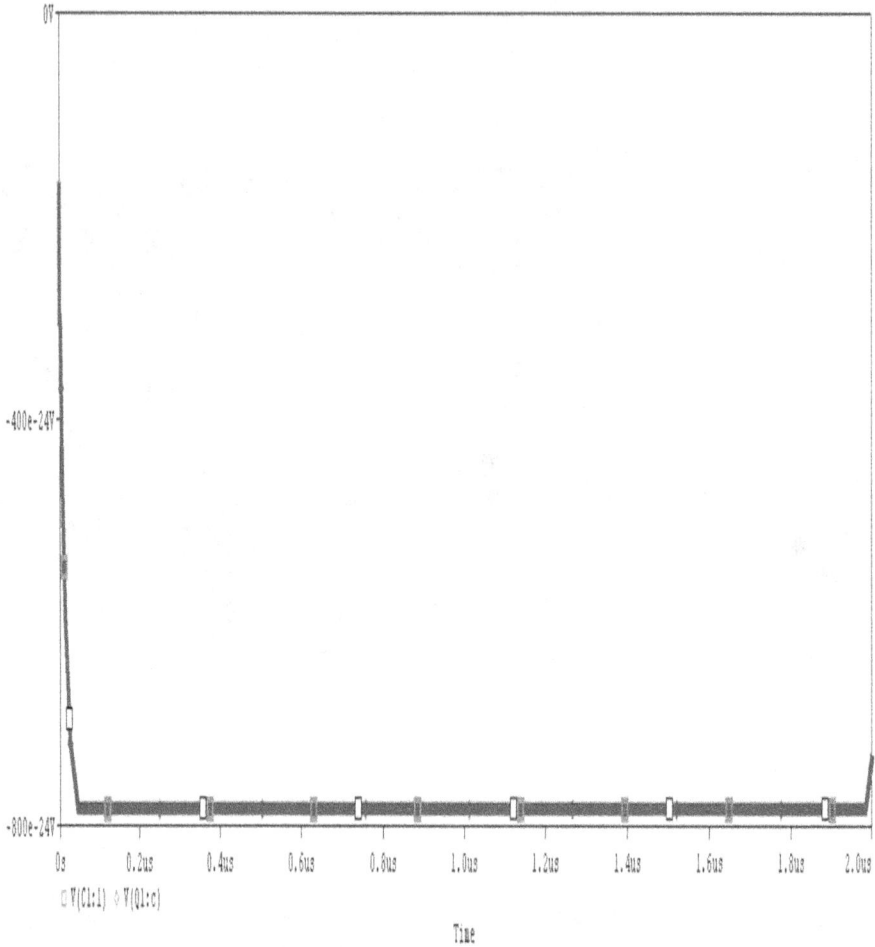

FIGURE 8.50   Output from transients.

8.1.14.1   Hierarchical Blocks

Typically, these hierarchical blocks are utilized for top-down design execution. Hierarchical blocks are placed in the top-level design, which is drawn before the lower-level design. The hierarchical block pushes from the top-level design to the lower-level design. In addition, they are local to a schematic design and are considered to be a component of the design database. Blocks are not kept in library files and hence are unavailable for use in other designs.

8.1.14.2   Pros and Cons of Hierarchical Blocks

Pros

- Hierarchical blocks can be used to develop the top-down designs.
- They can be created only in the schematic page and stored as schematic file.

Cons

- Drawback of using hierarchical blocks; they cannot be used for another design window as reference file.
- To implement the blocks, we desire to create this block for each design.

8.1.14.3   Hierarchical Symbols

Generally, hierarchical symbols are utilized in bottom-up designs. First, the lower-level schematic is drawn. The symbol is then positioned to symbolize a connection to a design at a higher level. As with regular/primitive symbols, hierarchical symbols are kept in symbol libraries and are available for use in other designs.

8.1.14.4   Pros and Cons of Hierarchical Symbols

Pros

- Hierarchical symbols can be used to develop bottom-up designs.
- They can be created from an existing schematic.
- They can be used in multiple designs.
- They can be stored in symbol library file, and they are also reusable.

Cons

- Main disadvantage in using hierarchical symbols is that they require schematic reference for tracking the file always while using.

8.1.14.5   Creating Hierarchical Blocks

Below are the steps to create and run the hierarchical block with the help of V_{Divide} circuit as example:

Step 1: In the schematic window, go to "place" menu and click on the Hierarchical Block as shown in Figure 8.51.

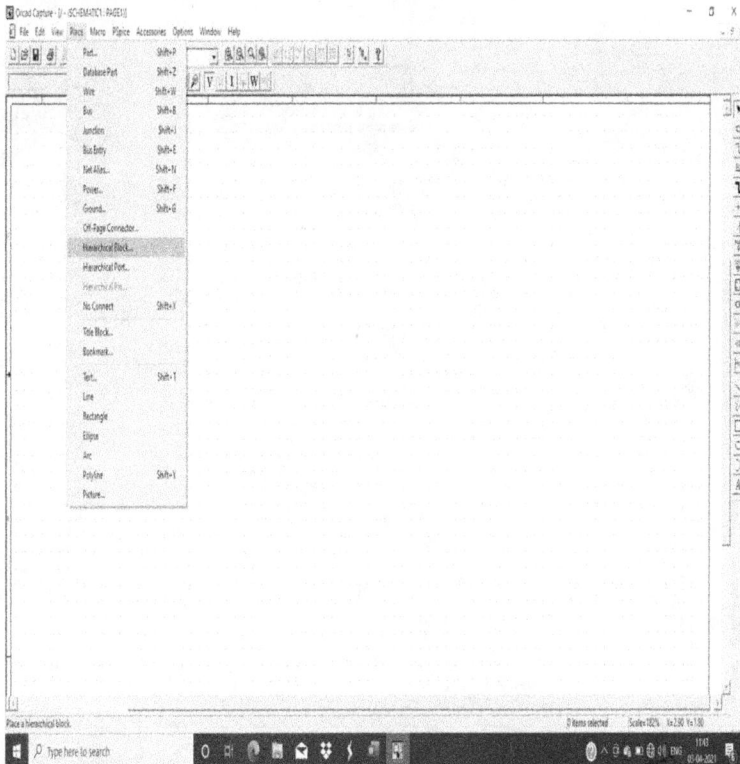

FIGURE 8.51   Hierarchical block selection.

Step 2: Place hierarchical blocks will appear. In that, five reference as HB1. In primitive option select No and give implementation type as schematic page. After giving implementation, name will enable. In that give VDivide(name of circuit) then press ok. Now Hierarchical block will be created, as shown in Figure 8.52.

Step 3: Select the hierarchical block that has been created and go to place option and click the hierarchical pin option shown in Figure 8.53.

Step 4: Place Hierarchical Pin window will open. In that, give name as IN and width as Scalar and type as input for input pin and press OK. Now place the input pin as shown in the Figure 8.54. In a similar way create another pin for output. Give name as Out and type as output and press okay.

After creating both the pins, the VDivide hierarchical block will appear as shown in Figure 8.55.

Step 5: Select the created hierarchical block and right click and select descend Hierarchy option as shown below. Now new schematic page for VDivide block will be created. Name the page name as VDivide and press OK, as shown in Figure 8.56.

Step 6: In the new schematic page, the pins are placed as shown on the left side of Figure 8.57. Now place the required components for VDivide circuit from the place part option and connect it with the help of wire.

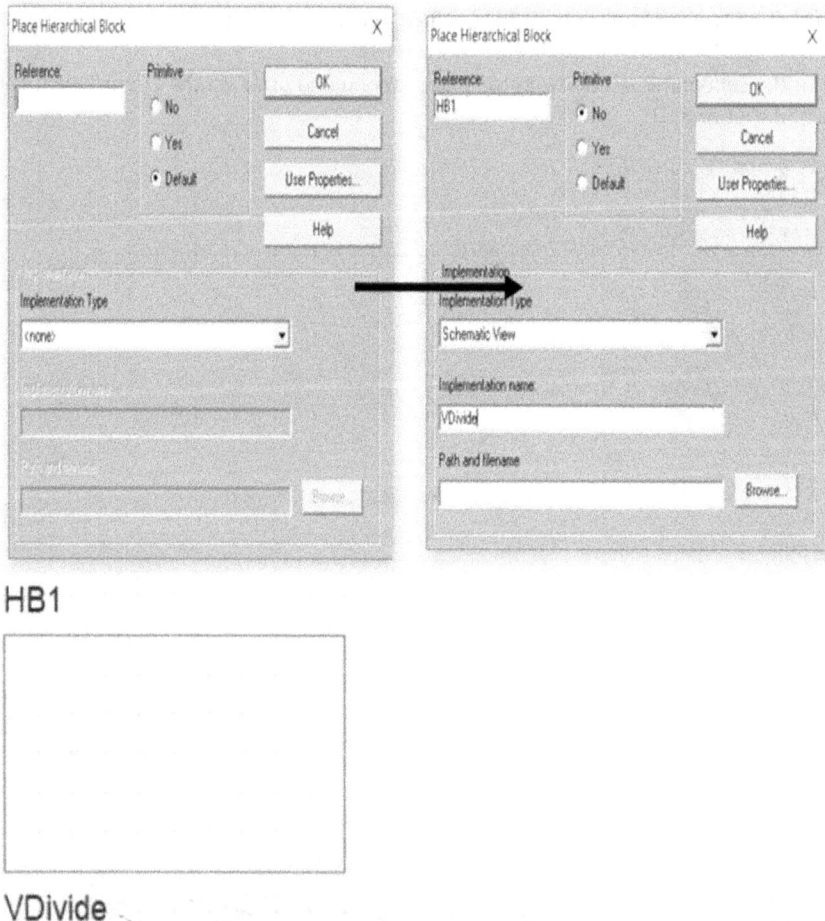

FIGURE 8.52 Hierarchical block creation.

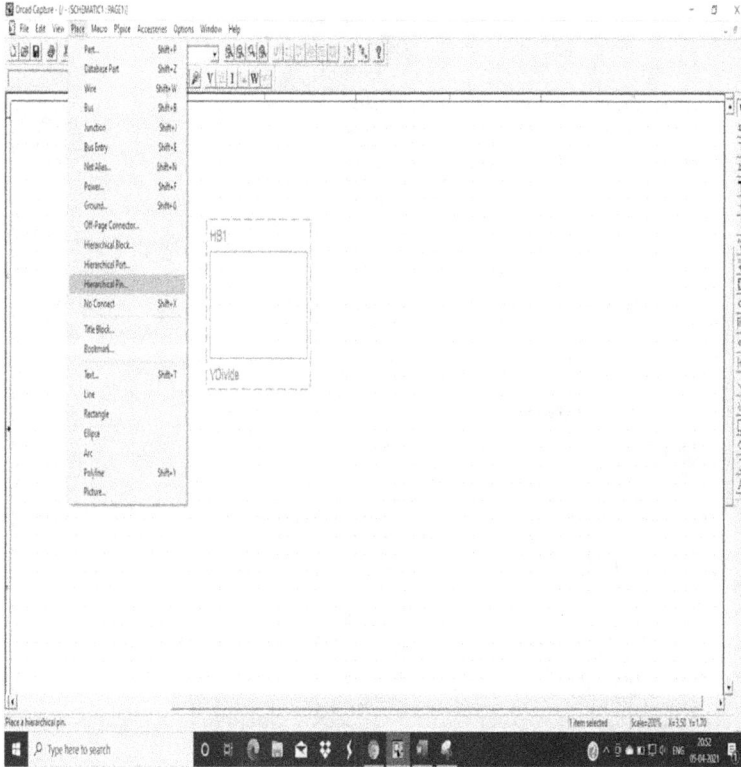

FIGURE 8.53 Hierarchical pin selection.

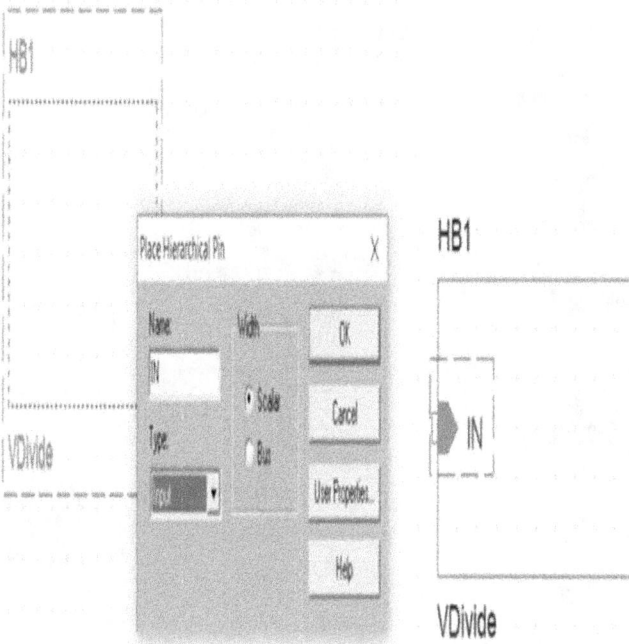

FIGURE 8.54 Hierarchical pin placement.

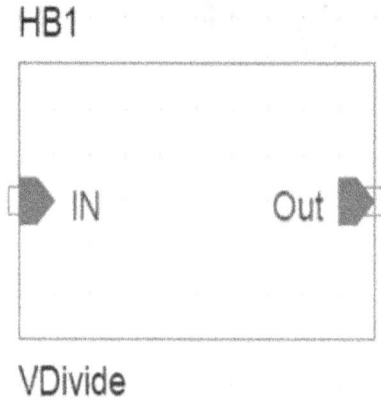

FIGURE 8.55 VDivide hierarchical block.

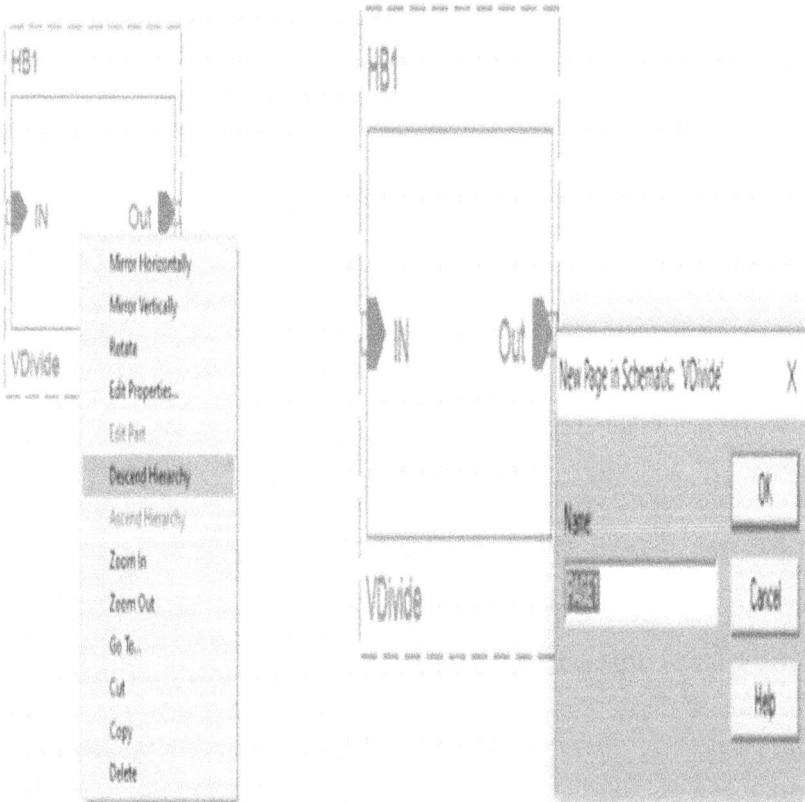

FIGURE 8.56 Hierarchical page name.

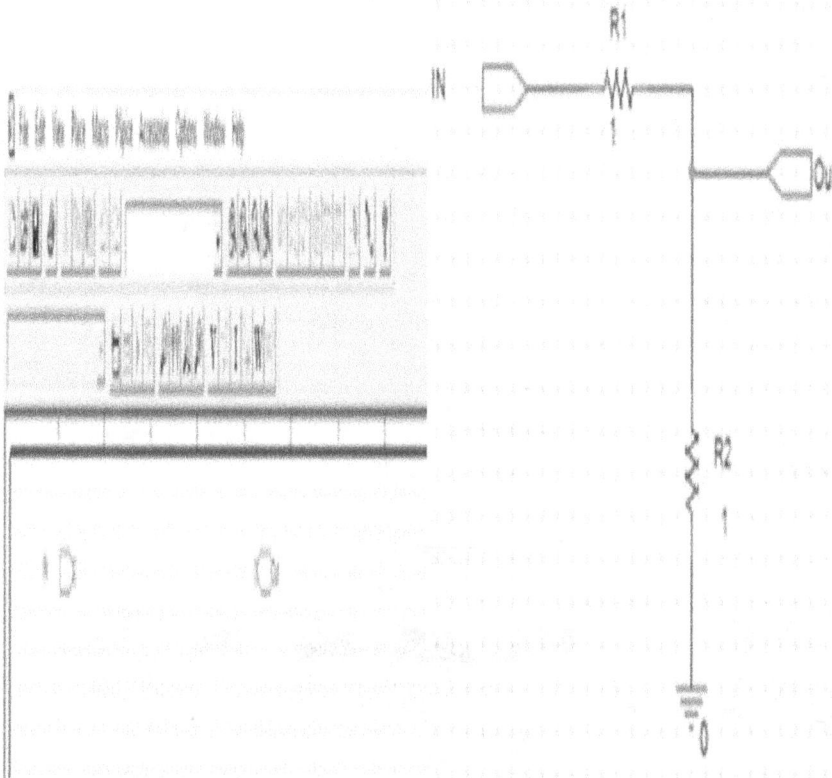

FIGURE 8.57 Hierarchical block placement.

Step 7: Now right click on the page and press Ascend hierarchy option, as shown in Figure 8.58. Now the screen will go to the hierarchy block which has been created early.

Step 8: In that Block, connect the voltage source, ground, and resistance as shown for simulation in Figure 8.59.

Step 9: Now save the created hierarchical block. In simulation setup select Bias point as analysis type as in the Figure 8.60 and press OK. Now run the simulation.

Step 10: Now in the output window, press file option on the left side and we will see the node voltage values in output file as in Figure 8.61.

This is how we can create hierarchical blocks for any of our circuits and save and run the blocks. This helps to show the simplified view of our circuit, which is required.

8.1.14.6   Creating Hierarchical Symbols

Creating hierarchical symbols is another way of creating a library file for easy and convenient usage. Here we going to see the steps for creating library files for that same voltage divider circuit as discussed.

Step 1: Create a Hierarchical block for voltage divider circuit name as VDivide.

Step 2: Go to file and New option. Click library as shown Figure 8.62. Now the library file will be created for the new project.

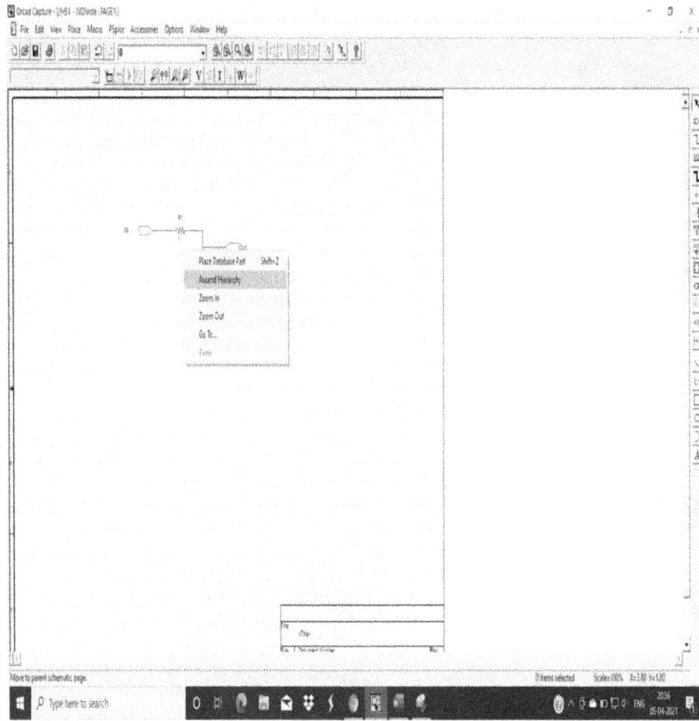

FIGURE 8.58   Ascend hierarchy.

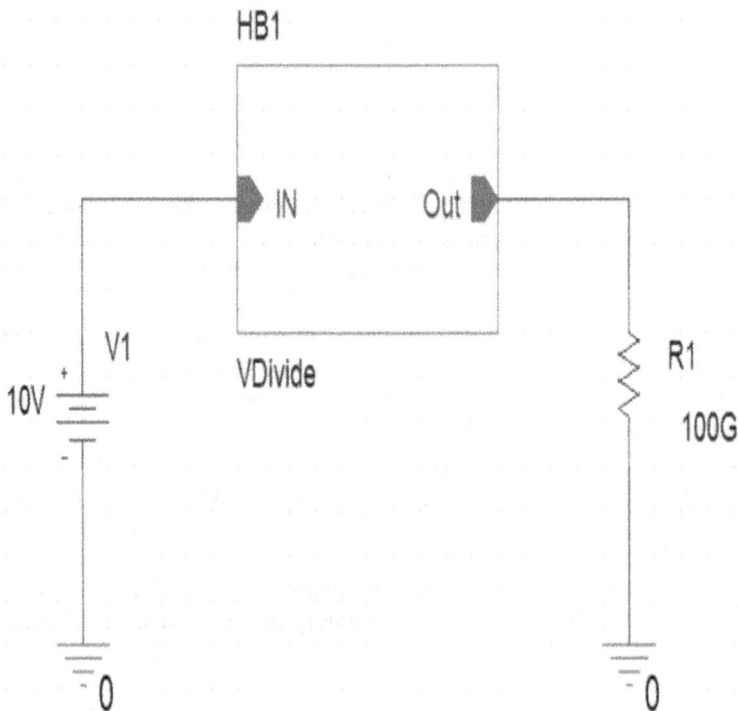

FIGURE 8.59   Hierarchical block.

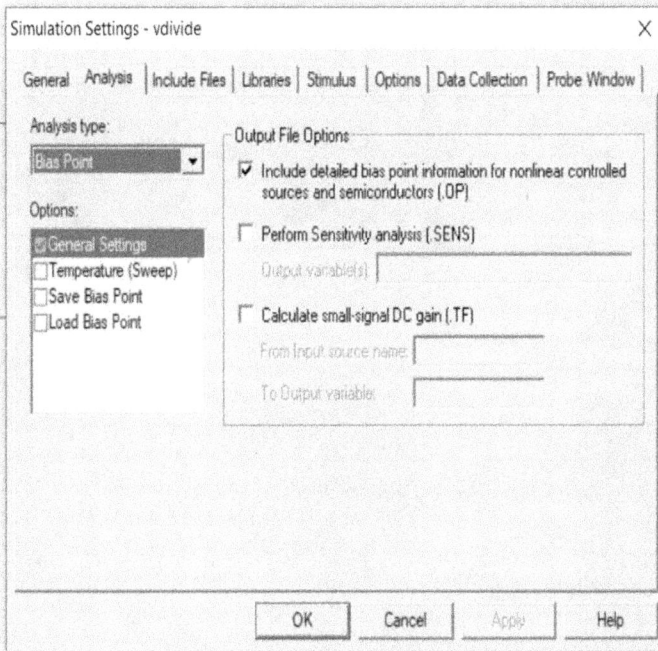

FIGURE 8.60   Simulation setting.

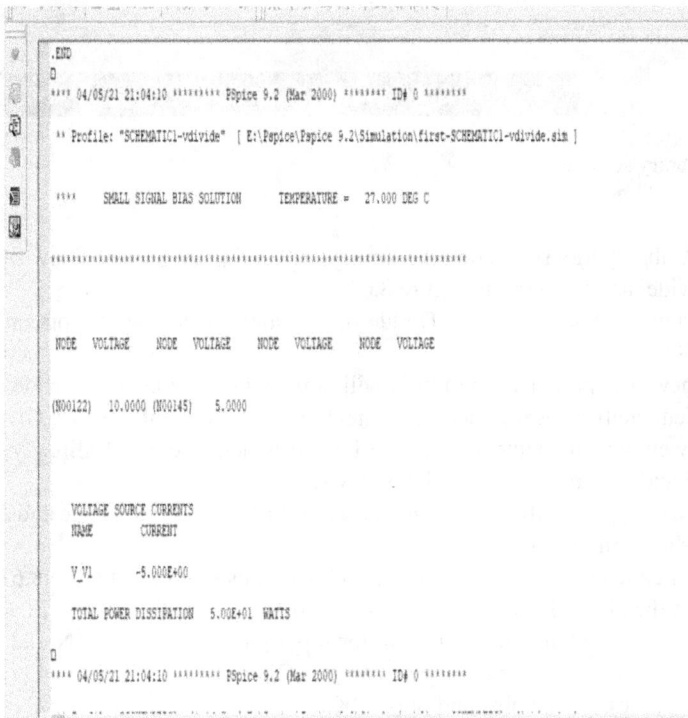

FIGURE 8.61   Output window.

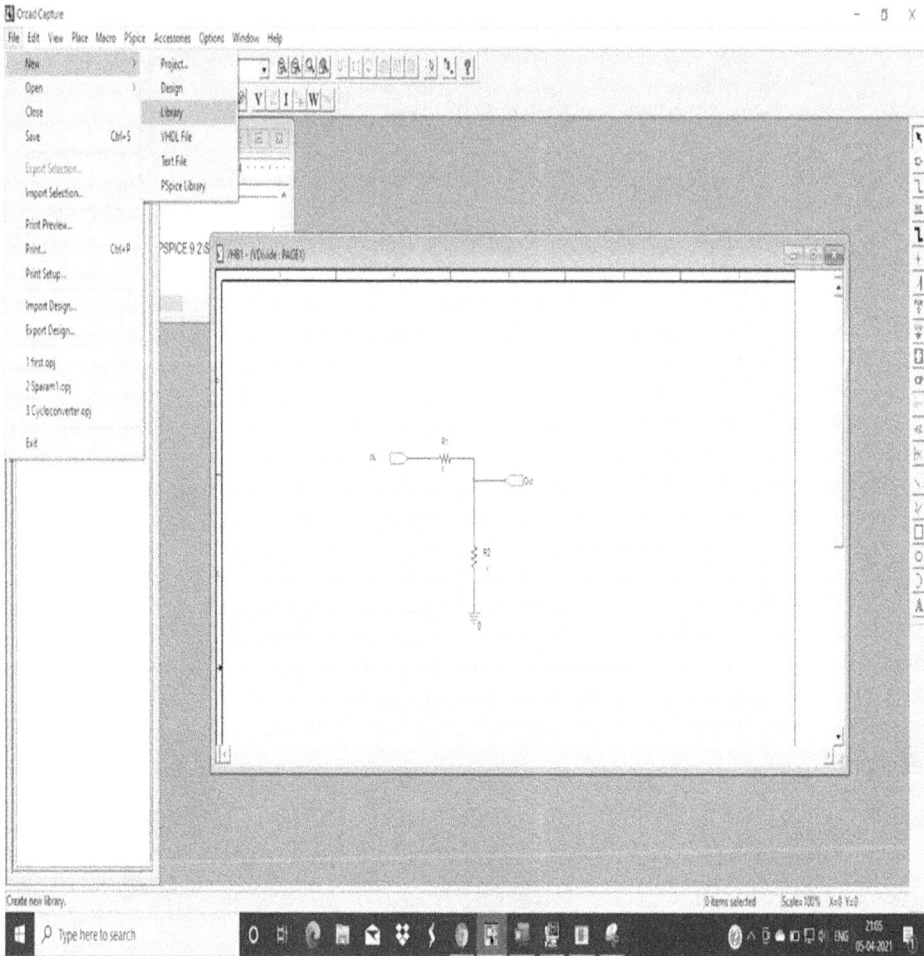

FIGURE 8.62   Library selection.

Step 3: In that library file, right click the library1.olb and press save as. Now save the libray file as VDivide, as shown in the Figure 8.63.

Step 4: Next right click the created VDivide library and select new part option as shown in the Figure 8.64.

Step 5: Now new part properties windows will appear. Give name as VDivide and in select attach implementation, now attach implementation window will appear. Give the type as schematic view and implementation as VDivide which is created allready as shown in Figure 8.65 and the press OK in both windows.

Step 6: A box will appears, after that go to place option and select rectangle and draw it inside the box as shown in Figure 8.66.

Step 7: Go to the menu "place" and select "pin" option as shown in Figure 8.67 or it can be also selected the pin in the left side toolbar also, then place pin window will appear, allows to create to pins as input and output pin, for input pin give name as IN and number as 1 and type as input and for output pin give name as OUT and number as 2 and type as output shape as short for both the pins and press OK.

Step 8: After creating each pins place the pin in the box as shown the Figure 8.68 and right click the Value and edit as VDivide (circuit name) for creating final library block.

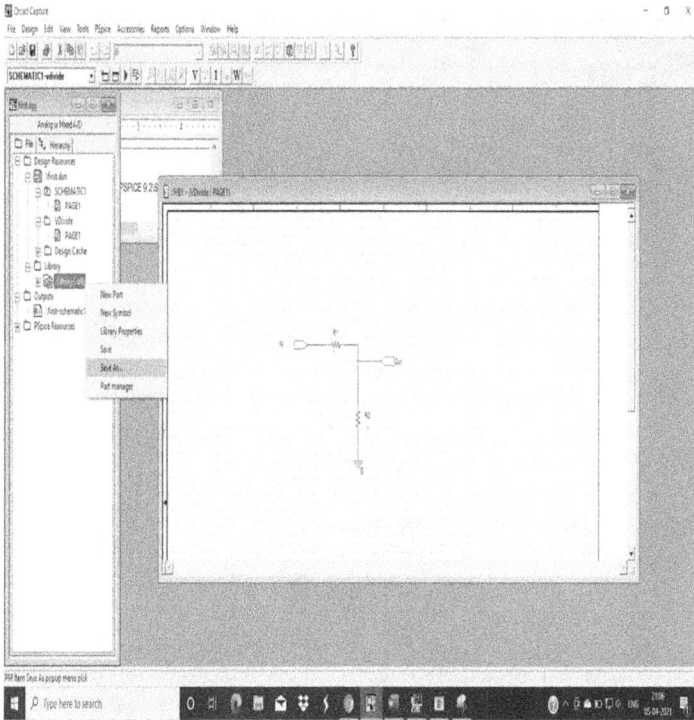

FIGURE 8.63 Library save.

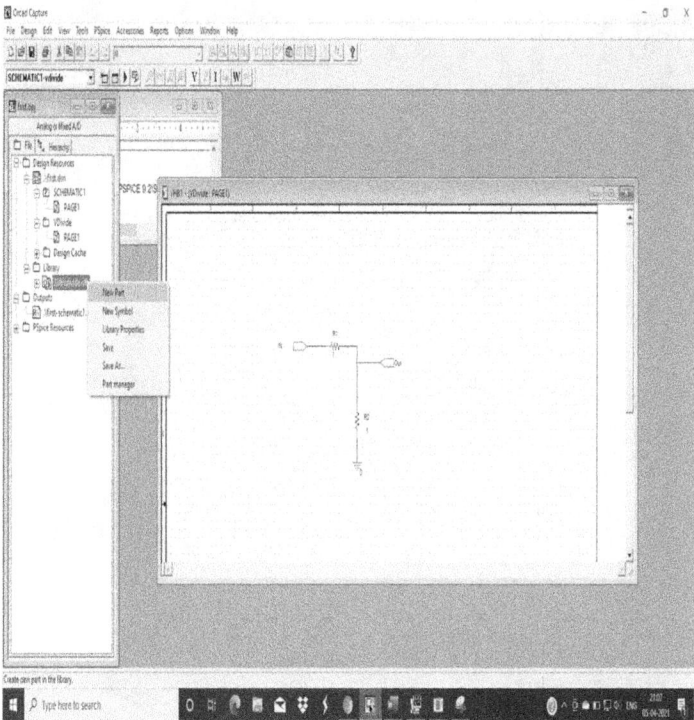

FIGURE 8.64 New port.

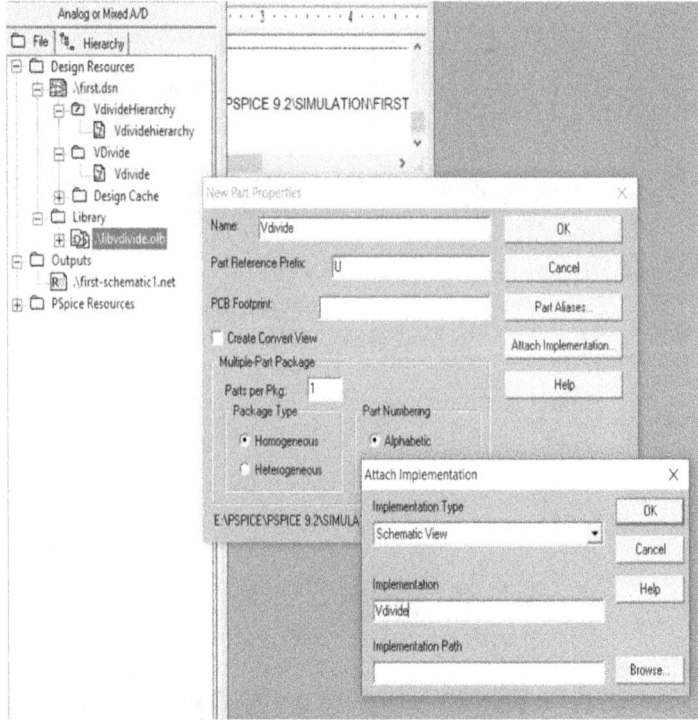

FIGURE 8.65　Attach implementation for new part properties.

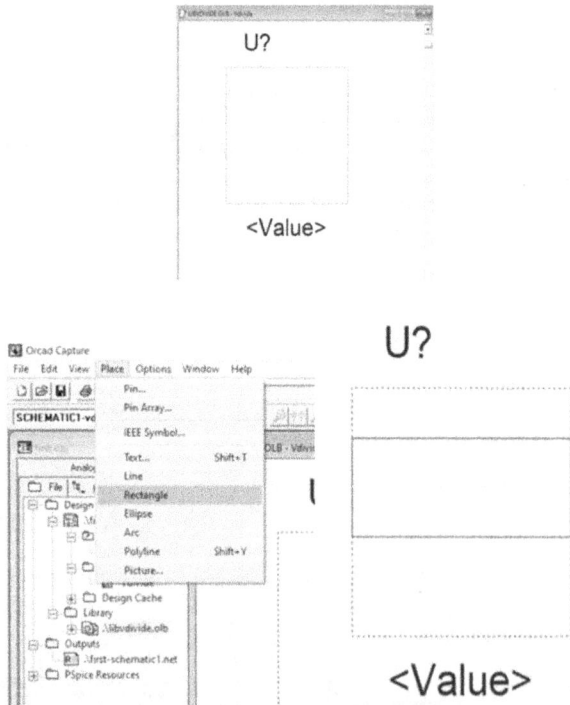

FIGURE 8.66　Part creation.

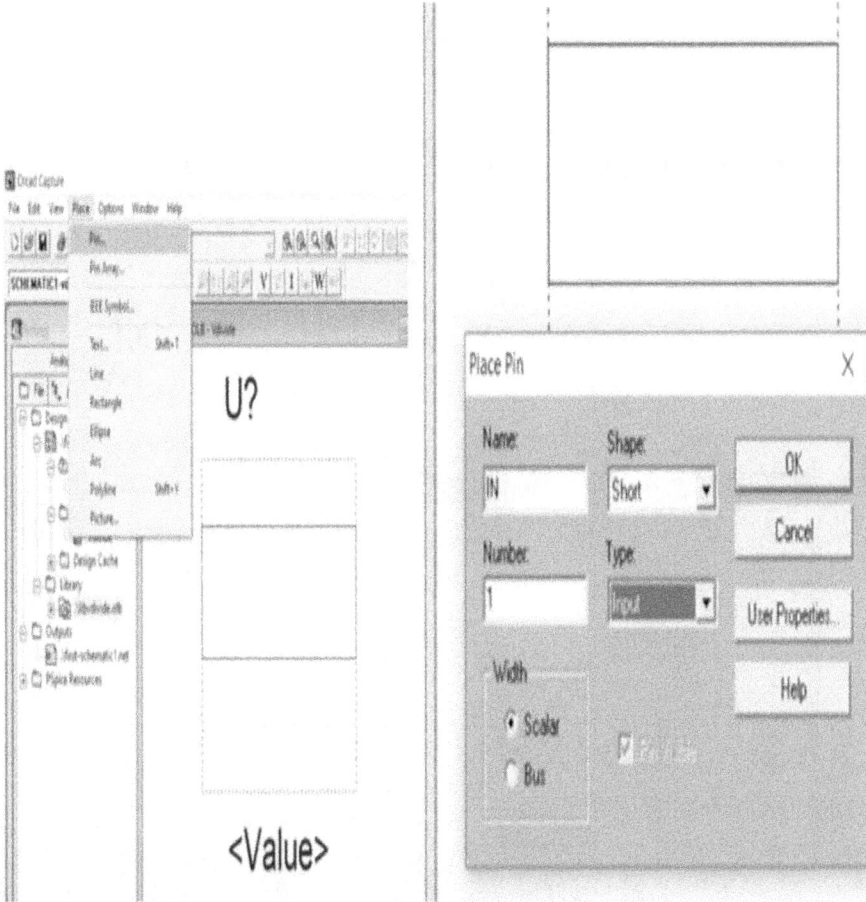

FIGURE 8.67 Place pin.

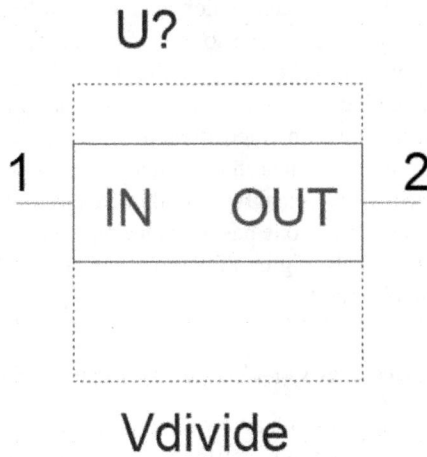

FIGURE 8.68 Block with pin.

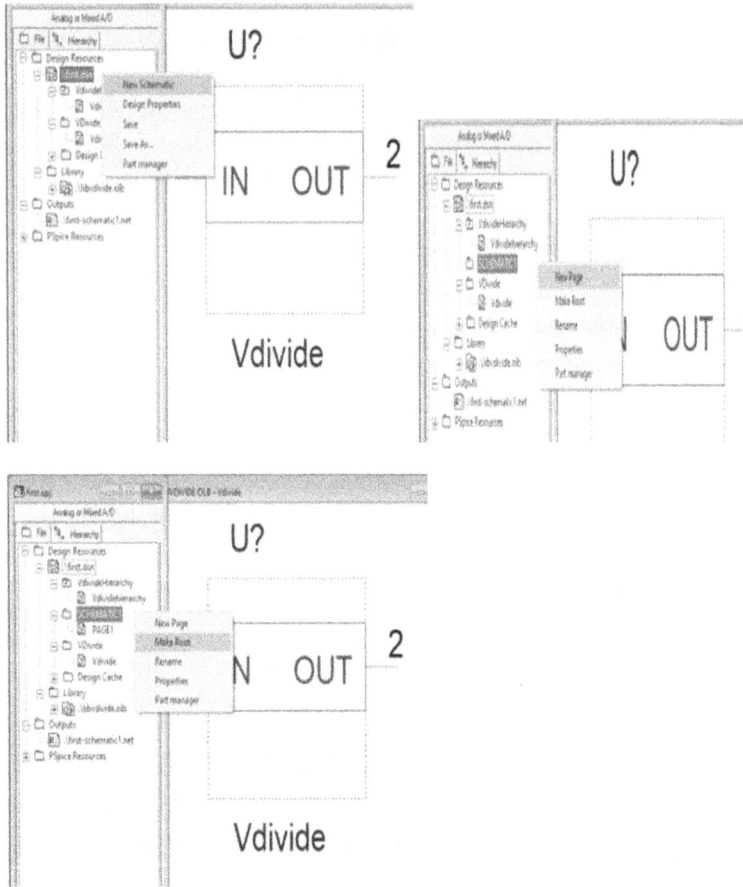

FIGURE 8.69   Library creation.

Step 9: Now go to.dsn file which you have been created right click it and go to new schematic, now right click on the new schematic page which been created then select new page, now save the entire project which have been done so for. After that right click the created schematic and select make root option as shown in the Figure 8.69.

Step 10: Now in new page open place part window now press add library option in that add the library file which you have been created in step 3, then press okay now the created VDivide circuit will appear as component in the schematic page as in the Figure 8.70.

Thus, how one can create a hierarchical block and make use of it in all projects by creating library files. And, for each circuit, one has to open a new project for creating hierarchical blocks and hierarchical symbols. Figure 8.71 shows the entire file creation for VDivide hierarchical block and symbol circuit.

8.1.15   Power Electronic Converters with EMI Filters using PSpice

It is clear that with the help of PSpice, creating a new schematic window for the implementation of the converter, interconnecting the components, getting the models of the components, tuning the simulation parameters for simulating the converter, building and running the implemented converter to get the resulting waveforms, and tuning the obtained waveforms in the data display window can be done.

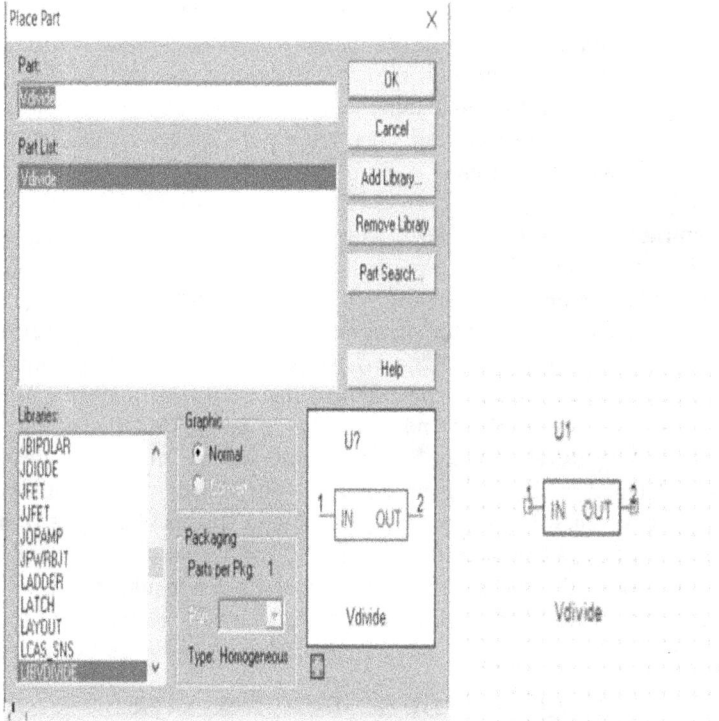

FIGURE 8.70 Place part.

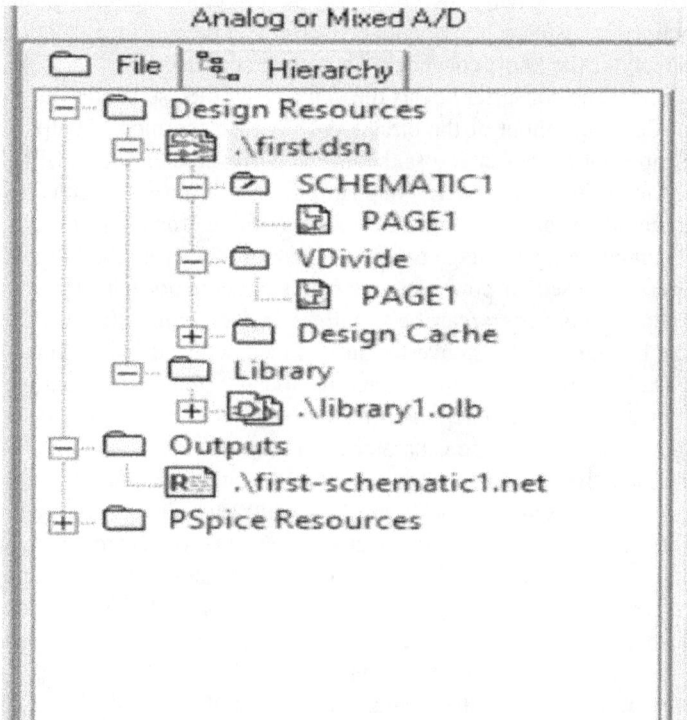

FIGURE 8.71 Design hierarchical block.

PSpice is used to explain the concept used for mitigating the electromagnetic interference (EMI) problems in the power electronic circuits. It has high power simulation tools that can be used for analysing the circuit. A detailed explanation on the PSpice is discussed with the step by step methods of implementing the circuits. Implementation of the S-Parameter and the conduction emission are also analysed. This chapter also discusses about the steps for measuring conducted emission using PSpice for the following converters.

i. **Bipolar Symmetric Output DC–DC Converter in Ultrasound Medical Imaging Systems**: This section describes a bipolar switching DC–DC converter with symmetric output for digital pulse generators in ultrasonic medical imaging systems. The suggested power stage architecture can operate in a time-interleaved manner while generating symmetric bipolar supply voltages. This converter separates the unipolar supply into positive and negative polar output, which is more efficient than a unipolar converter. The conducted emission of this circuit is simulated and eliminated by using EMI filters.

ii. **Diode Clamped Multilevel Inverter for Grid Application**: This section presents a three-phase, diode-clamped, multilevel inverter for grid-connected systems with a 2 kV output power and a 600 V renewable energy input. The simulation and analysis of a multilevel inverter using the ORCAD PSpice programme. For all converters to be highly efficient, conducted emission is the primary factor to consider. As a result, the filter reduces harmonics or conduction emission levels and the power converter's efficiency improves.

8.2   BIPOLAR SYMMETRIC OUTPUT DC–DC CONVERTER

8.2.1   Bipolar DC–DC Converter

The operation of bipolar DC–DC boost converter is similar to the simple boost converter. Bipolar output is obtained by connecting two boost converters by combining common ground point. Positive polar output is obtained by using a simple boot converter and the negative polar output is obtained by reversing the supply of the boost converter. These two converters are combined by connecting common point together. The converter boosts the voltage and then separates the output as negative and positive polar. So, if the input of the circuit is 10 V, then the bipolar output of the converter depends upon the capacitor converter across them. The positive polar output will always be double the capacitance voltage value and the negative polar output will always be equal to the capacitance value. The bipolar output power converter achieves a maximum efficiency of 83.4% at a load power of 187.5 mW, while maintaining a voltage balancing error of 2% across the full load range.

Bipolar converters are used to power unipolar pulse generators with the same peak-to-peak pulse amplitude. Bipolar converters can be unidirectional or bidirectional. In medical applications, unidirectional bipolar DC–DC converters are utilized, whereas bidirectional bipolar DC–DC converters are utilized in microgrid applications. A bipolar symmetric output switching DC–DC converter is used for digital pulse generators in ultrasound medical imaging systems. To promote non-invasive imaging, the digital pulse generator is an essential component of ultrasound systems. A commercial ultrasonic medical imaging system employs either unipolar or bipolar pulse generators. Unipolar pulse generators are capable of producing ultrashort, extremely high voltage unipolar pulse impulses. They are utilized in a range of applications due to their relative usability and large bandwidth. For high-frequency applications, such as Doppler and B-mode imaging, bipolar pulse generators are preferred due to their high sensitivity and improved signal-to-noise ratio. Bipolar pulse generators also have better energy efficiencies, as the digital signals used to electrically excite the piezoelectric transducers have a lower unwanted DC component.

Moreover, as the peak-to-peak pulse voltage can be double the voltage rating of the coaxial cables and transducers, the system's size and cost can be drastically reduced, particularly for arrays of many transducer elements.

8.2.2 Circuit Topology

The circuit topology for Boost, Unidirectional Bipolar DC–DC converter, and Bi-Directional DC–DC converter is shown Figures 8.72 to 8.74.

8.2.3 Applications of Bipolar Output DC–DC Converters

1. Medical imaging systems
2. Microgrid applications
3. Ship power systems
4. Photovoltaic applications
5. Silicon micro-displays
6. Bipolar DC bus

From Figure 8.75, the unipolar DC supply is directed to the bipolar DC–DC converter so that bipolar output can be produced. Bipolar DC–DC converter converts unipolar DC source to positive and negative output polarity and its behaviour is identical to that of a simple DC–DC boost converter. This bipolar output is provided as input to the digital pulse generator. The digital pulse generator is an integral component of ultrasonic devices for non-invasive imaging. In order to electrically stimulate piezoelectric transducers, digital pulse generators utilize a variety of pulsing techniques. Consequently, digital pulse generators require bipolar pulses as input. The digital pulse generator supplies pulses to transducers that generate acoustic waves with varying phase delays and amplitudes. The ultrasound medical imaging systems receive the output from the transducers. Furthermore, these supply voltages must be symmetric in order to produce homogeneous properties such as the amplitude, phase, and timing of the digital pulses. This is essential for the ultrasound

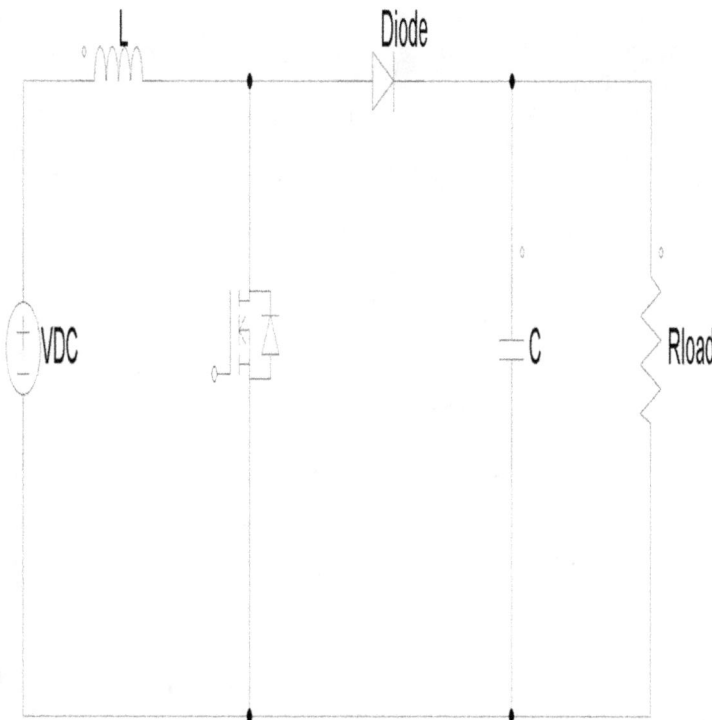

FIGURE 8.72 Boost converter.

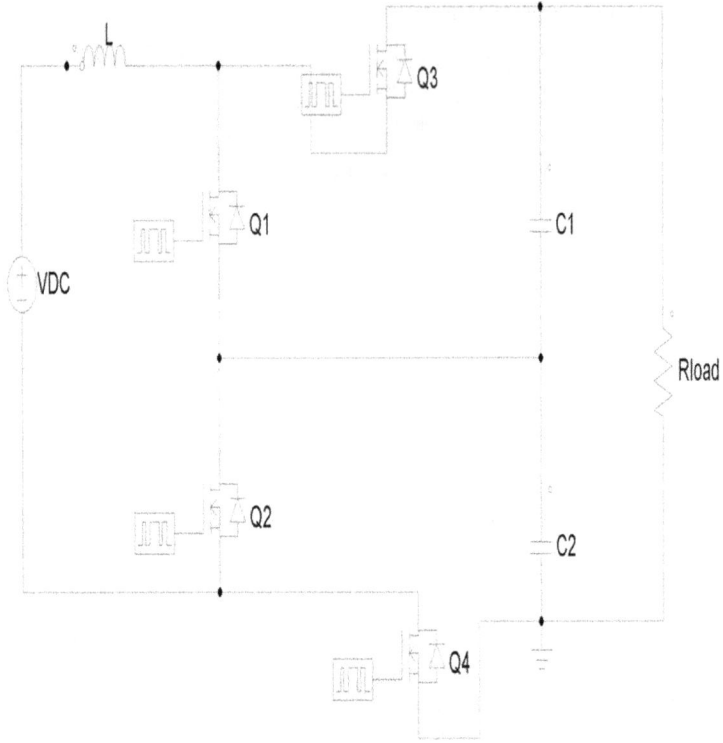

FIGURE 8.73 Uni-directional bipolar DC–DC converter.

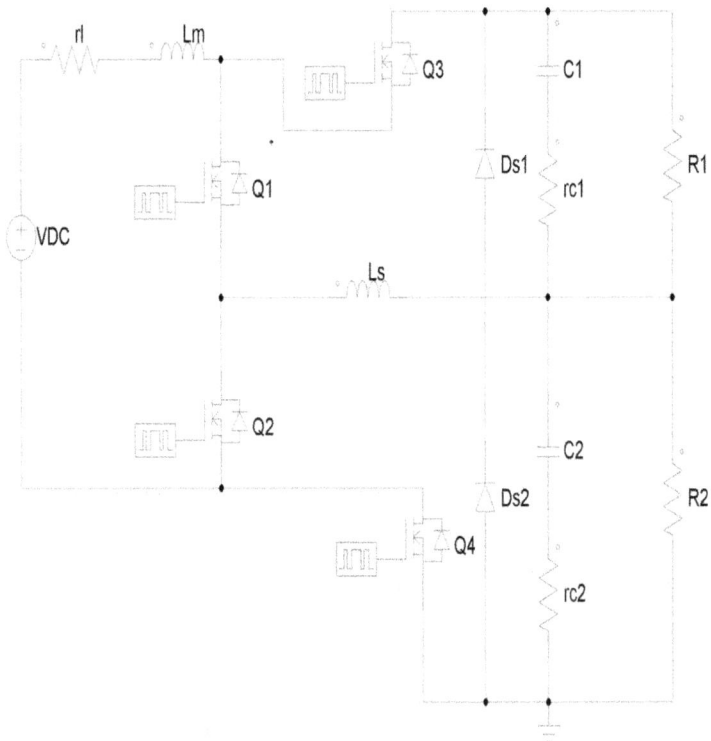

FIGURE 8.74 Bi-directional bipolar DC–DC converter.

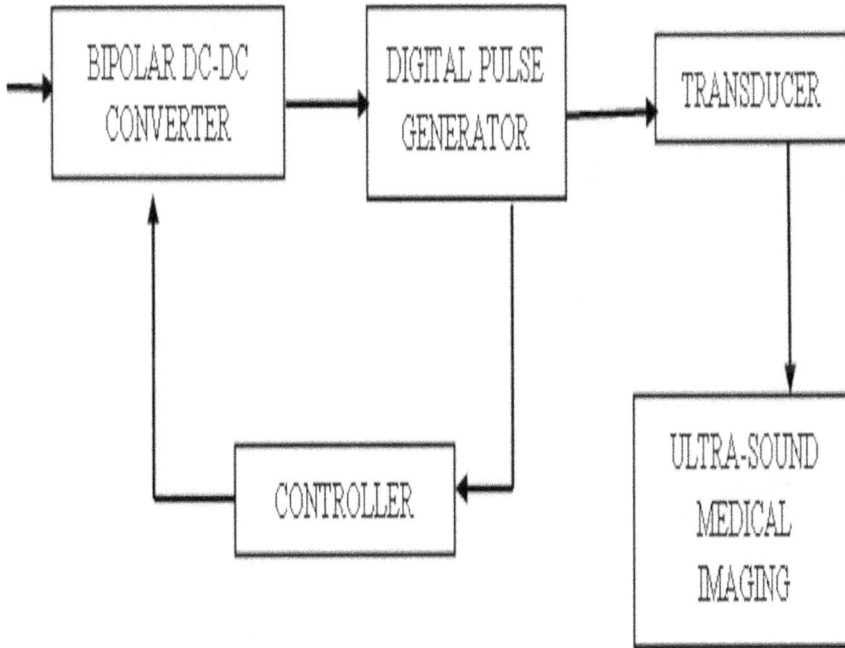

FIGURE 8.75 Bipolar symmetric output DC–DC converter-block diagram.

transmitter to attain a good sensitivity and signal-to-noise ratio. In addition, for portable ultrasound applications where the imaging system's small size is of utmost importance, a compact implementation of the bipolar output power supply is very desirable.

Consider the simplest method for providing both positive and negative supply voltages while developing an alternative architecture for a bipolar power stage. A conventional boost converter is used to provide a positive voltage V_{op}, whereas a second boost converter with its input power supply V_{in} inverted can be used to provide the necessary negative voltage V_{on}. Nevertheless, as explained above, this necessitates the use of two cumbersome power inductors, several pins, and I/O pads, which is undesirable for portable ultrasound imaging applications. In order to create a single structure capable of delivering bipolar supply voltages, it is preferable to mirror the negative boost converter with regard to its positive counterpart, while keeping the polarity of V_{in} and L.

The circuit shown in the Figure 8.76 is the positive polar boost DC–DC converter. Here, two switches are used to produce the positive polar output. The output voltage is equal to the 2×Vc1. The outputs are taken from ×1, Vc1, and V_{op}. The negative polar output can be obtained by reversing the supply of the positive polar boost DC–DC converter shown in Figure 8.77. In the negative polar DC–DC converter, the output is taken from the V×2, Vc2, and V_{on}.

Here, the positive polar DC–DC converter is kept as such and it is connected together with the reversed negative polar DC–DC converter. The converters are connected together by combining the common ground points. Here the output voltage will be twice the capacitor voltage. The bipolar DC–DC converter consists of four switches, as shown in Figure 8.78. The switches Sw1 and Sw2 provide positive polar output and the switches Sw3 and Sw4 provide negative polar output.

Moreover, if the two power stages are operated identically throughout the charge phase $\emptyset_{-ch}$ and discharge phase $\emptyset_{-dch}$, the series-operating inductors and input V_{in} can be combined into a single element. This results in a bipolar output power stage topology. An interleaving operation scheme is used to control the switching actions in the power stage. From Table 8.9, for the phase $\emptyset_{1ch} = 1$, during the charge period, S1 and S3 are turned on and S2 and S4 are turned off, causing the inductor to be energized at a rate of V_{in}/L. During the discharge period $\emptyset_{1dch}$, S2 and S3 are switched

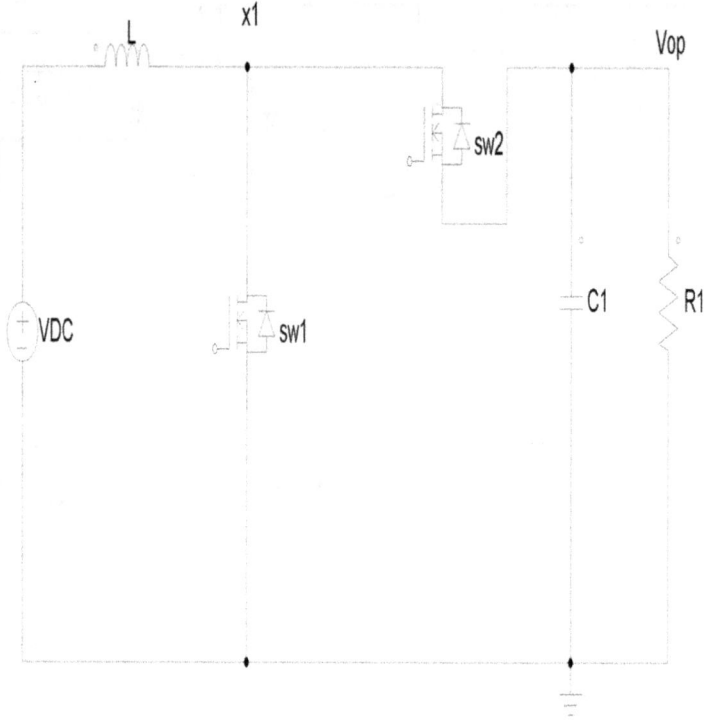

FIGURE 8.76　Positive polar boost DC–DC converter.

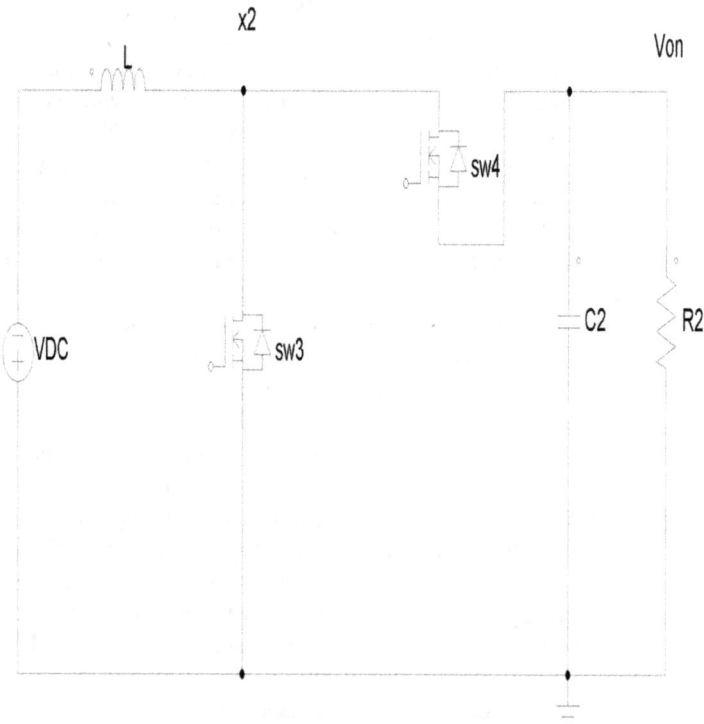

FIGURE 8.77　Negative polar boost DC–DC converter.

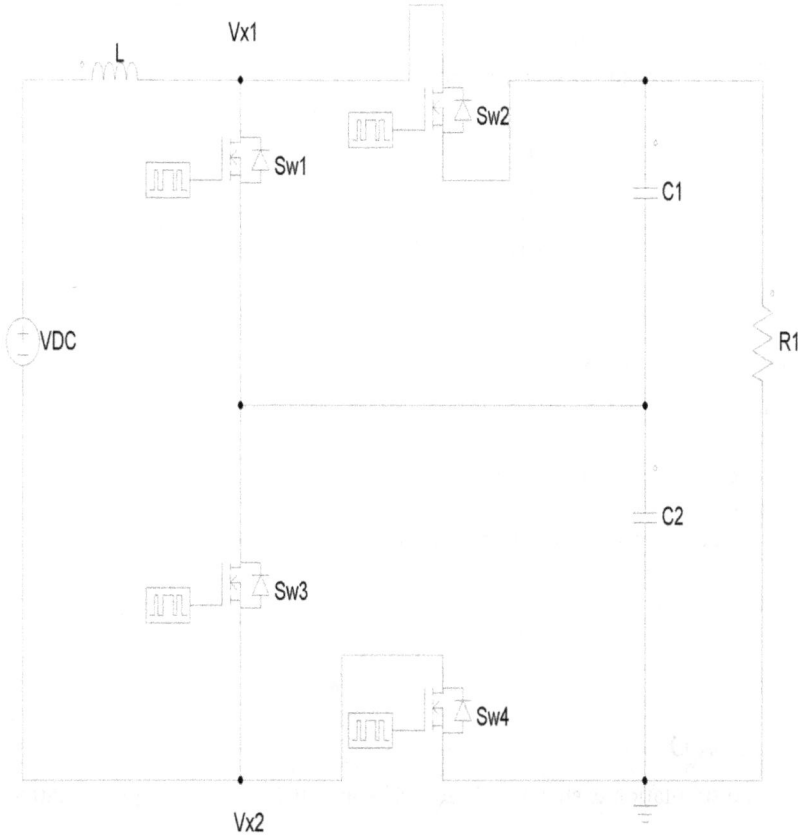

FIGURE 8.78 Bipolar DC–DC converter.

TABLE 8.9
Switching Action Table

Phase	S1	S2	S3	S4
$\varnothing_{1ch}$	ON	OFF	ON	OFF
$\varnothing_{1dch}$	ON	OFF	OFF	ON
$\varnothing_{2ch}$	ON	OFF	ON	OFF
$\varnothing_{2dch}$	OFF	ON	OFF	ON

off, while S1 and S4 are turned on. This causes the node V×2 to be shorted to the negative output, allowing the inductor L to charge C2 at a rate of $(V_{in} - |V_{on}|)/L$. For phase $\varnothing_{2ch} = 1$, similar switching processes ensure that the positive output V_{op} and capacitor C1 are energized. In such systems, the power stage of the bipolar output power converter can be reconfigured as a single output power converter capable of generating a voltage level of $V_{out} = 2V_{op}$. The interleaving operating scheme of the power stage is also maintained, such that C1 and C2 are electrified in two complementary phases two and one, respectively. As a result, Vc1 = Vc2 = V_{on}, and $V_{op} = 2$Vc1 = 2Vc2, where Vc1 and Vc2 are the voltages across C1 and C2, respectively. This results in a power converter architecture that can effectively power both unipolar and bipolar pulse generators. Table 8.10 shows the required components.

TABLE 8.10
Components Required

S. No.	Component	Model No.	Quantity
1.	CAPACITOR	–	2
2.	INDUCTOR	–	1
3.	MOSFET	IRF320	4
4.	DIODE	D1N4001	1

8.2.4   DESIGN SPECIFICATIONS

1. Inductance = $V_s*D/(f_s*$change in $I_0)$
   $L = 120$ uH
2. Capacitance = $I_0*D/(f_s*$change in $V_0)$
   $C = 96$ uF
3. Switching Frequency = 1/(total time period)
   =1/(8 us) = 125 kHz
4. Duty Cycle = ton/T
   0.75 for positive waveform
   0.5 for negative waveform

8.2.5   SIMULATION OUTPUT

The circuit for the simulation is shown in Figure 8.79 and its result from Figures 8.80 to 8.83.

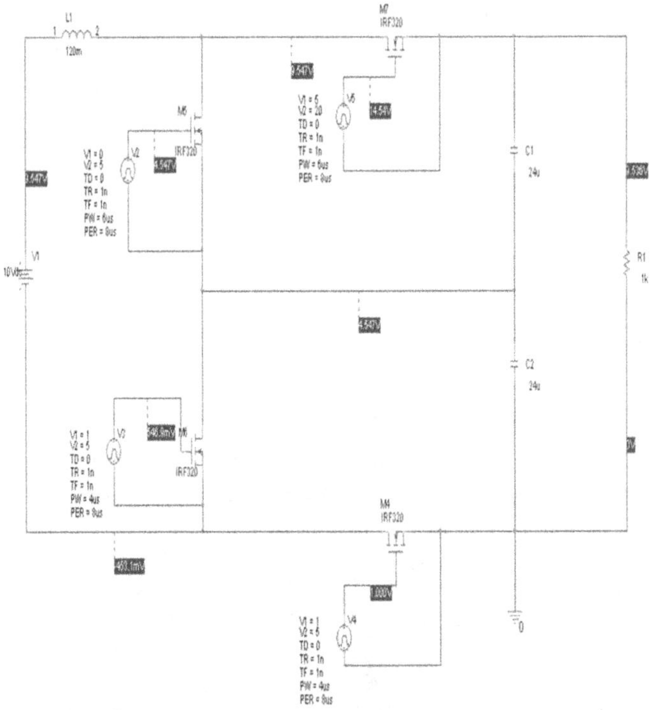

FIGURE 8.79   Simulation circuit.

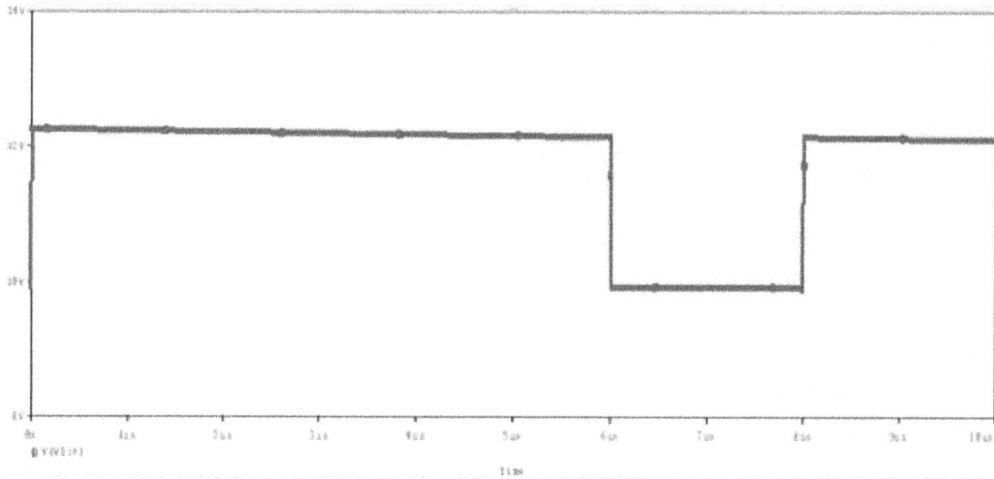

FIGURE 8.80 Supply voltage simulation output.

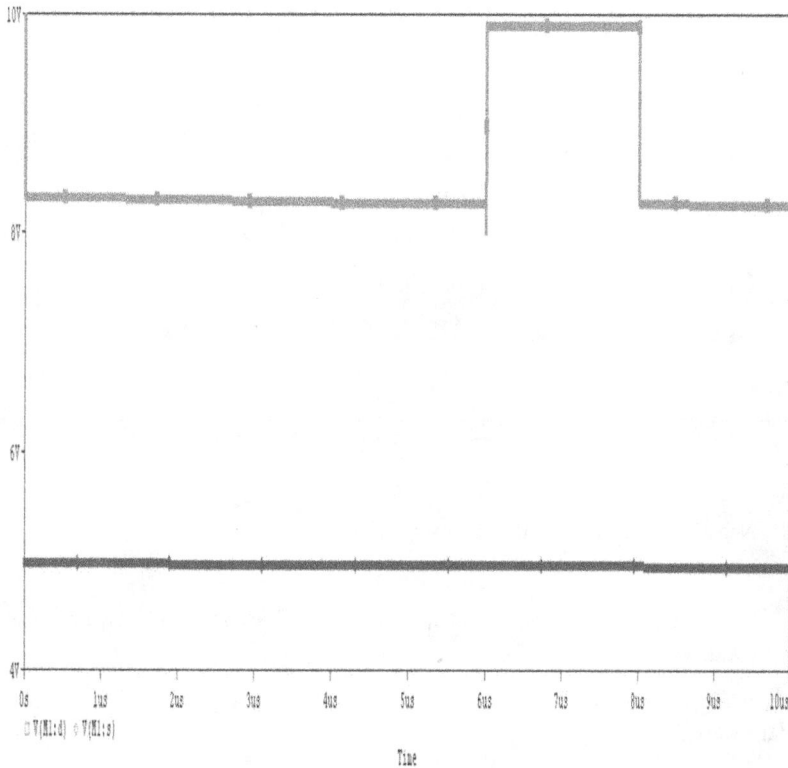

FIGURE 8.81 Positive waveform simulation output.

The supply has to rise and fall to $V_{in} - V_{on}$. Here the supply is 10, and it rises to 12.1 and falls to 12.1 − V_{on}, where V_{on} = 2.1 V. Therefore, it falls to 10 V.

In Figure 8.81, this is a positive waveform, which is obtained from the Terminal ×1. When it is triggered, it reaches to 2×Vc1 which is 10 V.

This is negative waveform obtained from the terminal V×2. When it is triggered, it reaches to zero. The voltage rating can be raised by using a body diode parallel to Sw4.

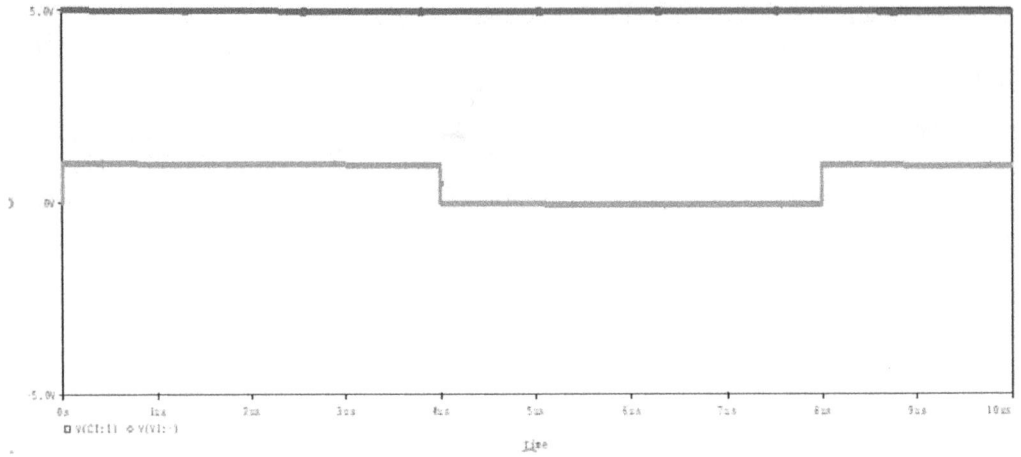

FIGURE 8.82   Negative waveform simulation output.

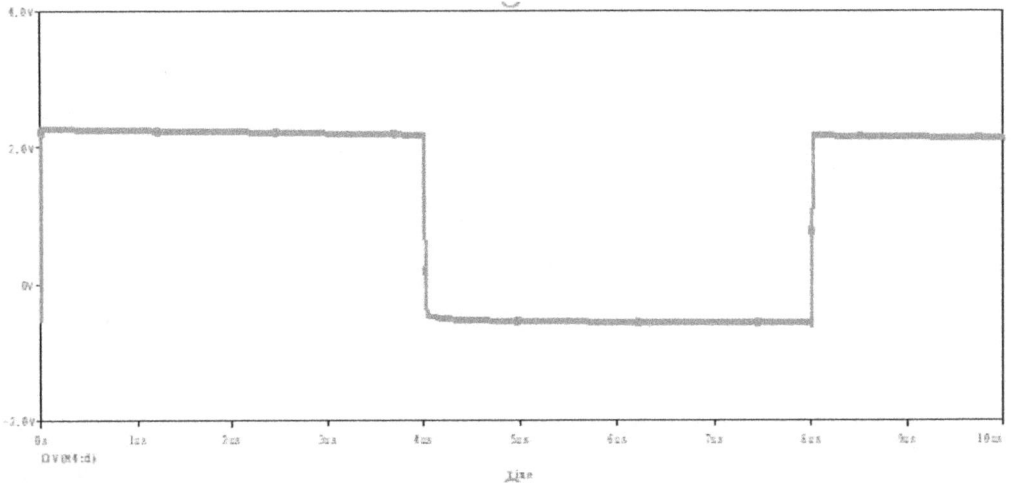

FIGURE 8.83   Negative waveform obtained by using body diode.

The above negative is obtained by connecting a diode parallel to the Sw4. By doing so, the output voltage is improved.

8.2.6   S-PARAMETER

The representation of circuit network in the harmonic generation, the parameters such as Z, Y, H, are used for low and medium harmonic generation. These parameters are not used for the high and very high frequency range because the open and short circuit conditions to determine the voltage and current are difficult to achieve. On low frequency range, the impedance and the admittance matrix are widely used, but for the high frequency range it cannot be used, therefore this forms the basis of the scattering mechanism. The Scattering parameters can be measured in a power line by two port networks connected to the vector network analyser (VNA). This analyser can measure the Scattering parameters over a high frequency range. The understanding of Scattering parameters is very important for the applications having high frequency range.

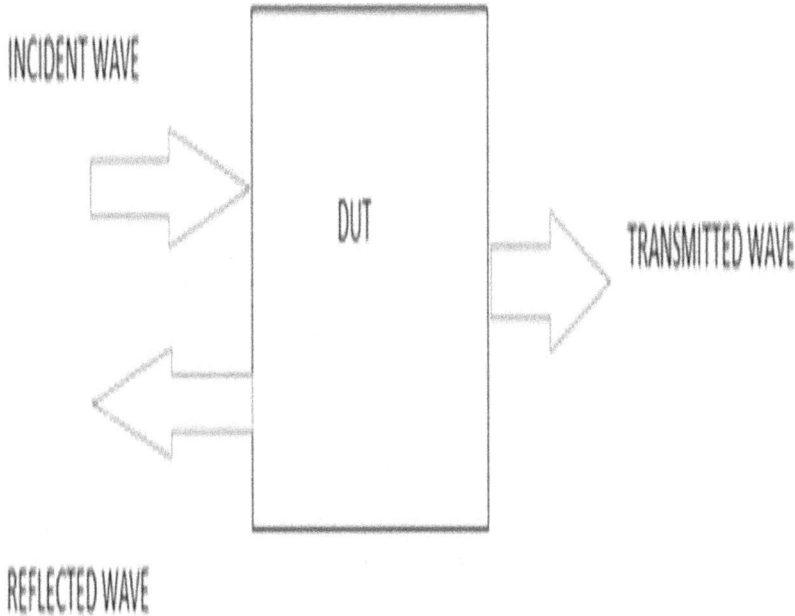

FIGURE 8.84   Scattering parameter.

a. **Scattering Parameter**

For higher frequencies, it is impossible to measure the total voltage, current, and power waves. It can be measured in terms of reflected, transmitted, and incident wave. Therefore, the parameter used is called Scattering parameters, shown in Figure 8.84.

The scattering parameter is defined as the ratio between the normalized reflected wave at ith port and the normalized incident wave at jth port.

$S_{ij} = b_i/a_j$

where, b_i = normalized reflected wave at ith port.

a_j = normalized incident wave at jth port

$b_i = V^+/\sqrt{z}0$

$a_i = V^+/\sqrt{z}0$ where $\sqrt{z}0$ is the characteristics impedance of line.

b. **S-Parameter in PSPICE**

How to find S-Parameter in PSPICE?

To find S-parameter in PSpice, two hierarchical blocks are required. They are:

1. XMITS
2. REFLEX

   XMITS

   The XMITS circuit shown in Figure 8.85 is used to measure the

   – Forward transmission coefficient, S21
   – Reverse transmission coefficients, S12

   Since the output load and input load are identical, the transmission coefficients equal the output voltage multiplied by 2. Gain of the E device, E1, is 2. The circuit (CKT) port is for connecting to the external circuit. The synchronous transmitter (STR) port is a concealed pin. OrCAD Capture will generate a unique net if it is left unconnected in a schematic. Alternately, a specific net can be named for the connection by modifying the PSpiceDefaultNet attribute value of the XMITS component instance shown in Figure 8.85; in this case, the STR pin will have a known label (S21) while evaluating simulation results in the Probe window.

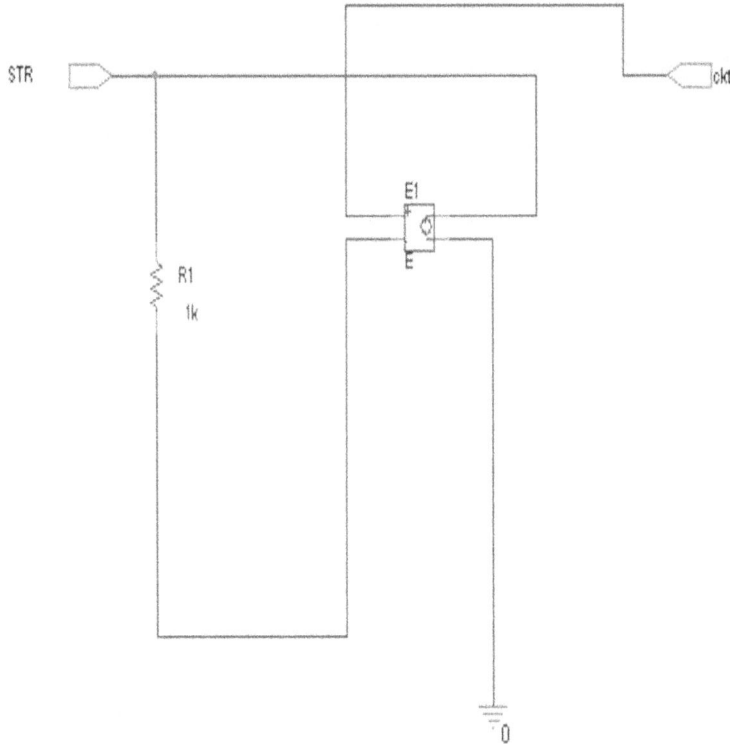

FIGURE 8.85 XMITS schematic circuit.

c. **REFLEX**

The REFLECTS circuit shown in Figure 8.86 is utilized to measure the
- Input reflection coefficient, S11
- Output reflection coefficient, S22

The coefficients of reflection equal the input voltage multiplied by 2 minus AC unity. Gain of the E device, E1, is 2. As with the circuit for measuring transmission coefficients, the CKT interface pin is utilized to connect to the external circuit. Synchronous receiver (SRE) is a hidden pin similar to STR, which was discussed previously. In addition, the REFLECTS sign has a DCbias property. On active circuits, the DC level of voltage source V1 can be adjusted by modifying the DCbias property of the REFLECTS instance in OrCAD Capture. This property is set to zero by default.

d. **Implementation of XMITS and REFLEX**

These sub-schematic circuits are given to a two-port hierarchical block. These circuits are connected outside the block by connecting XMITS to the input and the REFLEX circuit to the output of the block. By connecting XMITS to the input, we can get S12 and by connecting REFLEX to the output, we can get S22. By connecting XMITS to the output, we can get S21 and by connecting REFLEX to the input, we can get S11.

The sub-schematics are applicable to both passive and active circuits. These sub-schematics can also be used to generate and implement design symbols. This application note refers to a design example that employs XMITS and REFLECTS symbols derived from these sub-schematics as in Figure 8.87. Refer to the Building and Editing Components section of the OrCAD Capture User Guide for information on the converter whose S-parameter is placed inside the block and the XMITS port is given to the input side and the REFLEX is given to the output side. The S-parameter is taken from the sub-schematic circuits. XMITS gives S12 in the input side and S21 in the output side and REFLEX gives

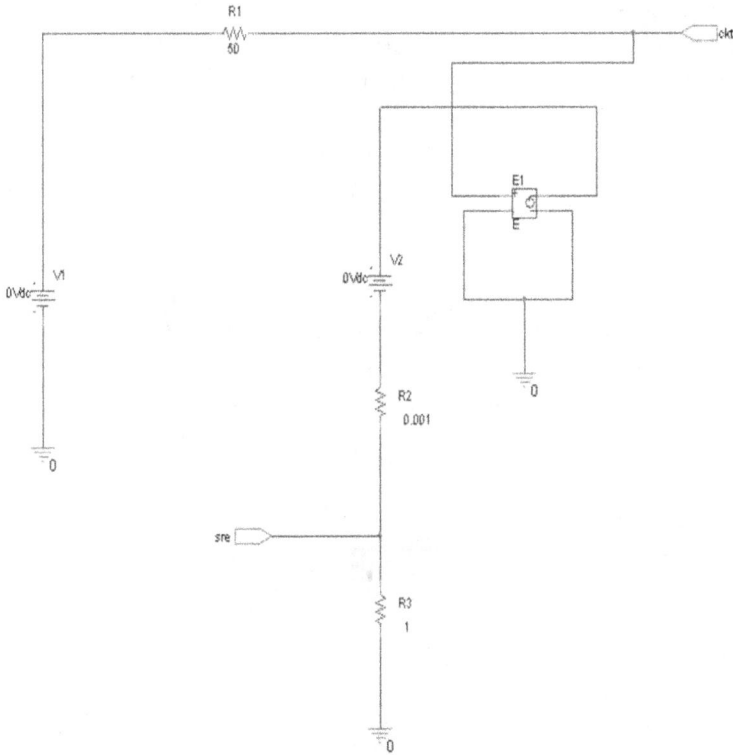

FIGURE 8.86 REFLEX schematic circuit.

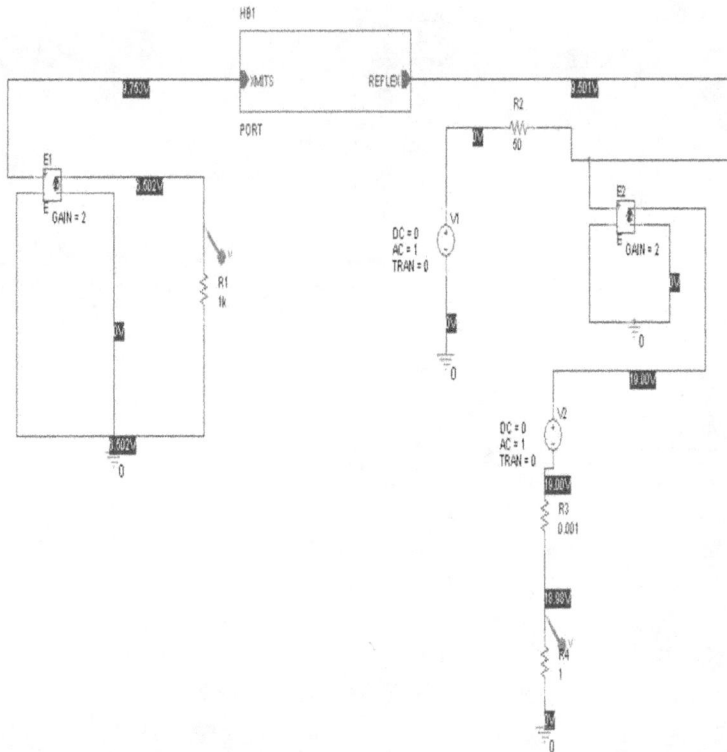

FIGURE 8.87 Port hierarchical block.

S11 in the input side and S22 in the output side. The output obtained by XMITS and REFLEX will be in voltage and these outputs have to be converted into decibels and then S-parameter is obtained. The circuit connections are given inside the port block shown in Figure 8.88.

Result of S-Parameter is shown in Figures 8.89 and 8.90.

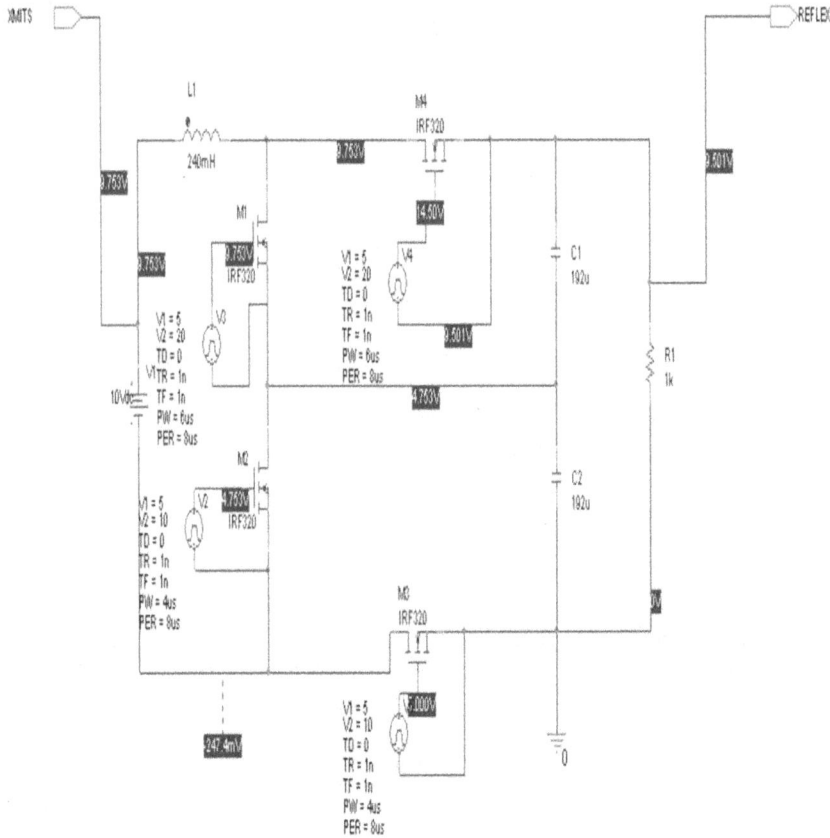

FIGURE 8.88 Circuit inside the block.

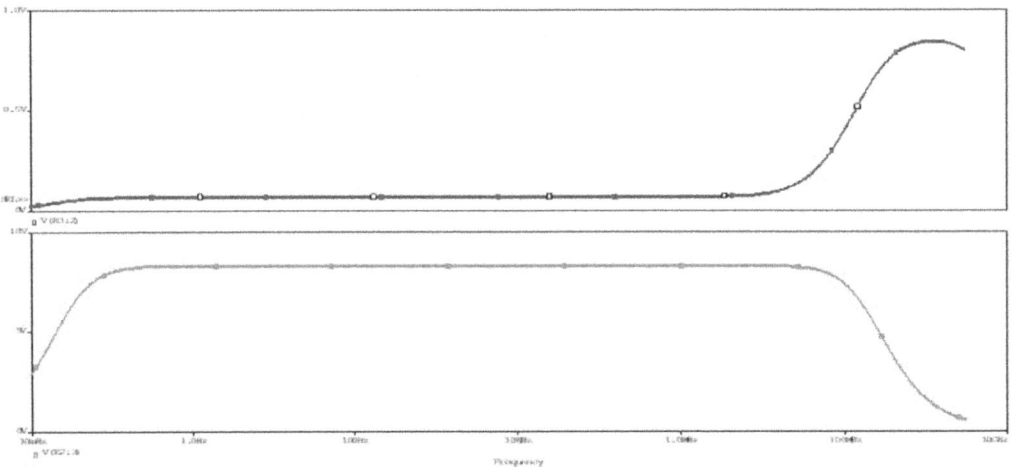

FIGURE 8.89 S12 and S22 waveforms.

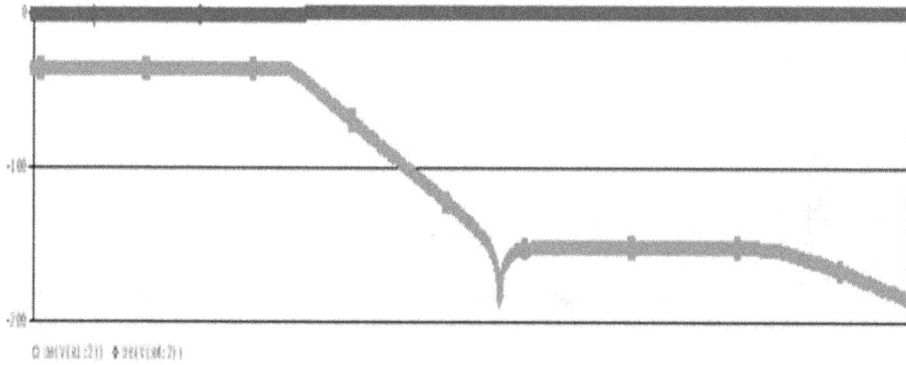

FIGURE 8.90 S21 and S11 waveforms.

8.2.7 CONDUCTED EMISSION

Line impedance stabilization network (LISN) is connected to the positive and negative leg of the circuit as shown in Figure 8.91 and the connected emission is taken by using a summing amplifier.

Conducted Emission

The measurement results are shown in Figures 8.92 and 8.93.

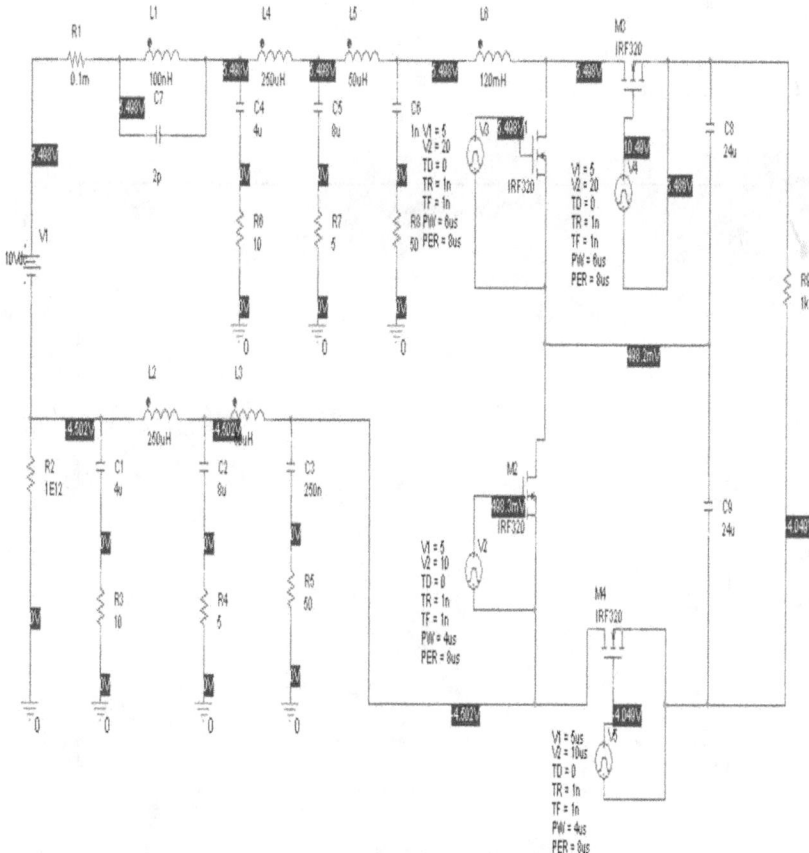

FIGURE 8.91 Circuit with LISN.

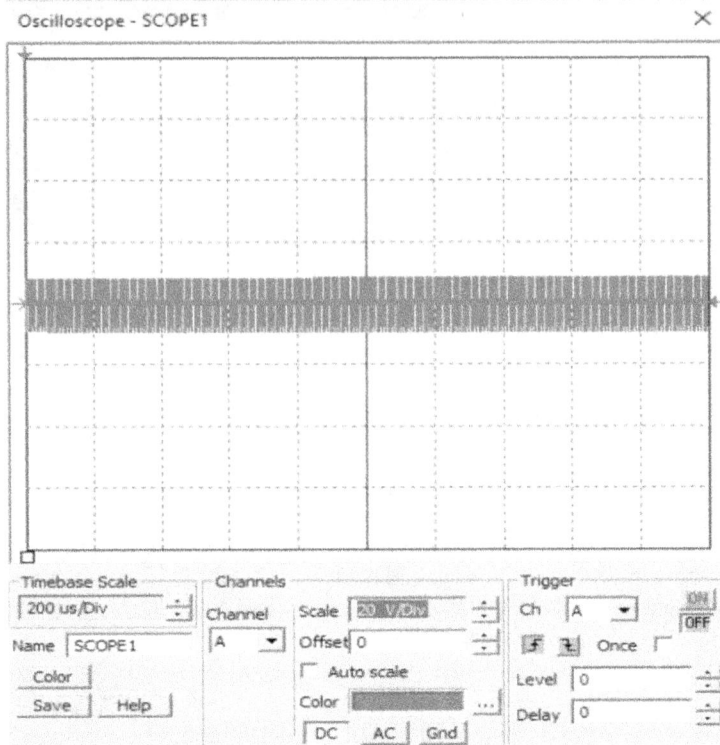

FIGURE 8.92   Conducted emissions.

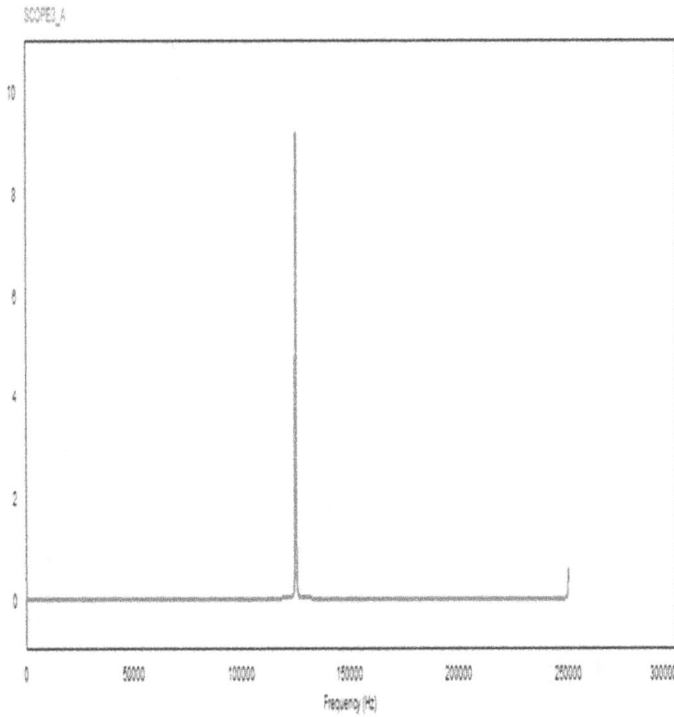

FIGURE 8.93   Emission without filter.

The conducted emission in this circuit can be reduced by using EMI filters. The EMI filters are of three types:

1. *L*-type
2. *π*-type
3. *T*-types

These filters are capable of reducing the conducted emission in the circuit. The EMI filters are connected to the summing amplifiers so that after summing the conducted emission from the positive and negative leg, the output of the summing amplifier is given to the EMI filter. The filter reduces the attenuation of the conducted emission. Here two *L*-type filters are connected back-to-back and the reduced emission is taken from the filtering, which is given resistance.

8.2.8 Circuit with EMI Filter

In Figure 8.94, EMI filter is inserted and ready for measuring the conducted emission.

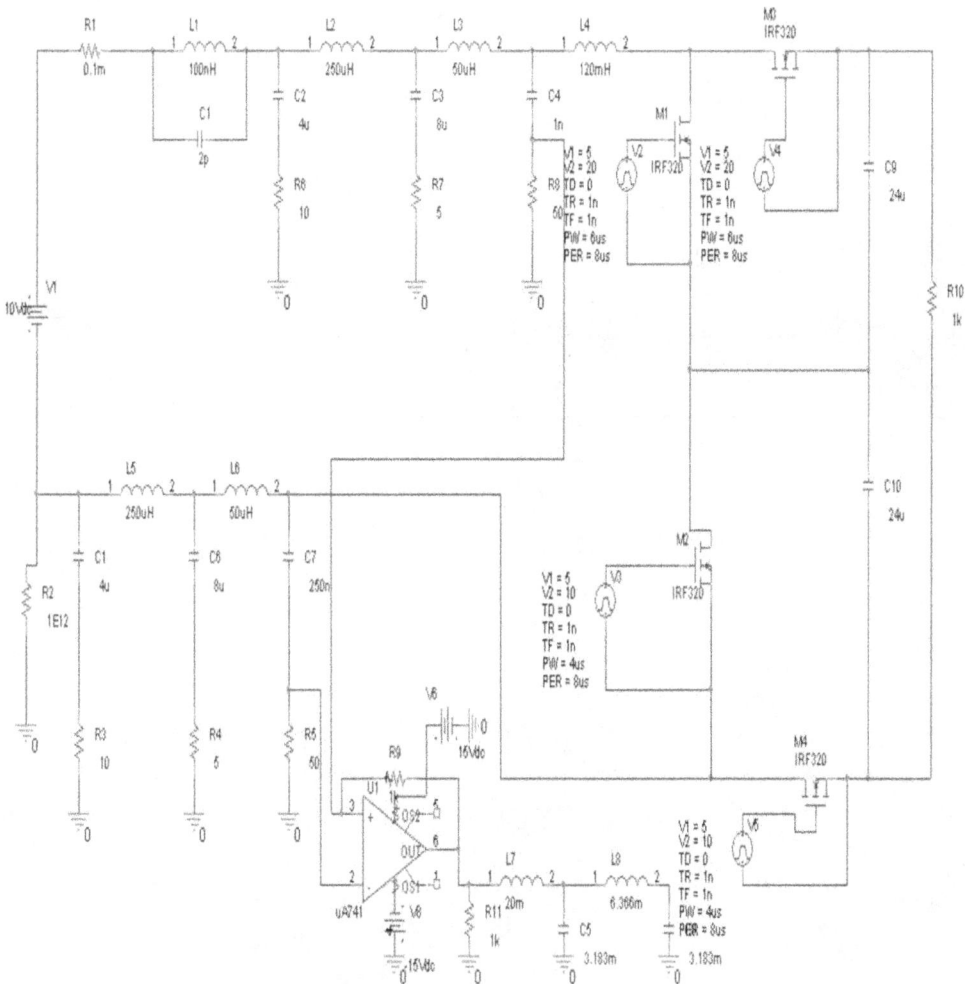

FIGURE 8.94 Circuit with EMI filter.

8.2.8.1 Emission Output with Filter

The conducted emission measurement results are shown in Figures 8.95 and 8.96. The conducted emission's attenuation has been reduced to a great extent by using EMI filters.

8.2.9 Conclusion

In medical systems, bipolar DC–DC converters can be implemented. In such a case, the output of these converters must be precise and dependable. The bipolar output power converter achieves a maximum efficiency of 83.4% at a load power of 187.5 mW, with the voltage balance error maintained under 2% across the whole load range. By using the output of the bipolar converter as the input to the digital pulse generator, beam formation occurs in the medical imaging systems. The S-parameter for this is found using two sub-schematic circuits XMITS and REFLEX. The output of XMITS and REFLEX should mirror image of each other. Thus, the coefficients of S11, S12, S21, and S22 are obtained and mentioned in the chapter. The conducted emission of the circuit can be found by connecting LISN in the positive and negative leg of the supply and the output of this is summed by using the summing amplifier. The obtained conducted emission is reduced by implementing EMI filters. The conducted emission of this circuit is taken and its attenuation is reduced to some extent by using EMI filters.

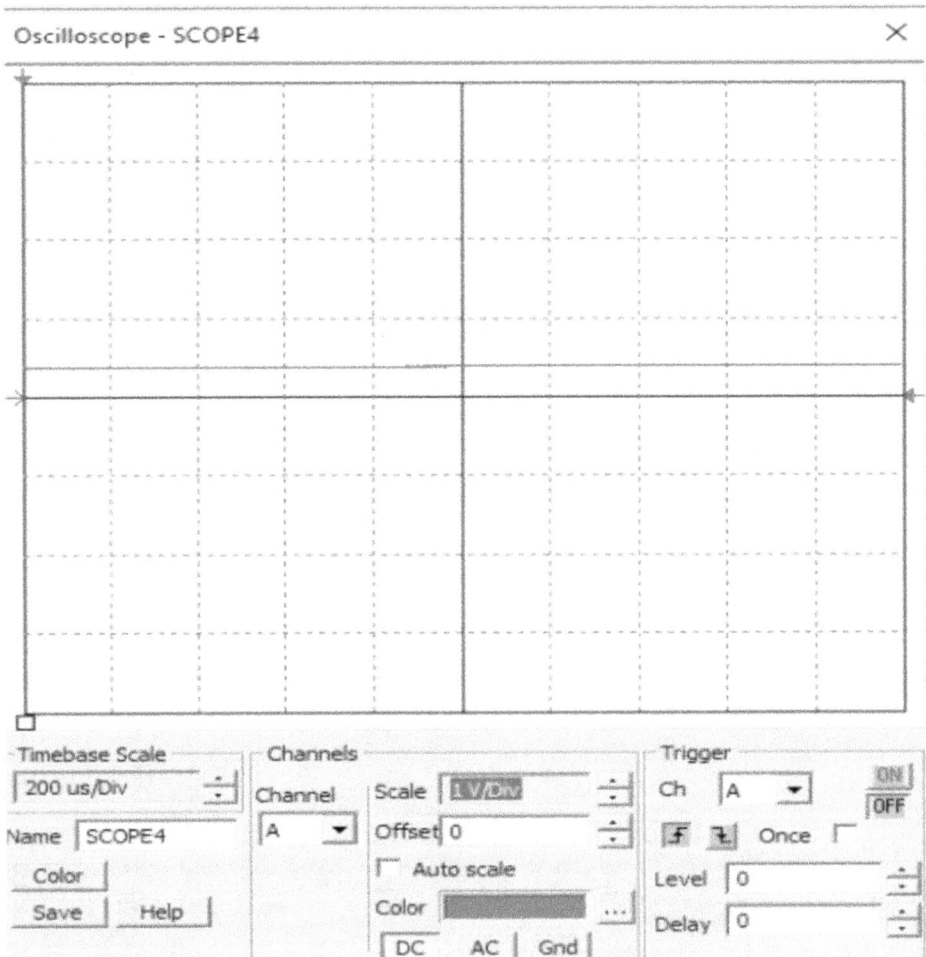

FIGURE 8.95 Reduced emissions due to EMI filter.

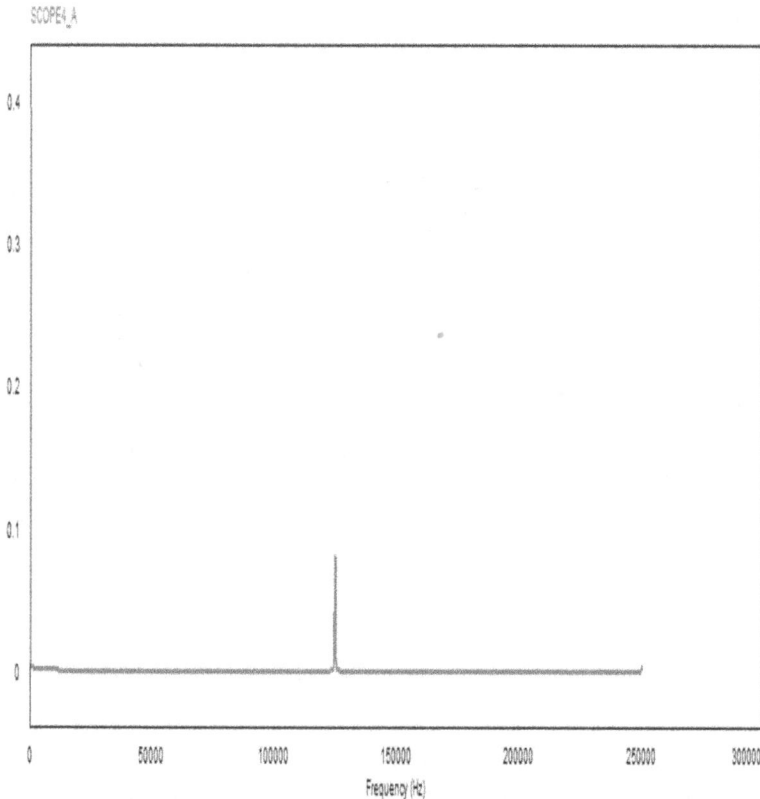

FIGURE 8.96   Emission with filter.

8.3   MLI FOR GRID APPLICATION

Power electronics deals with the conversion and control of electric power. The main task of the power converters in power electronics devices is to process and control the flow of electric charges by supplying the current and voltages for suited loads. The recent survey about the main task of the power converters is to extend the care for consuming power and current consumed by the converters and the output load voltage to meet the required electricity. Inverters are said to be semiconductor-based power converters. MLIs are said to be an array of power semiconductor devices. Compare with conventional converter and multilevel inverter for produce the sinusoidal output waveform and observed that harmonics also reduced. MLI will generate n level of output waveform. By increasing the number of voltage level, we will achieve the sinusoidal output waveform at the output AC side. In this section, three-phase three-level diode clamped multilevel inverter for grid connected system of output power 2 kV is considered with the renewable energy of 600 V as the input source. The proposed model of multilevel inverter is simulated and analysed with the help of ORCAD PSpice software. Conducted emission is the major consideration for the entire converter to give high efficiency. LISN is used to measure the conducted emission and according to the emission results L filter is constructed. The filter results have reduced harmonics or reduced conduction emission values, resulting in the efficiency of the power converter being increased.

8.3.1   MULTILEVEL INVERTER

Inverters are said to be semiconductor-based power converters. MLIs are said to be an array of power semiconductor devices. They generate multiple step voltage waveforms with variable and

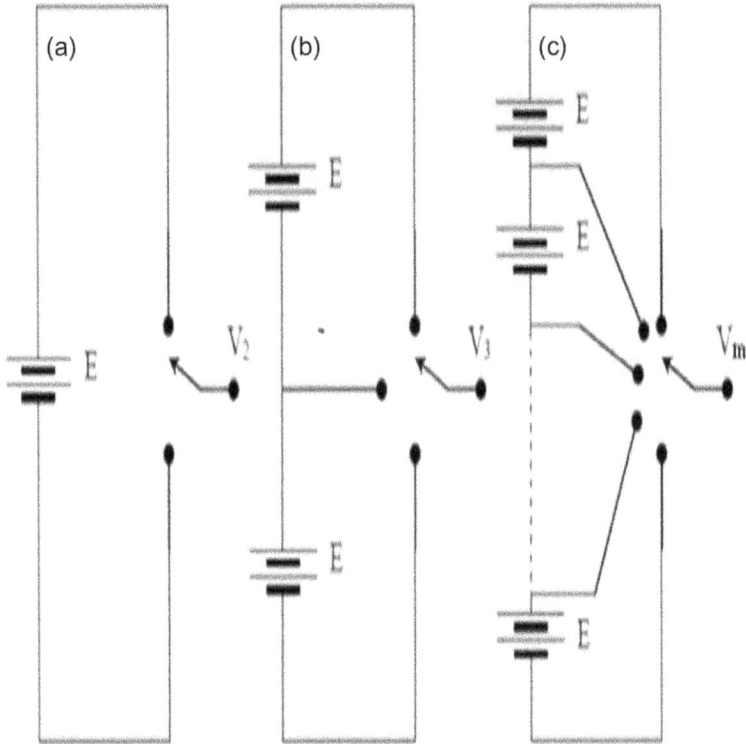

FIGURE 8.97   Multilevel inverter (MLI) topology—two level, three level, n level.

controllable frequency, amplitude, and phase. In a practical way, the pure sinusoidal waveform is not possible, but with the use of multi-level inverter one can produce a pure sinusoidal output waveform. MLIs have n number of levels (i.e., single pole n throw switch) Figure 8.97 shows the multilevel inverter topology for second level, third level, n level respectively. One can make use of MLIs depending upon the required applications. It has n number of real-world applications. Some of them are wind energy conversion, uninterruptible power supplies, and electric and hybrid vehicles. The purpose of multilevel converter is to give high power output from the medium voltage sources such as batteries, super capacitors, and solar panels.

Output voltage of the two-level inverter can be either high voltage or zero voltage, but here in the case of multilevel inverter the output voltage can be obtained in the form of multiple voltage levels achieved by dividing the input DC voltage. Hence, depending upon the desired number of voltage level requirements for the applications, voltage level in multilevel inverter topology can be selected. A normal DC–AC inverter will convert a DC) into an AC at various frequency and various output voltages. The same topology can also be applied for multilevel inverter. Generally, homes, office, and factory building all are supplied with AC electricity at various frequencies and various voltage ranges such as 50 Hz/100 V and 60 Hz/220 V. Frequency range can be depending upon the climatic conditions; for example, 50 Hz frequency is suitable for India, whereas 60 Hz frequency is suitable for the United States. In order to drive the electrical appliances efficiently, there is a need to convert the main supply into optimal voltage and frequency range. Inverters are used for a wide range of applications; the most specific, widely used one is VFD and UPS the so-called emergency power systems. Here the inverter will convert the AC main power into DC main power and the DC power will store in the battery. In case of power failure, the stored DC power in the UPS is converted back into AC power source as needed. This becomes one of the main emerging uses of inverter application.

8.3.2 MLI Topology

MLIs have become the preferred choice in all industries for high power and high voltage applications. Multilevel converters come under the inverter topology, where inverters are mainly classified as CSI and VSI, in that MLIs come under VSI category with more than two-level voltage sources. Where CSI and VSI are used for low medium voltage source application, MLIs are used for high power and high voltage source applications. The three main topologies of MLIs are as shown in Figure 8.98.

8.3.3 Power Converters

Power electronics deals with the conversion and control of electric power. The main task of the power converters in power electronics devices is to process and control the flow of electric charges by supplying the current and voltages for suited loads, as depicted in Figure 8.99. Range of power scale is about milliwatts (mW), megawatts (MW), and gigawatts (GW). Power electronics is said to be interference between the electronics and power. The recent survey about the main task of the power converters is to extend the care for consuming power and current consumed by the converters and the output load voltage to meet the required electricity.

Table 8.11 shows the classification of Power Converters.

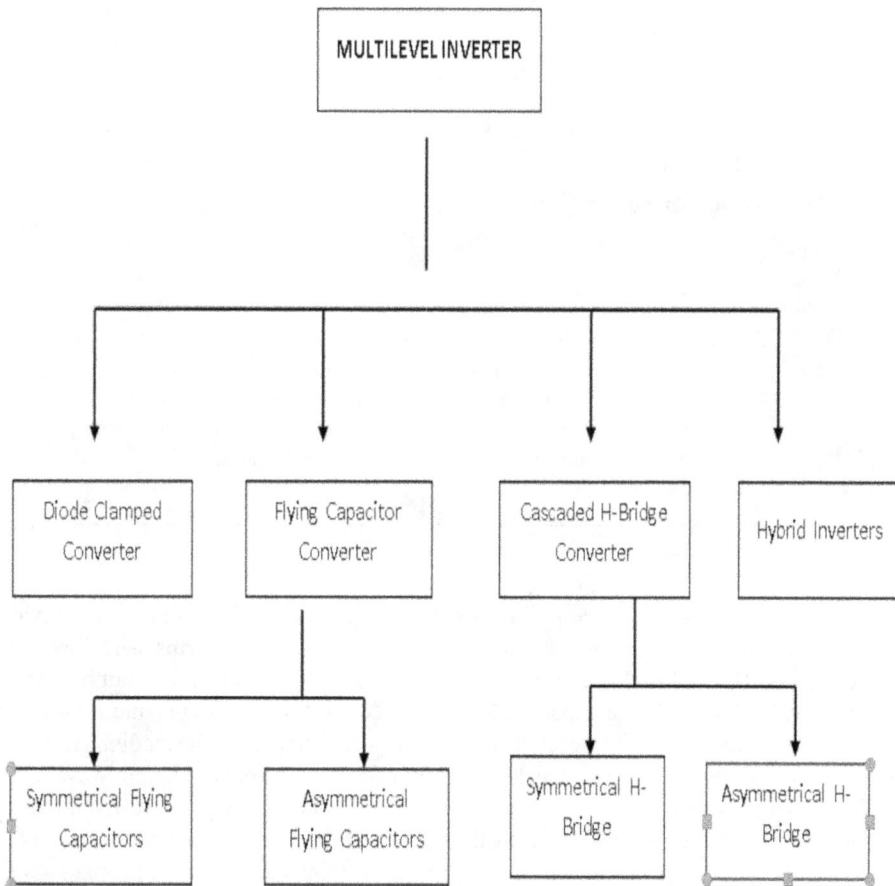

FIGURE 8.98   Multi level inverter (MLI) topology.

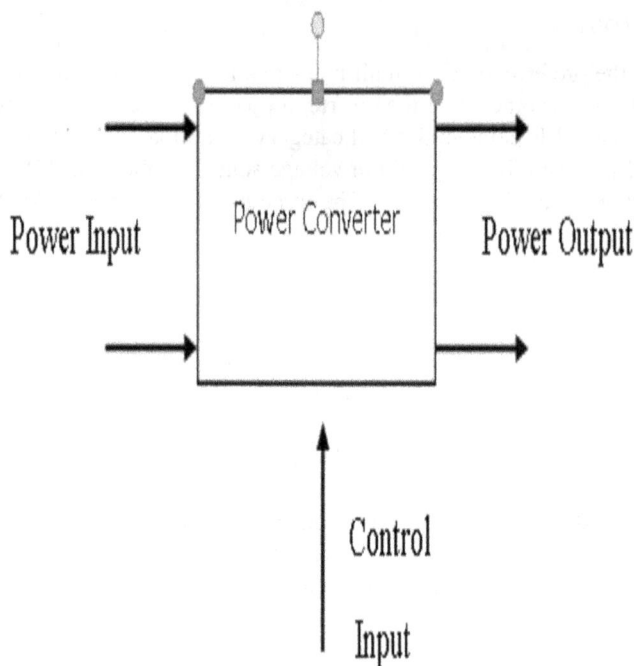

FIGURE 8.99 Block diagram of power converter.

TABLE 8.11
Classification of Power Converters

Output Power Input Power	DC	AC
AC	Alternating current (AC) to direct current (DC) converter (Rectifier)	AC to AC converter (Fixed frequency: AC controller Variable frequency: Cycloconverter)
DC	DC to DC converter (Chopper)	DC to AC converter (Inverter)

In past few years, the significance of MLIs has increased. MLIs are commonly employed in high-power, high-voltage applications due to their ability to mix waveforms with low total harmonic distortion (THD) and high harmonic spectrum (HS) (i.e., neutral point clamped MLI, cascaded H-bridge (CHB) MLI, flying-capacitor MLI, etc.). Neutral point clamped multilevel inverters (NPCMLIs) are used in a wide range of industrial applications that cover the medium voltage range due to their numerous advantages, such as low-harmonic content, improved output waveforms, and high-efficiency, among others. In NPCMLI, an exclusive DC source can be converted into any level of AC output. By increasing the level while utilizing NPCMLI, nearly sinusoidal AC output can be obtained, but THD will drop. The discrepancy in neutral-point voltage is the primary issue with NPCMLI. Another important disadvantage of NPCMLI is that it is difficult to manage the real power flow of each isolated converter in a multi-level converter, and a large number of clamping diodes are required. There are numerous systems used to build neutral point clamped (NPC) topology, including PSpice, ADS, and MATLAB.

Typically, the generation, transmission, distribution, and consumption of electric power involve the conversion of DC to AC. Inverters are the general name for DC-to-AC power converters, which play a crucial part in variable-frequency drives, uninterruptible power supply, induction heating, high-voltage DC power transmission, electric vehicle drives, AC transmission systems, and energy storage systems. Inverters may be square wave inverters, quasi-square wave inverters, two-level PWM inverters, or multilayer inverters, depending on the shape of the output waveform (MLIs). Due to their capacity to operate in all settings, including low, medium, and high voltage, MLIs have attracted a great deal of attention. The fundamental idea underlying an MLI is to generate a staircase waveform, following a sinusoidal route, by combining numerous lower voltage DC levels using semiconductor switches.

Multiple input DC levels typically include capacitors, batteries, or other traditional storage devices, and renewable energy voltage sources. Rearranging and regulating the power switches complete the aggregate of these many DC input levels and supplies a high output voltage. Thus, in MLIs, the operating voltage is much higher than the stress voltage on a power switch because the voltage source to which the switch is connected determines its rated voltage. MLIs offer numerous benefits, including a better harmonic profile, reduced dv/dt stress resulting in a significant reduction of filter components, and operation in both medium and high voltage applications with low voltage rated switches. Neutral point clamping (NPC), flying capacitor (FC), and CHB are the most prevalent and traditional MLI topologies, which have been widely implemented in numerous energy storage and electrical conversion systems.

In this section, diode clamped MLI stimulates grid application, where renewable energy is used as the source for the converter and the output of the converter is connected to the grid system. Construction of three-phase three-level diode clamped multilevel inverter is done. For output power of around 2 kW, assume that the renewable source form the solar or wind energy system as 600 V in the input DC side. At the end, Scattering parameter and conducted emission is calculated and filter circuit design is implemented for the proposed neutral clamped inverter.

8.3.4 Advantage of MLI

Conventional inverters, the so-called single-phase/three-phase, two-level voltage source inverters (VSIs) are widely used for low and medium power applications. And, current source inverters (CSIs) are used extensively for the medium voltage range and high-power range applications. Whereas MLIs are mainly concentrating on high power range of about 1 MW to 6 MW and high voltage range of about 3.3 kV to 6.6 kV in general. Higher number of voltage levels is not possible in conventional two-level inverters, but in MLIs one can generate n number of voltage levels depending upon the requirements. Because of this advantage in MLI, ideal pure sinusoidal output voltage is practically possible. The fact here is that the smoothness of the waveform is directly proportional to the voltage level of the converters. Conventional single-phase inverter requires its own DC power supply that can be derived from the low frequency transformer multi-winding or higher frequency. In this condition, balancing the series input DC voltage is difficult to maintain. So this will be overcome by using multilevel converters.Features of using MLIs are summarized as follows:

- Pure sinusoidal output voltage is possible by increasing the number of voltage levels.
- Aids in obtaining high-quality output voltage or current waveform.
- Easy to produce a high-power, high-voltage inverter with multi-level structure, and also device voltage gets reduced.
- Provides a high-power output from medium voltage sources including batteries, super capacitors, and solar panels.
- Operates at both fundamental and high switching frequency PWM.

- Better electromagnetic compatibility, higher power quality, staircase waveform quality, and higher voltage level.
- By increasing the voltage level, the harmonics content of the output voltage decreases significantly.
- Similarly, by increasing the voltage level, smoothness of the output voltage waveform gets increased and also satisfies the desired/required output voltage.

The overall advantage of MLIs is summarized as:

- Common Mode (CM) Voltage: The MLIs generate CM voltage, which reduces motor stress and prevents motor damage.
- MLIs can draw input current with minimum distortion.
- Switching Frequency: The multilayer inverter may work at both higher and lower switching frequencies, which are fundamental switching frequencies. It should be noticed that the reduced switching frequency results in decreased switching loss and increased efficiency.
- Reduced harmonic distortion. As a result of the selective harmonic elimination technique and multi-level topology, the THD in the output waveform is reduced without the use of a filter circuit.

8.3.5 Three-Phase Three-Level Diode-Clamped Inverter

Diode clamped inverter topology was first proposed in 1981. The input DC voltage source is divided by the capacitors; hence, they are also called as neutral point inverters. In the year 1992, more number of research works were published about this diode clamped converter. Diode clamped converter follows the single pole triple throw (SPIT) switch principle. In this converter, clamping devices are used in between the switches, where diodes are used as clamping devices. The input DC voltage is subdivided with the help of capacitors.

8.3.6 Principle—SPTT Switch

Two-level inverters are single pole double throw switch. Three-level inverters are single pole three throw switch. The pole in the converter connects in ways as $+V_{dc}$, $-V_{dc}$ and neutral. During positive cycle, throw1 (T1) switch is connected to the pole, during negative cycle throw2 (T2) is connected, and the pole is connected to neutral when the pole voltage is greater than the reference voltage. Figure 8.100 depicted the single pole triple throw (SPTT) switching configurations. Figure 8.101 shows the schematic circuit representation of diode clamped multilevel inveter circuit diagram (3 ph 3 level)

8.3.7 Components Requirement

In general, the number of componets used in all MLIs are based on the number of volatge levels choosen for construction. Table 8.12 classifies the number of voltage level depending on which number of clamping diode, capacitor, and active switches is needed for circuit construction, which is calculated based on formula.

8.3.8 Features of Diode Clamped Inverters

- Diodes are used as clamping devices.
- For n voltage level, there is a requirement of n-1 switch pairs; hence, for three levels, two switching pairs are required for each phase leg.
- Switching devices (e.g., MOSFET/IGBT) ideally block only the supplied DC voltage, whereas the clamping diodes help to share voltage DC source via neural point.

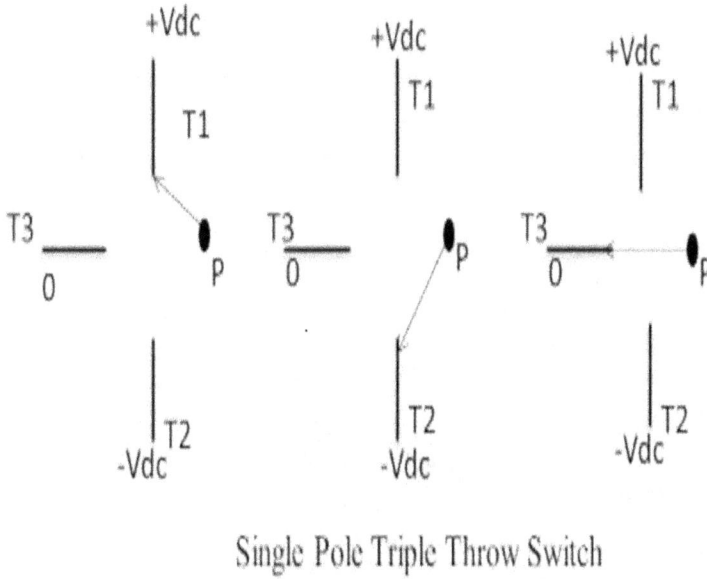

Single Pole Triple Throw Switch

FIGURE 8.100 SPTT switch.

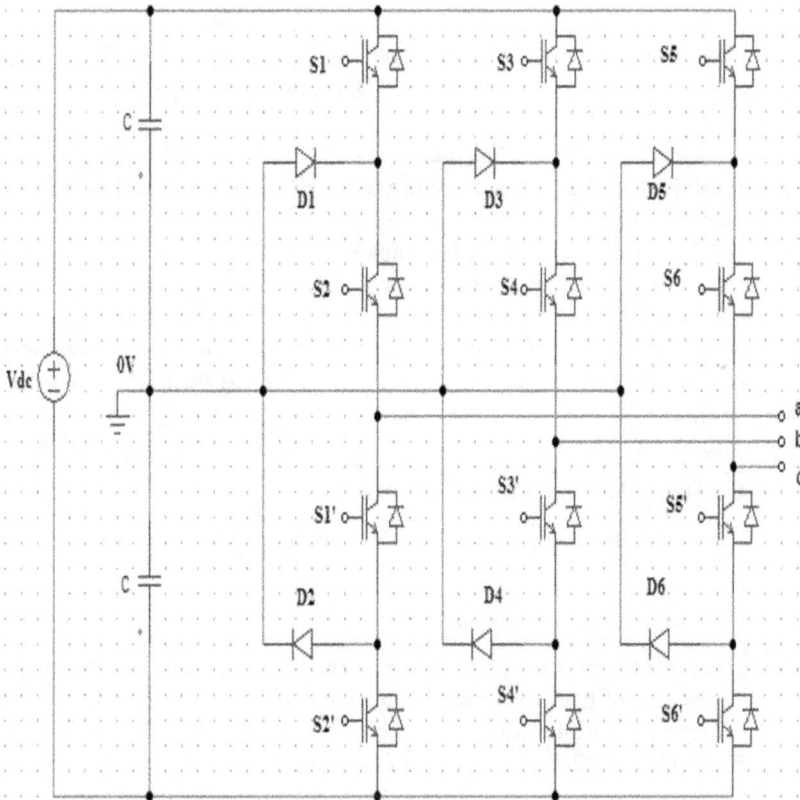

FIGURE 8.101 Diode clamped MLI circuit diagram (Three-phase three-level).

TABLE 8.12

Number of Component Requirement for Diode Clamped Multilevel Inverter

Voltage Level (m)	Active Switches 6*(m-1)	Clamping Diodes 3*(m-1) *(m-2)	DC Capacitors (m-1)
3	12	6	2
4	18	18	3
5	24	36	4
6	30	60	5
7	36	90	6

- Each diode must block a voltage equal to the number of switches above it multiplied by the DC voltage supplied.

a. Advantages
   - There is no dynamic Voltage Sharing Issue:
   - During commutation, each switch in an NPC inverter can withstand just half of the total input DC voltage.
   - **Static voltage balancing without extra components**: In NPC inverter, leakage current of the bottom and top switches is lower than the inner switches, hence the static voltage equalization is achieved
   - **Low THD and dv/dt**: MLI can increase the number of voltage levels. Hence, by increasing the voltage level the THD and dv/dt values are reduced
b. Disadvantages
   - Fluctuation occurs in the DC bus midpoint voltage.
   - By increasing the voltage level, there is a requirement of additional number of clamping diodes and switches.
   - Design of diode clamped inverter is complex.
   - Cause uneven loss distribution in each leg

This chapter explains about the MLI and diode clamped MLIs. From this, one can understand the basic introduction about the three-phase three-level diode clamped MLI. The converter simulation is done using the PSpice software.

8.3.9   DIODE CLAMPED MLI SIMULATION—PSPICE

Figure 8.102 represents the three-Phase three-level diode clamped MLI in PSpice simulation.

8.3.10   CONSTRUCTION

Diode clamped inverter contains three limbs, wherein each limb has four switches and two diodes; hence, totally it is made up of twelve switches with six diodes. In each limb, the four switches are divided into two upper switches and two lower switches. The centre of the limbs is connected to the terminals R, Y, B of AC load where centre of the upper and lower pairs of switches are connected with the diode and this connection is connected to the midpoint of the supply. Number of switches, diodes, and capacitors selected depends on the number of voltage levels.

Diode clamped inverter works on the principle of SPTT switch as discussed; in DC side DC bus capacitor is split into parts and also provided with neutral connection. Here during positive and

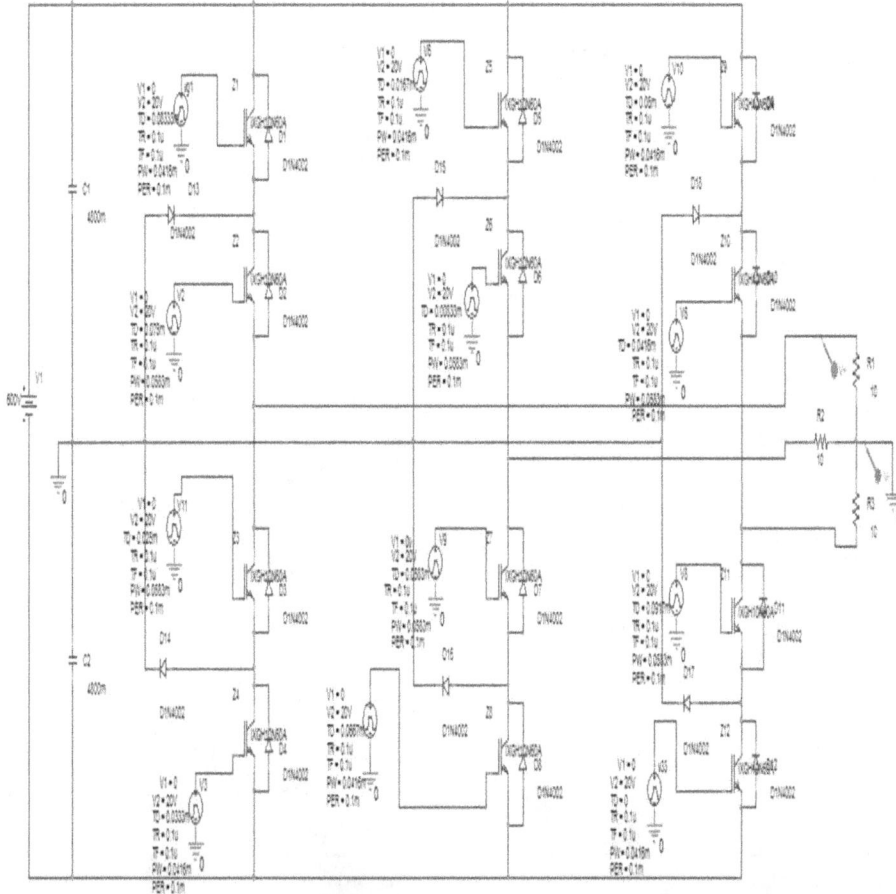

FIGURE 8.102   3 Phase three level diode clamped multilevel inverter in PSpice simulation.

negative modes of output, AC source, either the upper limb or lower limb switches, are in contact in each leg.

Let us take one single leg for analysing the detailed working process. If the top two switches S1 & S2 are in on state and bottom switches S3 & S4 are in off condition, which means that the pole gets connected to the top throw, the average voltage becomes $+V_{dc}/2$. Inverse of this condition where bottom two switches S3 & S4 are in on state and upper two switches S1 & S2 are in off condition, which means the pole gets connected to the bottom throw, the average voltage becomes $-V_{dc}/2$. Another connection is the switches S2 & S3 are in on state and S1 & S4 are in off state, which means the pole is connected to the neutral point throw where average voltage is zero. In this last case, switches S1 & S2 and S3 & S4 pairs are complementary in nature. The above three operations are happening depending upon the direction of the load current flow.

If the load current flow is normal, then the pole gets connected to upper throw and lower throw during positive and negative cycle mode, respectively. If the load current flow is in opposite direction, then the pole gets connected to the midpoint, which means neutral point of the DC side, where switch S2 is on for positive and switch S3 is on for negative cycle mode. This process is similar while adding legs; if two legs, it is said to be single phase AC supply. Similarly, three legs are connected for three phase AC supply (Figure 8.103). This is represented in Table 8.13.

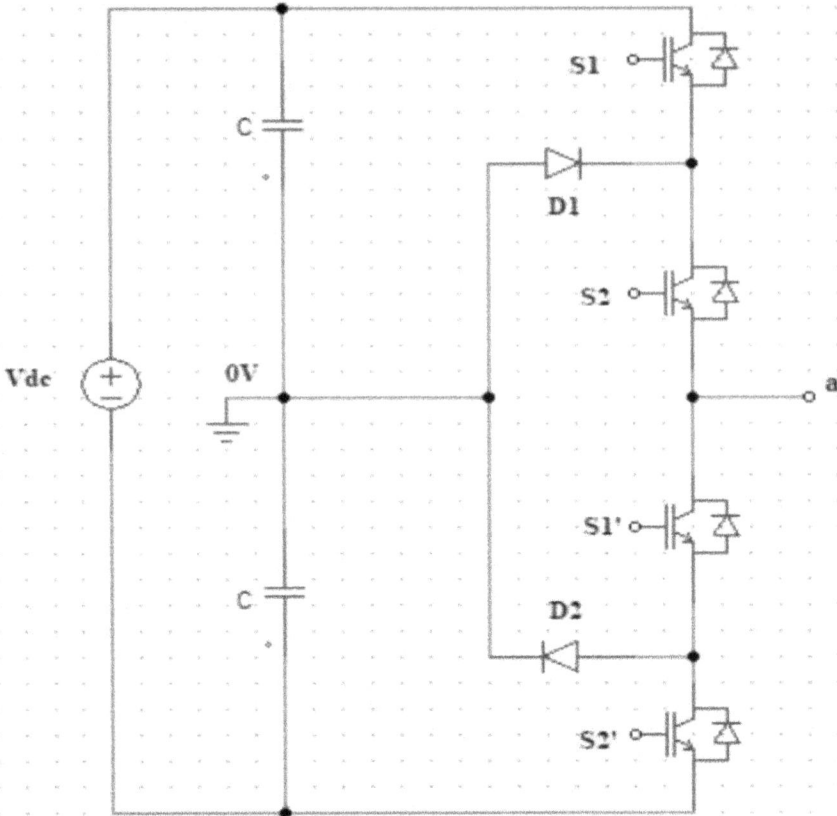

FIGURE 8.103 Single leg topology of diode clamped inverter (Three level).

TABLE 8.13
Switching Table for Single Leg of Diode Clamped Multilevel Inverter

S1	S2	S1′	S2′	Van
ON	ON	OFF	OFF	$\frac{+Vdc}{2}$
OFF	ON	ON	OFF	0
OFF	OFF	ON	ON	$\frac{-Vdc}{2}$

8.3.10.1 Switching Sequence

The three-level three-phase diode clamped MLI is obtained from the configuration of 12 switching devices. The switches S1, S2, S3, S4, S5, and S6 are complementary to S1′, S2′, S3′, S4′, S5′, and S6′. The voltage vectors are V1, V2, V3, V4, V5, V6, V7, V8, V9, V10, V11, and V12. Here one cycle is divided into 12 sectors with 30 degree each. The angle and its voltage vector are shown in Table 8.14.

TABLE 8.14
Voltage Vector

Sector	Angle	Voltage Vector
1	$0° < \theta \le 30°$	V1
2	$30° < \theta \le 60°$	V2
3	$60° < \theta \le 90°$	V3
4	$90° < \theta \le 120°$	V4
5	$120° < \theta \le 150°$	V5
6	$150° < \theta \le 180°$	V6
7	$-180° < \theta \le -150°$	V7
8	$-150° < \theta \le -120°$	V8
9	$-120° < \theta \le -90°$	V9
10	$-90° < \theta \le -60°$	V10
11	$-60° < \theta \le -30°$	V11
12	$-30° < \theta \le 0°$	V12

TABLE 8.15
Switching Sequence of Voltage Vector

Sector	Voltage Vector	S1	S2	S3	S4	S5	S6	S1′	S2′	S3′	S4′	S5′	S6′
1	V1	1	1	0	0	0	0	0	0	1	1	1	1
2	V2	1	1	0	1	0	0	0	0	1	0	1	1
3	V3	1	1	1	1	0	0	0	0	0	0	1	1
4	V4	0	1	1	1	0	0	1	0	0	0	1	1
5	V5	0	0	1	1	0	0	1	1	0	0	1	1
6	V6	0	0	1	1	0	1	1	1	0	0	1	0
7	V7	0	0	1	1	1	1	1	1	0	0	0	0
8	V8	0	0	0	1	1	1	1	1	1	0	0	0
9	V9	0	0	0	0	1	1	1	1	1	1	0	0
10	V10	0	1	0	0	1	1	1	0	1	1	0	0
11	V11	1	1	0	0	1	1	0	0	1	1	0	0
12	V12	1	1	0	0	0	1	0	0	1	1	1	0

From the table, one can infer that the switching table is formed using θ. For example, if θ lies between $0° < \theta \le 30°$, then it comes under sector 1, and voltage vector as V1, then the corresponding switching state as 1,10,000. The switching table for all the vector state are listed as shown in the Table 8.15.

The switches S1, S2, S3, S4, S5, and S6 are complementary to S1′, S2′, S3′, S4′, S5′, and S6′ switches.

8.3.10.2 Design Specifications
- Three-phase clamped diode MLIs have three legs with a shared DC bus.
- Through capacitors, this DC voltage is separated into switches. Therefore, $(m-1) = (3-1) = 2$ DC capacitors are required, where m is the number of voltage levels.

- For m-levels, m-1 switch pairs are necessary; so, two switch pairs are required in this instance. One of the switches in each pair must be activated. If one switch is activated, the other should be deactivated.
- Switching devices (e.g., IGBT) must block just the supplied DC voltage, while clamping diodes have an entirely other story to tell. Each diode must block a voltage equal to the number of switches above it multiplied by the DC voltage supply.

  For three-phase three-level diode clamped MLI,

  Voltage level = m = 3 levels

  Number of active switches = 6*(m−1) = 6*(2) = 12 active switches

  Number of clamping diodes = 3*(m−1) *(m−2) = 3*2*1 = 6 diodes

  Switching angle calculation, conduction angle(α) is 30°

8.3.10.3 Components Requirements

Components used here are IGBT, diode, Vpulse for gate signal (Table 8.16).

8.3.10.4 Simulation Results

a. **Input Side**

- The input DC source is 600 V waveform, and voltage across capacitor C1 and C2 as in the Figure 8.104

  Inference: DC source is 600 V, the input DC is divided by using the capacitors; hence, the voltage across the capacitors is 300 V.

b. **Gating and Switch Block**

- Gating signal g1, g2, g1′, and g2′ and voltage across the switches s1 and s1′ as in Figure 8.105.

  Inference: Gate signal g1 and g1′ are complementary; similarly switch s1 and s1′ are complementary to each other. When gate signal g1 and g2 are on switch s1 becomes on, when g1′ ad g2′ is on the switch s1 becomes off state, and switch s1′ is complementary to s1.

- Gating signal g3, g4, g3′, and g4′ and voltage across the switches s3, s4, s3′, and s4′ (leg two) as in Figure 8.106.

  Inference: Gating signal at leg one and leg two are at 120-degree phase shifted. Gate signal g3 and g3′, g4 and g4′ complement each other; similarly switch voltage s3 and s3′, s4 and s4′ complement each other. When gate signal g3 and g4 are on switch s3 become on, when g3′ and g4′ is on the switch s3 becomes off state, and switch s3′ is complement to s3. The same is for leg three.

c. **Output Side**

- Line to Line voltage as in Figure 8.107
- Load current as in Figure 8.108

TABLE 8.16

Components Used in the Simulation Circuit

Component	Name	Specification	Market Availability
IGBT	IXGH10N60A	VCE = 600 V, IC = 20 A, VCE (sat) = 2.5 V	Available
Diode	D1N4002	Reverse voltage = 50 V–100 V, Forward current = 1 A	Available
Capacitor	–	450 mF	Available
Vpulse	Gate Pulse	Fs = 10 kHz	Available

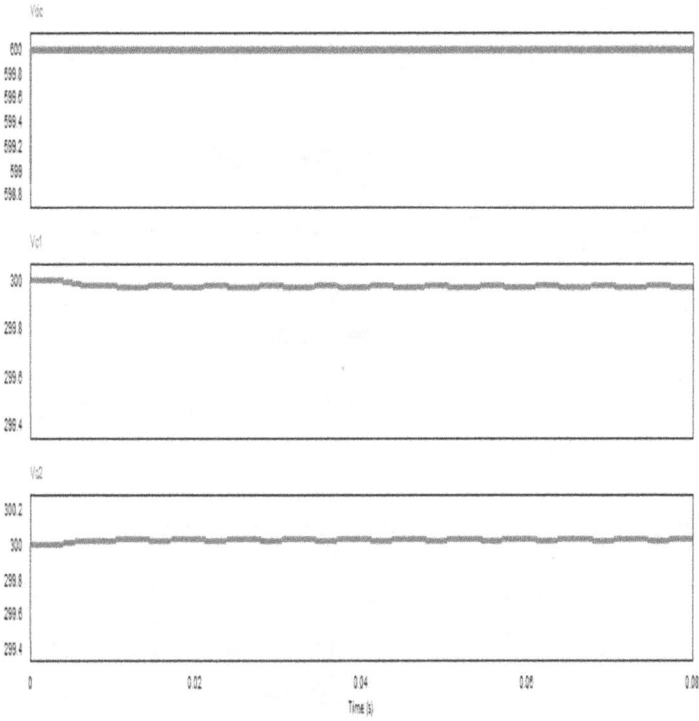

FIGURE 8.104 Input voltage, capacitor voltage waveform.

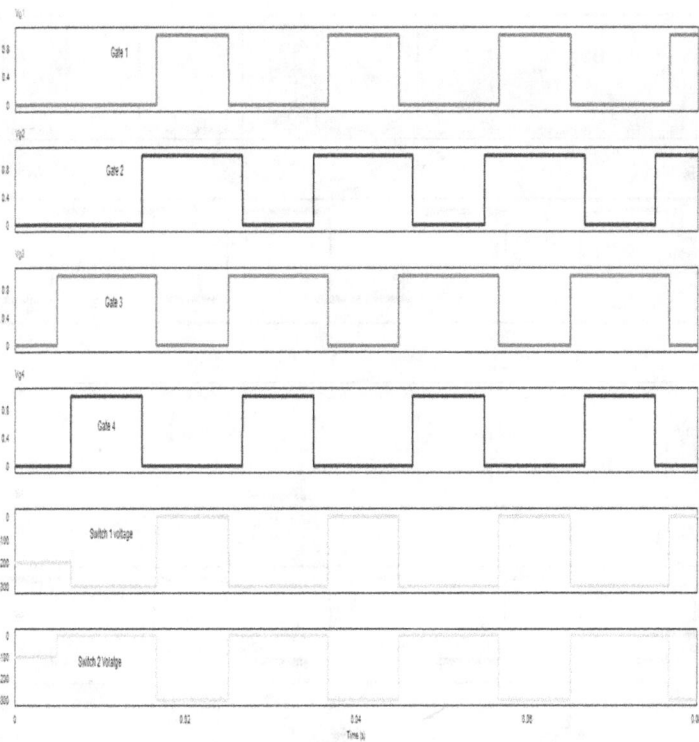

FIGURE 8.105 Gating signal and switch at leg one.

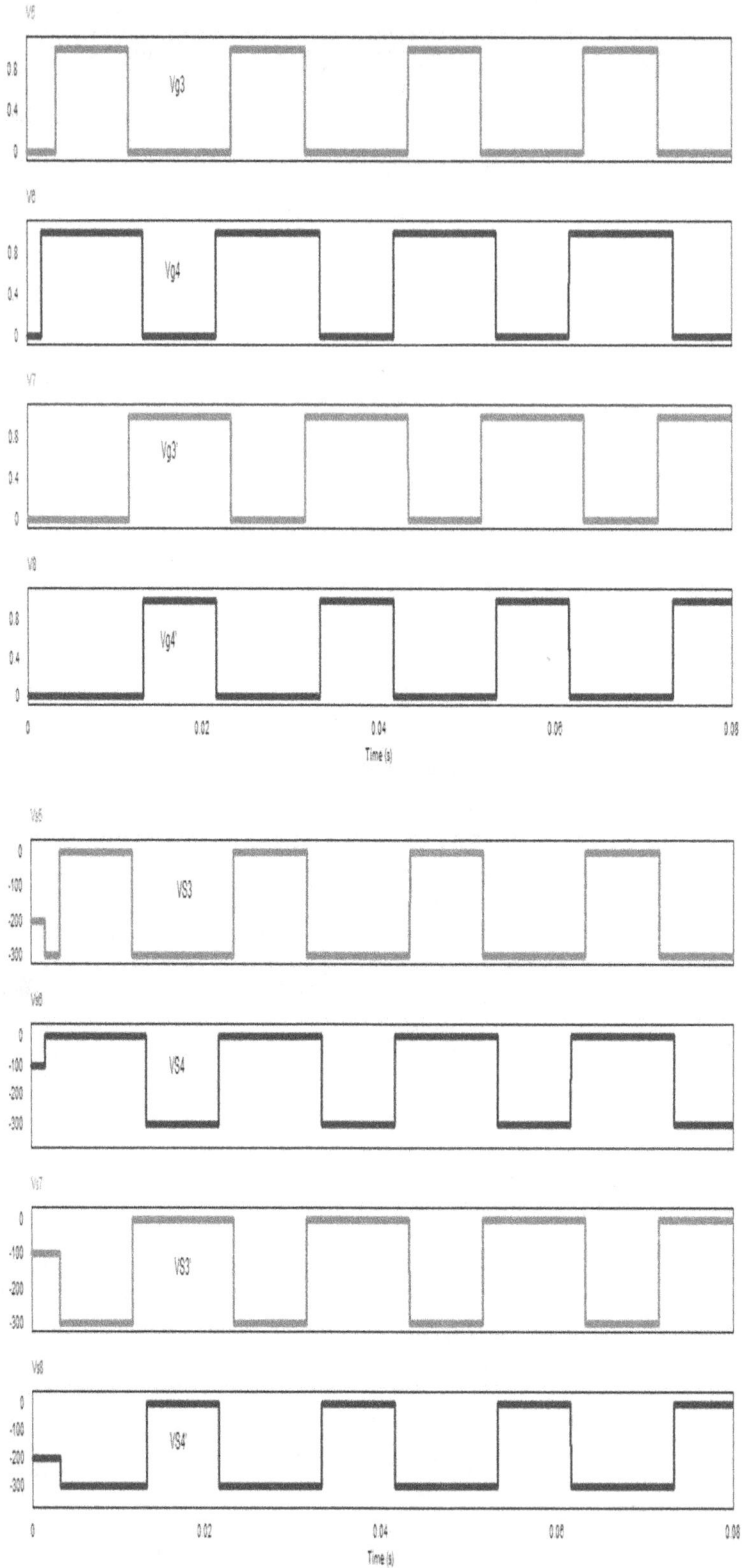

FIGURE 8.106　Gating signal and switch at leg two.

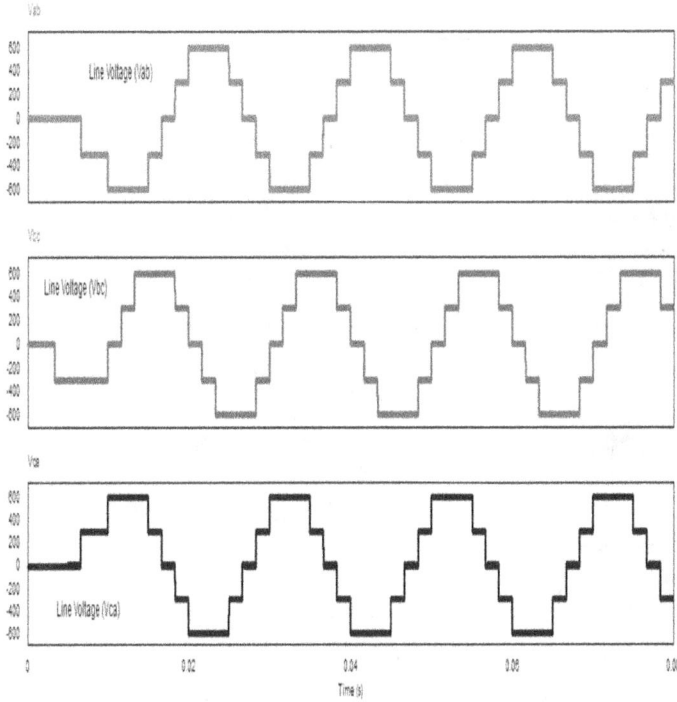

FIGURE 8.107 Line voltage of the three-phase three-level neutral-point-clamped multilevel inverter (NCMLI).

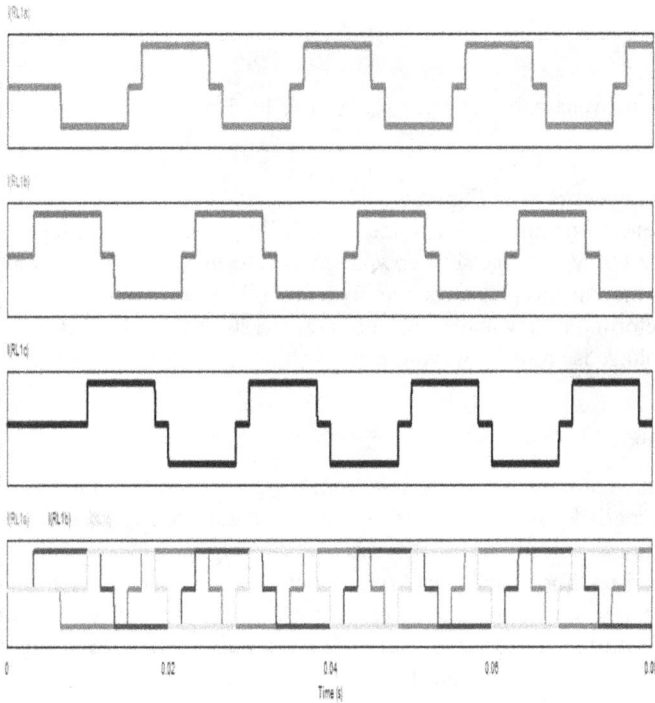

FIGURE 8.108 Load current of the three-phase three-level neutral-point-clamped multilevel inverter (NCMLI).

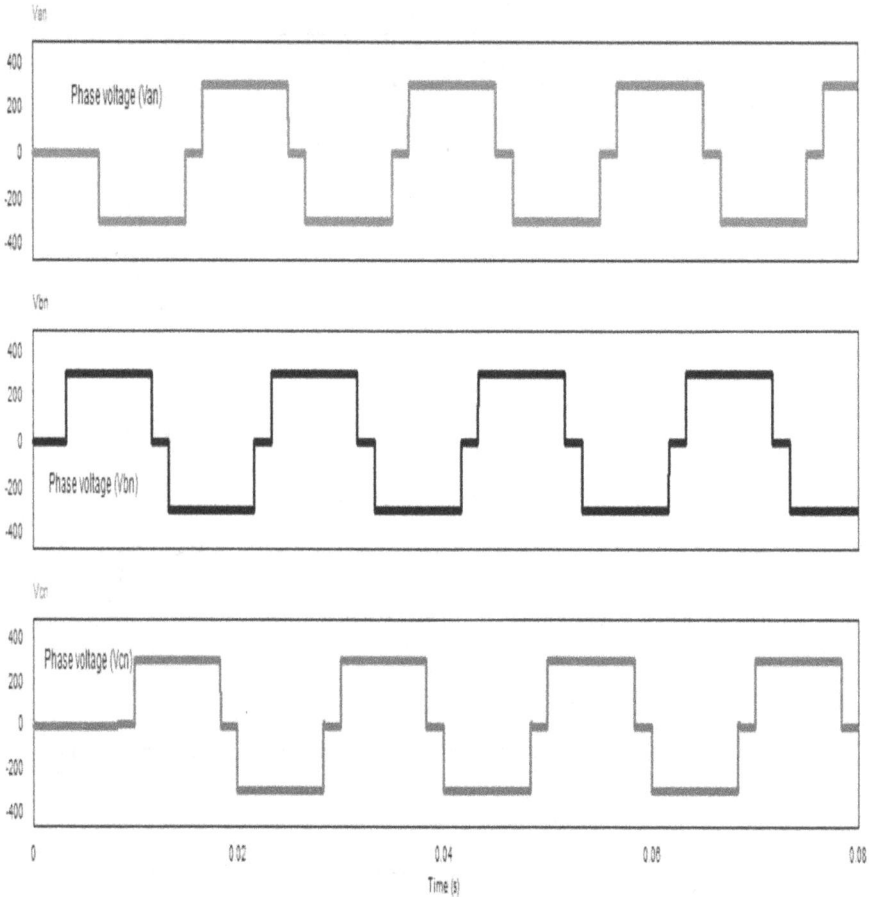

FIGURE 8.109 Phase voltage of the three-phase three-level neutral-point-clamped multilevel inverter (NCMLI).

- Phase voltages are as in Figure 8.109
  Inference: For line-to-line voltage, each line-to-line voltage has the input voltage of around 600 V and line six step voltage waveform. Load current is like square wave, which is the same as phase voltage. By using PWM or SPWM, we will get the sinusoidal waveform. Phase voltage, each phase is at 120-degree phase shift to each other and phase voltage is around 380-volt, phase voltage is the three-level voltage waveform.

8.3.11 S-Parameter Measurement

Scattering parameters are complex matrices; from S-parameters, one may evaluate reflection and transmission characteristics like amplitude or phase. Stimulus or reaction relates to both the VAN and TDR (time-domain reflectometers). Electrical scattering parameters are a "black box." It contains interconnected resistors, capacitors, and inductors that interface with other circuits through ports. The network is a scattering parameter matrix. It's described as an N-port network with an N-by-N matrix. Scattering parameters are in degrees and radians. Frequency, impedance, and number of electrical network ports must be given to determine scattering parameters, shown in Figure 8.110.

The scattering parameter is defined as the ratio between the normalized reflected wave at ith port and the normalized incident wave at jth port.

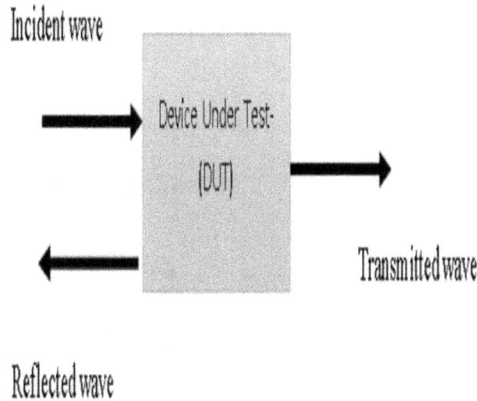

FIGURE 8.110 Block diagram of scattering parameter.

- $Sij = b_i/a_j$
  where, b_i = normalized reflected wave at ith port
  a_j = normalized incident wave at jth port
  $b_i = V^+/\sqrt{z}0$
  $a_i = V^+/\sqrt{z}0$ where $\sqrt{z}0$ is the characteristic impedance of line.

8.3.11.1 S-Parameter in PSpice

In PSpice XMITS and REFLECTS, sub-schematic circuits are needed to calculate the S-parameter for the network, where XMITS sub-schematic circuit is used to measure the forward transmission coefficient and reverse transmission coefficient, while REFLECTS sub-schematic circuit is used to measure the input reflection coefficient and output reflection coefficient. In general, XMITS are used to measure the transmission coefficient and REFLECTS sub-schematic circuit is used to measure the reflection coefficient. In PSpice, we want to create/connect the XMITS and REFLECTS sub-schematic circuits as shown in Figures 8.111 and 8.112.

8.3.11.2 Creating XMITS and Reflects Block in PSpice

Hierarchical blocks are created for XMITS and REFLECTS circuits. For S11 and S22 blocks, the internal circuit is REFLECTS sub-schematic circuit. Similarly for S12 and S21 hierarchical blocks, the internal circuit is XMITS sub-schematic circuit. For the S-parameter measurement, we want to connect the XMITS and REFLECTS blocks at the input and output sides of the main circuit (DUT) and before that convert the main circuit into a block.

The below are the steps to explain how to check and connect the S-parameter for active network.

Step 1: S11 and S21 hierarchical block as shown in Figure 8.113 for measuring input reflection and forward transmission coefficient.

Step 2: For active circuit in Figure 8.114, take the transistor or IGBT. Here IGBT is used, because MLI has IGBT switch. Now connect the S11 and S21 at the input and output of the component/DUT, respectively, to measure input reflection and forward transmission as shown in Figure 9.132 and the simulation results are in Figure 8.115.

Step 3: Similarly create S22 and S12 blocks for measuring output reflection and reverse transmission coefficient and connect the REFLECTS (for S22 measurement) in the output side and connect the XMITS block (for S21 measuremnt) in the input side and check the scattering parameter results (Figure 8.116).

Step 4: Simulation results of input and reflection coefficients are shown in Figures 8.117 and 8.118.

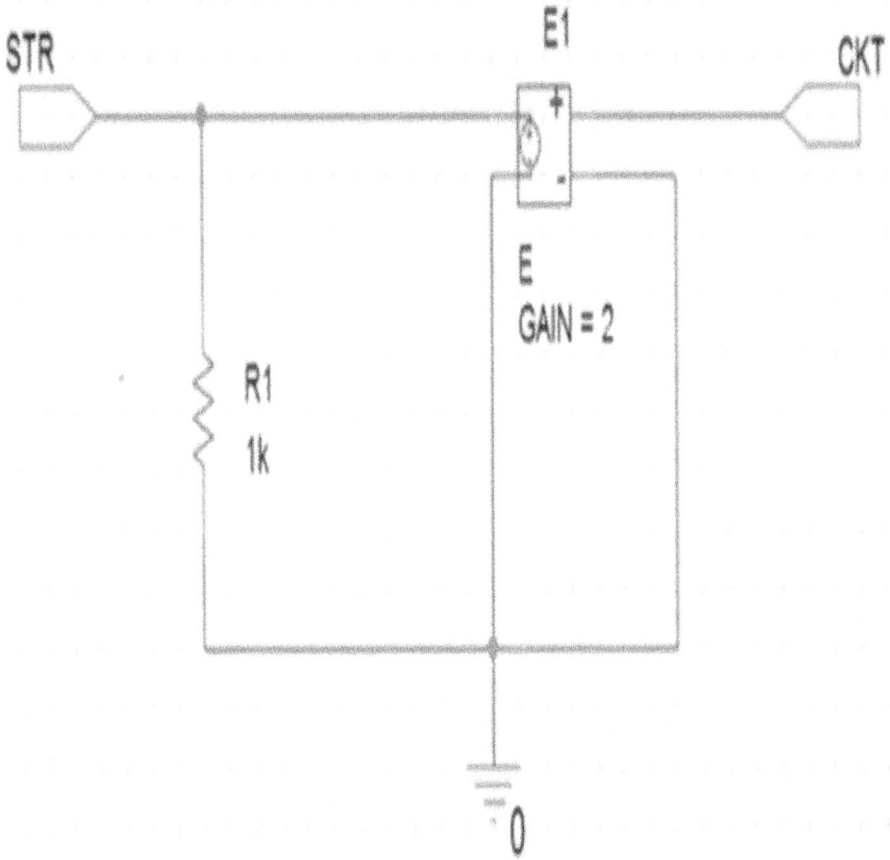

FIGURE 8.111 XMITS circuit.

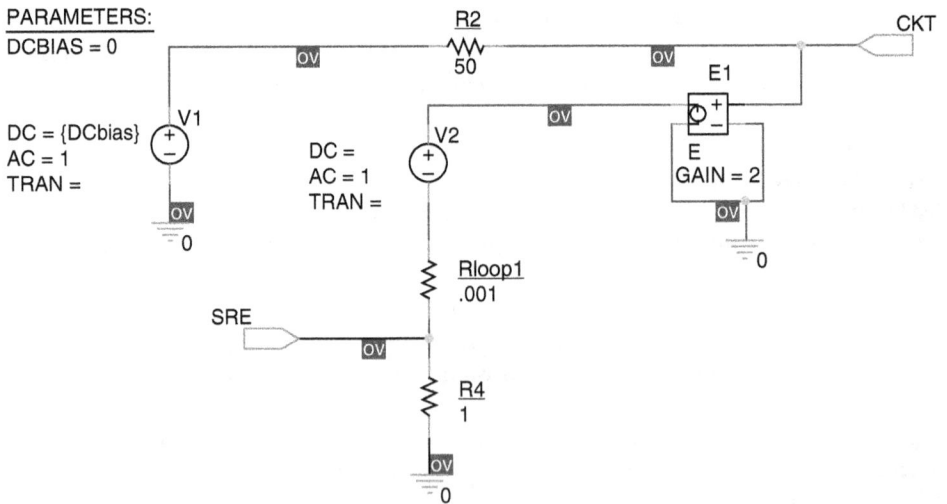

FIGURE 8.112 REFLECTS circuit.

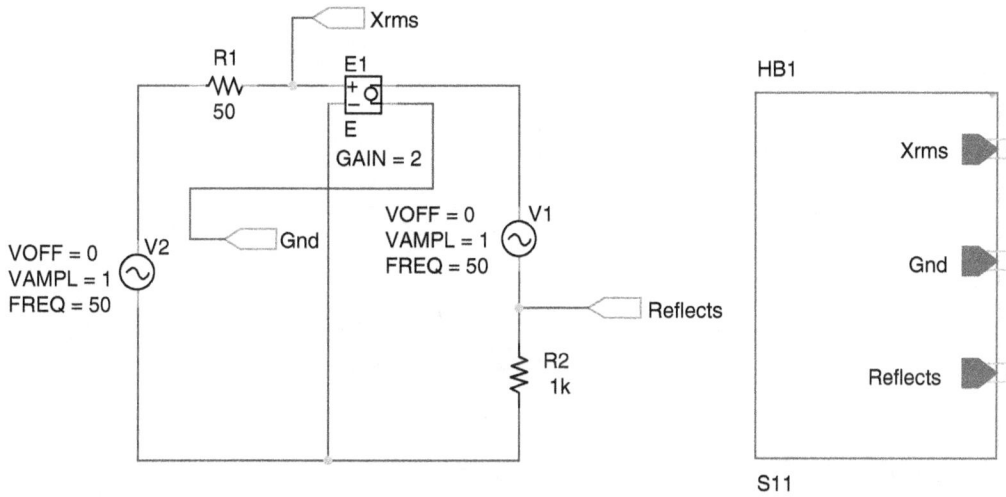

FIGURE 8.113 Inside of S11 block and REFLECTS hierarchical block for S11 measurement.

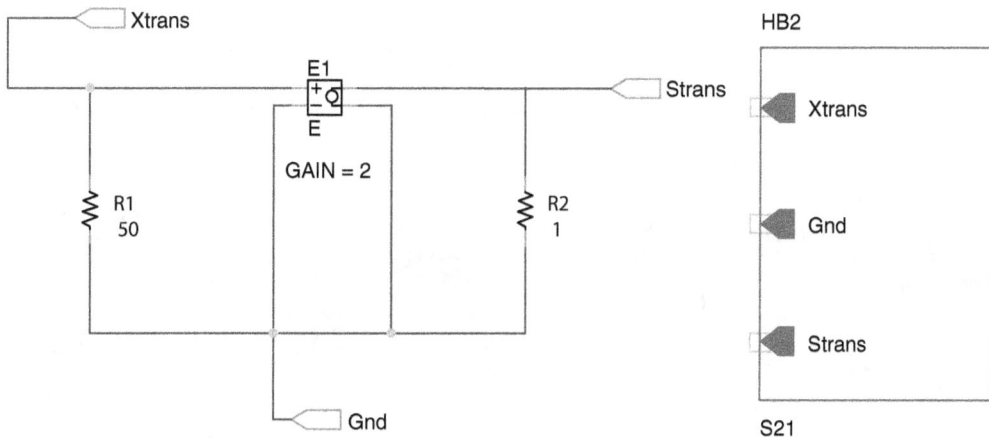

FIGURE 8.114 Inside of S21 block and XMITS hierarchical block for S21 measurement.

8.3.11.3 S-Parameter for MLI

Here the steps for measuring input reflection coefficient, output reflection coefficient, forward transmission coefficient, and reverse transmission coefficient for the three-phase three-level diode clamped MLI in PSpice is shown below.

Step 1: Create a hierarchical block for diode clamped MLI as shown in Figure 8.119.

Step 2: Create a hierarchical block for S11, S12, S22, and S21.

Step 3: Connect S11 (REFLECTS) and S21 (XMITS) at the input and output side of diode clamped MLI, respectively, as shown in the Figure 8.120.

Step 4: Connect S12 (XMITS) and S22 (REFLECTS) at the input and output side of diode clamped MLI, respectively, as shown in the Figure 8.121.

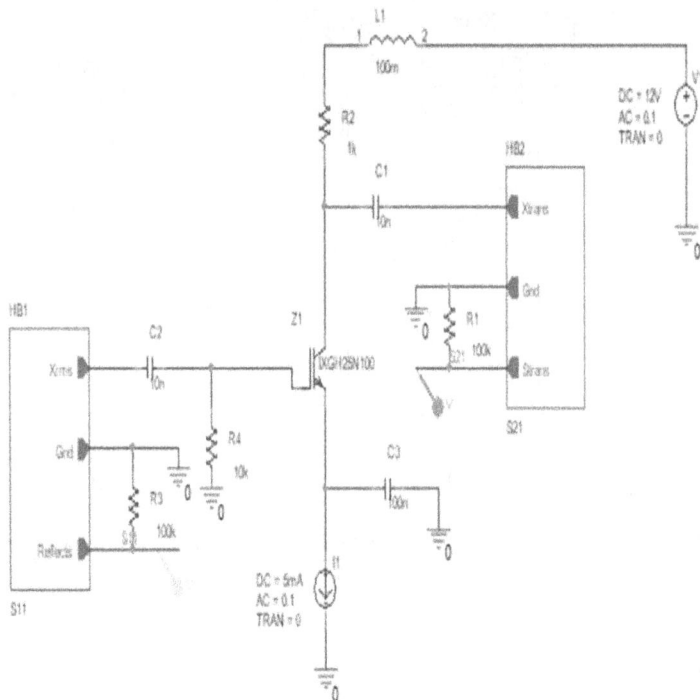

FIGURE 8.115   Circuit connection for measuring S11 and S21.

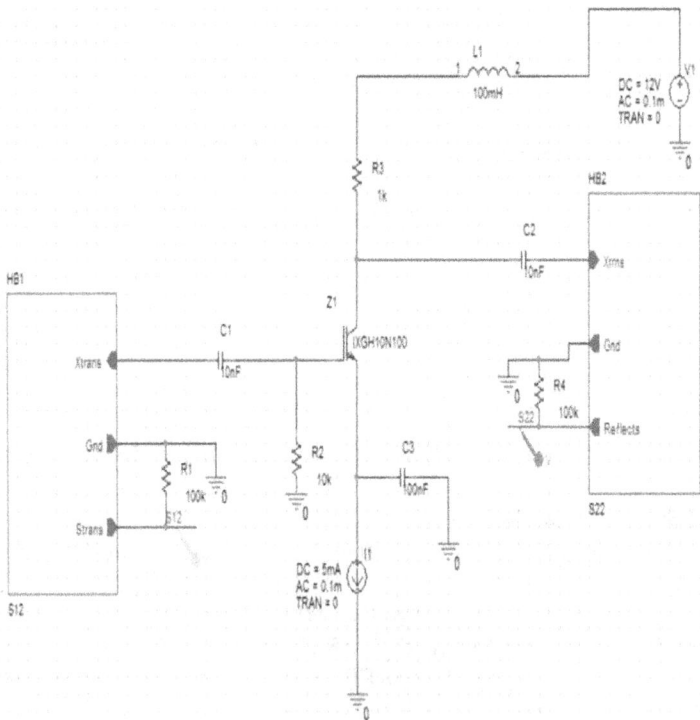

FIGURE 8.116   Circuit connection for measuring S12 and S22.

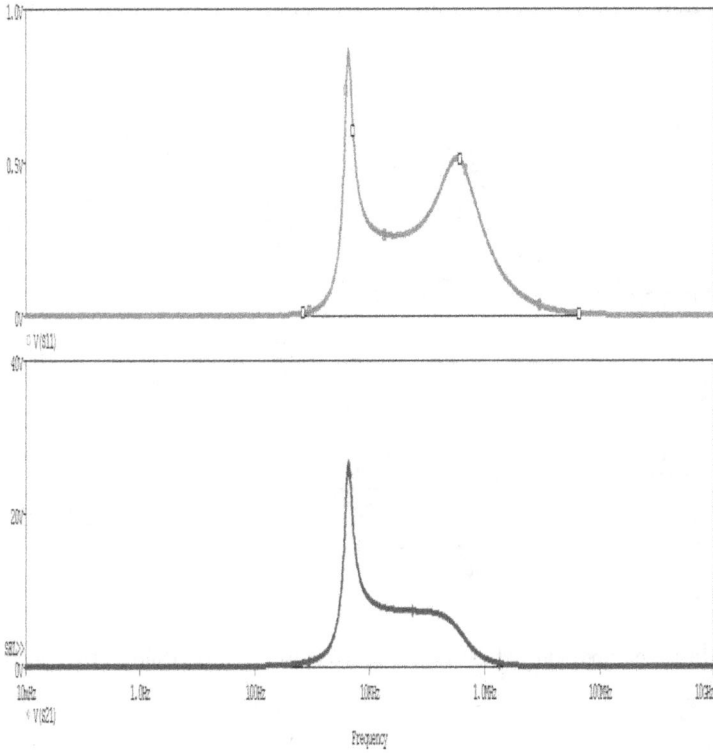

FIGURE 8.117 S11 and S21 graphs.

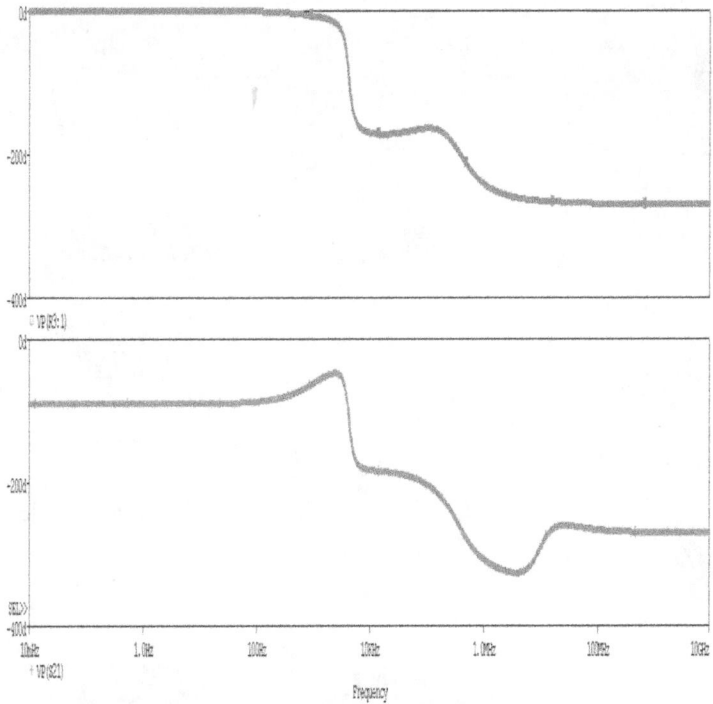

FIGURE 8.118 S12 and S22 graphs.

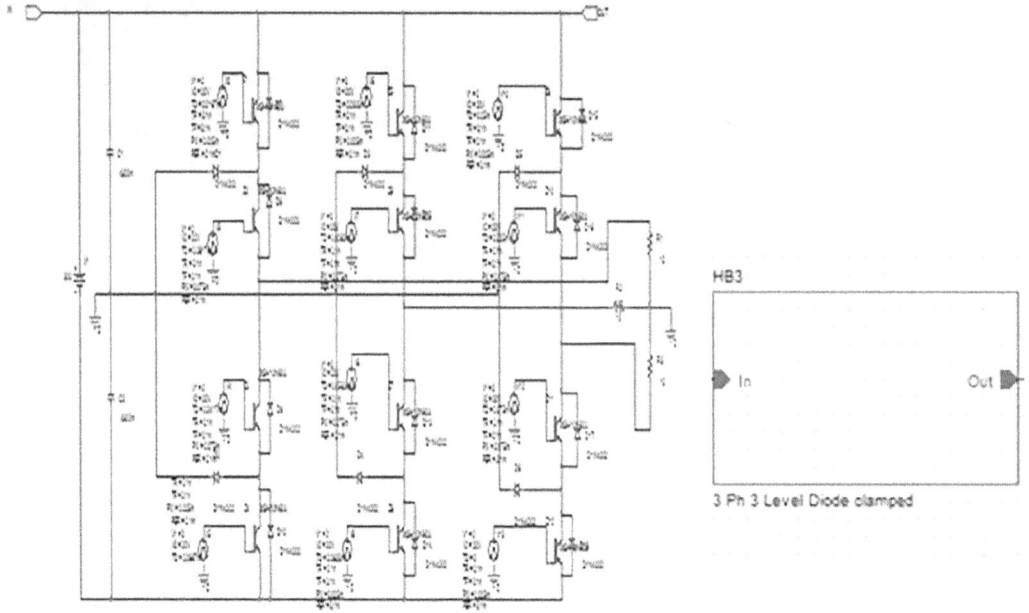

FIGURE 8.119   Inside of hierarchical block for multilevel inverter (MLI) and its DUT hierarchical block.

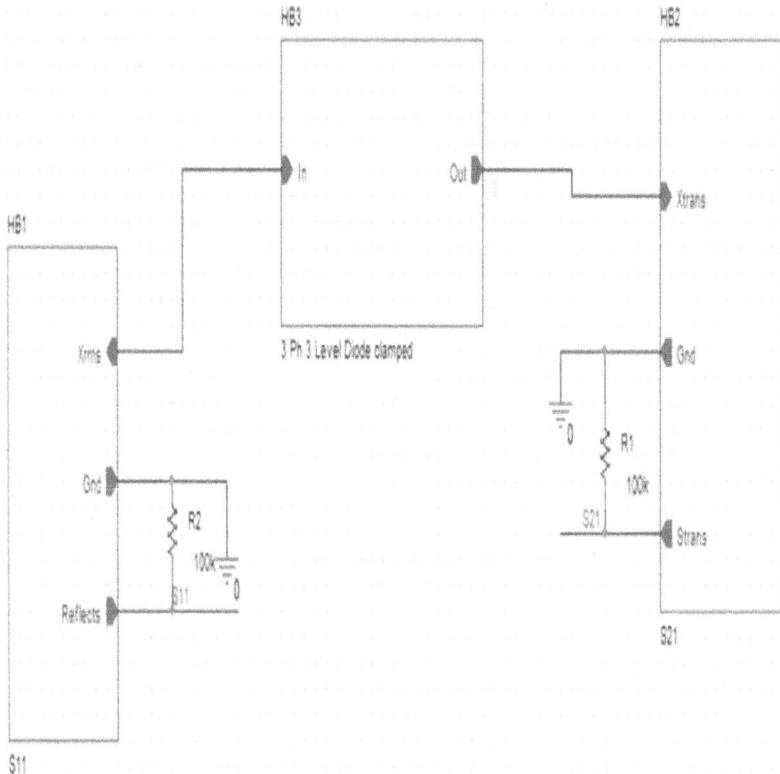

FIGURE 8.120   Multilevel inverter (MLI) circuit connection for S11 and S21 measurement.

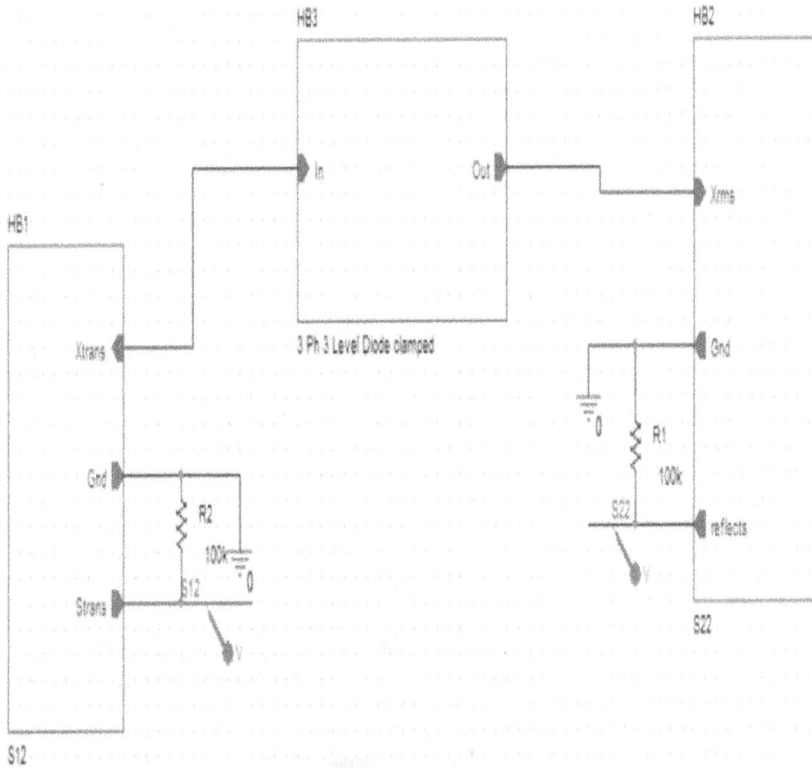

FIGURE 8.121 Multilevel inverter (MLI) circuit connection for S12 and S22 measurement.

8.3.11.4 Simulation Output for S-Parameter

Inference from the graph is shown in Figure 8.122. For S11 and S21, input reflection and forward transmission the waveform get changed after the frequency of about 500 kHz. For S22 and S12, out reflection and reverse transmission the waveform gets changes after 1 GHz.

Scattering parameter is one of the important parameters to find out reflection and transmission coefficient for n port network. In this chapter, we saw the Scattering parameter, uses of Scattering parameter, and how to implement the Scattering parameter in PSpice and finally implement for diode clamped MLI.

8.3.12 EMI Filter Design

8.3.12.1 Line Impedance Stabilization Network

LISN measures conducted emission. Conducted emission is monitored using current probes and voltage probes depending on the circuit application. An LISN is a low-pass filter installed between an AC or DC power source and the EUT (equipment under test) to produce a known impedance and offer an RF noise measurement port. LISNs can be used to forecast conducted emission for diagnostic and pre-compliance testing. Under changing test situations, the power supply's line impedance changes. Decoupling the device terminal from the power line during EMI noise measurements requires a LISN (Figure 8.123).

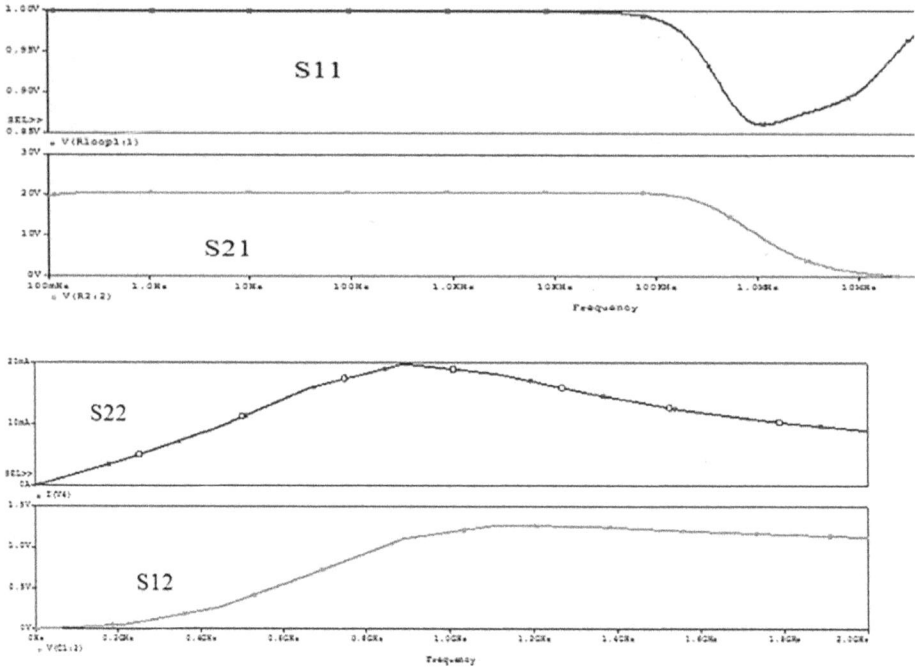

FIGURE 8.122   S11, S21, S22, and S12 waveform for the diode clamped multilevel inverter (MLI).

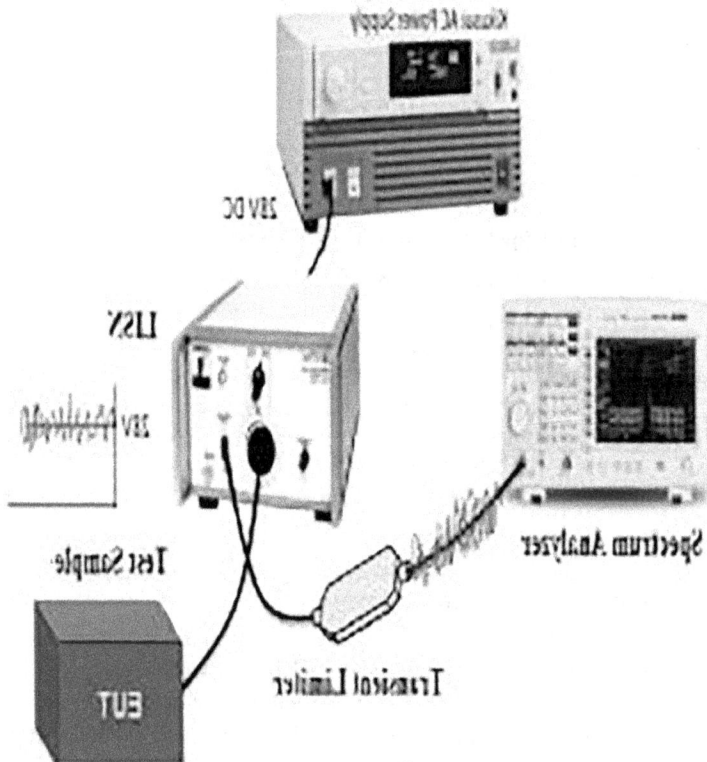

FIGURE 8.123   Basic setip for conducted emission test.

Using a LISN produces problems with input power line impedance (from short circuit to open circuit and also highly inductive input power line). Different converter types are compared. First, the converters' huge input capacitors may resonance with a LISN inductor. Second, the LISN inductor changes the operating point of converters using an input inductor. Producers commonly verify voltage drop limitations (between the source and the DUT), fidelity (I/O identical stable characteristics), and LISN reproducibility.

The LISN contains a low-pass filter for the input power line and a band pass filter for 0.15–30 MHz to stabilize its input impedance as depicted in the Figure 8.124. A CM and differential-mode (DM) noise separator is added to the LISN. A ferrite toroidal inductor makes a low-cost, small LISN.

The LISN contains a low-pass filter for the input power line and a band pass filter for the frequency range of 0.15–30 MHz to stabilize the LISN's input impedance in accordance with the CISPR-22 standard. For measuring both CM and DM noise simultaneously, an EMI separator is added to the LISN. A ferrite toroidal inductor is used to create a cost-effective and compact LISN.

8.3.12.2   Conducted Emission Measurement Using LISN—PSpice

PSpice is a robust general-purpose analogue and mixed-mode circuit simulator used to validate circuit designs and anticipate circuit behaviour. This is especially significant for integrated circuits. In PSpice, the LISN block is not included by default; therefore, to measure conducted emission for

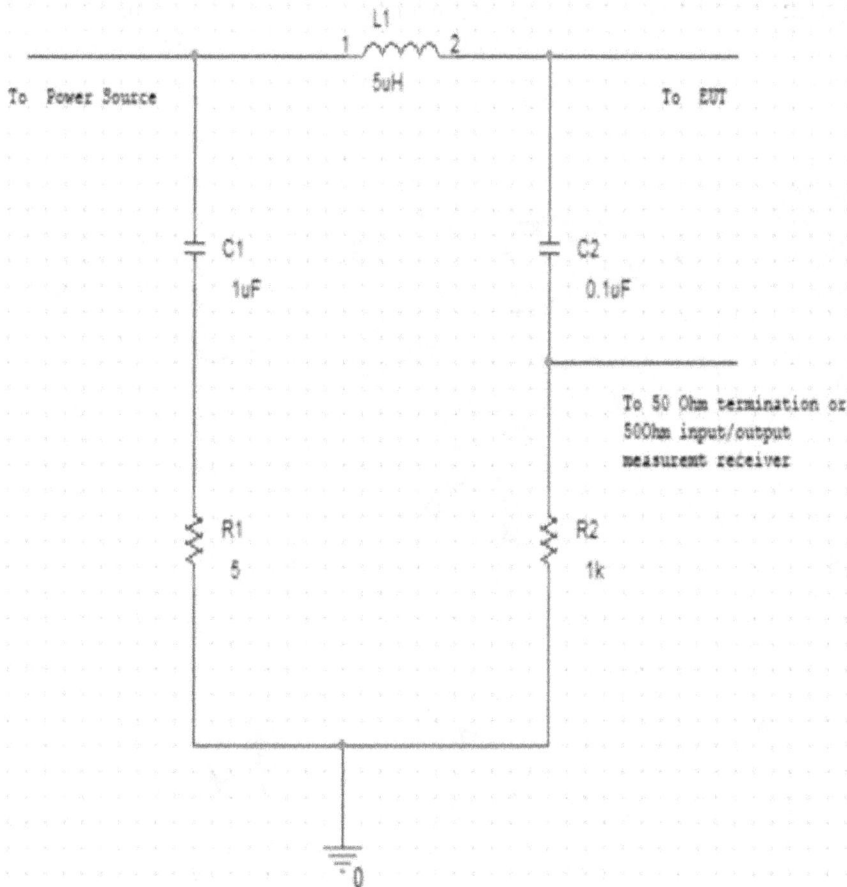

FIGURE 8.124   LISN block diagram.

a circuit, one must create the LISN block using the CISPR-22 standard, connect it to the circuit's input leg, and measure the conducted emission for the proposed network.

8.3.12.3 LISN in PSpice

Below is the procedure to create a LISN block in PSpice

Step 1: Create a hierarchical block for LISN by referring the CISPR 22 Standard (Figure 8.125).
Step 2: Create a library file or hierarchical symbol for LISN for easy usage (Figure 8.126).
Step 3: Save and run the simulation.

8.3.12.4 Conducted Emission Measurement for Diode Clamped MLI

Connected the LISN block in the input positive and negative terminals of the circuits and measured the conducted emission by connecting 50 Ohm resistor and checking the results by using scopes. The simulation circuits for conducted emission measurement for diode clamped MLI using LISN is shown in Figure 8.127 and corresponding emission waveform is shown in Figure 8.128.

Conduction emission output from the positive voltage and the negative leg of the DC input source are as follows.

8.3.13 SIMULATION CIRCUIT—EMI FILTER DESIGN

A network of electrical filters will filter out lower or higher frequency bands while allowing particular frequency bands to pass. Typically, the insertion-loss characteristic describes the fundamental

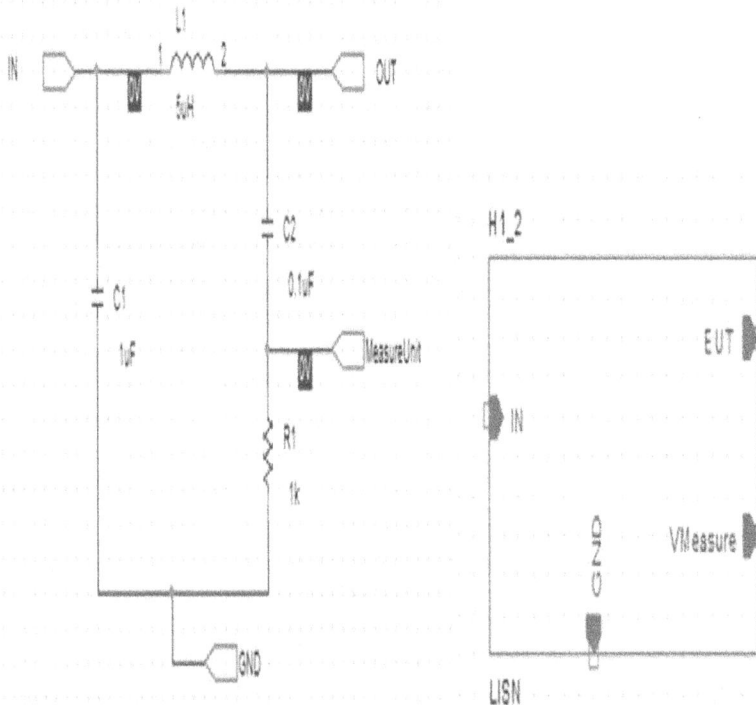

FIGURE 8.125 Inside of LISN hierarchical block and LISN hierarchical block.

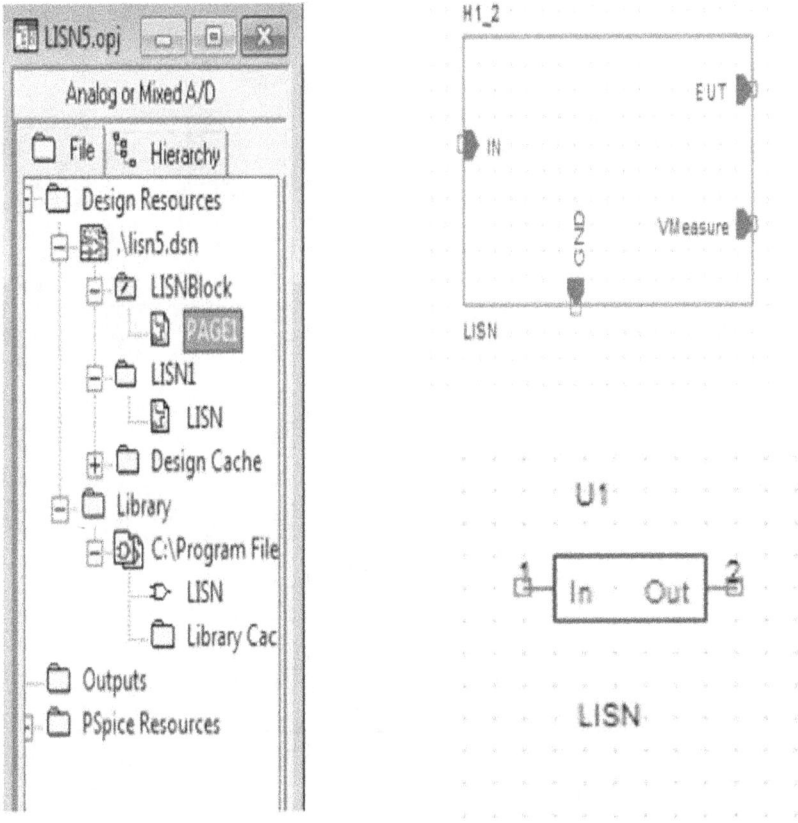

FIGURE 8.126　Inside of LISN hierarchical block and LISN hierarchical symbol – Pspice.

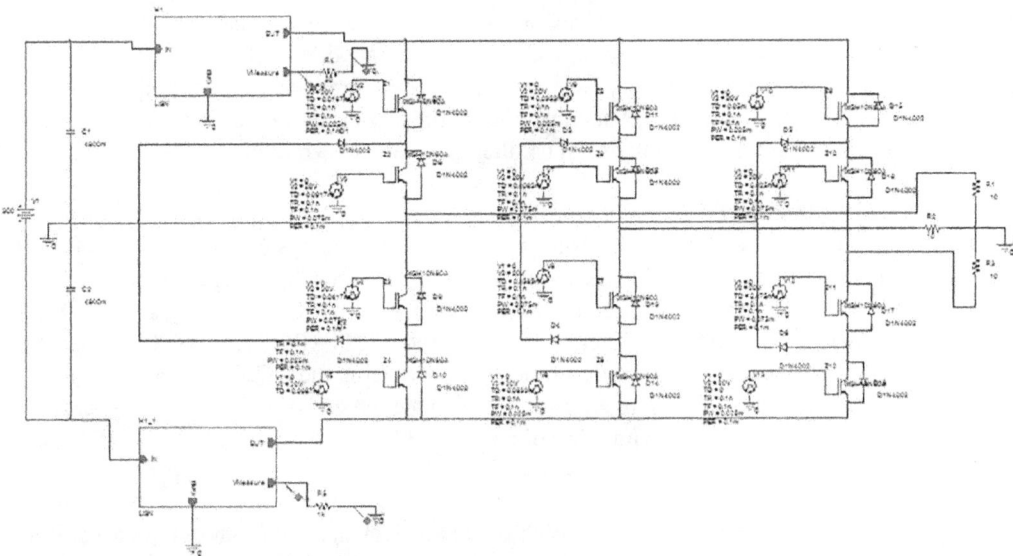

FIGURE 8.127　Diode clamped multilevel inverter connected with LISN.

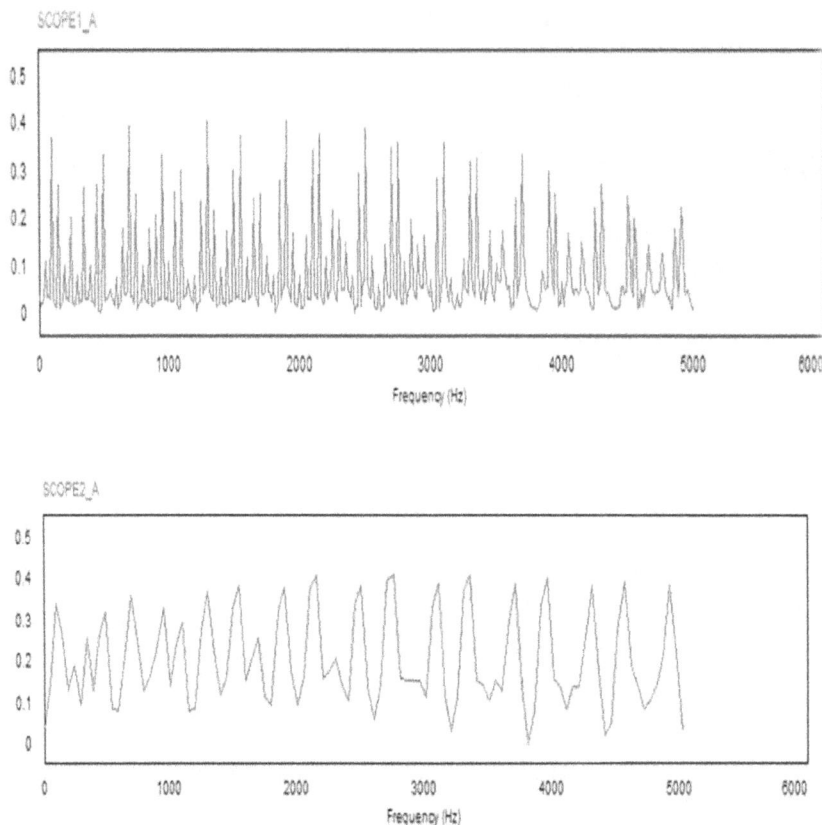

FIGURE 8.128 Line impedance stabilization network (LISN) output (V+) and (V−).

property of an EMI filter. This characteristic is often frequency-dependent and corresponds to the EMI filter's attenuation. Several factors complicate the measurement of insertion loss. EMI filter design engineers consider attenuation, insertion loss, and filter impedance, but a regular wave filter designer considers poles, zeros, group delay, pre-distortion, attenuation, and the order of the filter. Similar to the AC distribution network, DC distribution networks with intermittent renewable energy sources (RES) and variable load demands cause power imbalances and produce a voltage variation in the DC bus. By attaching a suitable filter to the input side, we may lower the imbalance and, consequently, the conducted emission. The filter types are shown in Figure 8.129.

8.3.13.1 DC Filters

The input side of the inverter is DC, as DC also has imbalance in the system, which affects the efficiency of the converter; hence, DC filter is designed for the converter. There are two types of filters available, namely, AC filter and DC filter, as shown in the Figure 8.130. In DC filters, there are two categories: one is APF (Active Power Filtering) and the other is PPF (Passive Power Filtering). APF may dynamically minimize DC voltage fluctuations by paralleling the DC side voltage converter. L filter is used in the simulation circuit to eliminate the conduction emission in the APF type. L filter will give more accuracy compared with other filters.

8.3.13.2 *L* Filter

The L type is the most common filter type. With only two elements, the L filter offers a 12 dB loss per octave. All of these figures represent the loss beginning above the cut-off frequency. If the load has switchers, a single L filter performs optimally in the DC mode. In a double L filter, as in

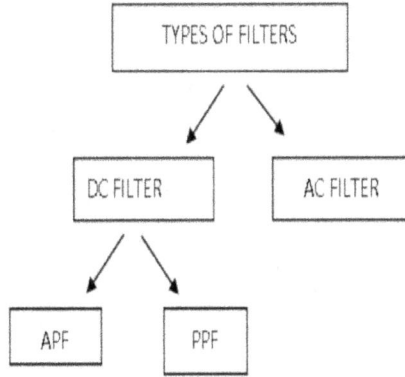

FIGURE 8.129 Types of filters.

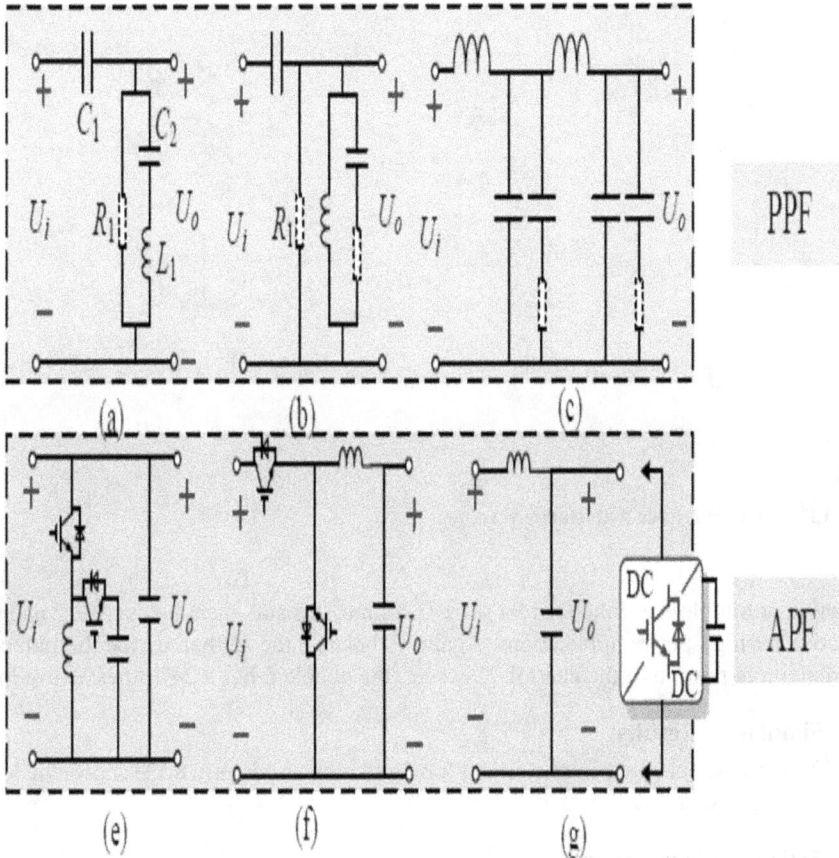

FIGURE 8.130 Basic direct current (DC) filters available.

Figure 8.131, the sum of the inductor and capacitor values required for the same amount of loss would be less than in a single L filter. A smaller output capacitor might not provide sufficient energy storage for the subsequent switcher.

The capacitor reactance must be less than the switcher's input impedance. As long as the drop is not extreme or the switcher frequency is high enough, the double L may be utilized. Even when

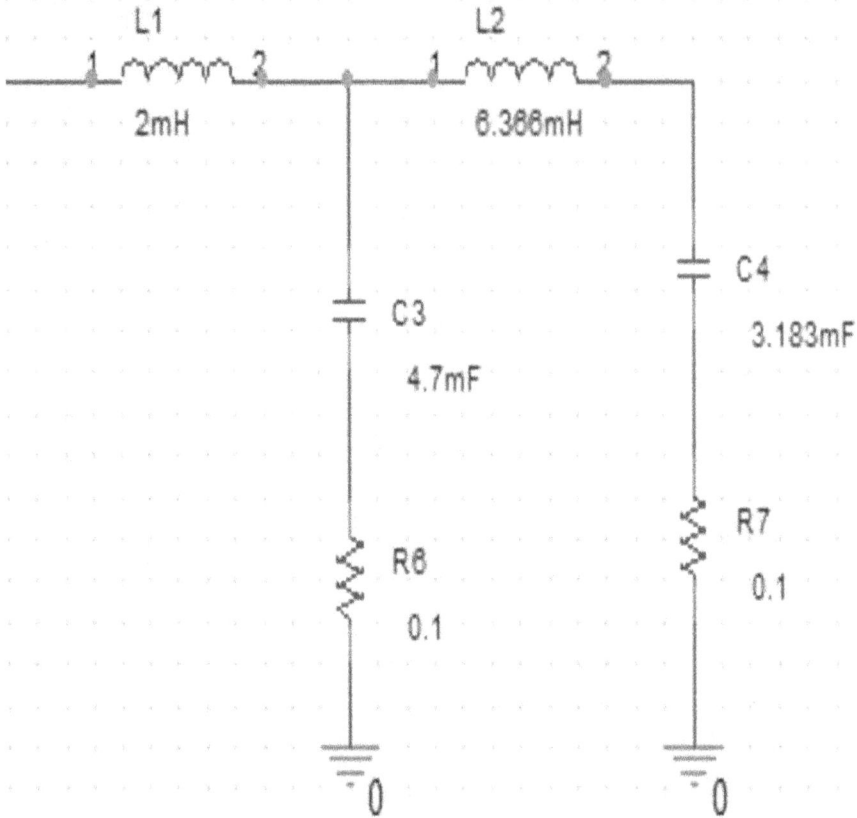

FIGURE 8.131 Double L filter with standard values.

the attenuation or filter loss is enhanced for the DC source, this statement holds true. L and multiple L are effective for high-power applications. Again, to balance the L filter, divide the inductors and place the remaining portion in the neutral. However, the double L has a 24 dB per octave loss.

8.3.13.3 Simulation Results

Figure 8.132 shows the entire schematic circuit for simulation and Figure 8.133 shows the schematic representation of L filter design.

8.3.13.4 Emission Measurement

a. **Without Filter**

Conducted emission waveform without filter—summing amplifier output is shown in the Figure 8.134.

b. **With Filter**

Output conducted emission waveform with filter—By adding L filter shown in Figure 8.135.

Output conducted emission waveform with filter—By adding Double L filter shown in Figure 8.136.

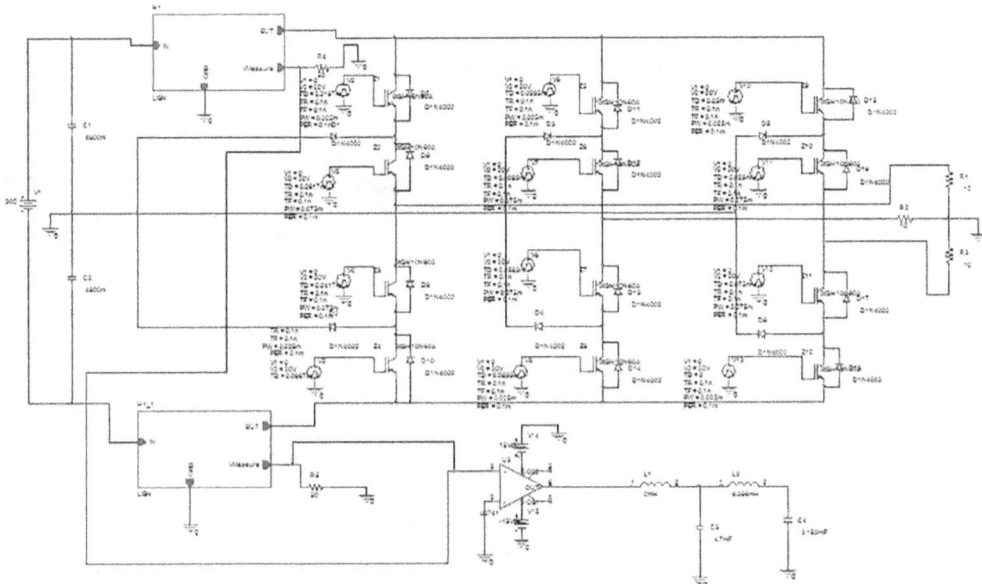

FIGURE 8.132 NPMLI with double *L* filter simulation circuit.

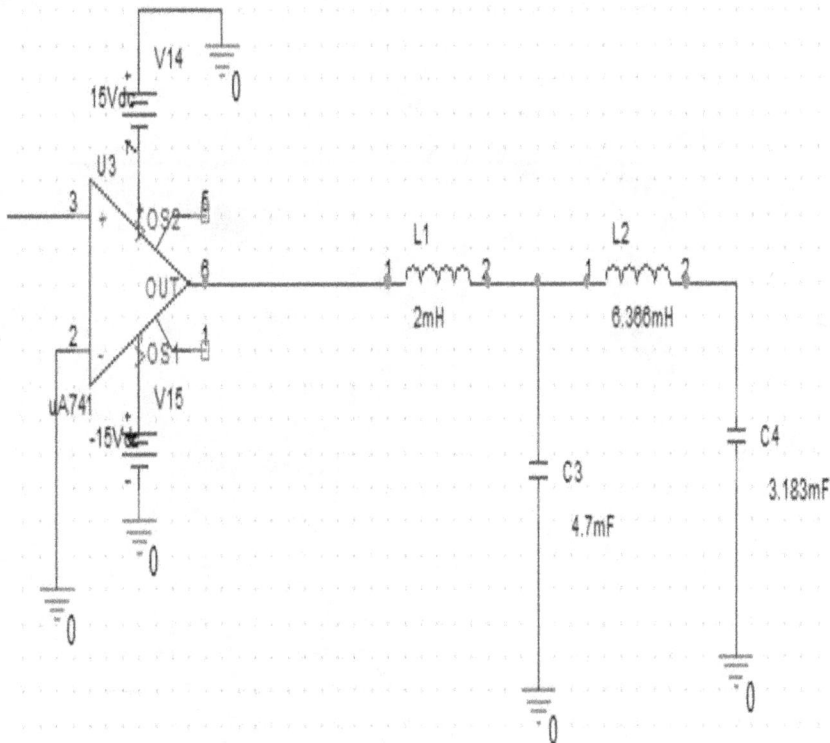

FIGURE 8.133 Double *L* filter design.

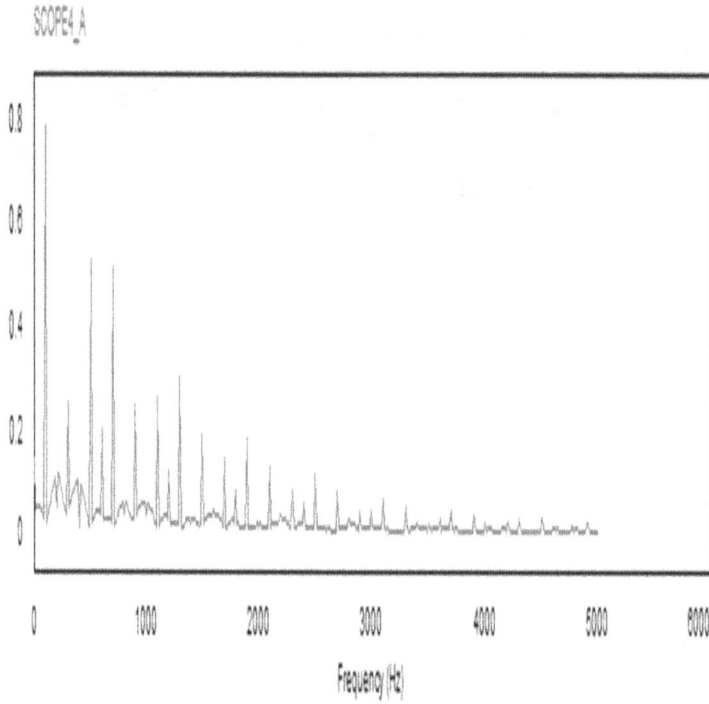

FIGURE 8.134   Conducted emission waveform without filter.

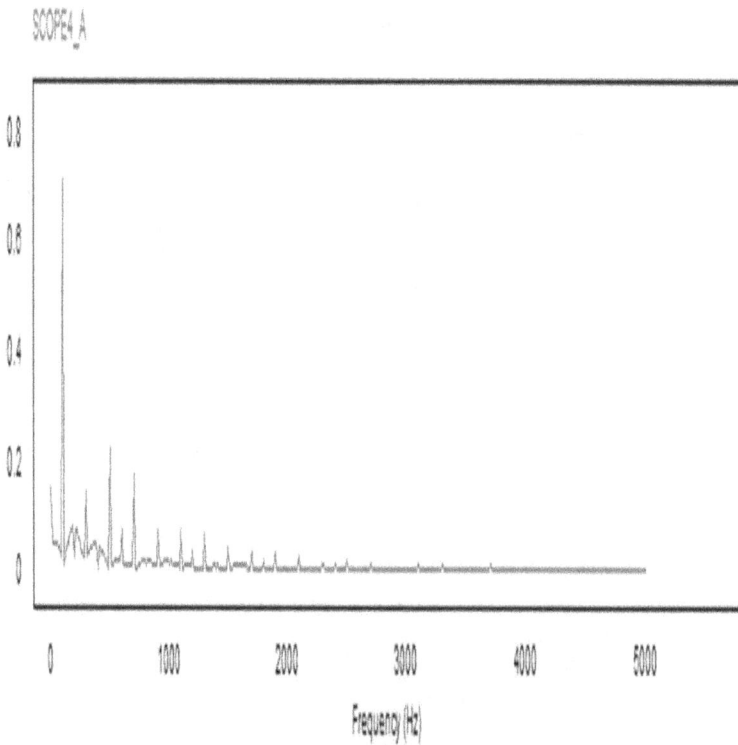

FIGURE 8.135   Conducted emission waveform by adding L filter.

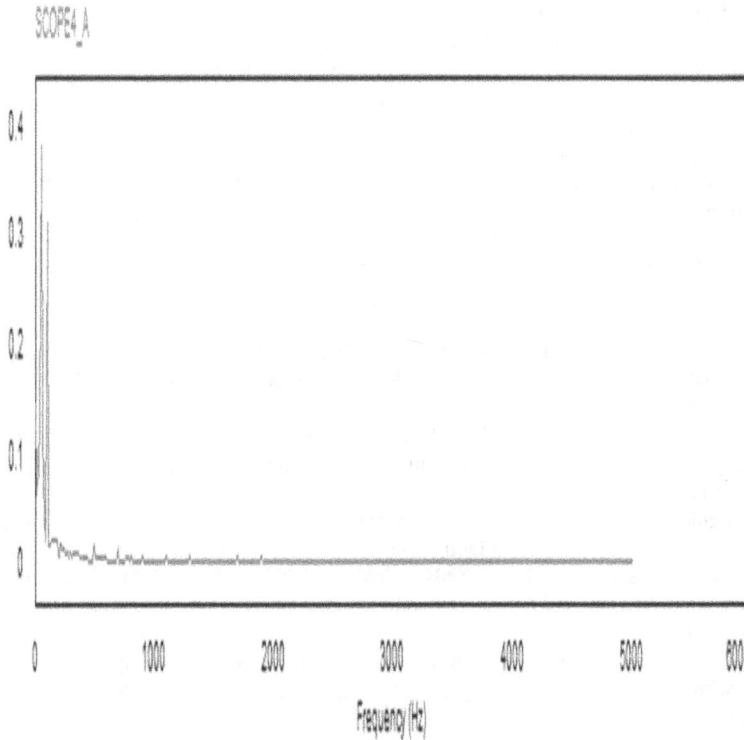

FIGURE 8.136   Conducted emission waveform by adding double L filter.

Inference: While comparing with and without filter, the attenuation in conducted emission by adding filter gets reduced. For higher frequency and for the circuit having more number of switches, double L filter is most suitable and preferred. Using Double L filter gives better results when compared with L filter. Hence, multiple L filter network is best choice while using increased number of voltage levels.

8.4   CONCLUSION

In this chapter, three-phase three-level diode clamped MLI of 600 V input DC source for grid connected application of 2 kw power is simulated in the PSpice software. The simulation results are analysed for every block in the inverter circuit. SPWM method is used; hence, sinusoidal waveform is obtained in the output side. For two port network analysis, Scattering parameter values are measured, and with the help of LISN conducted emission is measured. For the measured conducted emission, L filter is designed for the simulation circuit and reduces the conducted emission. And clearly explained about the S-parameters, the design and the implementation of the S-parameter setup to obtain the reflection and transmission coefficients of the S-parameter matrix. Conducted emission is measured using LISN and a spectrum analyser. Noise reduction techniques by means of filters and the implementation of such a designed topology to reduce the conducted emission and the resulting waveforms of all the topologies are simulated using the PSpice tool.

BIBLIOGRAPHY

1. MD. Halim Mondol, Mehmet Rida Tur, Shuvra Prokash Biswas, Kamal Hosain, Shuvangkar Shuvo, and Eklas Hossain, Compact Three Phase Multilevel Inverter for Low and Medium Power Photovoltaic Systems. *IEEE*. March 2020. DOI: 10.1109/ACCESS.2020.2983131.

2. Wanchai Subsingha, *Design and Analysis Three Phase Three Level Diode-Clamped Grid Connected Inverter*. 2016. IEEE. https://doi.org/10.1016/j.egypro.2016.05.019
3. R. Dharma Prakash and Joseph Henry, Analysis of Switching Table Based Three Level Diode Clamped Multilevel Inverter. *IJSER International Journal of Scientific & Engineering Research*, vol. 5, no. 4, April 2014, 102. ISSN 2229–5518.
4. https://www.pspice.com/resources/application-notes/obtain-s-parameter-data-probe-window
5. https://education.ema-eda.com/iTrain/PSpice163/pspiceTOC.html. PSpice application note by EMA Design automation company.
6. <A>S-Parameter Measurements application note by Keysight technologies. https://www.keysight.com/in/en/assets/7018-05986/application-notes/5992-2693.pdf
7. Richard Lee Ozenbaugh and Timothy M. Pullen, *EMI Filter Design*. Third Edition. CRC Press, US, Dec 2017.
8. Jianquan Liao, Niancheng Zhou, and Qianggang Wang, "Design of Low-Ripple and Fast-Response DC Filtersin DC Distribution Networks". Received: 28 September 2018; Accepted: 9 November 2018; Published: 12 November 2018.
9. Maddu Srinivasa Rao, D. Nagendra Rao, V. Usha Shree, "Design and Development of Compact Line Impedance Stabilization Network for Measurement of Conducted Emission". *IJERECE*, vol. 4, no. 11, November 2017.
10. PSpice user guide – by ORCAD. https://resources.pcb.cadence.com/i/1180526-pspice-user-guide/25?
11. Rajdeep Bondade, Yikai Wang, and Dongsang Ma, "Design of Integrated Bipolar Symmetric Output DC-DC Power converter for Digital pulse Generators in Ultrasound Medical Imaging Systems", Manuscript Submitted to *IEEE Transaction on Power Electronics*, IEEE, 2013. DOI: 10.1109/TPEL.2013.2264511.
12. X. Zhou, Y. Wang, L. Wang, Y.-F. Liu, and P. C. Sen, "A Soft-switching Transformerless DC-DC Converter with Single Input Bipolar Symmetric Outputs", *IEEE Transactions on Power Electronics*. DOI: 10.1109/TPEL.2020.3048230.
13. https://www.pspice.com
14. Zhi-cheng Wang, Xiaowen XuXu, Zhi-yong Lu, *The EMI Filter Design in Pspice*. IEEE-2013. DOI: 10.1109/ICMTCE.2013.6812471.
15. Shuo Wang, Fred C. Lee, and Williem Gerhardus Odendall, "Characteristics and Parasitic Extraction of EMI Filters Using Scattering Parameters". *IEEE*, March 2005. DOI: 10.1109/TPEL.2004.842949

9 EMI Using ADS

9.1 ABOUT THE SOFTWARE AND INTRODUCTION

Advanced Design System (ADS) is a premier high-speed design platform that is widely used for implementing and analysing Radio Frequency (RF) and Electro Magnet (EM) problems in power electronic circuits. It includes powerful simulation tools for analysing the circuit. This chapter explains the ADS software that is being used for simulation. It clearly showcases how to get started with the software, the simulation schematic window, component placement, connection techniques, several parameters required for simulation, and how to build and run the design to get waveforms on the output data window.

9.1.1 HISTORY

Advanced Design System is an Electronic Design Automation (EDA) tool by Keysight Technologies. For RF electronic products, including mobile phones, pagers, wireless networks, satellite communications, radar systems, and high-speed data lines, it offers an integrated design environment. To assist you in getting started more quickly, PathWave ADS provides integrated design guidance through templates. It is simple to find the part you need thanks to extensive component libraries. When creating schematic drawings, automatic sync with layout enables you to see the physical layout (Table 9.1).

9.1.2 GETTING STARTED WITH ADS

Figures 9.1–9.8 show how to start with the ADS software environment in a sequential manner after installing it.

If you are new to the ADS environment, you can choose the option "New to ADS," which proceeds further with a required tutorial about the placement of the components, connection procedure, and the simulation procedure. From this, one easily understand the basics of the work environment. The following figures show the basic work flow to simulate any topology (Figures 9.3–9.7).

If you are familiar in working with the ADS environment, you can choose the option "Familiar with ADS," which proceeds further to create a new workspace or to open an existing workspace. All the recent workspaces created will be displayed in the window, so that you can directly choose the workspace that you need to work with (Figure 9.8).

TABLE 9.1
About ADS

Developer	**Keysight Technologies PathWave Design**
Initial release	1985 (Microwave Design System (MDS))
Operating system	Windows, Linux, Solaris
Platform	ADS
Type	Electronic circuit simulation

FIGURE 9.1 Information window.

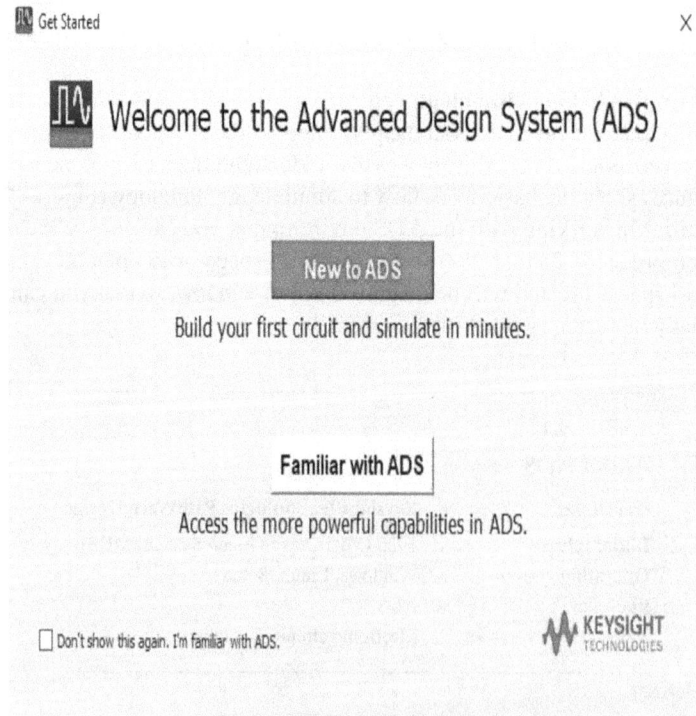

FIGURE 9.2 Tab to choose between new and familiar.

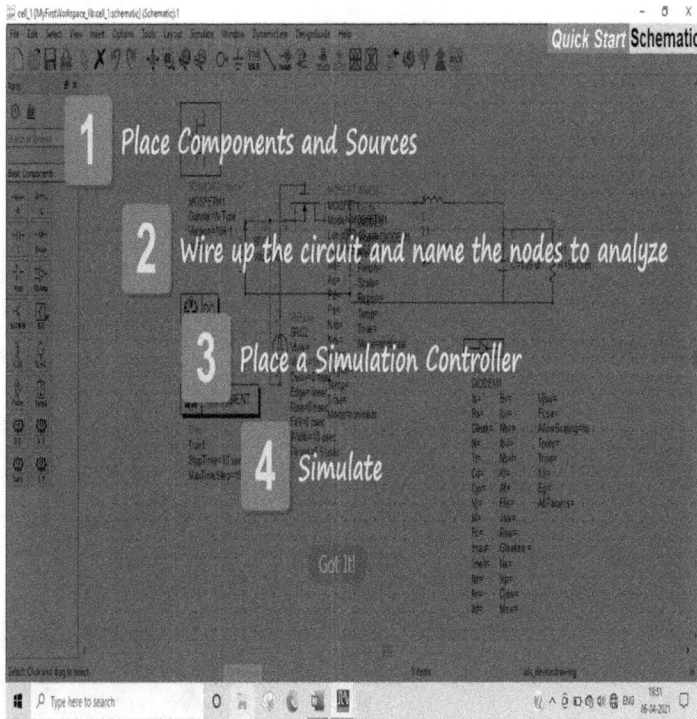

FIGURE 9.3 Basic workflow in the advanced design system (ADS) environment.

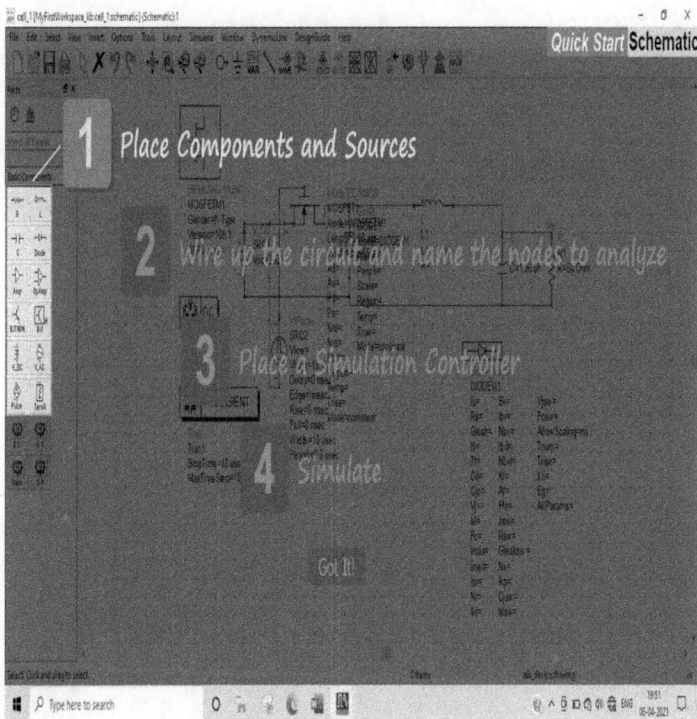

FIGURE 9.4 Palette to get the basic components.

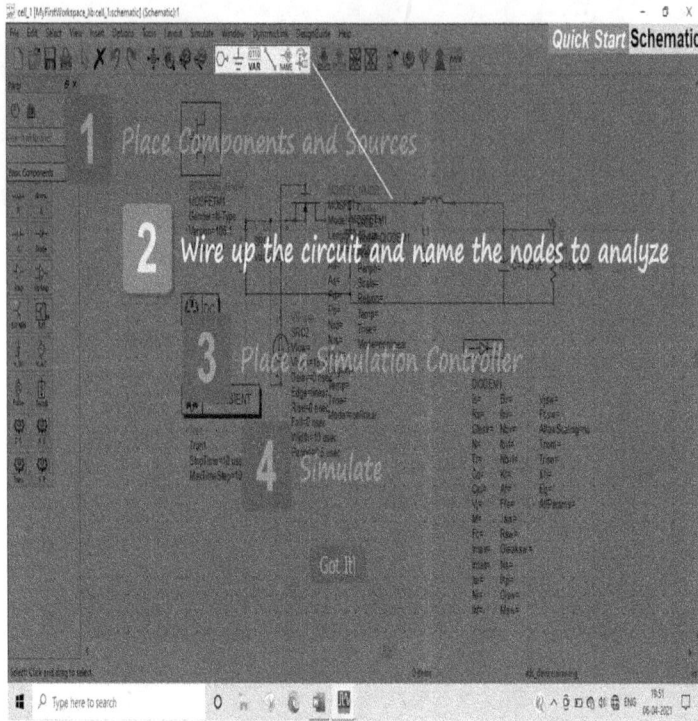

FIGURE 9.5 Tool bar for wiring and naming the components and nodes.

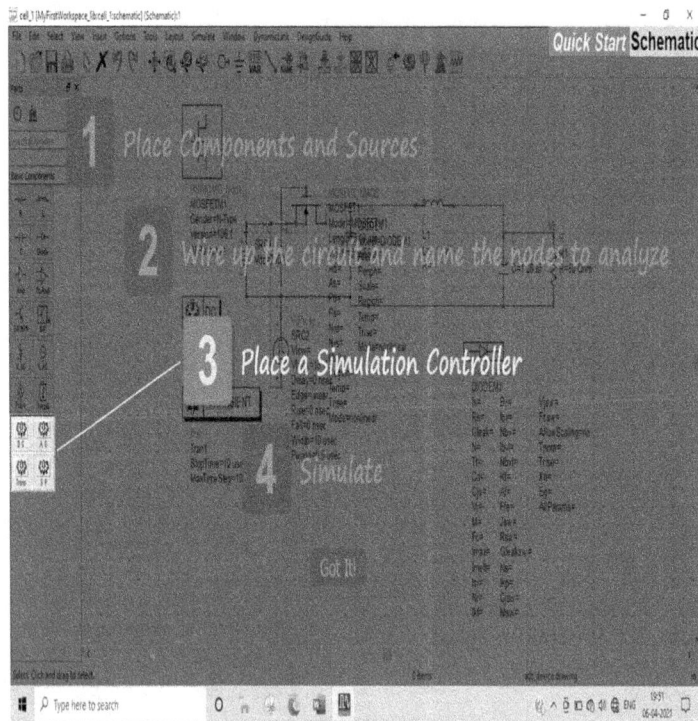

FIGURE 9.6 Palette to get the simulation controller.

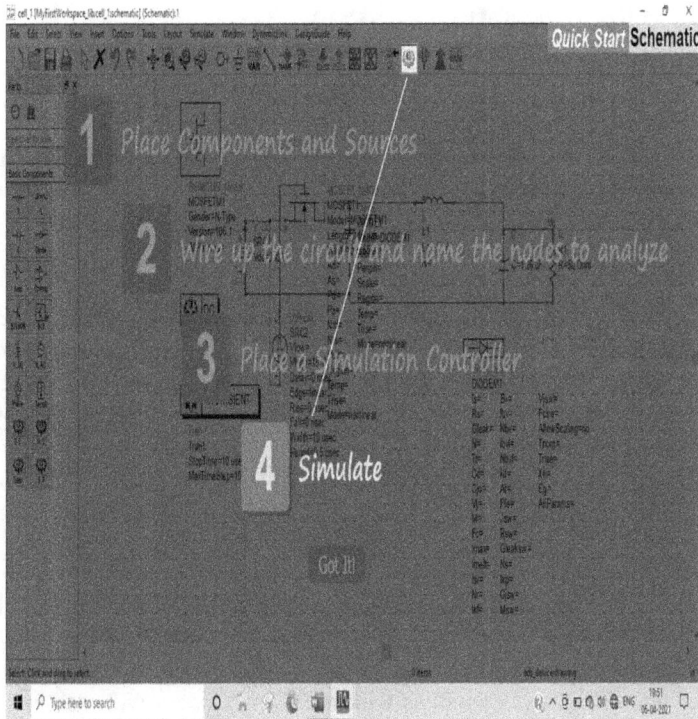

FIGURE 9.7 Icon to build and run the proposed topology.

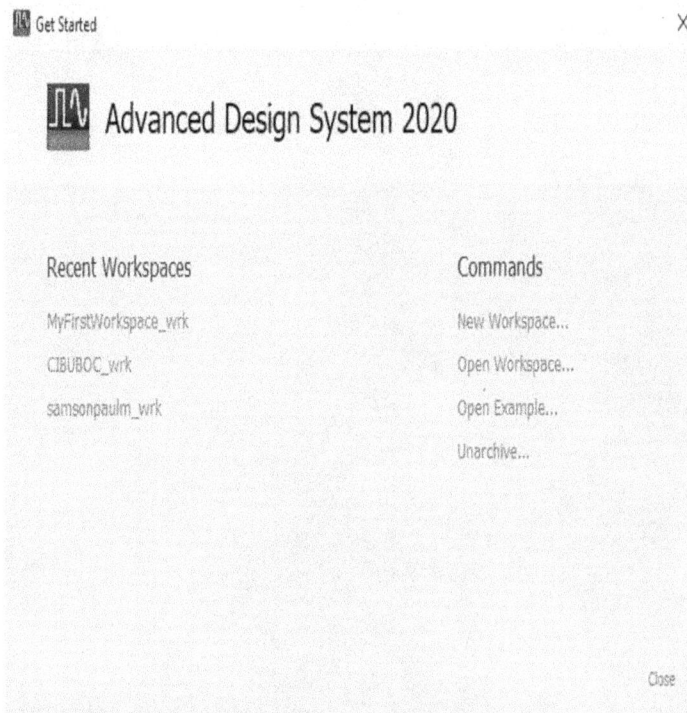

FIGURE 9.8 Tab to choose between new and existing workspace.

9.1.3 Creating a New Workspace

A workspace is one where the entire designed circuit is implemented, and it is simulated to get the desired results in a separate data display window. It contains the Schematic, Symbol, Layout and the Data Display Windows each having its own purpose.

Figures 9.9–9.12 show the steps involved in creating a new workspace for simulating a topology and opening an existing workspace.

9.1.4 Creating a New Schematic Window

A cell is one that contains the schematic window for implementing the designed circuit architecture and simulating the topology to obtain the results displayed in the data display window.

Figures 9.13–9.15 show the window, which is meant to create a new cell by clicking the "New Schematic Window" and the existing cells within the workspaces will also be displayed in the same window.

9.1.5 Creating a New Symbol

A Symbol is one that is created to represent the implemented circuit topology as a sub-circuit, literally a symbol, and it can be simulated in the same manner as mentioned earlier and the results could be viewed from data display window.

Figures 9.16–9.18 show the window that is meant to create a new symbol by clicking the "New Symbol Window" and the existing symbols within the workspaces will also be displayed in the same window.

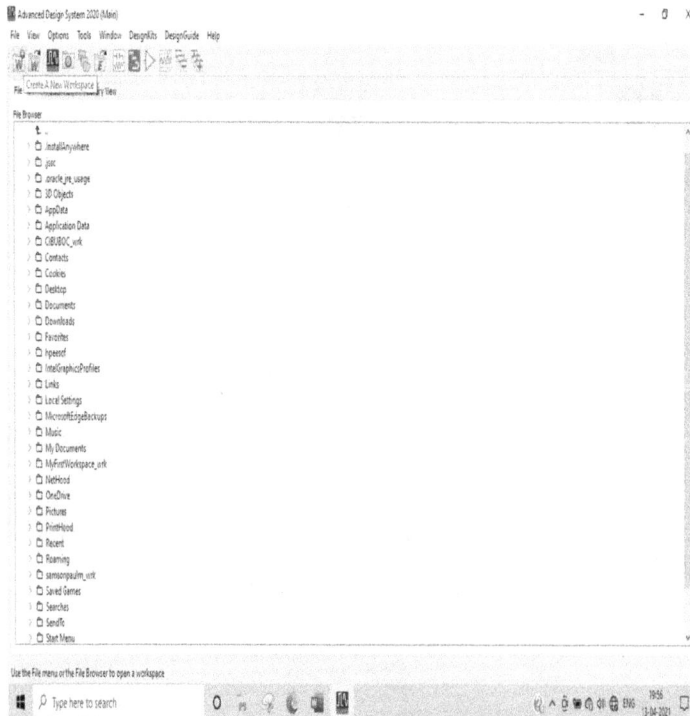

FIGURE 9.9   Icon to create a new workspace.

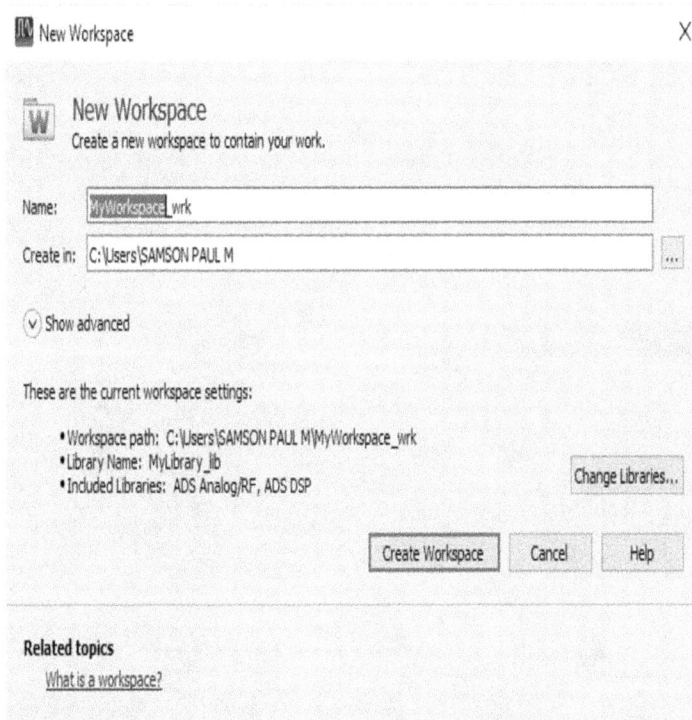

FIGURE 9.10 Tab to give a name to the new workspace.

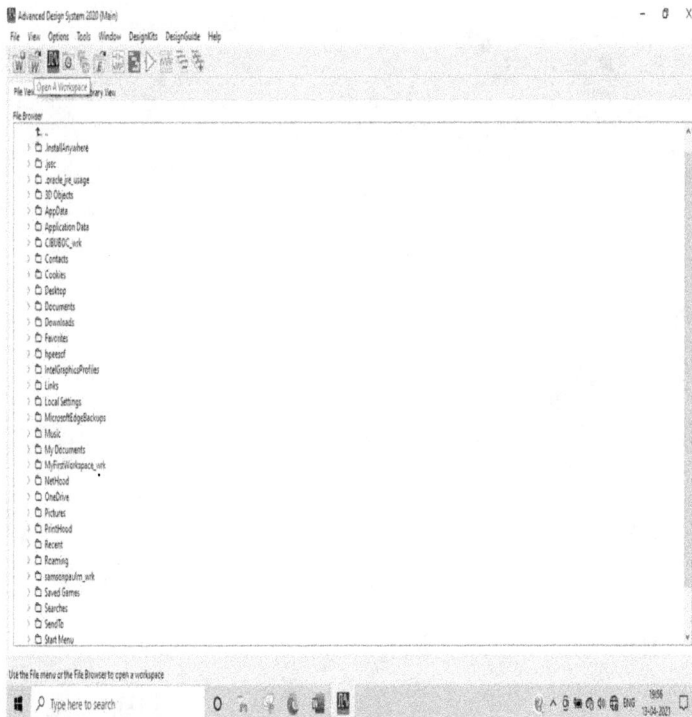

FIGURE 9.11 Icon to open an existing workspace.

FIGURE 9.12 New workspace.

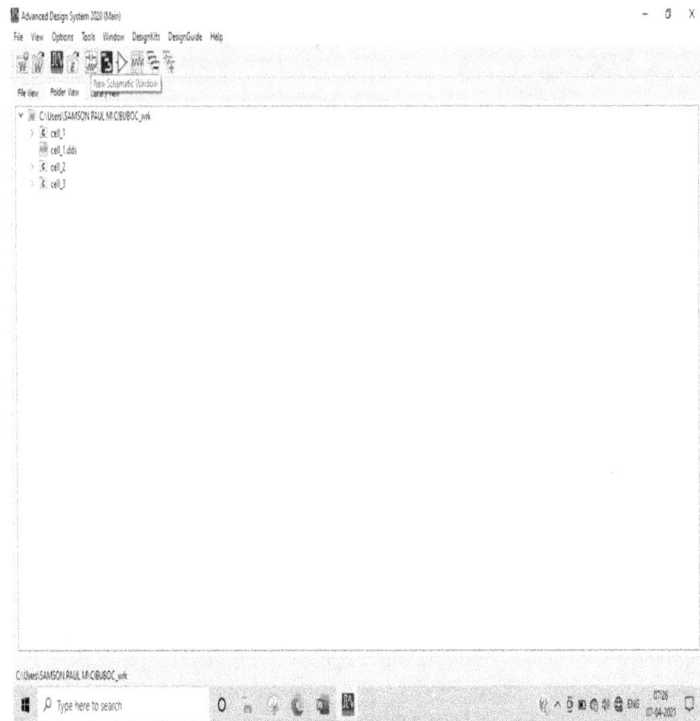

FIGURE 9.13 Tab to create a new cell or to open an existing cell.

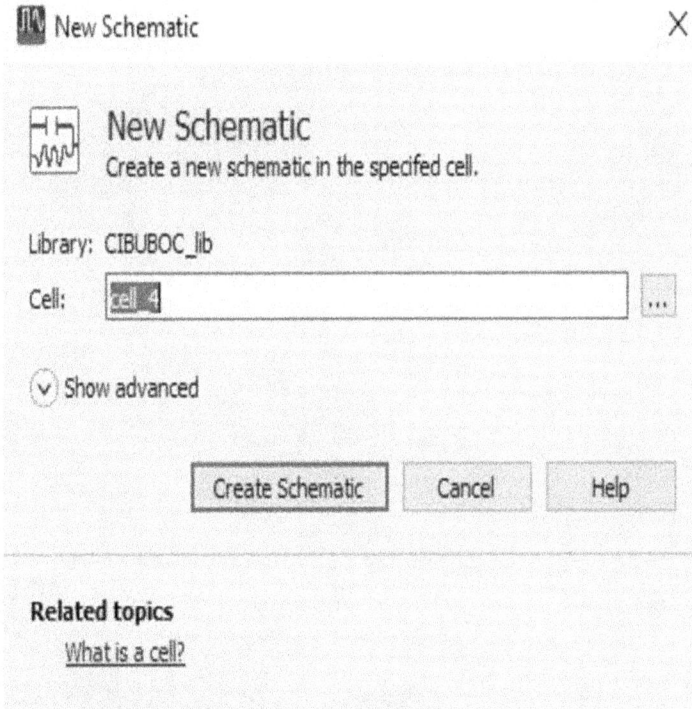

FIGURE 9.14   Tab to give a name to the new cell.

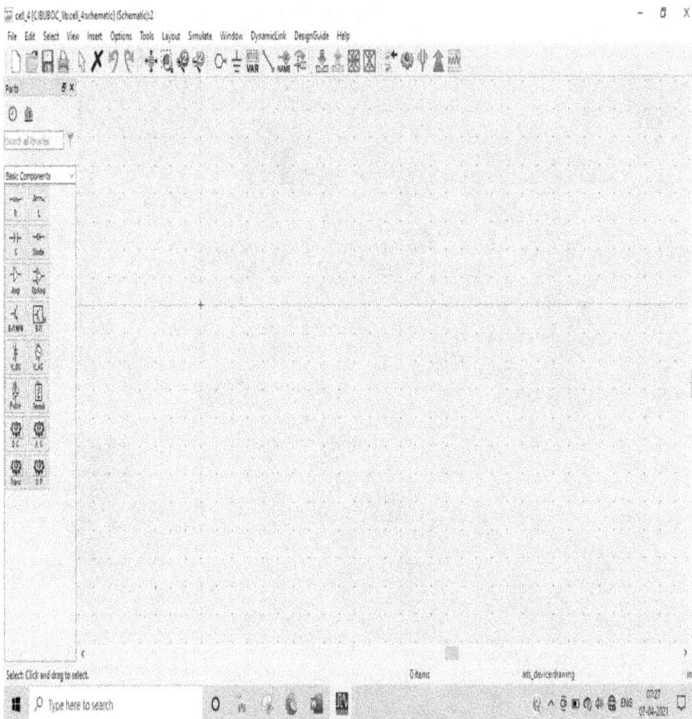

FIGURE 9.15   New schematic window.

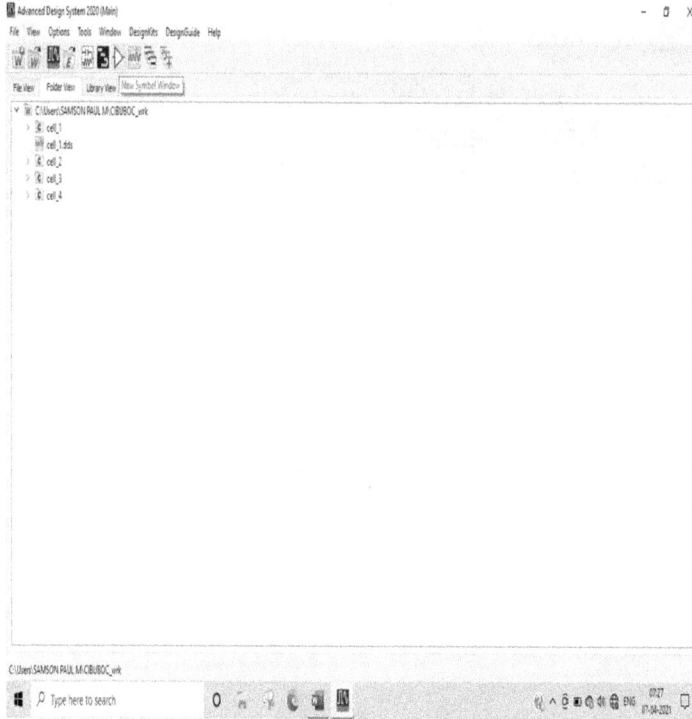

FIGURE 9.16   Tab to create a new symbol or to open an existing symbol.

FIGURE 9.17   Tab to give a name to the new cell.

FIGURE 9.18 New schematic symbol window.

9.1.6 LIBRARIES

The ADS LIBRARY contains several types of components and their equivalent models are grouped according to their functionalities and characteristics. Those components are available at the left end of the schematic window in a drop-down dialogue box. The dialogue box has the components grouped like basic components, lumped components, probe components, simulation-S_Param, Devices, Filter DG_All, etc. The group should be chosen based on the components required for the implementation of circuit.

Figure 9.19 shows the drop-down dialogue box containing the components group. Once the required group is selected, the components belonging to that group will be displayed in the same dialogue box and the respective component can be selected and placed.

The basic components like resistor, capacitor, inductor, alternating current (AC) and direct current (DC) voltage sources, diode, BJT, pulse generator, etc., are readily placed at the left side of the newly created schematic window shown in Figure 9.20.

There is also a search option in the library shown in Figure 9.21, which helps to easily search the required component to be placed in the schematic window. Based on the search component, all the related components and their equivalent models will be displayed on the left side of the schematic window.

FIGURE 9.19   Components palette in advanced design system (ADS).

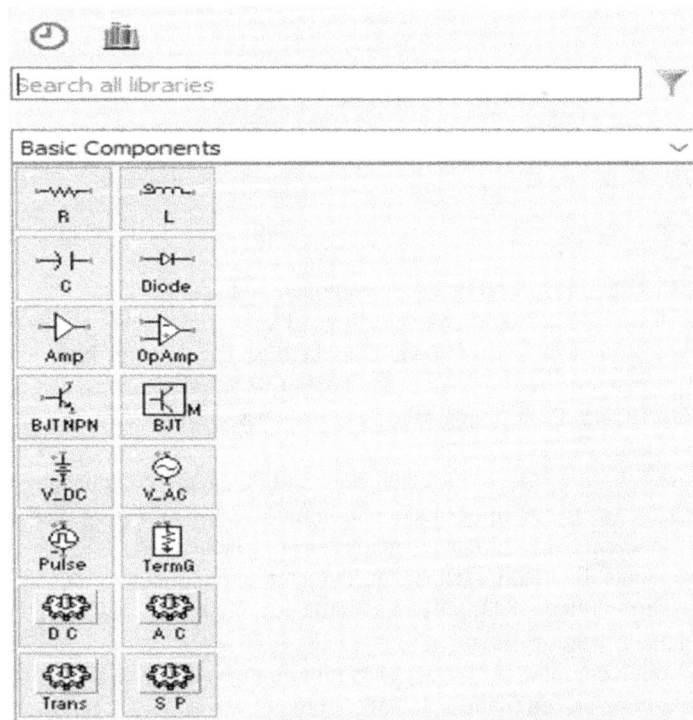

FIGURE 9.20   Basic components.

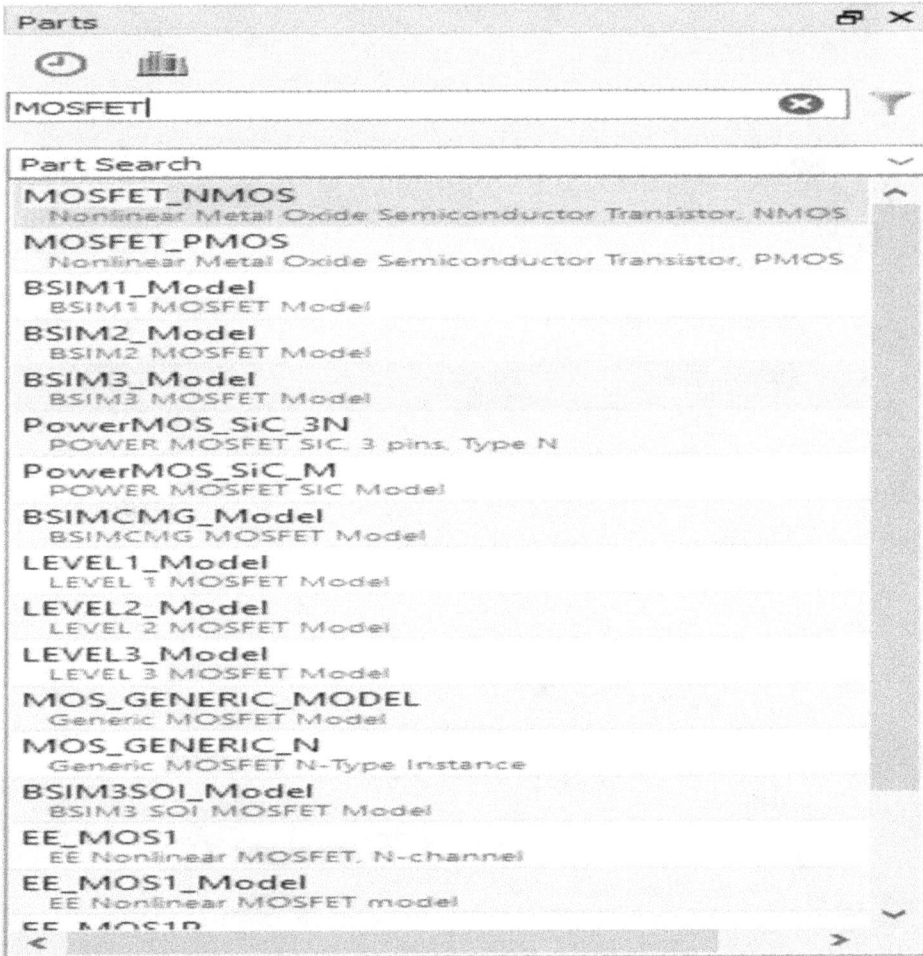

FIGURE 9.21 Components search library.

The Library Browser icon ![icon] is used to open the library browser of ADS, which contains all the components present in the ADS software and the cells created to implement the circuit.

Figure 9.22 shows the component library window opened after clicking the Library Browser icon. In the window that opened, there is an option named "Download Libraries," which can be used to download some new components unavailable in the library from the Key Sight Website and to add the downloaded component to the Netlist. Also, we can enter the values of some specific components based on the data sheet and save the components for further usage.

9.1.7 COMPONENT MODELS

The ADS environment allows you to place the equivalent models of the switching devices used in the implemented circuit, which enables the switching devices to function properly. These models are by default present in the ADS library. For some switching devices, the models may not be present by default. In such cases, the equivalent models can be downloaded from the Key Sight Website and can be included in the Netlist. Placement of the equivalent models is a very much important aspect in terms of simulating the topology. If an equivalent model for a switching device is not

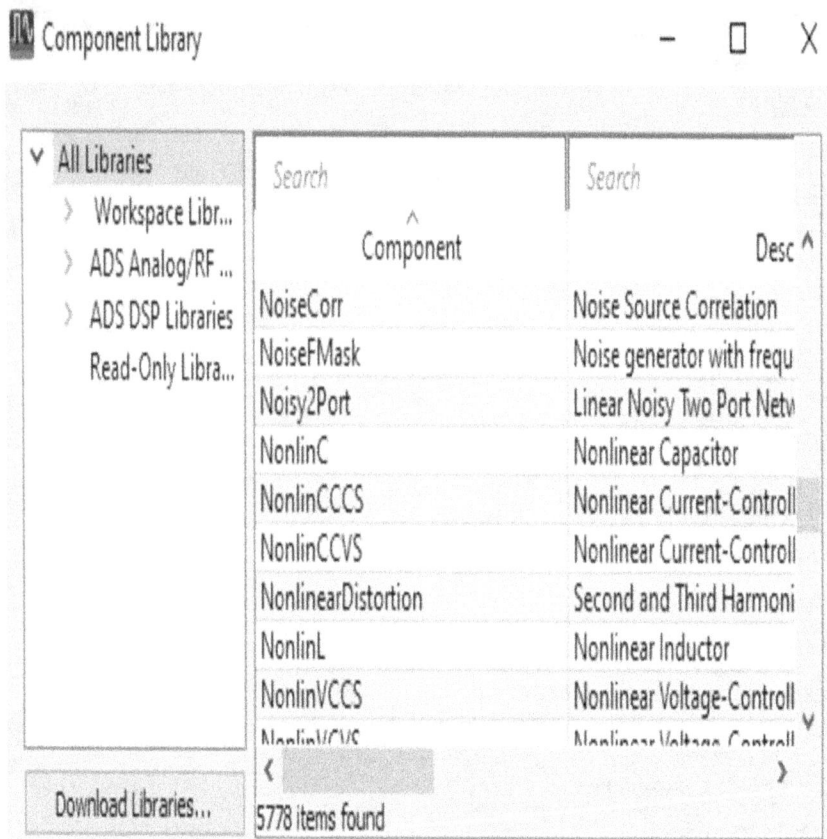

FIGURE 9.22   Component library.

placed, then the simulation throws an error message like "Instance of an Undefined Model". The name of the switching device and its equivalent model should be the same to match the device and its model during simulation.

Figure 9.23 shows an example of a switching device Metal-Oxide-Semiconductor Field-Effect Transistor (MOSFET) and its model present in the ADS library.

In the Instance parameter window, as shown in Figure 9.24, this can be obtained by double click-ing the device where we can see the instance name and the model name. The model name should be the same as that of the model you have added.

3. Similarly, Instance parameters of the insulated gate bipolar transistor (IGBT) model are obtained and in that there are different parameters like internal resistance, temperature, T_{rise}, etc. As in Figure 9.25, these data can be obtained from the datasheet of the specified rating of the device and the values can be entered.

4. Similarly, we have to enter the values of each component from the datasheet here. In resistor, the data can be entered from the datasheet as in Figure 9.26, depending on the current flow across the resistor.

5. Inductor values are considered because they are one of the key parameters of the filter. The values of the inductor as in Figure 9.27 should be obtained from the design and we also need to verify whether we have the inductor value in the market and we have to check the availability of the components. After that, we can enter the instance parameter values of inductor from the datasheet.

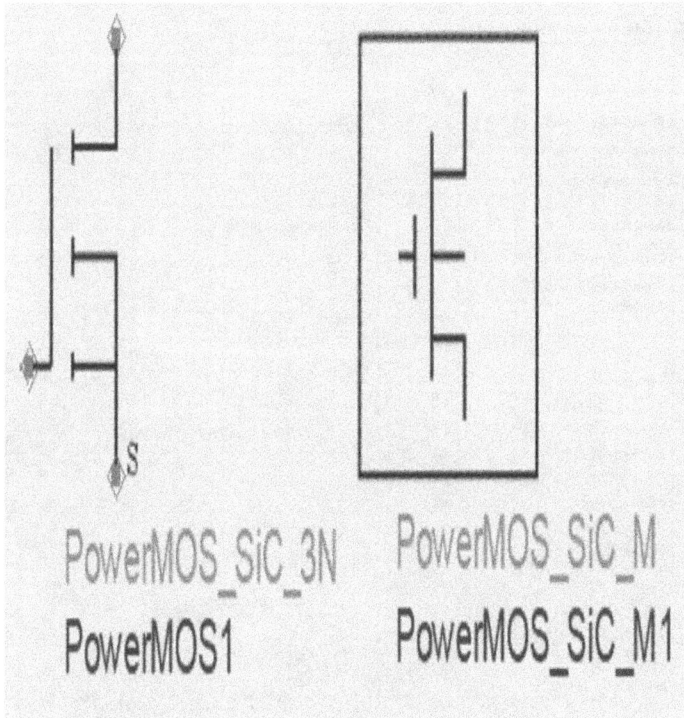

FIGURE 9.23   MOSFET and its equivalent model.

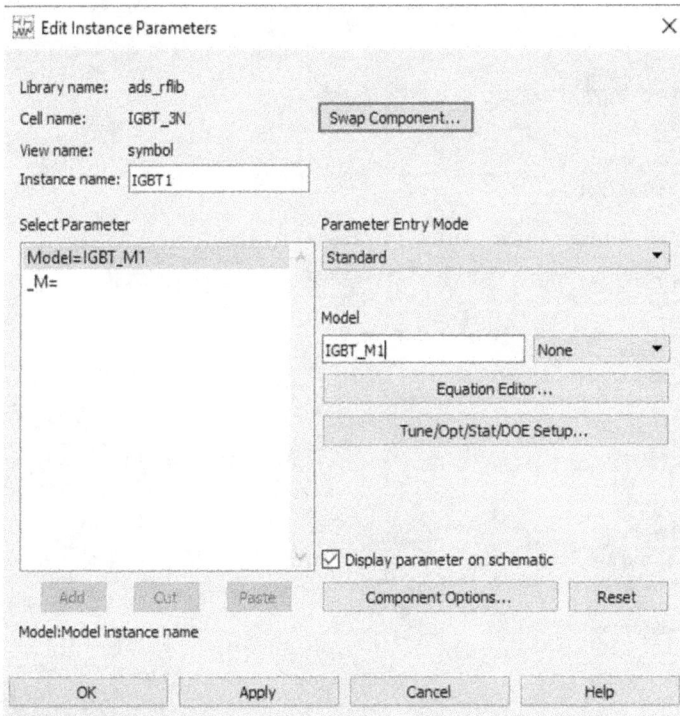

FIGURE 9.24   Instance parameter window.

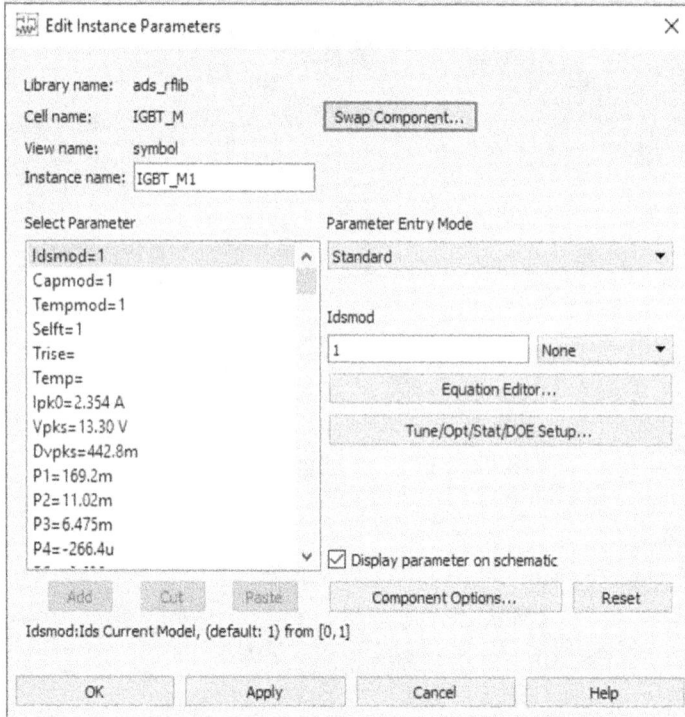

FIGURE 9.25   Instance parameter insulated gate bipolar transistor (IGBT).

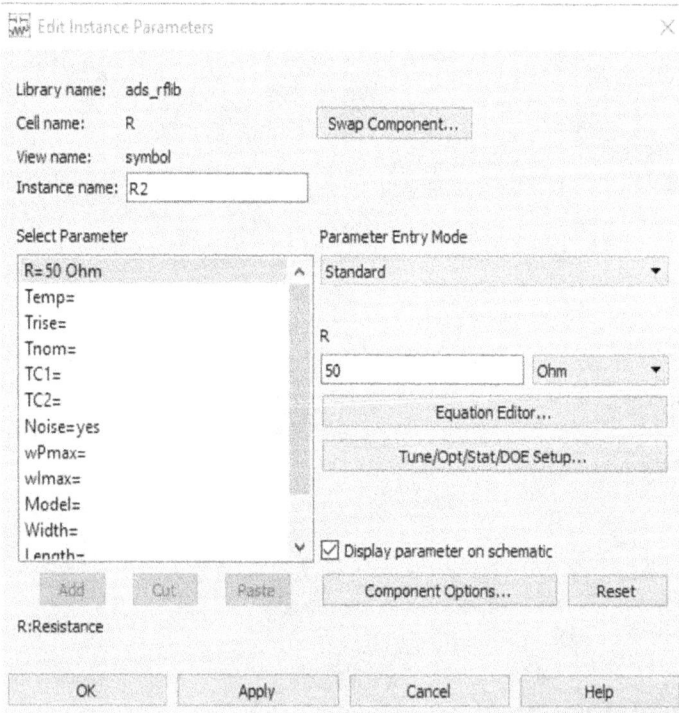

FIGURE 9.26   Instance parameter resister.

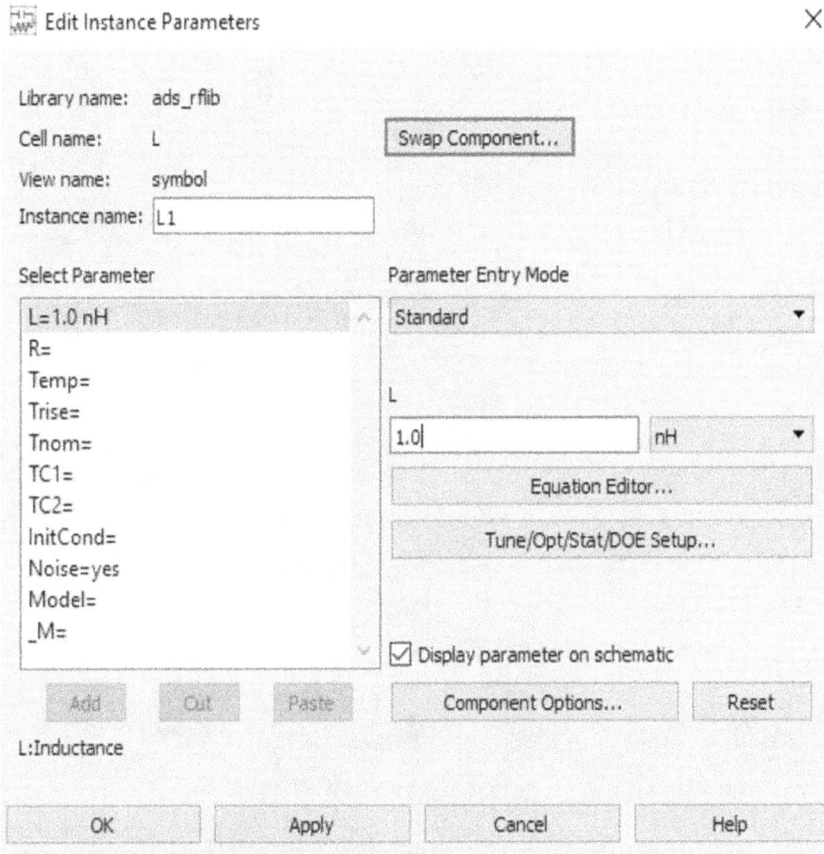

FIGURE 9.27   Instance parameter inductor.

6. Capacitor values are considered because they are the key parameters of the filter. The values of the capacitor as in Figure 9.28 should be obtained from the design and we also need to verify whether we have the capacitor value in the market and we have to check the availability of the components. After that, we can enter the instance parameter values of capacitor from the datasheet.

9.1.8 INSERTIONS

Various insertion options available in ADS are shown in Figure 9.29.

i. **PIN**
    The Pin icon is used to provide pins at the input and the output sides if required. The Pins are required to convert the implemented circuit into a sub-circuit used for creating a new symbol.
ii. **GROUND**
    The Ground icon is used to provide a common grounding to the circuit, i.e., the return path to circuit topology.
iii. **VARIABLE**
    The Variable icon **VAR** is used when a common parameter value is to be used for multiple component specifications.

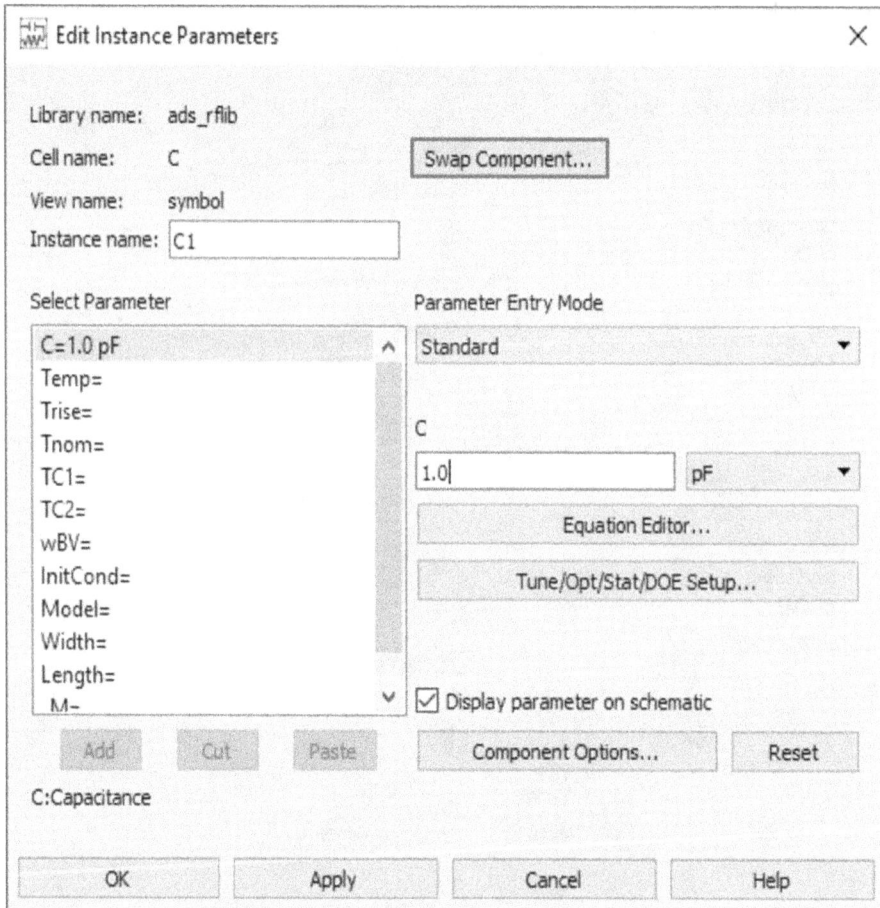

FIGURE 9.28   Instance parameter capacitor.

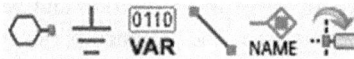

FIGURE 9.29   Toolbar for insertions.

iv. **WIRE**

The Wire icon ⟍ is used to connect the components placed, i.e., to wire the components. The wiring must be given in a proper way that there should not be any short circuits or open circuits between the components.

v. **NAME**

The Name icon NAME is used for providing names for the nodes. The Name icon is very much important to view the voltage at a particular node, i.e., the voltage at any node can be viewed only if that node is named using the icon.

vi. **ROTATE**

The Rotate icon is used for rotating the components during their placement such that the wiring becomes more convenient and free from shorts.

9.1.9 Template

The ADS environment makes available of default templates, i.e., default circuits along with the controllers such that they can be used wherever needed. This makes the simulation time lesser since the circuits are readily available. The simulation environment contains more number of templates. Figure 9.30 shows the toolbar to get the default templates readily available in ADS.

Figure 9.31 shows an example for default templates present in the software. The template shown below is a BJT Curve Tracer Circuit, which is used to generate and display the V-I characteristics of the Bipolar Junction Transistor. It can be used as it is wherever it is required, which makes the simulation easier.

9.1.10 Transient, AC and DC Models

The ADS software has several types of controller models to build and run the implemented circuit topology. Some of those controller models are transient model, AC Model, and DC Model. These models are used based on the requirement of the output at the data display window.

Figure 9.32 shows the transient model, which should be added inside the schematic editor to build and run topology. This transient model is used to determine the run time limit of the circuit simulation. It permits the simulation's start and end times, as well as the maximum time step required to achieve the desired outcome.

Figure 9.33a shows the AC Model, which should be placed inside the window to simulate an AC circuit and to get the AC voltages and AC currents at the output based on the circuit operation. It also enables start, stop, and step times for an AC circuit simulation to get the desired output.

Figure 9.33b shows the DC Model, which should be placed inside the window to simulate a DC circuit and to get the DC voltages and DC currents at the output side based on the circuit operation.

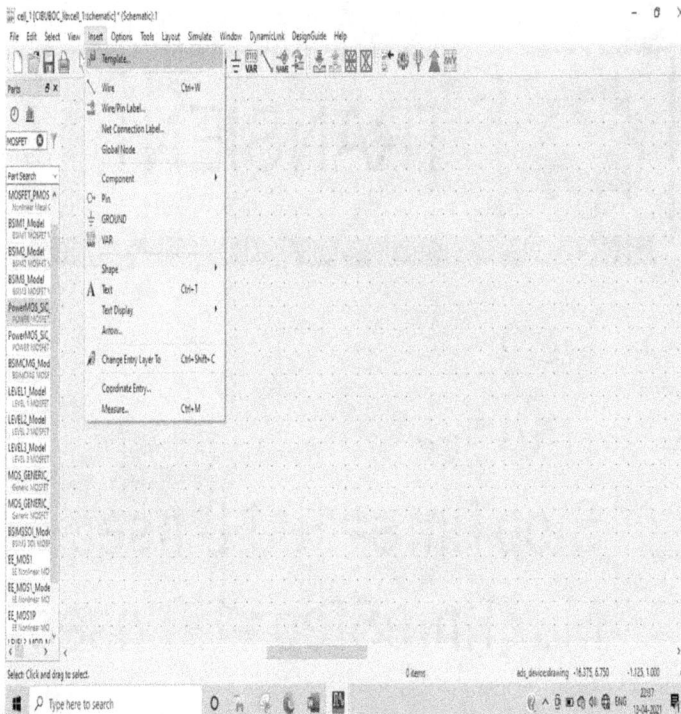

FIGURE 9.30 Template in advanced design system (ADS).

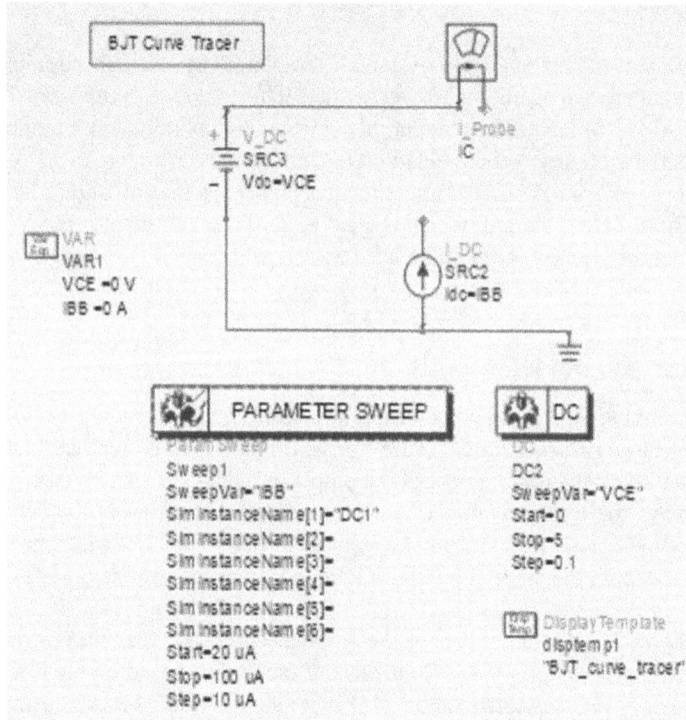

FIGURE 9.31 Template for BJT curve tracer.

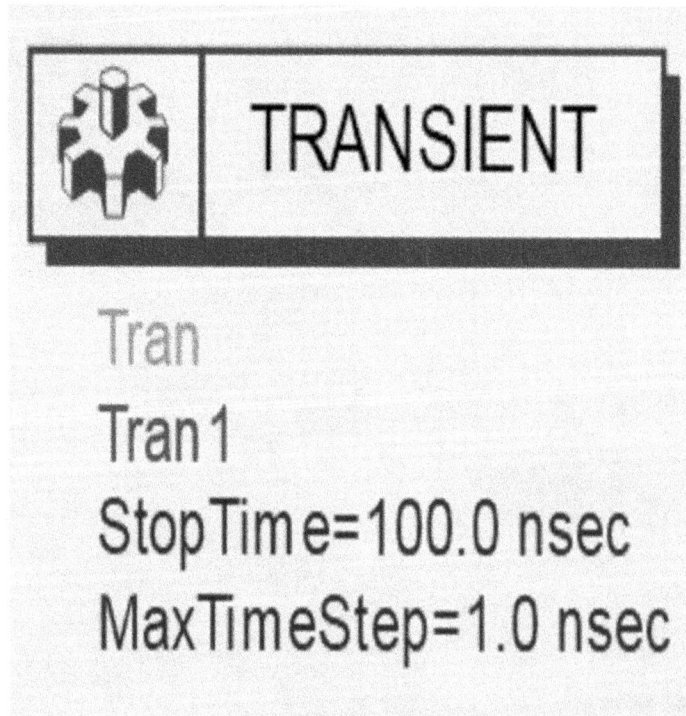

FIGURE 9.32 Transient model.

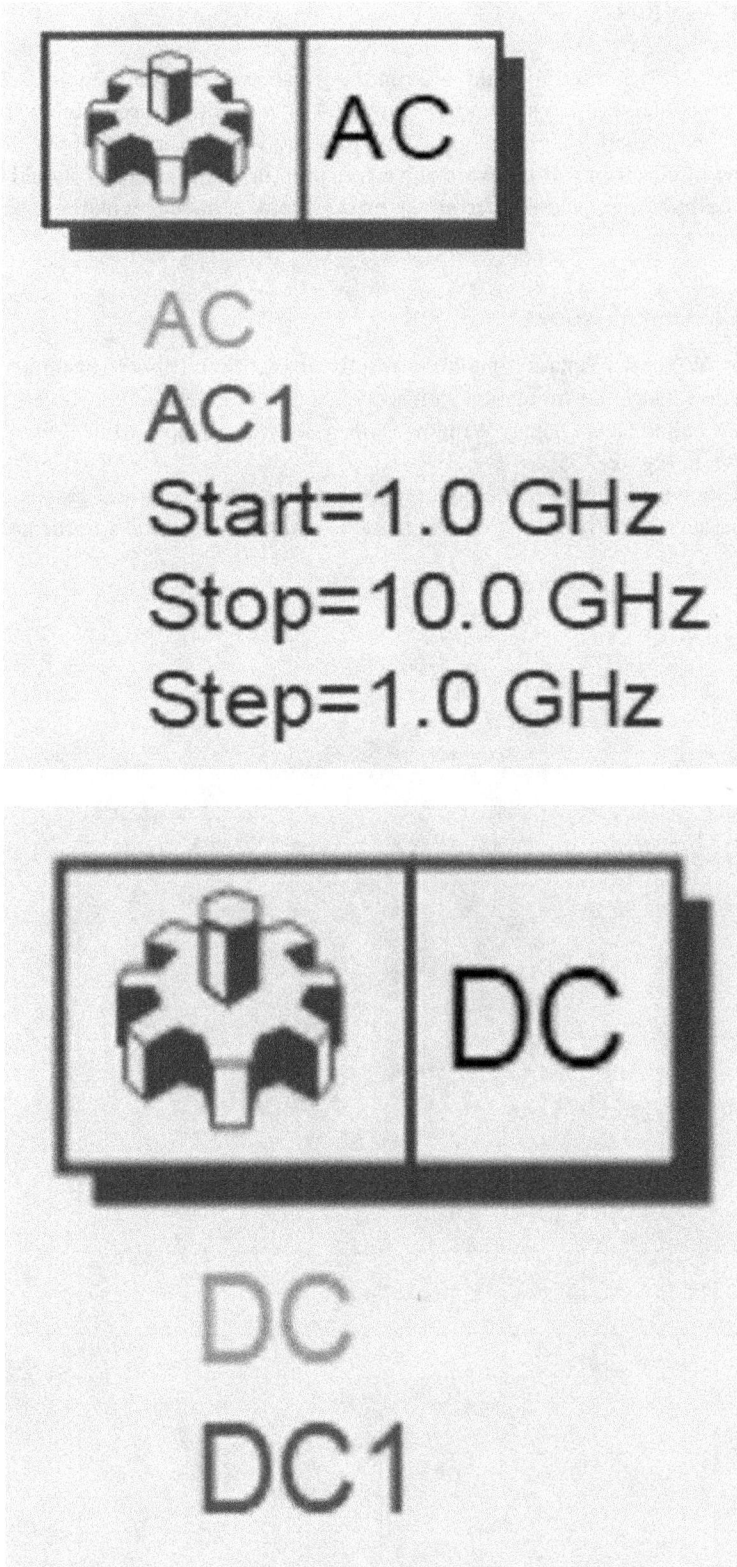

FIGURE 9.33 (a) Alternating current (AC) model. (b) Direct current (DC) model.

9.1.11   BUILD AND RUN

The Simulate icon ![icon] is used to build and run the implemented circuit topology. Once the icon is clicked, the error message tab shown in Figure 9.34 opens, which throws the warnings and the errors, if any, in the implemented topology. The errors and warnings may be from the connections or from the time limit settings. If there are any errors present, then the errors should be rectified to proceed to the output screen and if there are no errors, the data display window appears where the outputs can be viewed.

9.1.12   DATA DISPLAY WINDOW

A Data Display Window displays simulation results of the circuit topology implemented in the Schematic Window. Once the simulation icon is clicked and if the circuit encounters no errors with it, a new window called Data Display Window is opened with an empty white screen with a palette on the left shown in Figure 9.35.

The palette shown in Figure 9.36a contains several types of plots like graphs and equations such that the outputs from the circuit can be plotted in any of the formats in the palette based on

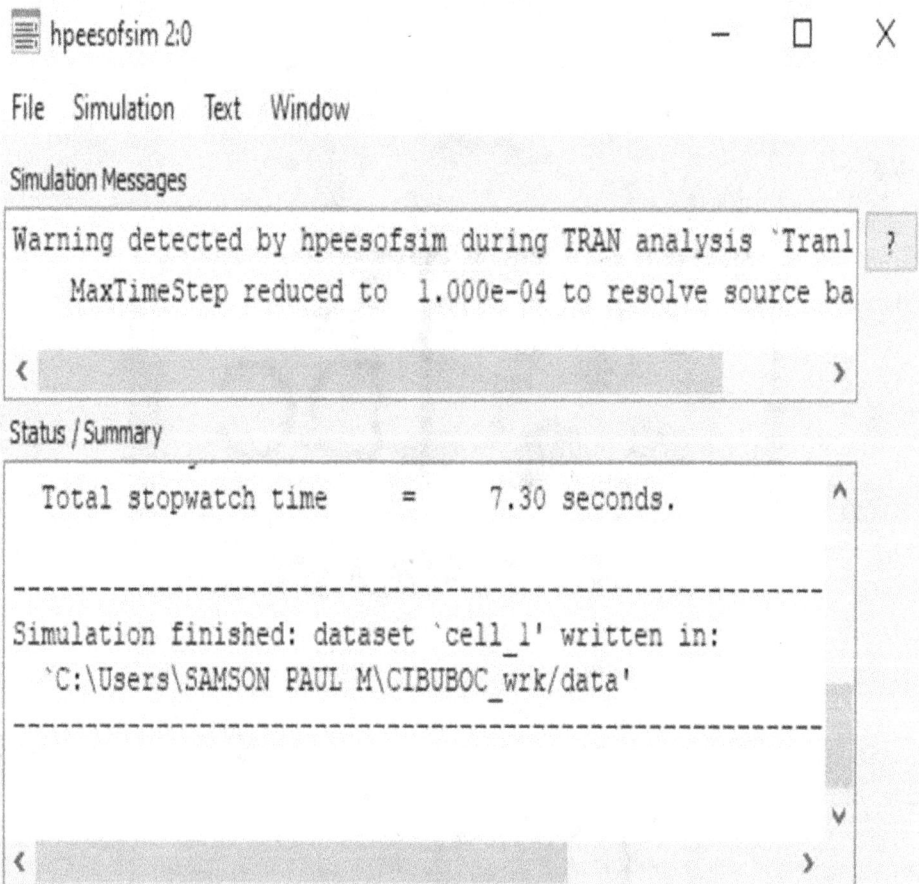

FIGURE 9.34   Error message tab.

FIGURE 9.35 Data display window.

the requirement. Most probably, the outputs will be displayed in the graphical format for making the analysis simple.

After selecting a required format in which the output is to be plotted from the palette displayed, a new window shown in Figure 9.36b appears asking for the parameters to be displayed in the selected plot.

The data display window icon ᴧᴧᴧ on the toolbar can also be used to open the same window in order to plot the required simulation results. The resulting waveforms can also be saved for further analysis.

9.1.13 DC Annotation

The DC annotation allows you to display the annotation values at particular nodes in the schematic window itself. It can be enabled from the simulate tool bar, as shown in Figure 9.37.

9.1.14 Tuning of Parameters

The Tuning icon ⍾ facilitates the modification of design parameters and their values. It is feasible to easily observe the influence of circuit topology on output due to the change made without

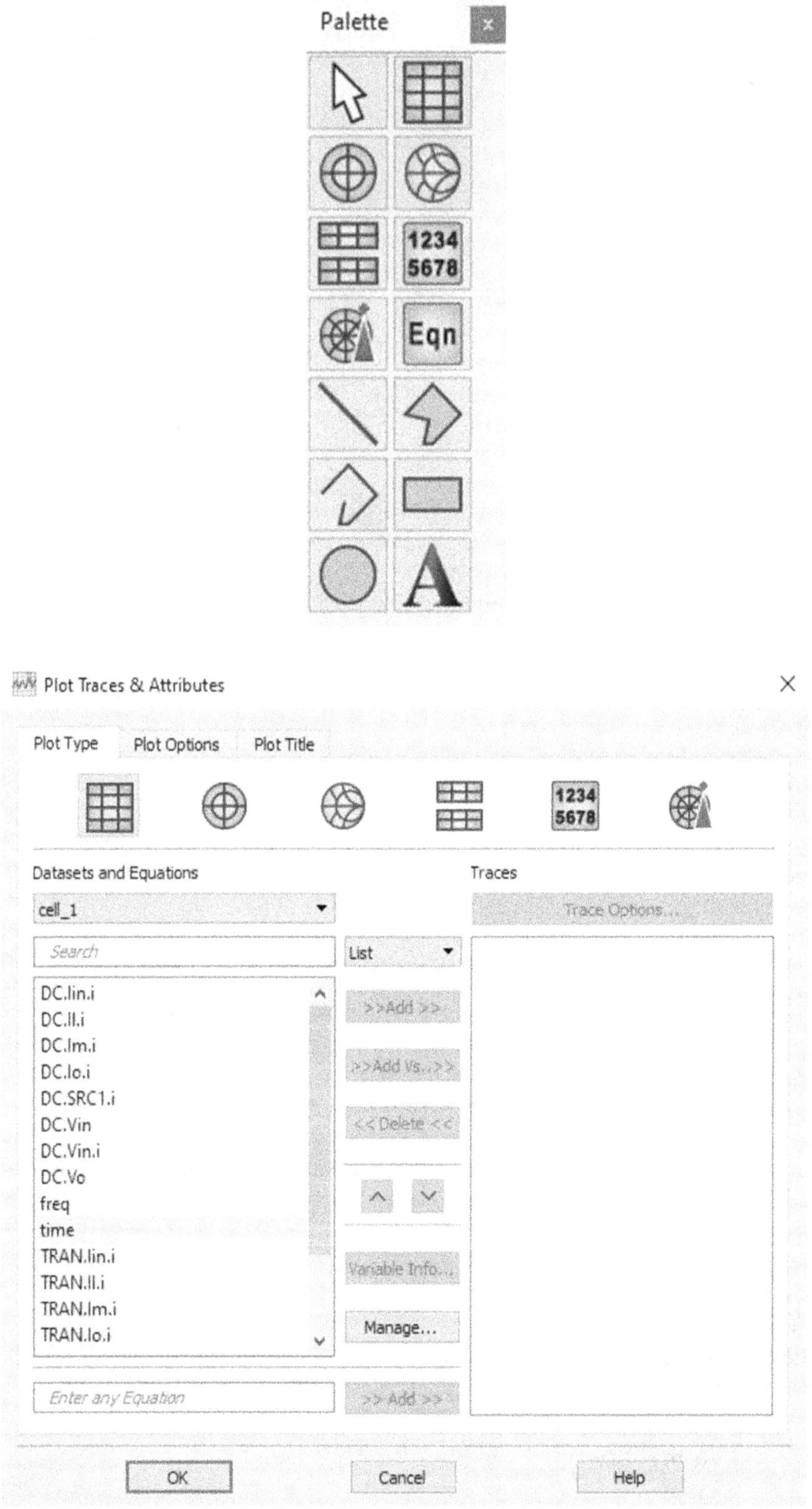

FIGURE 9.36 (a) Palette for different output plots. (b) Tab to select the parameters to be plotted.

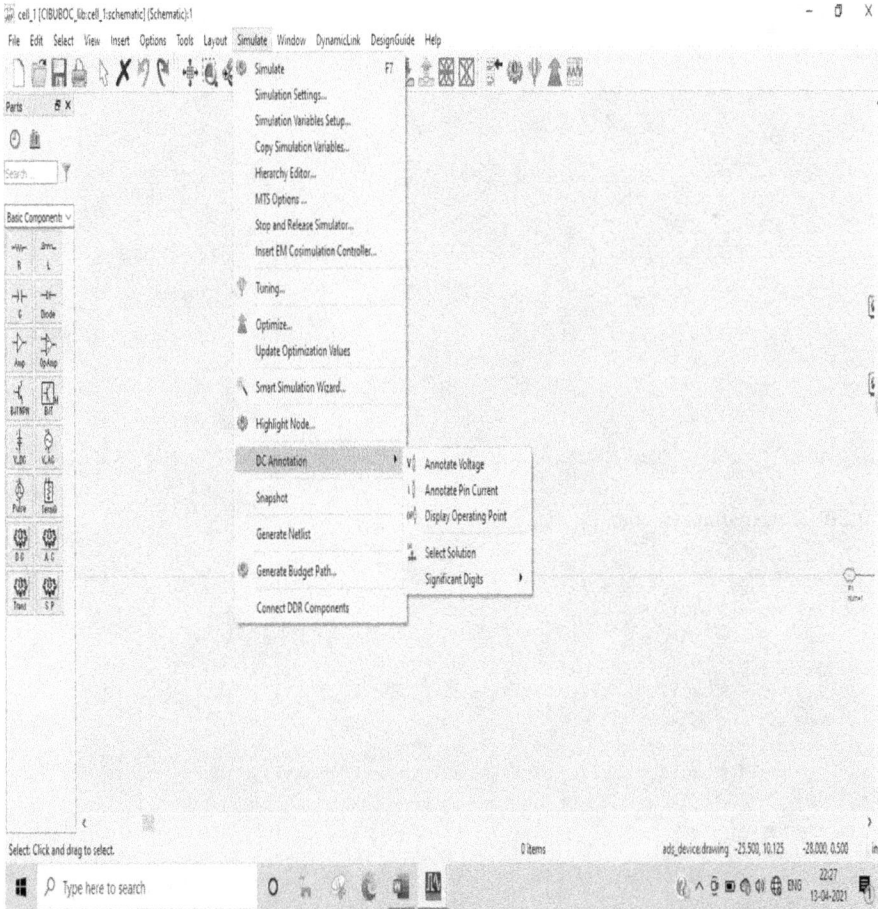

FIGURE 9.37 Enabling direct current (DC) annotation.

re-simulating the entire circuit implemented in the schematic window. After tuning the parameters, multiple plots will be generated from trials and they can be overlaid in the Data Display Window, which will be very much helpful for analysing the circuit.

Figure 9.38 shows the tune parameter tab in which the parameter to be tuned is selected and the results regarding the parameter selected is tuned and displayed in the data display window.

1. If the schematic design is hierarchical (has sub-networks), you can tune the parts within the sub-networks without having to leave the command.

    The Schematic window's View menu or the Push into Hierarchy icon can be used to choose Push into Hierarchy when the Tune Parameters mode is on. Choose the subnetwork that will be tuned. In the Schematic window, you can see how the subnetwork is set up. Parameters can now be changed.

2. To open the Enable/Disable Parameters dialogue box, click the Enable/Disable button within the Tune Parameters dialogue box, as shown in Figure 9.39. This dialogue lists all of the parameters in the hierarchy that are enabled or disabled for adjustment (Table 9.2).

3. Click the **Store** button to create your trace data and store parameter settings to memory. The *Store Traces and Values* dialogue box appears. Figure 9.40 shows the Tune Parameter Save window.

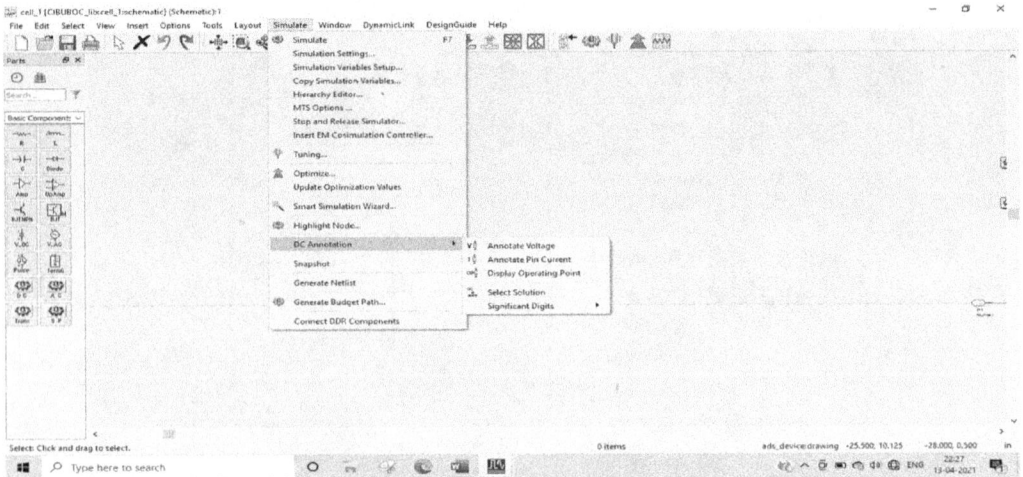

FIGURE 9.38   Tune parameters tab.

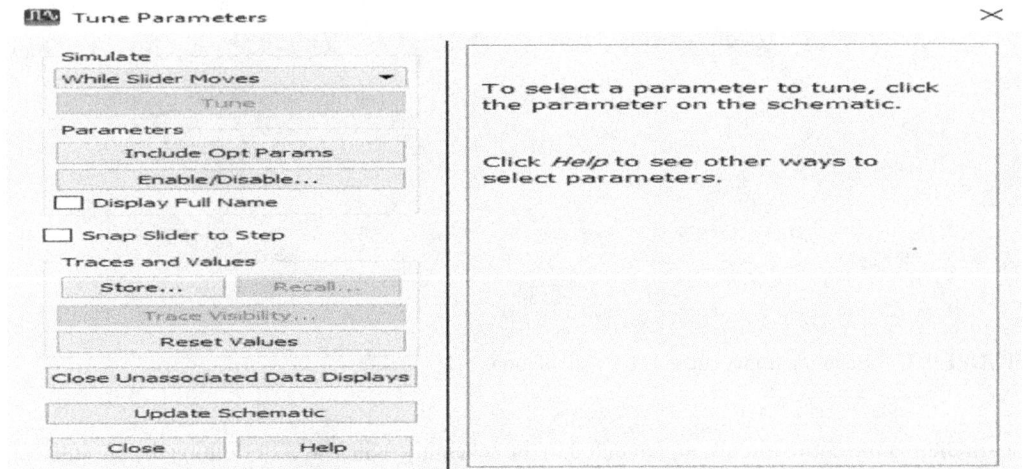

FIGURE 9.39   Tuning parameter—enable/disable.

- Give your tuning state a distinct name and a brief description.
- To preserve your memory trace and resume your tuning session, click OK.
- Click the Store button once more to save your updated trace data and parameter settings to memory if you want to store a different state after doing some more fine-tuning.
- A new instance of the Store Traces and Values dialogue box appears.
- Rename your new tuning state, add a comment, and click OK.
4. The Trace Visibility button in the Tune Parameters dialogue box permits the display or concealment of one or more stored memory traces in the Data Display window. To reveal or conceal a trace of stored memory.
  - Click the Visibility of Trace button. The Visibility of Trace dialogue box appears.
  - When deselecting a memory trace in the Trace Visibility dialogue box, the memory trace is no longer visible in the Data Display window after clicking Apply or OK. The memory trail remains accessible. Simply, it is not visible.

TABLE 9.2
Setting Up Parameters in Tuning

Simulate	After Clicking the Tune	Only after clicking the tune button performs an analysis. This option can be used for single modifications as well as several changes.
	Following Each Change	Analysis happens on each change after it has been made.
	During Slider Movement	Performs continual analyses while moving the slider. This option is similar to the "After Each Change" option, except that it is continuous.
	Tune	Tunes the design. This button is active only when the "After Pressing Tune" option is selected.
Parameters	Include Opt Parameters	Includes parameters that are enabled for optimization. If the optimization-enabled parameter does not already have a tuning setup, the optimization setup will be used for tuning.
	Enable/Disable	Launches the Enable/Disable Parameters dialogue box. This dialogue box is used to enable disabled parameters and disable enabled parameters.
	Display Full Name	Displays full names of the tune parameters in the tuning dialogue box. Otherwise displays the short names of the parameters. The short names are up to the last six characters of the tune parameters.
	Snap Slider to Step	For parameters tuned in a linear scale, the slider moves in increments of the step when "Snap Slider to Step" is selected. Otherwise, the slider moves continuously
Traces and Values	Store	Stores the tuned parameter values in temporary storage and creates a memory trace for each trace in the data display. Note that when you close tuning, all of the stored traces and values are deleted.
	Recall	Restores the parameter values for a specified, previously stored state. Note that if you have changed which parameters have been tuned since the state was originally stored, you may need to choose between the original values at the time the state was stored and the current values
	Trace Visibility	Lists all of the stored states that enable you to specify whether a memory trace is visible.
	Reset Values	Resets the tuned parameters to their nominal values.

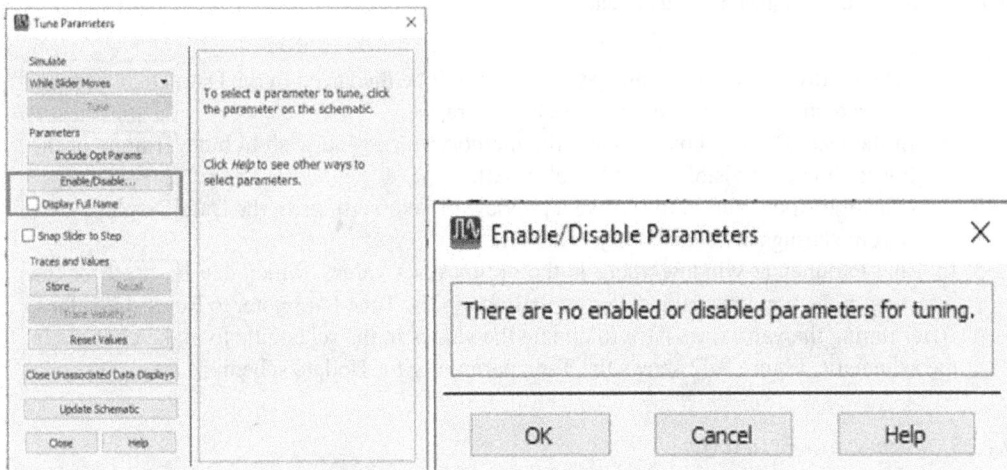

FIGURE 9.40 Tune parameter save.

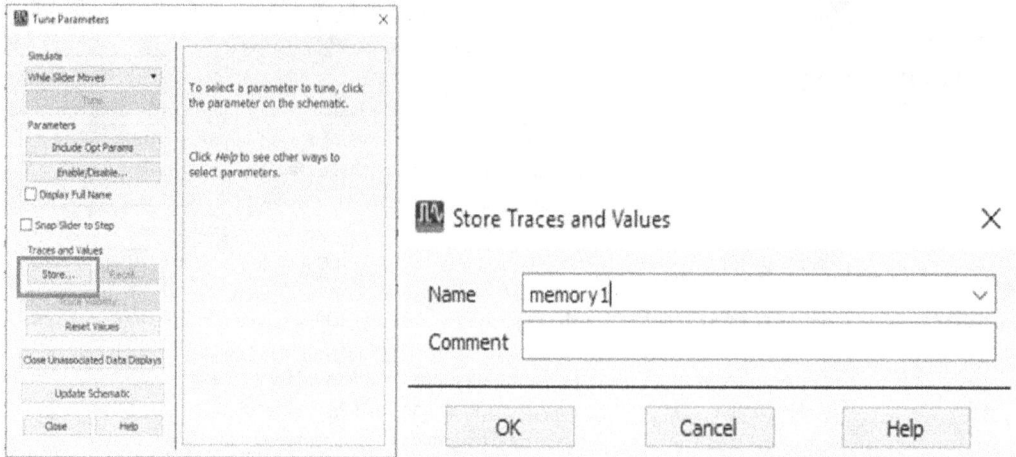

FIGURE 9.41 Tune parameter-reset values.

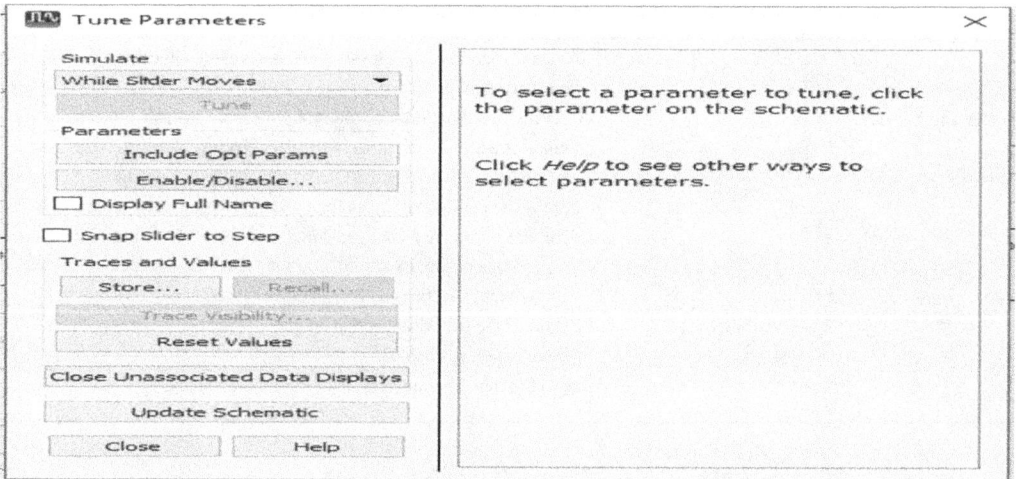

FIGURE 9.42 Tune parameter—update schematic.

- Select individually the memory traces that will be displayed in the Data Display pane. To make all traces visible, select Selected Traces.
- In the Data Display box, deselect the memory traces you wish to hide. If none of the traces should be visible, click Deselect All.
- Click the Apply button to receive a preview of your settings in the Data Display.
- If your settings are satisfactory, click OK.
5. In Tune Parameters window, there is the option reset values, which can be used for the resetting to the previous values. Figure 9.41 shows the Tune Parameter to Reset the values.
6. After tuning the values, we have to update the values in the schematic to make changes in the schematic. Figure 9.42 shows the Tune parameter for Update schematic.

9.1.15 SAVING THE SCHEMATIC

The implemented circuit topology in the schematic window can be saved for later use. The Save icon in the toolbar shown in Figure 9.43 allows you to save the schematic.

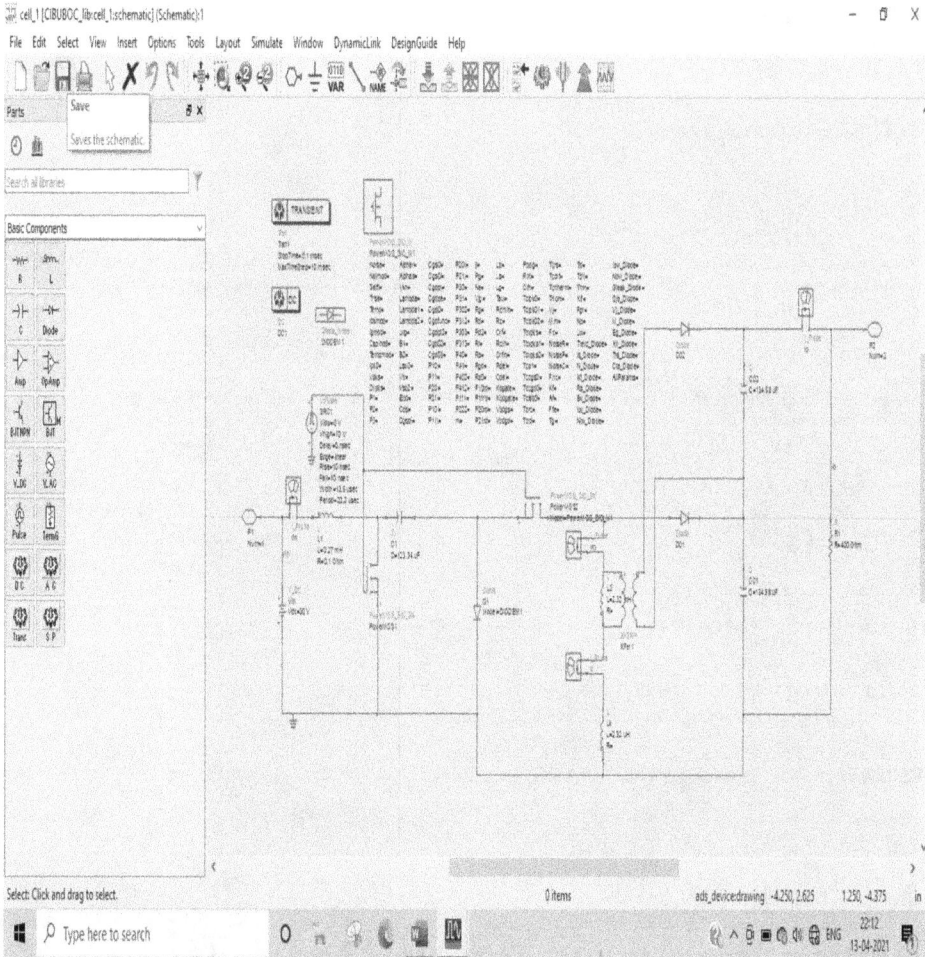

FIGURE 9.43   Icon to save the schematic cell.

9.1.16 Working with Examples

1. ADS has examples where it is possible to run the files and view the output. Figure 9.44 shows the page for ADS example.
2. Getting Started and Tutorial Figure 9.45 is for beginners where you will be guided with some ways to do the action.
3. Go to Simulation Examples, where you can find the list of sub-examples. Each sub-example consists of examples circuit with blocks and the specified output. Figure 9.46 shows the simulation example page.
4. Design flow examples consist of the demonstration-based design for different applications. Figure 9.47 shows the design example page.
5. Training Example consists of how we can use the ads window, and how we have use the other software which can be used for future research works. Figure 9.48 shows the training example page.

Design Guide will help us to find the different ways of designing; EM is used for analysing S-Parameters.

Open Example   —   □   ✕

←   →   🏠   🔍 Search...   ✕

ADS Examples

Home

Getting Started and Tutorials
Demonstrates the basic capabilities and concepts of ADS to help you design and simulate circuits for use in applications such as RF, Microwave, and High Speed Digital (HSD).

Simulation Examples
Demonstrates the simulation capabilities of ADS to help you set up and use the simulation options.

Design Flow Examples
Demonstrates various design flow examples such as RF Board, MMIC, RFIC, and Multi-technology Module (MTM).

Training Examples
Demonstrates the advance functionalities of ADS.

All Examples (in alphabetical order)
Example workspaces listed in alphabetical order.

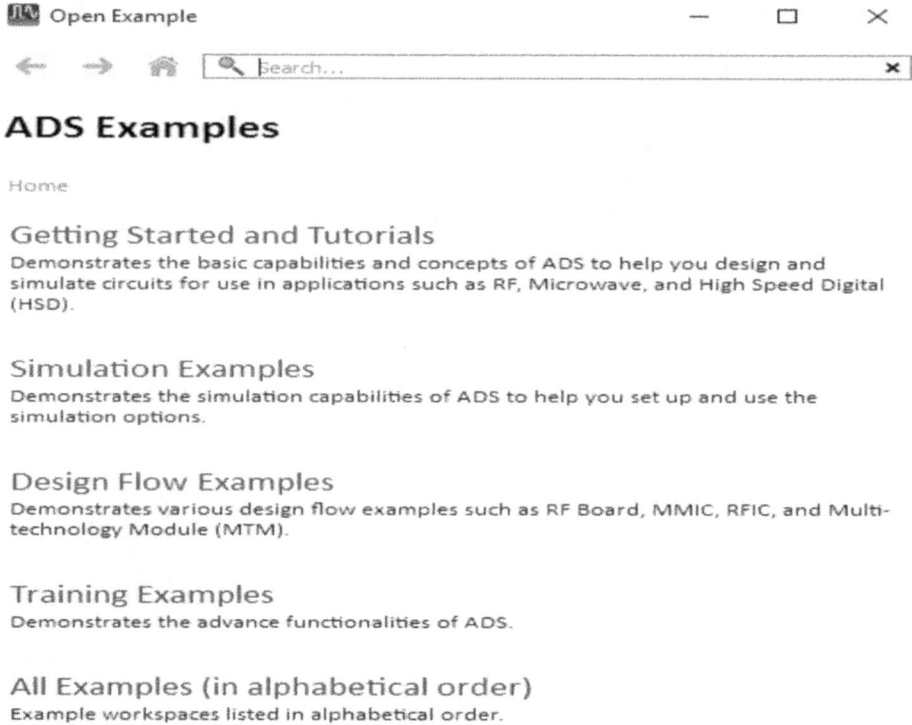

FIGURE 9.44   Advanced design system (ADS) examples.

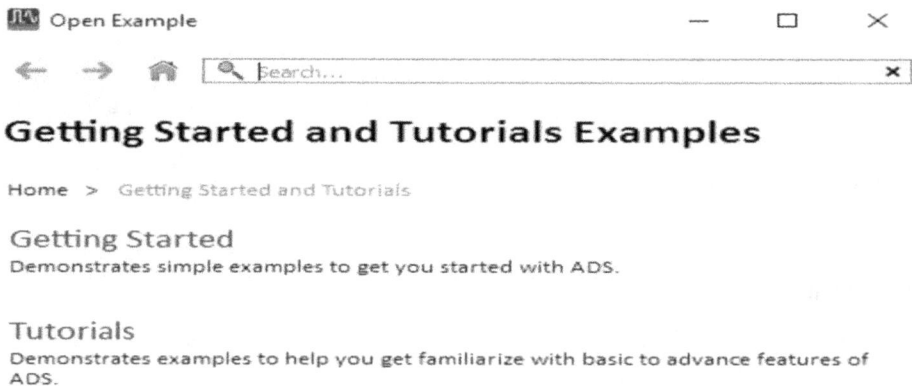

Open Example   —   □   ✕

←   →   🏠   🔍 Search...   ✕

Getting Started and Tutorials Examples

Home > Getting Started and Tutorials

Getting Started
Demonstrates simple examples to get you started with ADS.

Tutorials
Demonstrates examples to help you get familiarize with basic to advance features of ADS.

FIGURE 9.45   Getting started.

9.1.17 WORKING WITH MODELS

1. As discussed before in ADS, we need a Model for certain switches and the components, which are more powerful consider the example model EE Mos, Mos Generic, MosfetM. These are the specific models which are available for the certain switches. Figure 9.49 shows the EEMOS model setting.
2. The model consists of certain parameters where we have to enter the values from the datasheet of the specific components, where in ADS it has the most latest switch-like GaN. SiC devices are available, which can be used, and the model can be downloaded from the keysight website or we can include Net list.

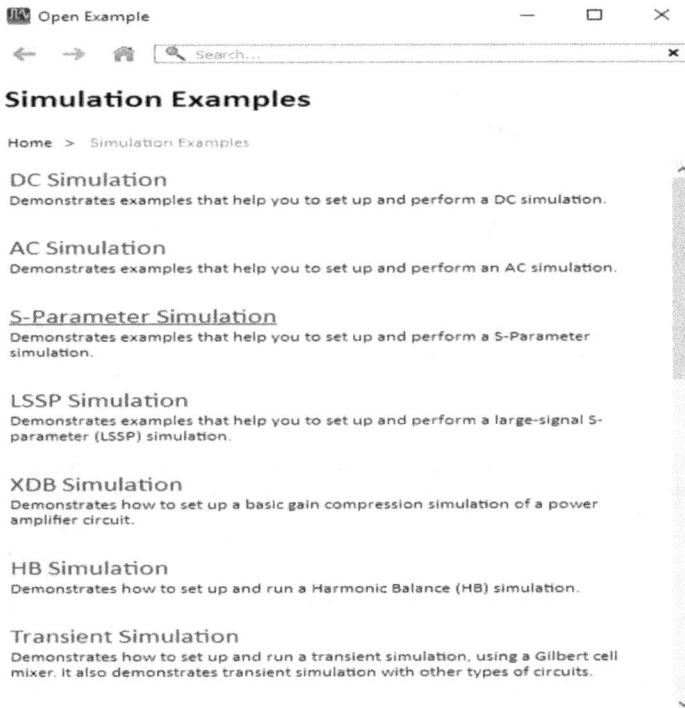

FIGURE 9.46 Simulation example.

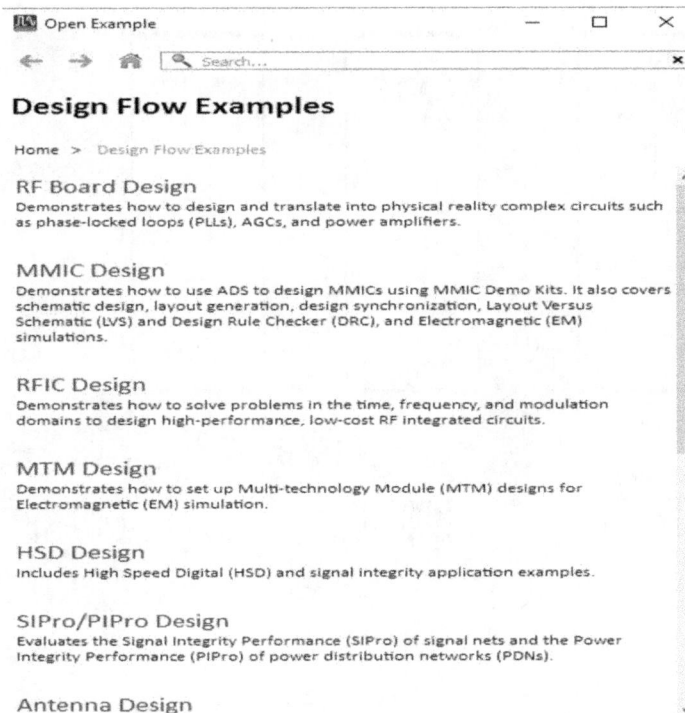

FIGURE 9.47 Design examples.

FIGURE 9.48   Training example page.

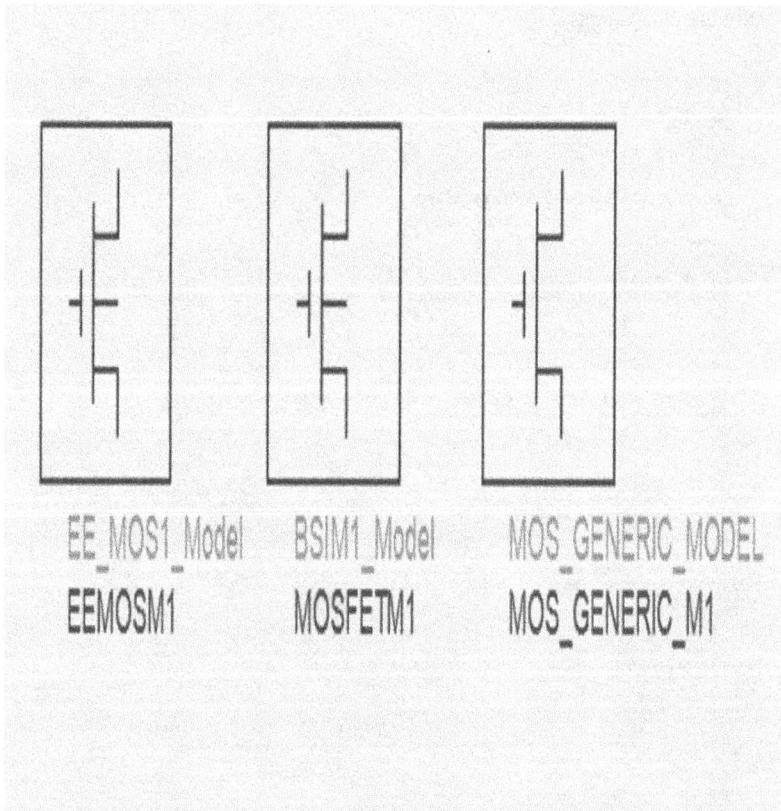

FIGURE 9.49   EEMOS model.

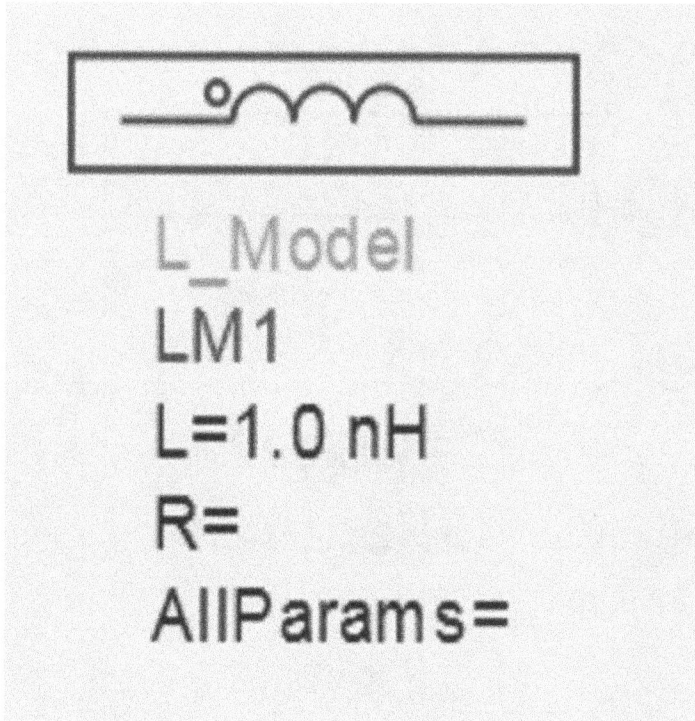

FIGURE 9.50 Inductor model.

3. When you are using inductance in the circuit you need to use the L Model in ADS software. This model as in Figure 9.50 has the standard values, which are taken from the datasheet. The parameters to be inserted in the model are discussed below.
4. When you are using capacitance in the circuit you need to use the C Model in ADS software. This model has standard values that are taken from the datasheet. The parameters to be inserted in the model as in Figure 9.51 are discussed below.

    When you have not used this model and you use high value of capacitance, the ADS will show error message like C value is unusually very high. This leads to the error and the output waveforms cannot be obtained unless you are using the C model.
5. When you are using resistance in the circuit you need to use the R Model in ADS software. This model has standard values that are taken from the datasheet. The parameters to be inserted in the model are discussed below. Usually R model is not necessary when you use low values of resistance as in Figure 9.52.

9.1.18 Data Display Window

1. Data display window will be popped up when you run the simulation without errors. You will get a dialogue box like this where you can find the different combinations of the graphs, as shown in Figure 9.53.
2. Traces of plot; in data display window, there are different plots available. One is a rectangular plot. In this, on the left hand side, there are the output functions, which can be selected for displaying the data, as shown in the Figure 9.54.
3. After inserting the datasets in traces, you can be viewing the output waveforms. If waveform is with respect to time we have to select the transient parameter. If the response should be in frequency then we can select the AC domain parameter where there will be

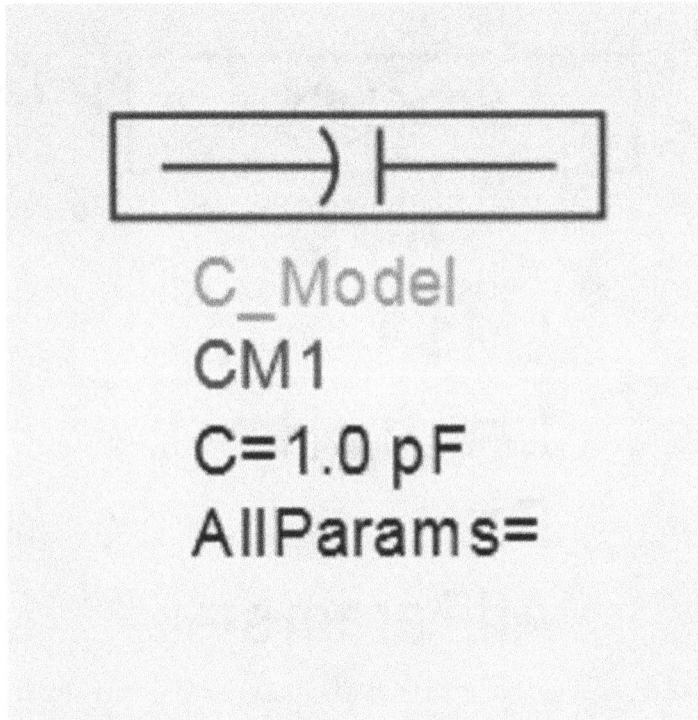

FIGURE 9.51   Capacitor model.

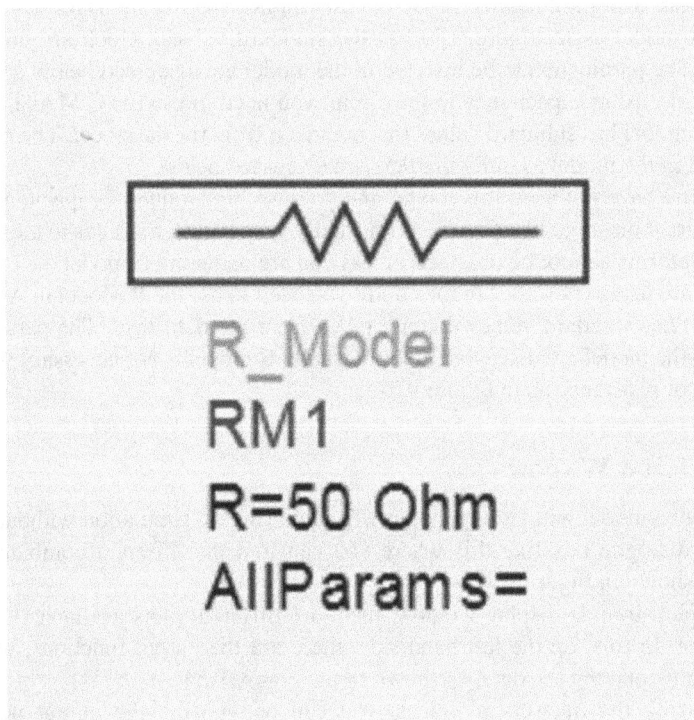

FIGURE 9.52   Resistor model.

FIGURE 9.53   Data display window.

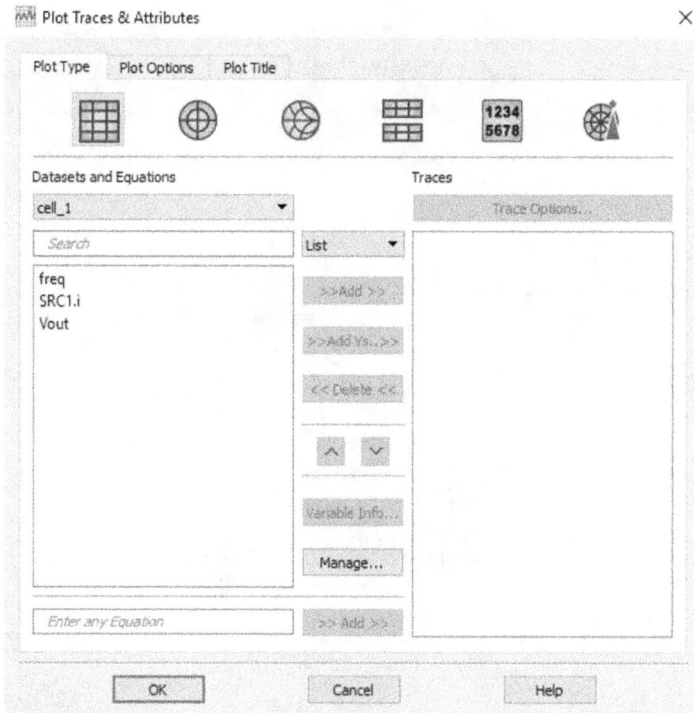

FIGURE 9.54   Plot-type parameter.

a message popping up for selection of the parameter of the curve like dB, mag....we can select the parameter we are in need of.

4. There are other plots like polar, smith plots where each can be selected and the output waveforms can be analysed.

5. We can change the plot that is displayed in the top of the plot traces and attributes window as in Figure 9.55.

6. To add the value, we need to select the parameter and then we need to click add. If we have to have versus plot, then select the X-axis parameter and then click Add Vs. Now add the versus parameter. You can change in data display window by pressing the plot as depicted in Figure 9.56.

7. To change the plot type:
   • Double-click on the plot.
   • On the Plot Type tab of the Plot Traces & Attributes dialogue box, choose a new plot type.

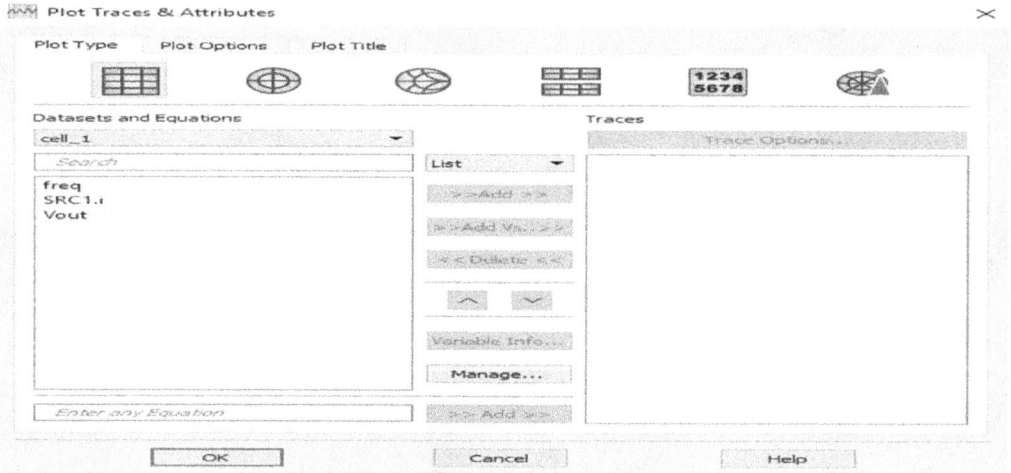

FIGURE 9.55 Plot styles.

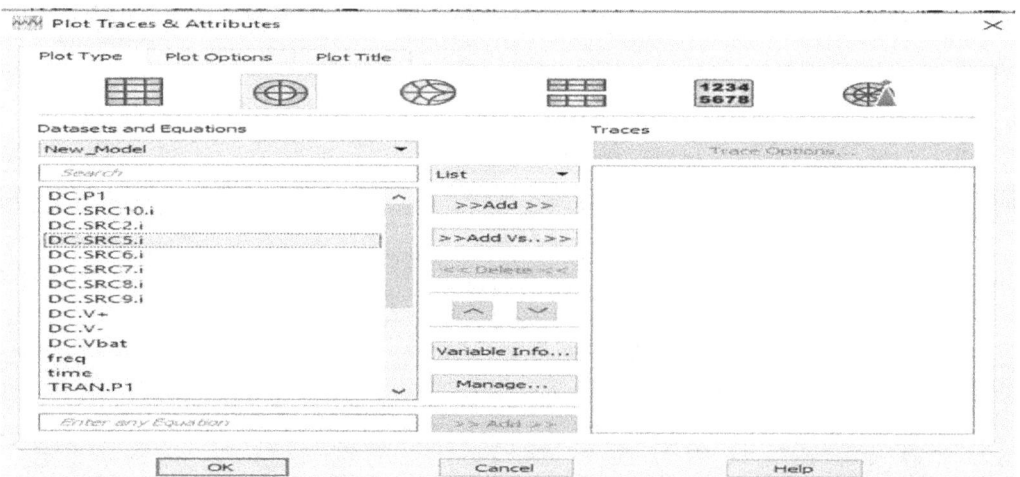

FIGURE 9.56 Plot traces and attributes.

To edit various plot characteristics:
- Double-click the desired plot or select Edit Item Options. Then click the Plot Options tab as illustrated in Figure 9.57.
- When finished with adjusting plot characteristics, click OK to save the modifications.

8. Plot title is used for naming the plot. Figure 9.58 shows the Plot title window.
9. Data display drawing facilities, as shown in the Figure 9.59, are used to draw line, rectangle, polygon, polyline, circle text, which can be used for describing the curves and the plots.
10. There are other functions like stacked rectangular plot, antenna, list equation, which have specific usage for different types of plot as in Figure 9.60. Where list can be used, we need to analyse the performance of a device for every change we can using list. Similarly, equation can be used when you need to take the output of specific parameter and you need to calculate the parameters.

12 NetlistInclude

1. NetlistInclude is a component that allows the ADS simulator to utilize an external file. NetlistInclude is the preferred way for including external files; however, deprecated components may continue to function. As depicted in Figure 9.61, the NetlistInclude component can directly read a Spectre file using the netlist include controller.

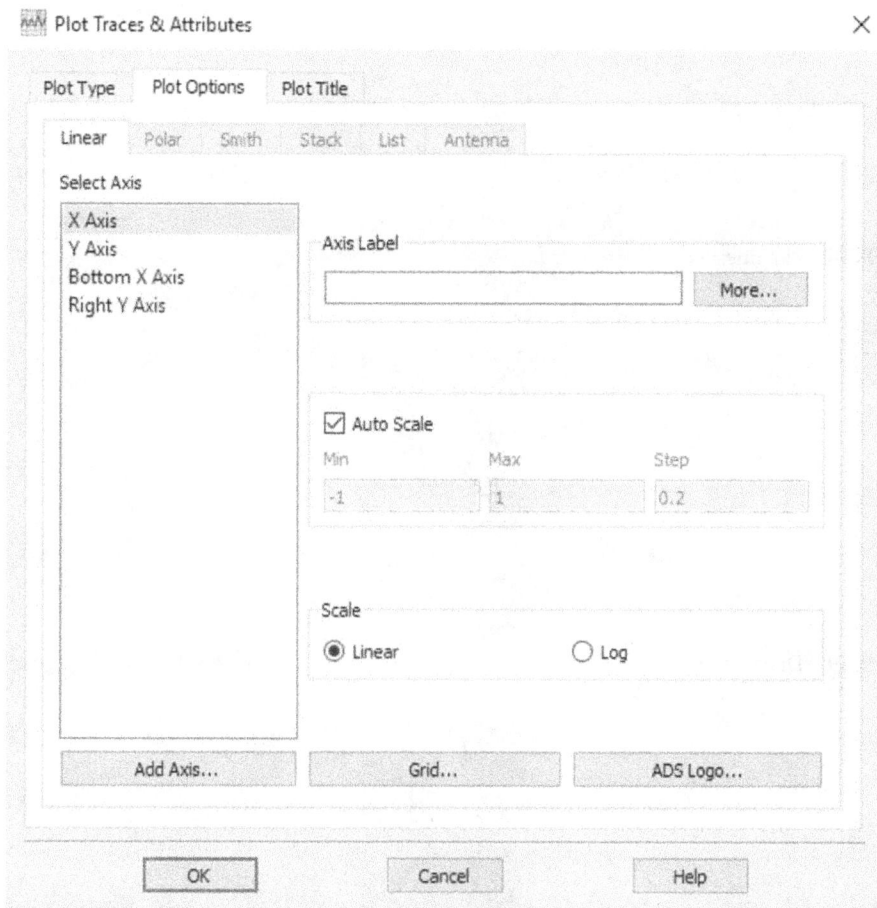

FIGURE 9.57 Plot setting.

FIGURE 9.58   Plot title.

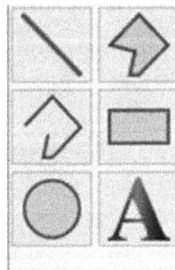

FIGURE 9.59   Drawing facility.

FIGURE 9.60   Plot types.

FIGURE 9.61   Netlist including controller.

2. The IncludeFiles argument allows you to specify a list of netlist files to include in the simulation. As demonstrated in Figure 9.62, use the Add button to include several files with a single NetlistInclude component.
3. The NetlistIncludeList component allows you to specify a list of Netlist include files to scan. This component can only be used with the Batch Simulation controller, as seen in Figure 9.63.
4. To use NetlistIncludeList component:
   - Insert a Netlist inclusion list component into the schematic. NetlistIncludeList can be found on the Simulation-Batch palette.
   - Double-click on the component to populate it with Netlist include files. Specify the appropriate Netlist type in the Parameter entry mode. In addition to the native ADS format, the SPICE format can be specified. Figure 9.64 shows the Edit Instance Parameter setting.

9.1.19   S-PARAMETER

At a particular frequency, the S-Parameter controller defines the signal-wave response of an n-port electrical element. It is a sort of small-signal AC simulation that is typically used to define a passive RF component and determine the small-signal properties of a device at a particular bias and temperature.

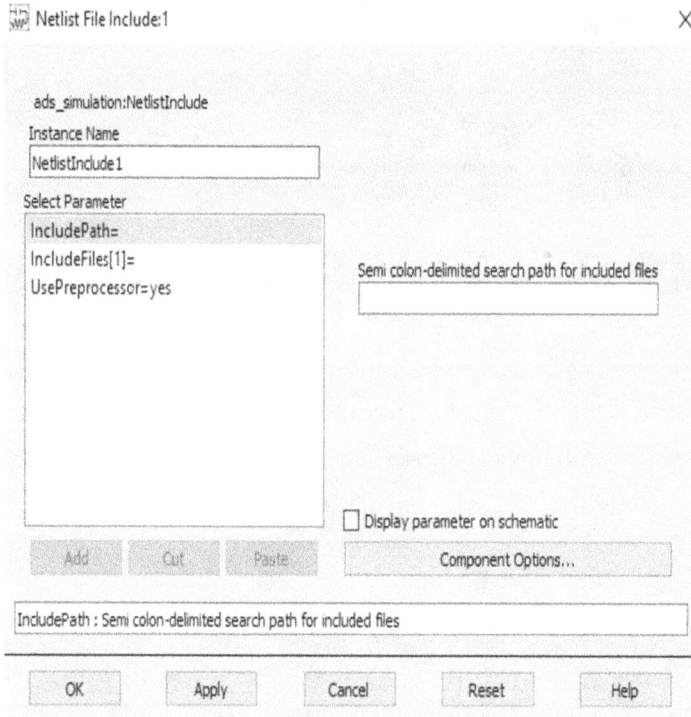

FIGURE 9.62   Netlist build.

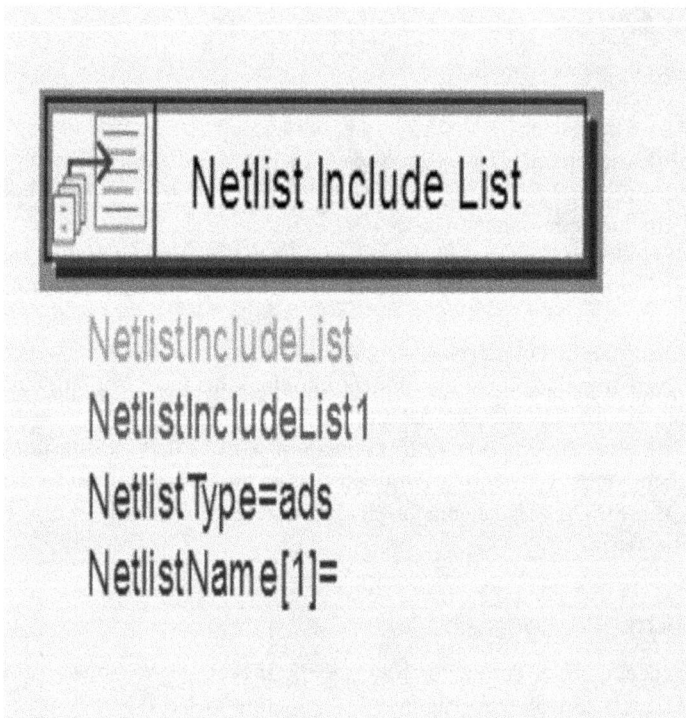

FIGURE 9.63   Netlist including list controller.

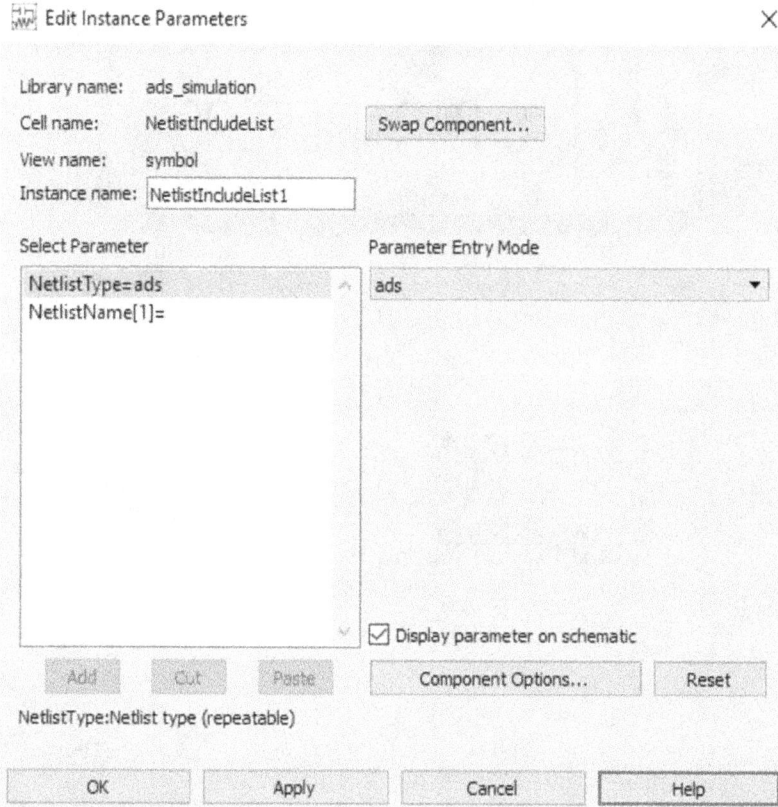

FIGURE 9.64 Edit instance parameter.

Figure 9.65 depicts the S-Parameter controller which is used to:

- Acquire the scattering parameters (S-parameters) of a component or circuit and transform them into Y- or Z-parameters.
- Plot, for instance, the fluctuations in swept-frequency S-parameters in relation to another variable that is changing.
- Simulate group delay. Simulate linear noise. Simulate the impact of frequency conversion on the S-parameters of tiny signals in a mixer-equipped circuit.

9.1.20 HARMONIC BALANCE

Simulating analogue RF and microwave circuits is ideally suited to the harmonic balance controller. It is a frequency-domain analysis method that simulates distortion in non-linear circuits and systems. Figure 9.66 displays the setting for frequency harmonic balancing.

In the context of simulation of high-frequency circuits and systems, harmonic balance provides the following advantages over standard time-domain transient analysis:

- It directly captures the steady-state spectral response.
- Numerous linear models are best represented at high frequencies in the frequency domain.
- In many real situations, the frequency integration necessary for transient analysis is prohibitive.

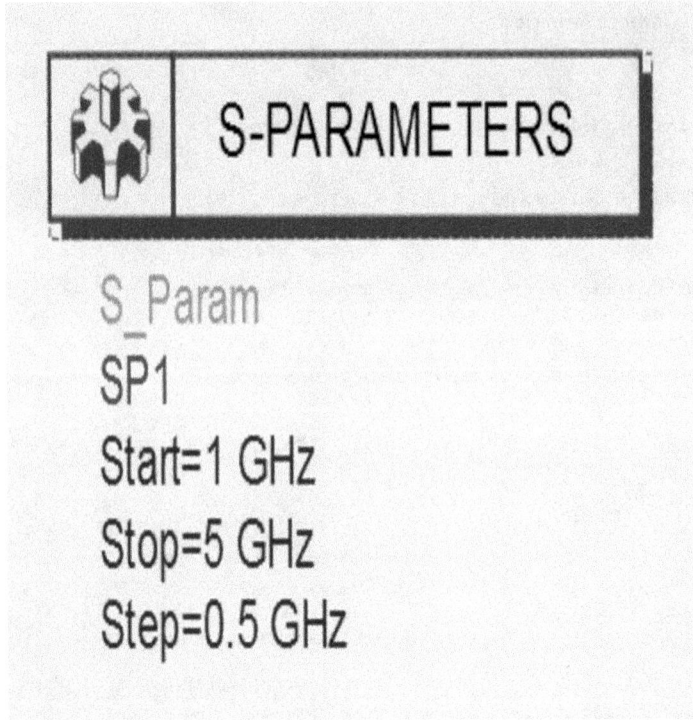

FIGURE 9.65   Scattering parameter.

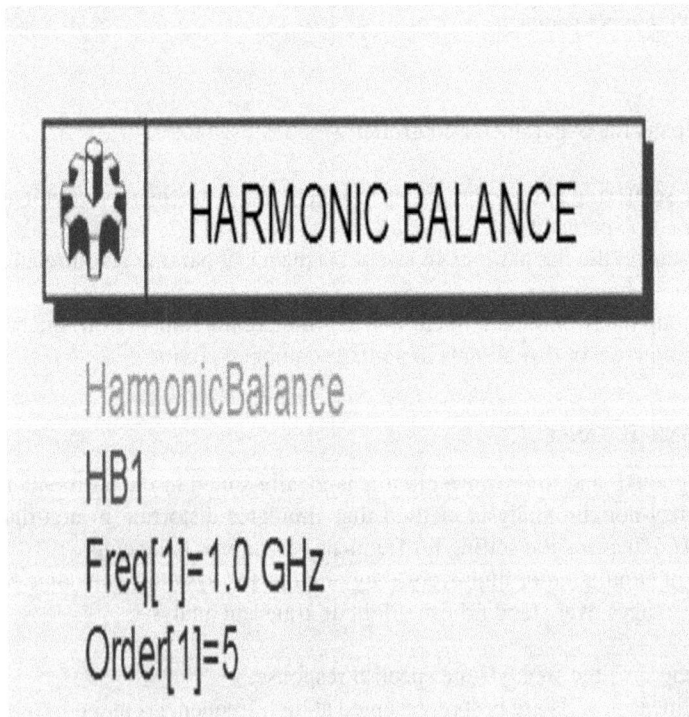

FIGURE 9.66   Harmonic balance—frequency.

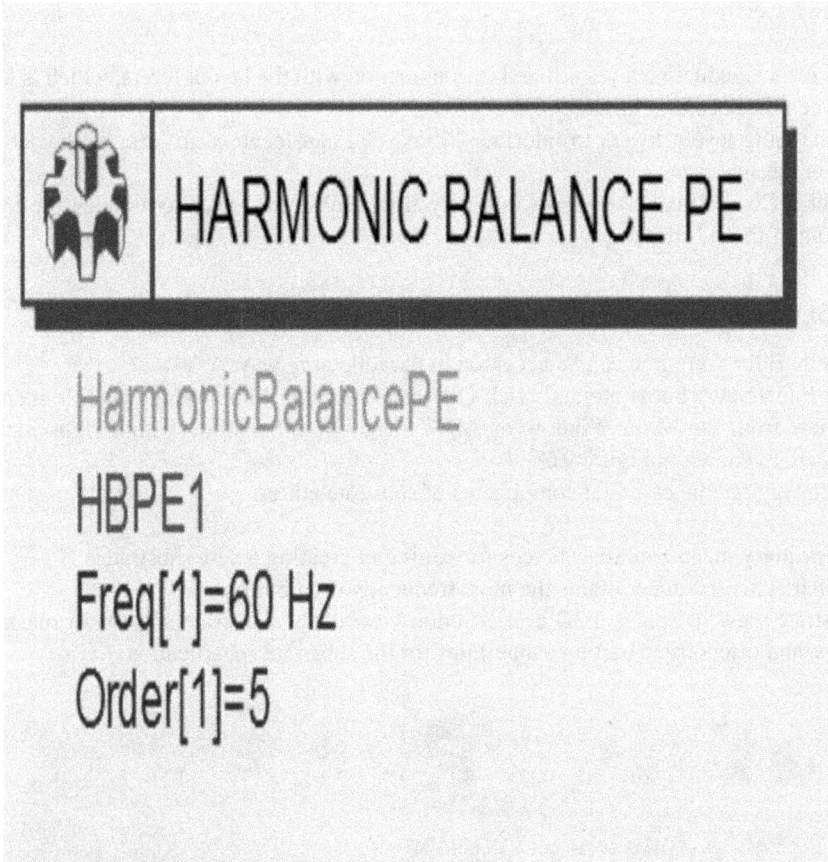

FIGURE 9.67 Harmonic balance PE instance.

There is separate harmonic balance circuit for the power electronics, which can be used for implementing the power electronics designs as in Figure 9.67.

Use the harmonic balance controller to:

- Determine the spectrum composition of voltages or currents using the harmonic balance controller.
- Compute quantities including intercept points of the third order, total harmonic distortion, and intermodulation distortion components.
- Perform load-pull contour analysis on power amplifiers. Perform a non-linear analysis of noise.

9.1.21 EM SIMULATION

In EM simulation, a substrate describes the medium in which a circuit exists. An example is the substrate of a multilayer circuit board, which is composed of metal traces, insulating material, ground planes, vias that connect traces, and surrounding air. A substrate specification allows you to specify attributes for your circuit like the number of layers in the substrate, the dielectric constant, and the height of each layer. A substrate is composed of the following alternating items: Substrate Coating: This layer defines the dielectric media, ground planes, coverings, air, and other layered substances.

Interface Layer:

- This is the conductive layer utilized in conjunction with the layout layers, which is located between the substrate layers.
- By mapping layout layers to interface layers, you can locate your circuit's layout layers within the substrate.
- Either a Cover (Interface) or an infinitely thick Substrate Layer terminates the top and bottom of the substrate.

9.1.22 SUBSTRATE EDITOR

The Substrate Editor window can be accessed in the following way:

Choose File > New > Substrate and click OK from the ADS Main Window's File menu. Select Library View from the Main Window of ADS. Select New Substrate with a right-click on any library or cell, as shown in Figure 9.68.

The following are the essential components of substrate editor:

1. The primary menu contains choices for editing or creating a new substrate.
2. Toolbar: This section contains the most frequently used buttons.
3. Substrate view: Displays a 3D cross-section view of the substrate stack with mask mappings, and offers basic editing capabilities for the substrate specification.

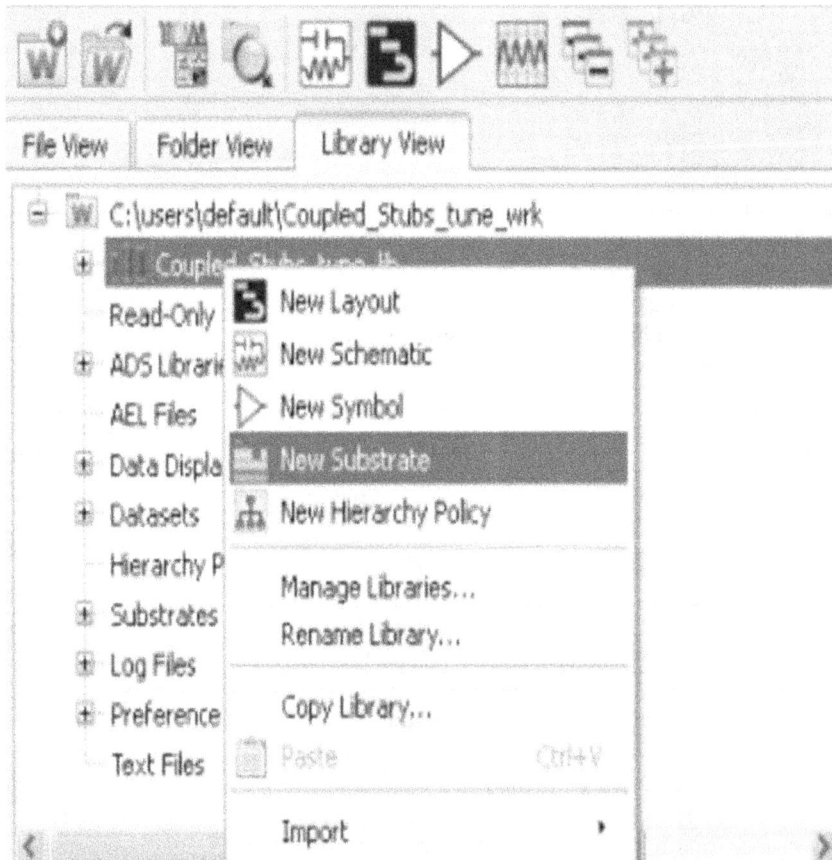

FIGURE 9.68 Substrate editor window.

4. Notifies the user of any warnings or faults for the substrate.
5. Attributes panel: This panel, located on the right, permits the modification of the properties of the currently selected item on the substrate.

To create a new substrate:

- Choose File > New from the Substrate window or File > New > Substrate from the ADS Main window.
- From the New Substrate window, select the library where you want to create the substrate.
- Enter the substrate's name in the File Name field and then click OK.

To open a predefined substrate

- Select File > Open from the Substrate Window or File > Open > Substrate from the ADS Main Window to open a substrate.
- Select the substrate from the Open Substrate window and click OK. The selected content will open in a separate window.

Substrate editor provides three options to save:

- Save: This option stores the current substrate's modifications.
- Save As: The Save As command allows the current substrate to be saved under a new name.
- Select the library from the Library drop-down menu and enter the substrate's File Name. The requested substrate is produced and shown in the selected library.
- Save a Copy as: The Save a Copy As command saves a copy of the current substrate.
- Select the library from the Library drop-down menu and enter the substrate's File Name. In the selected library, a copy of the present substrate is made.

9.1.23 Inserting, Moving, and Deleting Items

- The Substrate View allows you to see the substrate stack and perform simple editing operations.
- To add or remove an item from the substrate, right-click on the substrate view and choose an option from the menu that appears.
- After selecting the required action, the corresponding properties are displayed in the Substrate Editor's right panel.

To copy a design

- Launch a Workspace, such as Workspace A.
- In the ADS Main Window, navigate to File > Manage Libraries.
- Click Add Library Definition file and select the lib.defs file from a different workspace, for example, Workspace B. Select Close.
- Within the ADS Main Window, choose the Folder View tab.
- Right-click the desired cell and choose Copy Cell. In the Copy Files dialogue box, select the target library and click OK. This transfers the entire cell to the destination specified.

To open a design

- Launch ADS and launch an already workspace.
- Select File > Open > Schematic from the ADS Main window to open the Open Cell View dialogue box.

- Select the type from the list provided.
- If you want to open a built-in ADS design (in read-only mode), select the Show ADS Libraries checkbox to display a list of all available libraries under Library.
- Under Library, select the name of the Library containing the design.
- Select the name of the cell under Cell.
- Under View, choose a view type, such as symbol, schematic, or layout.
- To open the selected design, click Accept.

9.1.24 POWER ELECTRONIC CONVERTERS WITH ELECTROMAGNETIC INTERFERENCE (EMI) FILTERS USING ADS SOFTWARE

It is clear that with the help of ADS, creating a new schematic window for the implementation of the converter, where to get and how to interconnect the components, how to get the models of the components, how to tune the simulation parameters for simulating the converter, how to build and run the implemented converter to get the resulting waveforms, and how to tune the obtained waveforms in the data display window are possible.

ADS software is used to explain the concept used for mitigating the EMI problems in the power electronic circuits, it has high power simulation tools that can be used for analysing the circuit. A detailed explanation on the ADS software is provided with the step by step methods of implementing the circuits. Implementation of the S-Parameter and the CE are also analysed. This chapter also discusses about the steps for designing the EMI filter for different applications. The concept is explained using the following converters:

i. Power Electronic Circuits for grid-to-vehicle technology: Here the results of the system is done using the ADS simulation software and the results are analysed, and also the S-parameter measurements and the CE measurements are done and T filter is implemented for solving EMI issues.

ii. Power electronic converter for permanent motor direct current (PMDC) motor has been examined. The converter is used for hybrid energy storage system.

iii. Coupled-Inductor Buck-Boost Converter (CIBuBoC) provides a semi-quadratic ultra-high step-up with low input current ripple and low output voltage ripple with positive polarity used to enhance the power level of Solar Photovoltaics (PV) power generation in single stage and the enhanced power can be used exclusively for residential and industrial applications.

9.2 INTERLEAVED BUCK-BOOST CONVERTER FOR SINGLE-PHASE ON-BOARD CHARGER FOR ELECTRIC VEHICLE

To increase the usage of electrical vehicles (EVs), the charging system should be familiar to the user. The on-board charging system is easy for the user to charge and it is user friendly as it is operated in the AC grid voltage. As the number of EVs continues to rise, utility firms and suppliers are beginning to worry about the power quality of the utility grid network. Peak electricity use may necessitate the construction of a new electric power plant. Concerning the power quantity issue, simultaneous power consumption by EVs may result in insufficient power supply and raise peak demand. EVs will be produced in increasing numbers and will eventually dominate future mobility. In order to preserve the quality and quantity of the utility grid, an energy management system for EV consumption and ecosystem is essential. Grid-to-vehicle (G2V) and vehicle-to-grid (V2G) capabilities are essential elements that should be incorporated into vehicles and charging systems in order to improve the energy management of the utility grid. A bidirectional converter/inverter is required for bidirectional power transfer in this setup. In contrast to conventional on-board chargers,

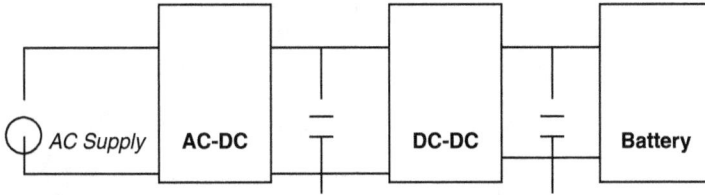

FIGURE 9.69 Converter block.

bidirectional on-board chargers can convert AC grid electricity to DC voltage for charging EV batteries and converting DC voltage from EV batteries to AC voltage for grid connectivity. Moreover, the bidirectional on-board charger can also handle power quality functions such as voltage regulation and power factor adjustment. Conducted EMI is created by the physical contact between conductors, whereas radiated EMI is caused through induction. At lower frequencies, EMI is caused by conduction, but at higher frequencies, it is caused by radiation.

The single-phase on-board charger consists of the AC–DC converter and DC–DC converter, bidirectional bridge where Switch used for the experiment analysis is insulated gate bipolar transistor (IGBT) and the topology of the circuit and the application of the on-board charger is used widely in the day to day life as it uses the AC grid voltage as the source as depicted in the Figure 9.69. The block diagram of the setup describes the overview of the circuit operation and the outline of the working of the converter.

9.2.1 IGBT (Insulated Gate Bipolar Transistor)

IGBT consists of BJT (bipolar transistor) and MOS (insulated gate field effect transistor) composite voltage-driven power semiconductor devices of the fully controlled, voltage-driven kind. IGBT combines the benefits of these two technologies; the driving power is minimal, and saturation occurs at a low voltage drop. Very suitable for the DC voltage of 600 V and above are variable flow systems such as AC motor, inverter, switching power supply, lighting circuits, and traction drive. IGBT is a switch, either in on or off state. By the gate-source voltage, when the gate-source plus +12 V (greater than 6 V, generally take 12 V to 15 V) the IGBT conducting. If gate-source voltage is not applied or is applied negative, IGBT turn-off, a negative pressure purpose is reliably shut off. No amplified voltage IGBT function, turn on can be seen as the wire, other as open circuit when disconnected.

9.2.2 AC–DC Rectifier

Rectifiers convert AC to DC; these controlled rectifiers are also referred to as AC–DC converters. To produce controlled output voltages, phase-control IGBTs are utilized. By changing the delay or firing angle of IGBT, the output voltage of IGBT rectifiers can be adjusted. In the case of a very inductive load, the phase-control IGBT is shut off by firing another IGBT during the negative half-cycle of the input voltage. These phase-controlled rectifiers are easy to use, less expensive, and highly efficient. Since these rectifiers convert AC to DC, they are also known as AC–DC converters and are widely employed in industrial applications, particularly variable-speed drives with power levels ranging from fractional horsepower to megawatts.

9.2.3 DC–DC Chopper

Chopper will transform a DC supply with fixed voltage into one with changing voltage. A DC–DC converter immediately converts DC to DC and is referred to as a DC converter. The DC equivalent of an AC transformer with a continuously changeable turns ratio is a DC converter. It can be used

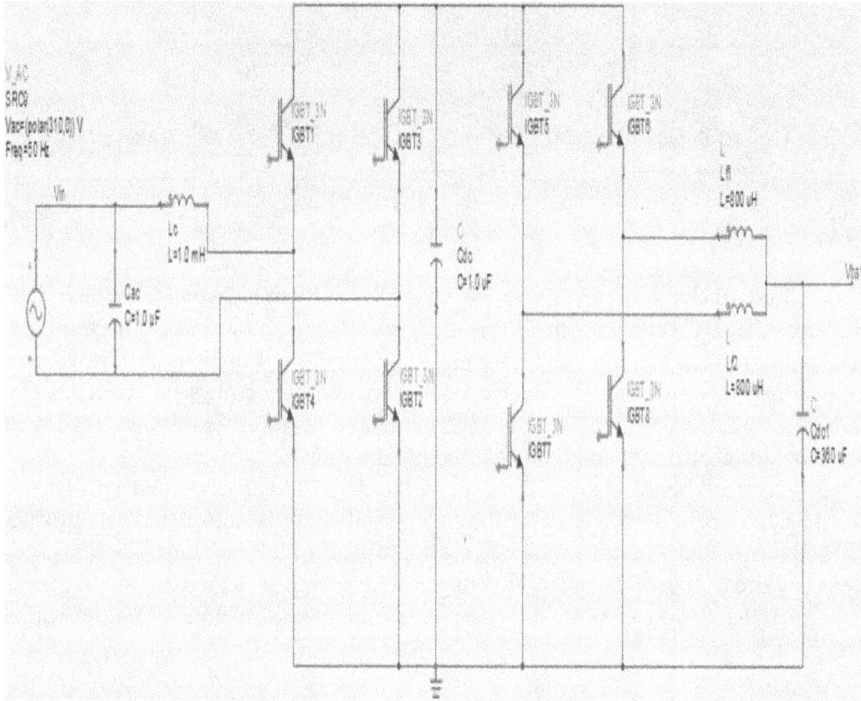

FIGURE 9.70 Circuit diagram.

to scale down or step up a DC voltage source, similar to a transformer. From a fixed or variable DC input voltage, the chopper can generate a fixed or variable DC output voltage. Ideally, the output voltage and input current of a DC – DC converter would be pure DC, but in practice, the output voltage and input current contain harmonics or ripples. Only when the converter links the load to the supply source and the input current is discontinuous does it draw current from the DC source.

This block diagram consists of two converters and two DC link capacitance, 230 Vrms voltage fed from the supply is rectified and the DC out of 310 V is stored in DC link and the 310 V DC is step down using buck converter reduced to 72 V. DC link capacitance is used between the chopper and the battery and the converted voltage is used to store in the battery, as shown in the Figure 9.70.

Implementation of the proposed circuit diagram is done in the ADS software by Key sight Technologies, which is the power full for analysis of EMI and CE. Here the Source voltage is given in the polar form and the output of the circuit is taken as Vbat.

9.2.4 Circuit Topology

The EVs on-board chargers typically include two-par topologies with cascaded: (1) AC–DC inverter and (2) DC–DC converter, as shown in Figure 9.71.

The grid is supplied with an input voltage of 220 Vrms and a frequency of 50 Hz; the system is designed for residential applications; the grid's current limit should not exceed 16A, and the charger power of the OBC is set at 1 kW. The grid to vehicle (V2G) is an AC–DC rectifier voltage from the grid before transmitting a DC–DC buck converter to charge the battery at a lower voltage. The AC–DC rectifier will function using pulse width modulation (PWM)-controlled switches S1, S2, S3, and S4 for charging the capacitor's DC link. The DC–DC buck converter employed both switches S5 and S7 to convert DC link voltage by reducing grid voltage from 310 to 72 V in order to charge the battery. The vehicle to grid (V2G) is a DC–DC boost converter that increases battery voltage

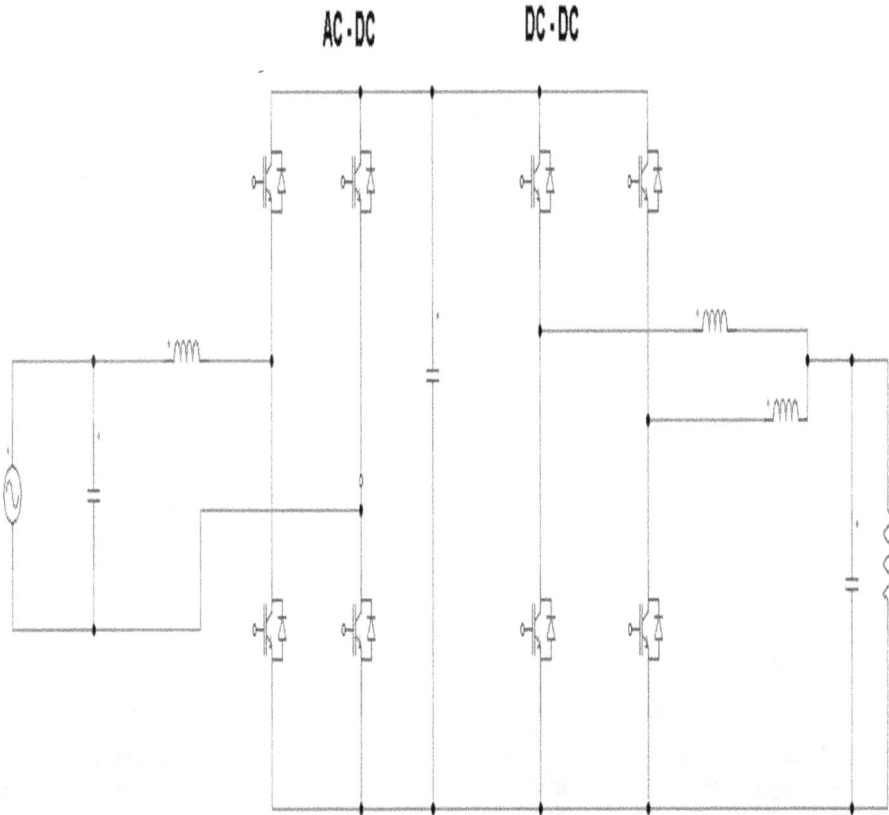

FIGURE 9.71   Circuit topology.

to DC link. The DC–DC boost converter that uses both switches S6 and S8 will operate alternately to prevent ripple while converting 72 V of battery power to 310 V for charging a DC -link with a voltage of 310 V.

9.2.5   DESIGN OF CONVERTER

Inductor Ripple Current

$$\Delta IL = \frac{V_s}{L}DT = 13\%$$

Output Voltage Ripple

$$\Delta V_o = \frac{I_o DT}{C} = 2\%$$

Filter Inductor

$$L = \frac{V_o}{\Delta IL\, fsw\, I_o}D^2(1-D) = 820\,\mu\mathrm{H}$$

Filter Capacitor

$$C = \frac{I_o \, D}{\Delta V_o \, fsw} = 330 \, \mu F$$

Boost Mode

$$VO = \frac{Vs}{1 - D}$$

$$VO = 310, \ Vs = 72$$

$$D = 0.77$$

Buck Mode

$$VO = VsD,$$

$$VO = 72, \ Vs = 310$$

$$D = 0.23$$

Experimental analysis describes about the design, model of the components, the Block diagram, and circuit topology. It also describes about the switch used in the circuit and the working and the gating of the switch are discussed (Tables 9.3 and 9.4).

Simulation Circuit

The circuit diagram is implemented in ADS software, as shown in Figure 9.72, and the Pulse used is PWM signal for the IGBT and the respective simulation tools are used for the software.

9.2.6   SIMULATION TOOLS

a. **Transient Simulation**

Transient block is used to run the stimulation, as shown in Figure 9.73; the maximum time and step size are to be entered in this tool.

TABLE 9.3
Component Value

Parameter	Symbol	Value
Power	Pout	1 kW
Grid Voltage	Vs	220 Vrms
Coupling Inductance	Lc	1 mH
DC Link Voltage	V DC	310 V
DC Link Capacitance	C DC	1,000 uF
Switching Frequency	Fs	30 kHz
Filter Inductance	Lf1 and Lf2	800 uH
Filter Capacitor	Cf	330 uF
Battery Voltage	Vbat	72 V

TABLE 9.4
Component Model

Parameters	Symbol	Value	Market Model	Manufacturer
IGBT Transistors	SW	560 V, 29 A	FGB3056-F085	ON Semiconductor / Fairchild
Coupling Inductance	Lc	1 mH	SRF1260A-102M	BOURNS
DC Link Capacitance	C DC	1,000 uF	ALS70A102DE400	KEMET
Filter Inductance	Lf1 and Lf2	800 uH	PM3602-200-RC	BOURNS
Filter Capacitor	Cf	330 uF	ALA7DA361CC400	KEMET
Battery	Vbat	72 V	BP72V28RT3U	TRIPP LITE

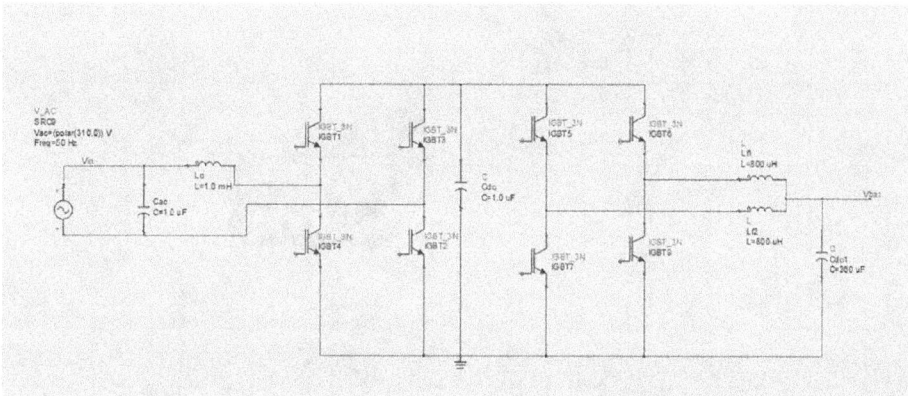

FIGURE 9.72   Circuit diagram.

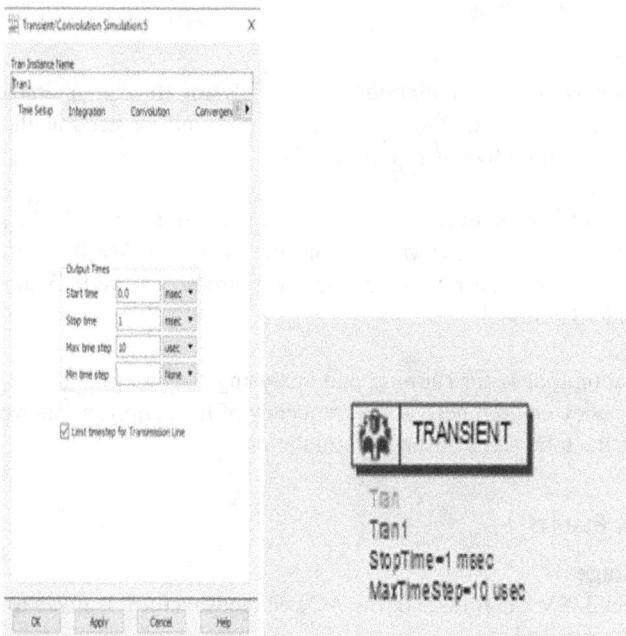

FIGURE 9.73   Transient block and parameter.

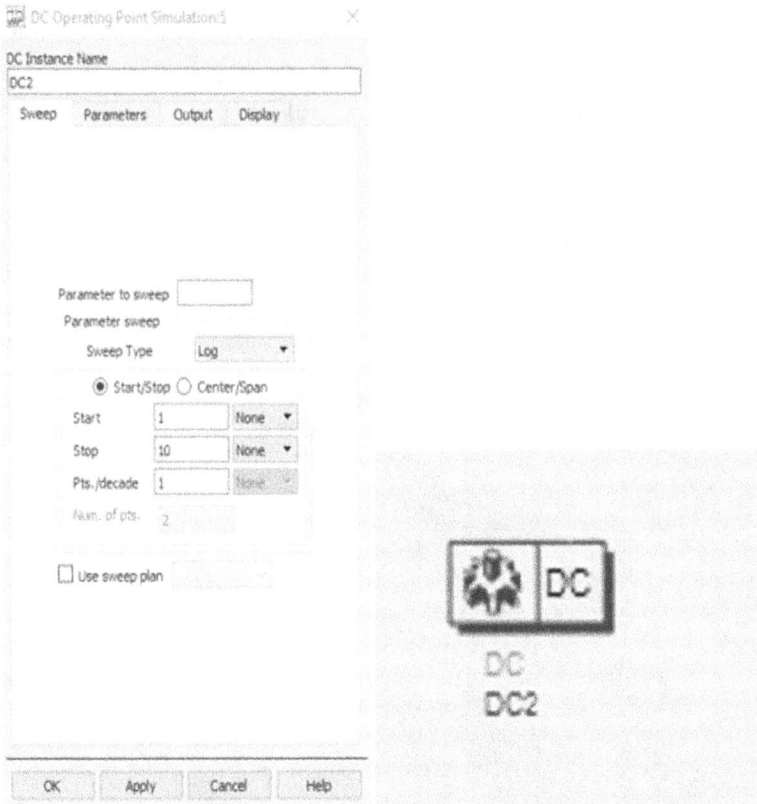

FIGURE 9.74 DC simulation block and parameter.

b. **DC Simulation**

DC simulation is used as simulation tool for analysing the DC parameters. The parameters to be entered in the simulation are the terms which we need as the output is to be entered in this simulation tool as in Figure 9.74.

c. **IGBT Block**

This IGBT block is used to control and operate the IGBT switch in the safe mode. Here the values to be entered are the values from the datasheet. We have to take the values based on the voltage and current under the short circuit. Figure 9.75 shows the Instance Parameter and IGBT Model.

d. **Simulation AC**

This simulation tool is for running and analysing the AC parameters available at the input; the frequency entered here is the frequency of the supply, as shown in Figure 9.76. The stepsize is the total steps of the AC simulation.

9.2.7 Simulation Results

a. **Input AC Voltage**

For the Input 325 V Magnitude of the AC voltage waveform is given as in Figure 9.77

b. **PWM Pulse Generated**

PWM pulse is generated using 12 V as the peak voltage and PWM waveform given of frequency 30 kHz, edge time 20 ns, and dead time 30 ns, as shown in Figures 9.78 and 9.79.

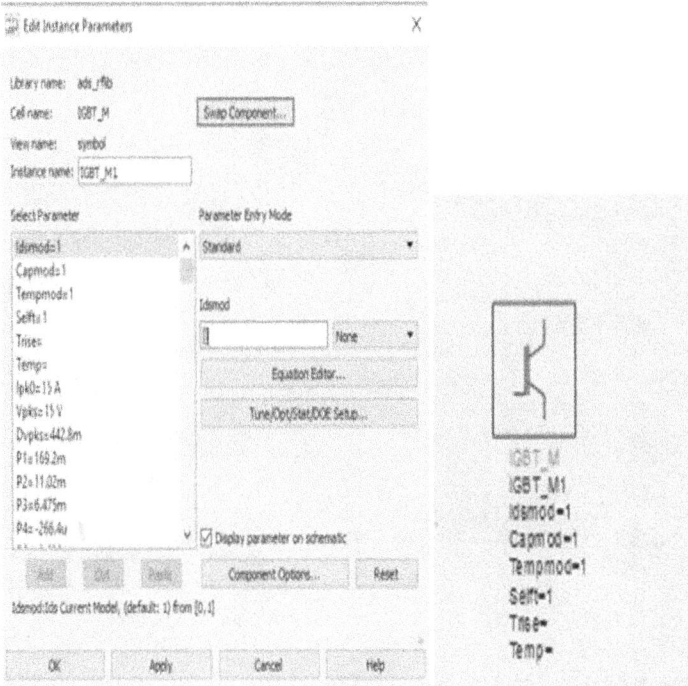

FIGURE 9.75 Instance parameter and insulated gate bipolar transistor (IGBT) model.

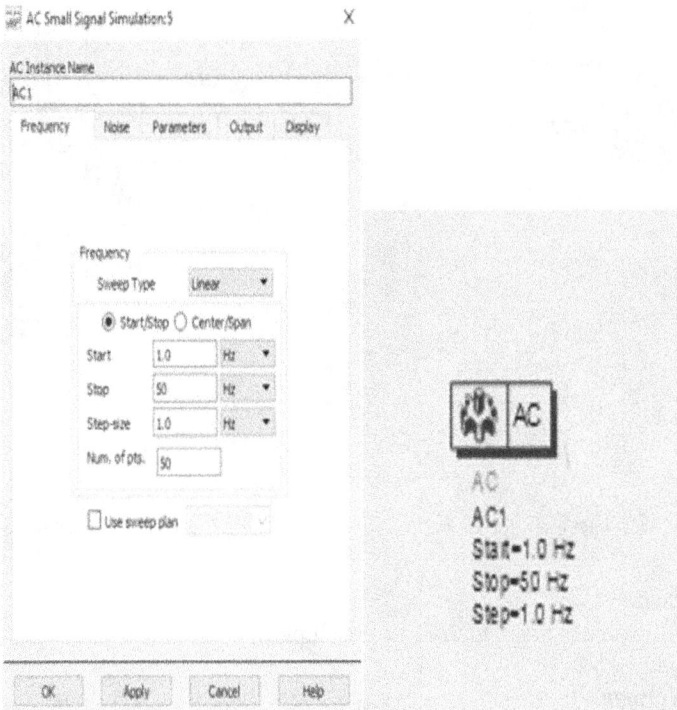

FIGURE 9.76 AC simulation tool and AC parameter.

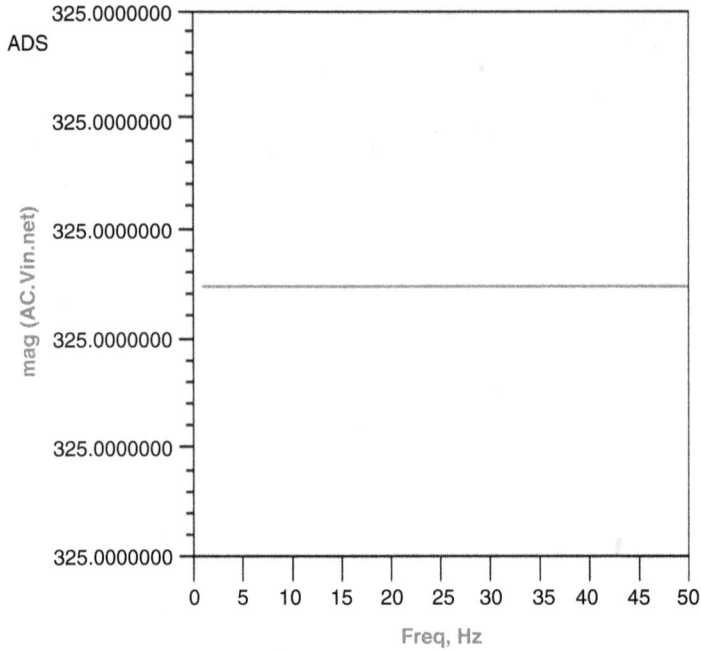

FIGURE 9.77   Input voltage.

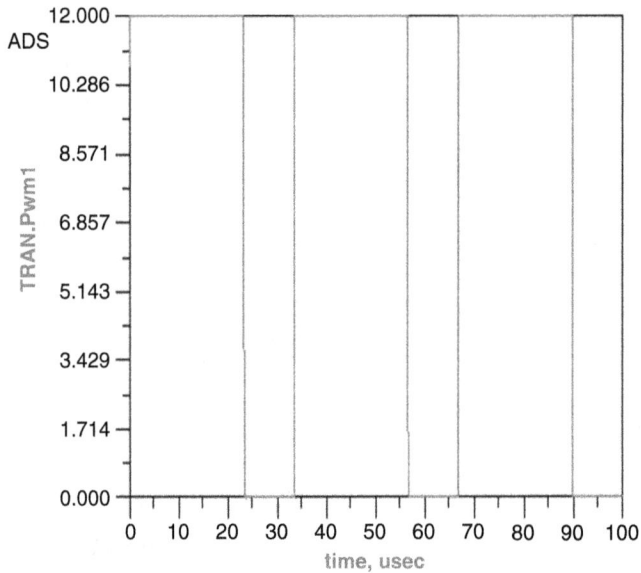

FIGURE 9.78   PWM 1 waveform for 20 ns.

c. **Signal to Switch**

PWM pulse gate is given to the gate signal block where the output of the gate signal is given in Figure 9.80.

d. **DC Link Voltage**

AC voltage is converted into DC voltage and maintained in the DC link capacitance as shown in Figure 9.79.

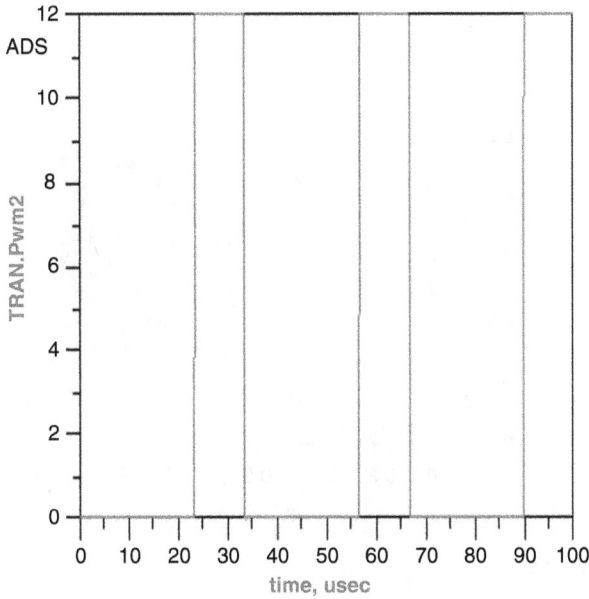

FIGURE 9.79 PWM 1 waveform for 30 ns.

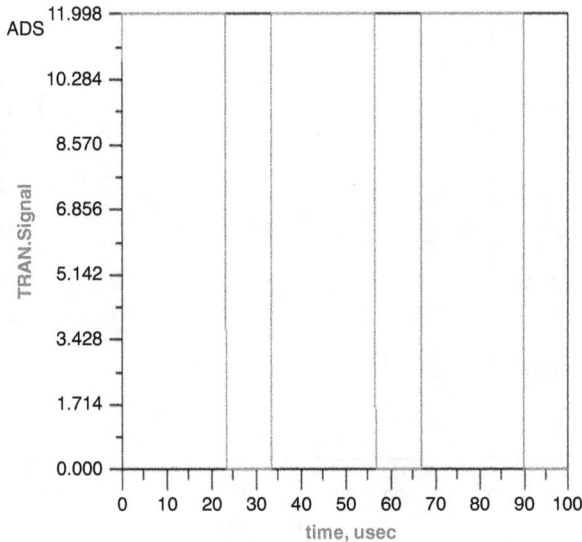

FIGURE 9.80 Output of gate signal.

e. **Output Battery Voltage**

Lumped Circuit (LC) filter is used at the output and the battery is connected for charging. Figure 9.82 shows the output battery voltage.

9.2.8 S-Parameter Implementation

S-parameters characterize the reaction of an N-port network to any or all incident signals. The first number indicates the replying port, while the second indicates the incident port. Therefore, S_{21} indicates the reaction at port 2 in response to a signal at port 1. One-port and two-port networks

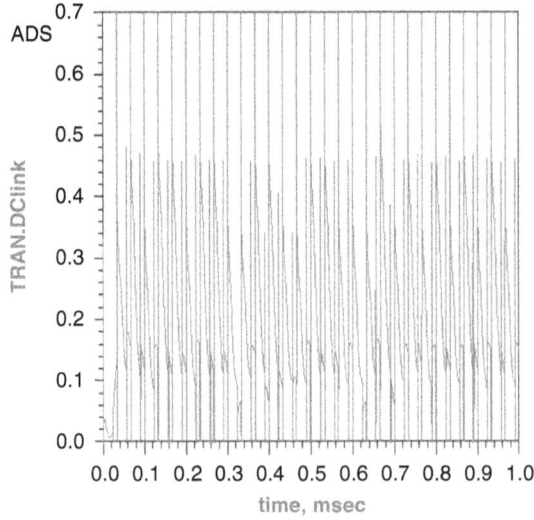

FIGURE 9.81 DC link voltage.

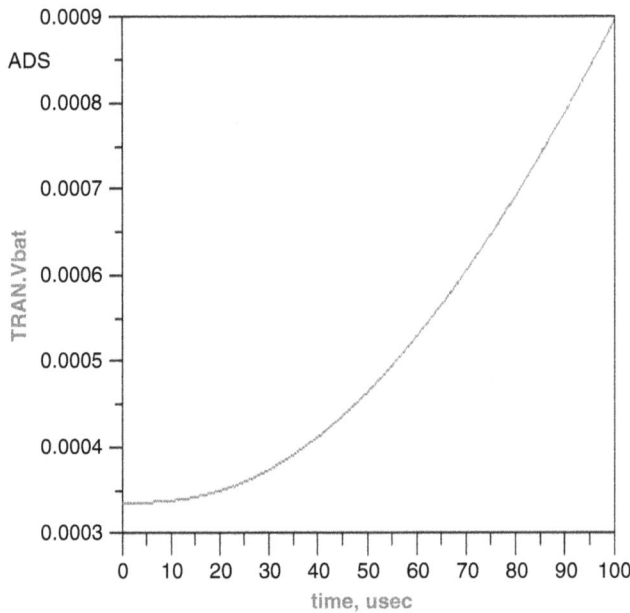

FIGURE 9.82 Battery voltage.

are the most prevalent "N-port" networks in microwaves, as depicted in Figure 10.83. Modelling three-port network S-parameters with software such as ADS is straightforward; however, accurate measurements of three-port network S-parameters are exceedingly challenging. Typically, vendors offer measured multi-port S-parameters for amplifiers and other devices, but you should always verify that your answers are realistic.

Let's analyse a network with two ports. Consider the signal at a port, port 1, as the superposition of two waves travelling in opposite directions. By convention, each port is depicted as two nodes in order to assign a name and value to these waves travelling in opposite directions. The variable ai represents an incident wave at port I while the variable bj represents a reflected wave at port j.

Generalized two-port
network, characteristic
impedance Z0

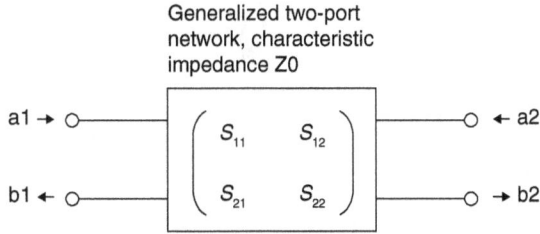

FIGURE 9.83 S-Parameter network diagram.

Don't obsess over how two signals can simultaneously occur at the same node! The magnitudes of the *ai* and *bj* variables can be viewed as voltage-like variables that have been normalized using a given reference impedance. This is quite convenient, as the square of these magnitudes equals the wave's power level. Remember that S-parameters are meaningless unless you know the value of the reference impedance, often known as Z_0.

$$S_{11} = \frac{b1}{a1}$$

$$S_{12} = \frac{b1}{a2}$$

$$S_{21} = \frac{b2}{a1}$$

$$S_{22} = \frac{b2}{a2}$$

The aforementioned equations for S_{11} and S_{21} are determined from network analysis or measurements by setting the incident signal's amplitude to 0 and solving for the ratios of S-parameters as a function of a_1. In a similar manner, S_{12} and S_{22} are calculated by setting $a_1 = 0$ and solving for the other ratios.

9.2.8.1 S-Parameter in ADS
S-Parameter symbol is created using the Pins P1 and P2 at input and output, respectively. This is used to create the sub-circuit of the input and output, as shown in Figure 9.84.

9.2.8.2 Circuit Symbol
After the creation of Pins, we have to create the symbol, and we have to add the ports and the symbol for the circuit which will be created as shown in Figure 9.85.

9.2.8.3 S-Parameter Sub-Circuit
The symbol is created and then with the symbol we have to create a sub-circuit with Z termination at the input and output ports for the calculation of the Z and S-parameters. We have to create a schematic and then we have to link the symbol with the schematic for creating the sub-circuit, as shown in Figure 9.86.

9.2.8.4 Results of Z-Parameter
The above waveforms (Figures 9.87–9.89) are taken from the S-parameter block where the Z1 and Z2 input and output port Z-parameters are analysed and the results are given above.

9.2.8.5 Results of S-Parameter
The above results (Figures 9.90–9.93) are the S-parameter outputs. These are obtained from the S-parameter simulation tools. These waveforms show how much of Scattering has occurred in the circuits and this can be calculated using these S-parameters.

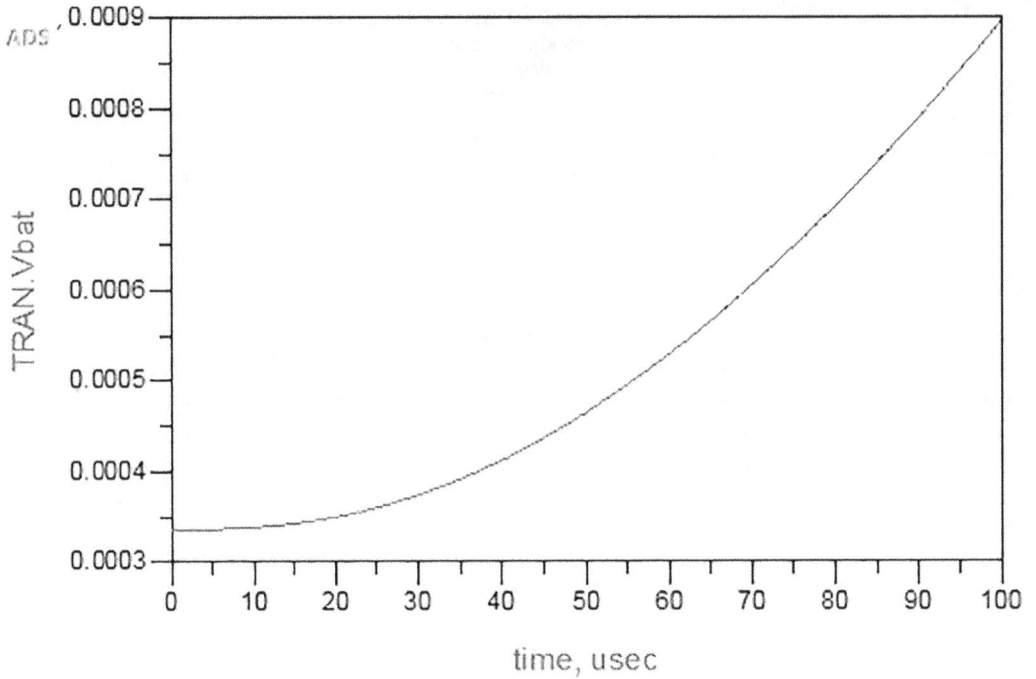

FIGURE 9.84 S-Parameter symbol.

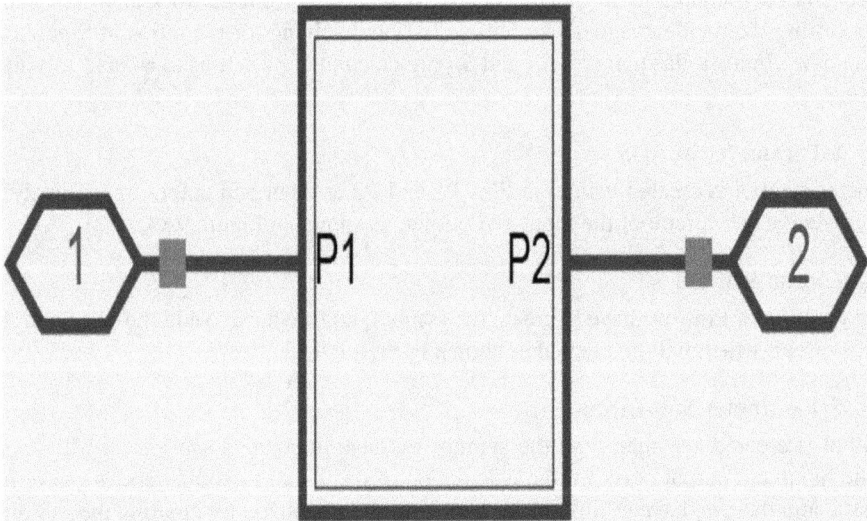

FIGURE 9.85 Circuit symbol.

9.2.9 CONDUCTED EMISSION (CE) CALCULATION

CEs are the noise currents produced by the device-under-test (DUT) that travel via the power cord or harness to other components/systems or the power grid. These noise currents can be measured utilizing either the voltage or current technique. Controlling and avoiding CEs should be simpler than radiated emissions. Being lower frequency, it is less susceptible to parasitic interference than higher frequency issues. Nonetheless, they remain an issue that must be considered.

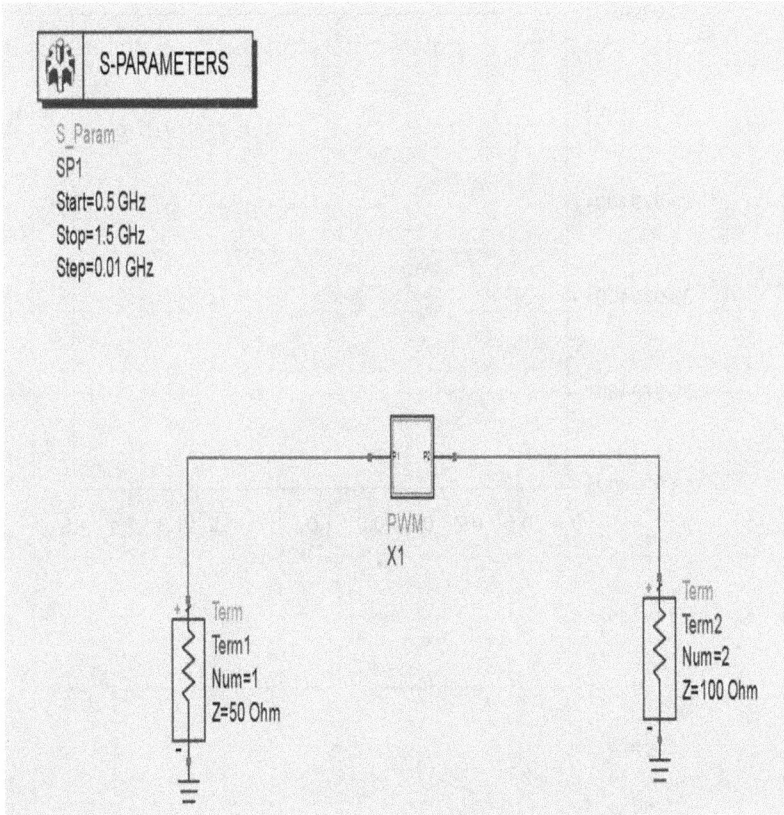

FIGURE 9.86   Sub-circuit.

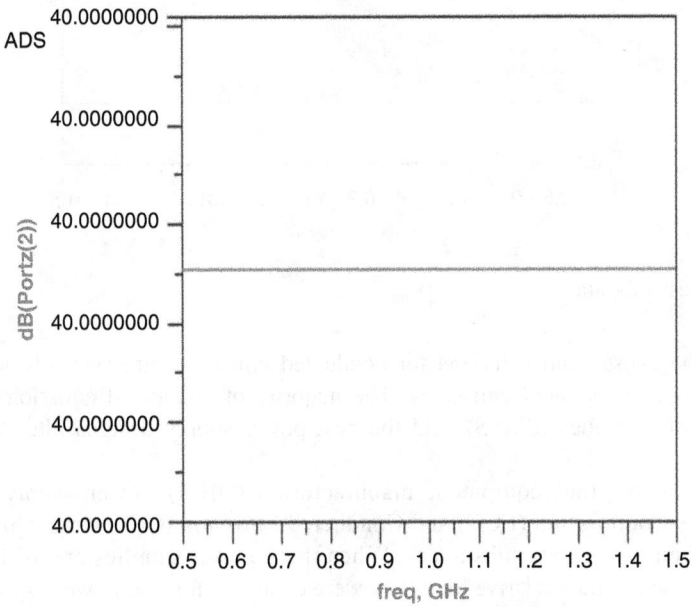

FIGURE 9.87   Result Z2 parameter.

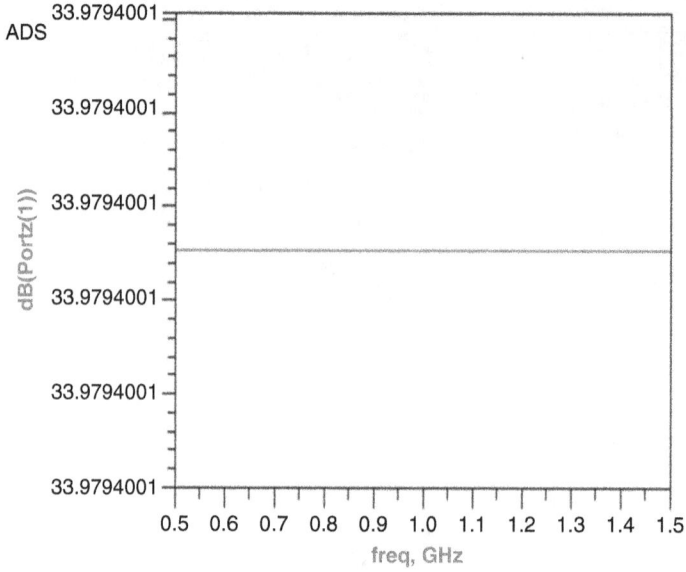

FIGURE 9.88   Result Z1 parameter.

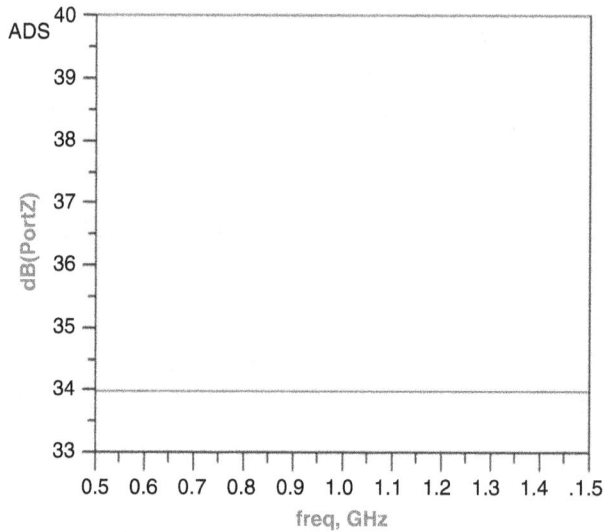

FIGURE 9.89   Result Z-parameter.

Consequently, the causes and solutions for conducted emissions are typically simpler to comprehend than those for radiated emissions. The majority of conducted emissions are caused by switch-mode power supplies (SMPS), and the best power supply designs filter the power input adequately.

However, many original equipment manufacturers (OEM) power supply carry Federal Communications Commission (FCC) and Conducted Emission (CE) marks while being poorly constructed and emitting toxic substances. When these power supplies are loaded with a reactive load, as opposed to the resistive load they were designed for, the power supply may become unstable or noisy, and additional measures are often required to keep it in compliance. In addition, with modern products, it is possible for higher frequency harmonics to contaminate the system power supply and leak out through the filter and back into the power line.

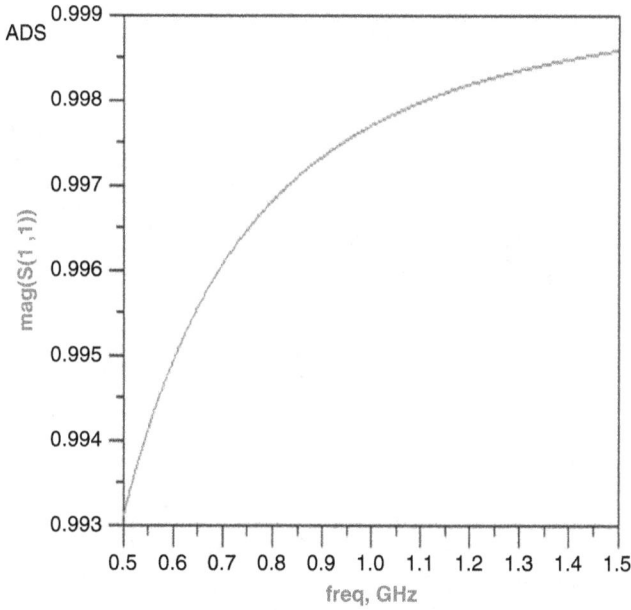

FIGURE 9.90 S_{11} output.

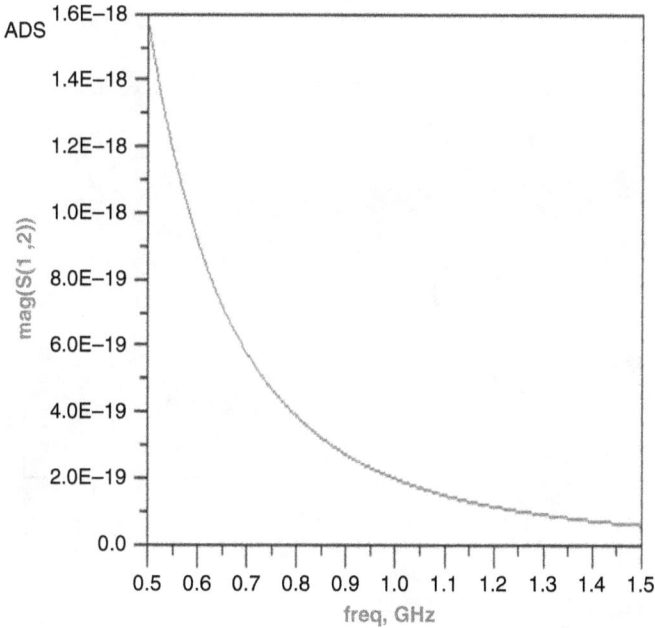

FIGURE 9.91 S_{12} output.

Always be on the lookout for circumstances in which the filter is compromised, either by design or by system design faults, such as inadequate internal cable routing, filter or power supply placement, or bad chassis or signal returns connection. Typically, the product's failure modes are low, but excessive emissions can disrupt sensitive measuring equipment or communications receivers in the vicinity or linked to the same power line circuit.

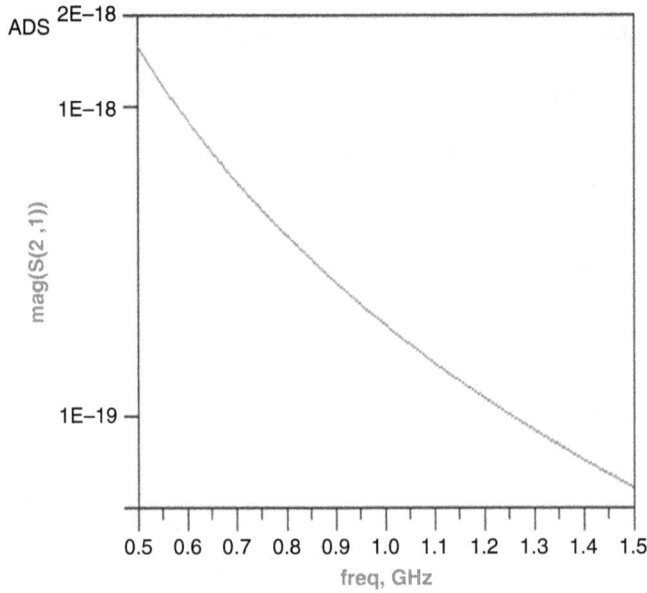

FIGURE 9.92 S_{21} output.

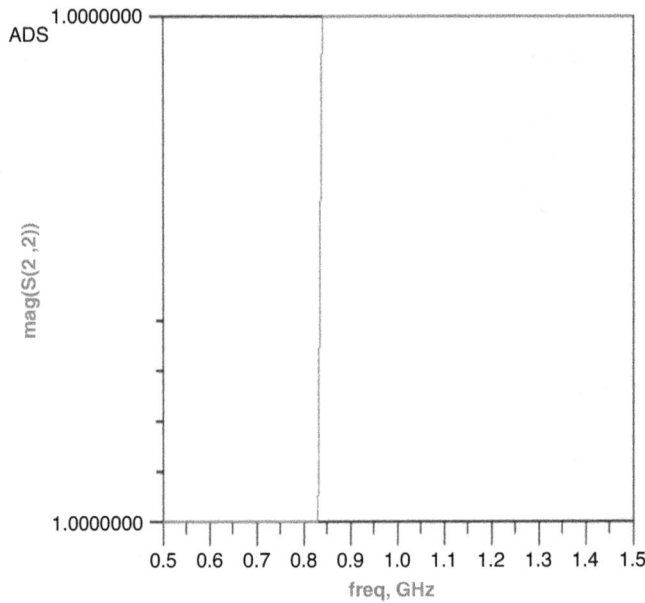

FIGURE 9.93 S_{22} output.

Testing was performed to estimate or determine the noise caused by a sudden change in voltage or current in the equipment's circuitry and emitted through the power cable into peripheral or loaded devices. The unwanted noise may be lethal to the connected devices and cause them to malfunction. Popular techniques for evaluating conducted emissions include:

• Super heterodyne EMI receiver
• Line impedance stabilization network (LISN)

The primary need for CEs is the placement of the equipment under test (EUT) in relation to the ground plane and the LISN, as well as the mains cable and earth connection. Placement influences the stray coupling capacitance between the EUT and the ground reference, which is part of the common mode coupling circuit, and must thus be closely managed; in most situations, the standards require a distance of 0.4 m.

Cable connections must have a controlled common mode inductance, which requires a defined length and the least amount of ground plane coupling as possible.

9.2.10 CIRCUIT DIAGRAM

LISN is added to the circuit where the EUT is connected to the output and V_{bat} is connected to the input as shown in Figure 9.94.

9.2.11 LISN MODEL

As depicted in Figure 9.95, LISNs are used to convert interference currents to voltages that can be measured by EMI receivers, as well as to ensure constant, stable, and standardized measuring conditions of conducted interferences introduced into the mains by the investigated EUT.

9.2.12 ABOUT LISN CISPR 25

This standard is intended to protect receivers from disturbances caused by a vehicle's conducted and radiated emissions. The provided test methodologies and limitations are meant to offer provisional control over vehicle-radiated emissions, as well as component/module conducted/radiated emissions of long and short durations.

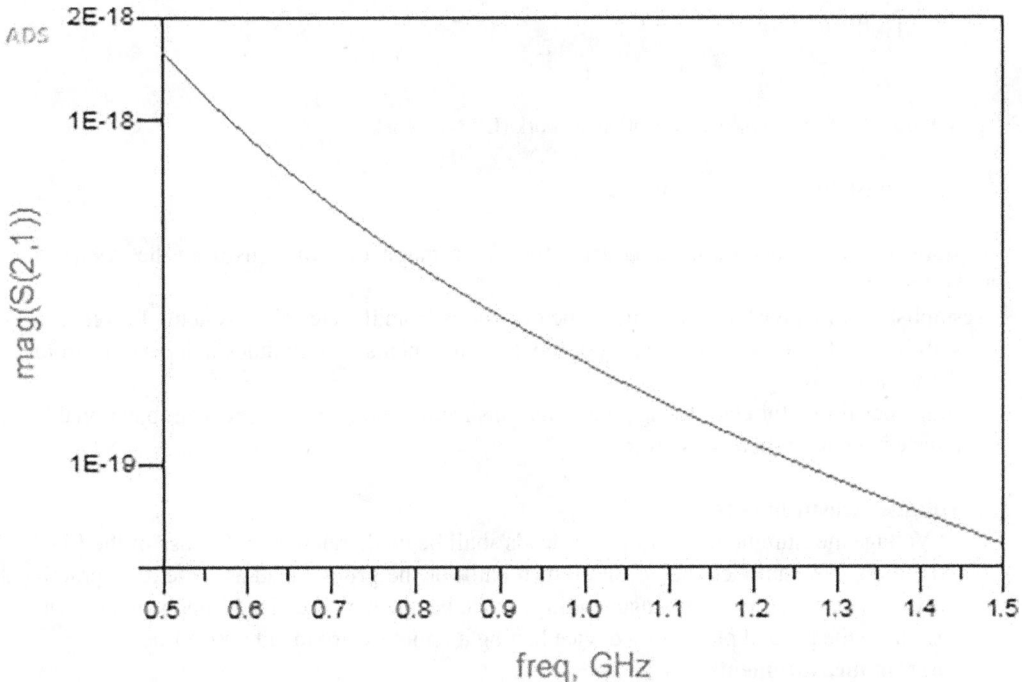

FIGURE 9.94 Line impedance stabilization network (LISN) implemented circuit.

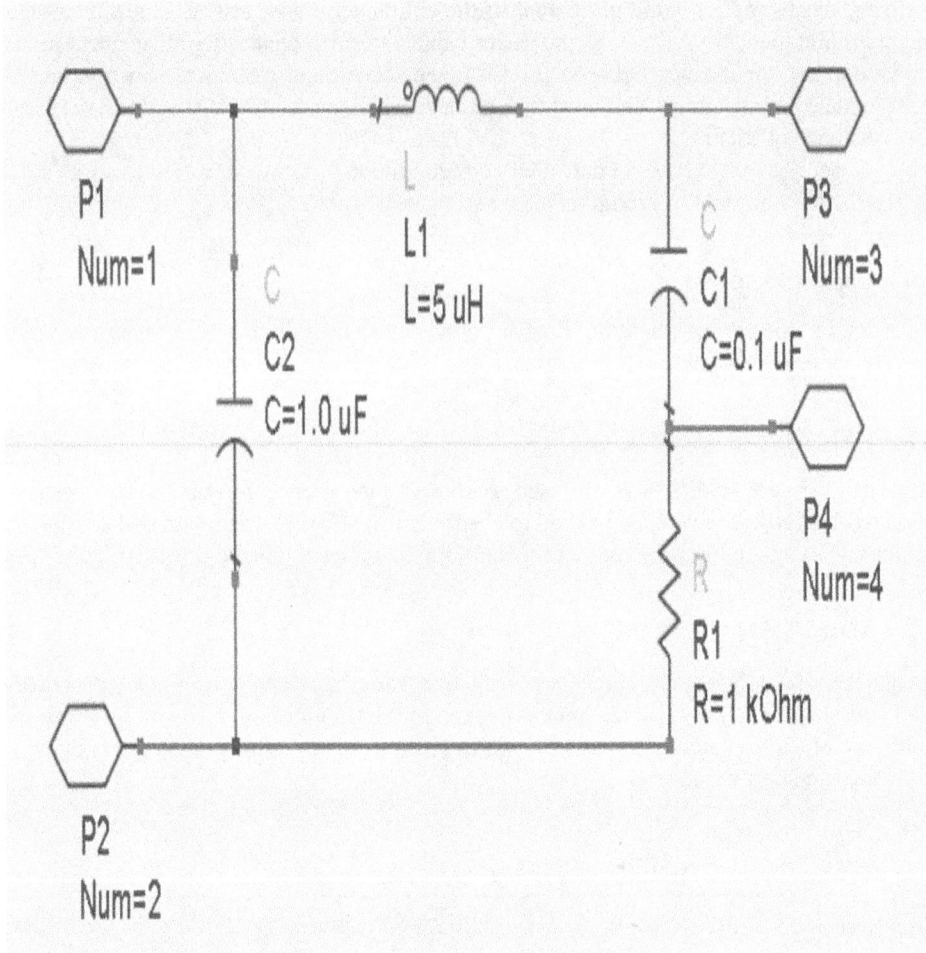

FIGURE 9.95   Line impedance stabilization network (LISN) model.

To accomplish this end, this standard:

– establishes a test method for measuring the electromagnetic emissions from the electrical
  system of a vehicle;
– establishes limits for the electromagnetic emissions from the electrical system of a vehicle;
– establishes a test method for testing on-board components and modules independent from
  the vehicle;
– establishes limits for electromagnetic emissions from components to prevent objectionable
  interference to on-board receivers;

a. **Voltage measurements**

   Voltage measurements on all power leads shall be made relative to the case of the EUT
   (where the case supplies the ground return path) or the ground lead as close as is practically possible to the EUT. Voltage readings are to be taken on each lead (supply and return)
   relative to the ground plane for a device having a remotely grounded return line.

b. **Current measurements**

   Current probe measurements must be performed on the control/signal leads as a single cable or in sub-groups commensurate with the current probe's physical dimensions.

The length of the test harness shall be nominally 1.5 m (or as specified in the test plan) and 50 mm above the ground plane. Unless otherwise indicated in the test plan, the test harness wires shall be nominally parallel and contiguous. Place the current probe 50 mm away from the EUT connector and then measure the emissions.

To ensure that the highest level is detected for frequencies greater than 30 MHz, the current probe must be positioned in the subsequent positions:

a. 500 mm from EUT connector;
b. 1,000 mm from EUT connector;
c. 50 mm from antenna (AN) terminal.

In the majority of circumstances, the position of greatest emission will be as close as possible to the EUT connector. If the EUT is equipped with a metal shell connector, the probe must be attached to the cable immediately adjacent to the shell, but not the shell itself. The EUT and other test setup components must be at least 100 mm from the edge of the ground plane.

9.2.13 LISN DESIGN

As per International Special Committee for Radio Interference (CISPR) standard, the value of C1 is 0.1 uF, C2 is 1 uF, L1 is 5 uH, and R is 1 kOhm as shown in Figure 9.96.

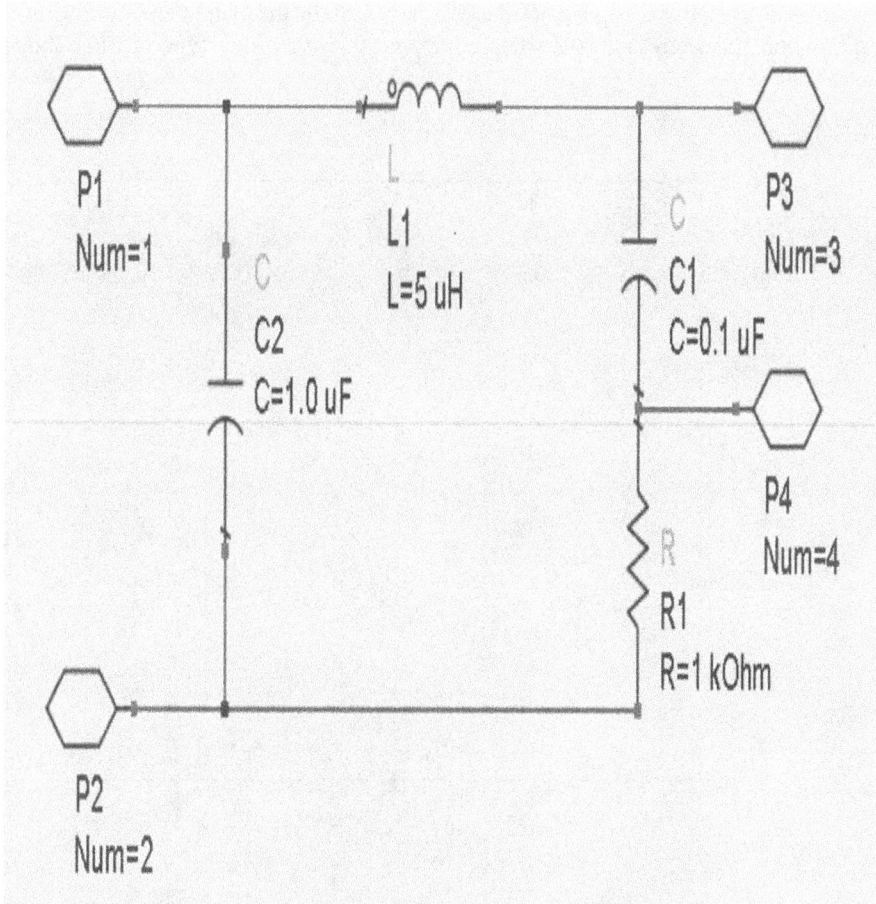

FIGURE 9.96   Values of components in line impedance stabilization network (LISN).

a. **Results of LISN**

The above LISN outputs shown in Figures 9.97 and 9.98 are taken from the LISN block available at the ADS software where these are taken at the $V_{measurement}$ terminal in the LISN block; these measured voltages are fed into the filter for amplification of the voltage signal obtained from the LISN.

T-Model Filter Design

$$L = \frac{Z_0}{\pi f_c} = 2\,\text{mH}$$

$$C = \frac{1}{Zo\,\pi\,f_c} = 80\,\text{uH}$$

$$f_c = \frac{1}{\pi\sqrt{LC}} = 10\,\text{kHz}$$

The cut-off frequency is calculated based on 40% of the switching frequency where with those values, the inductor and the capacitor values are calculated and implemented in T filter for the amplification where the amplifier used here is the operational amplifier (OP AMP).

T-Model Filter Circuit

T Filter is used after the analysis made on the other type of filter where T filter is best suitable for the AC input and also has less loss in the output waveform analysed in the digital spectrum analysed when compared with the other type of filter shown in Figure 9.99.

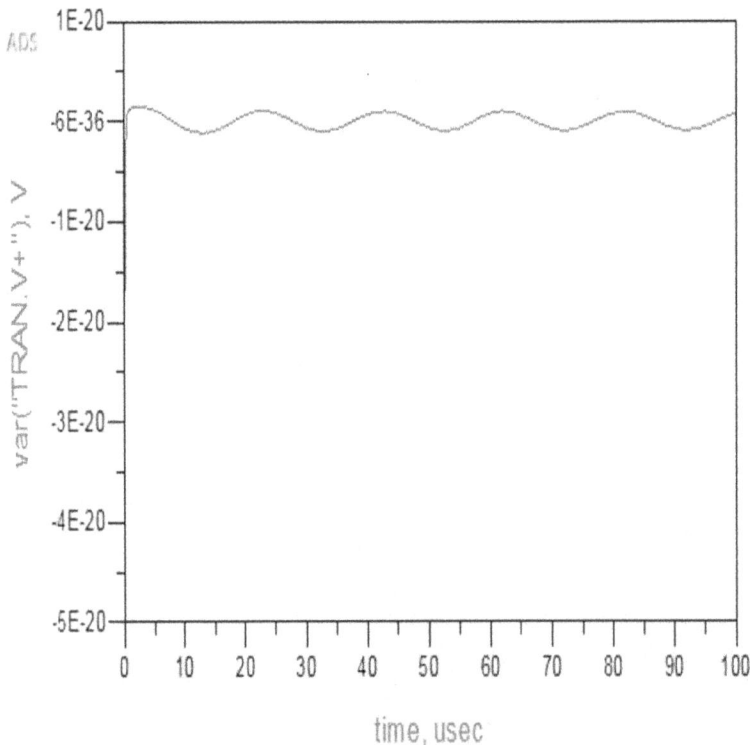

FIGURE 9.97   Line impedance stabilization network (LISN) V+ output.

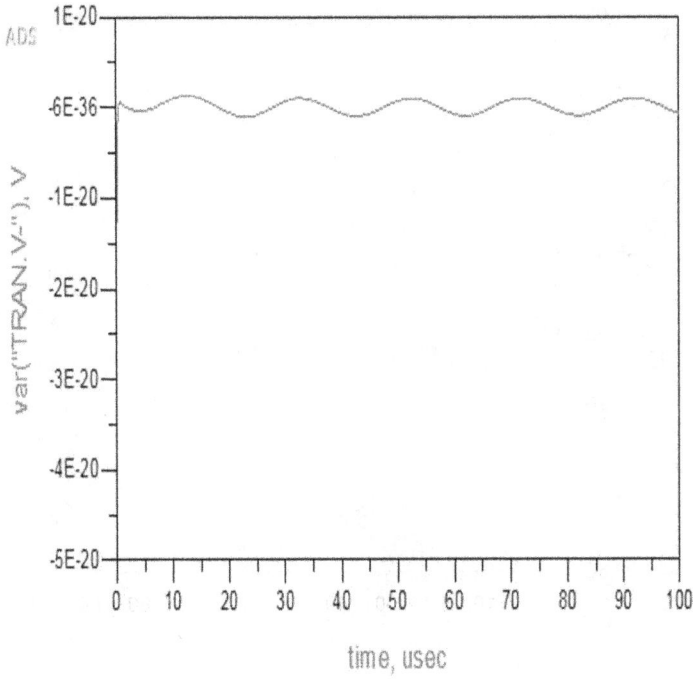

FIGURE 9.98   Line impedance stabilization network (LISN) V– output.

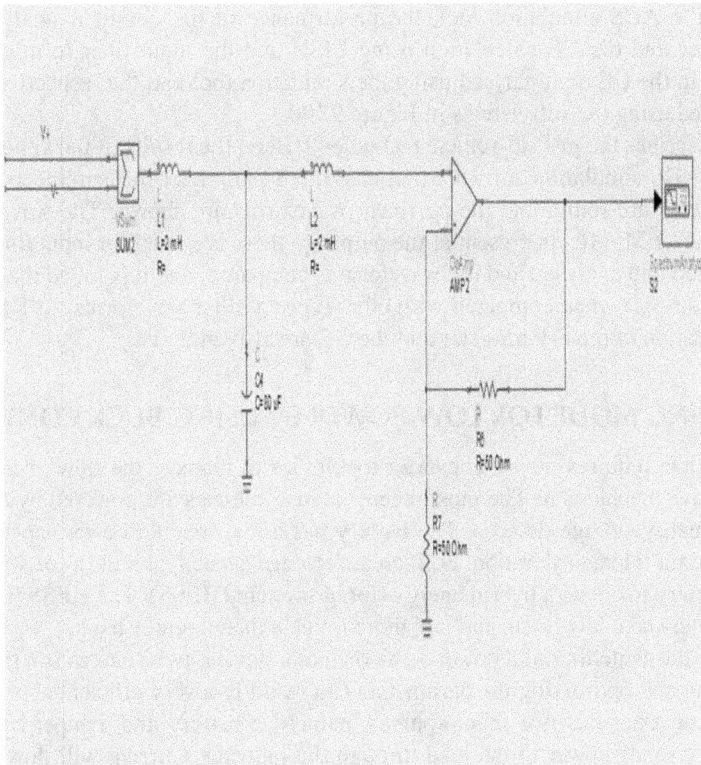

FIGURE 9.99   T-model filter.

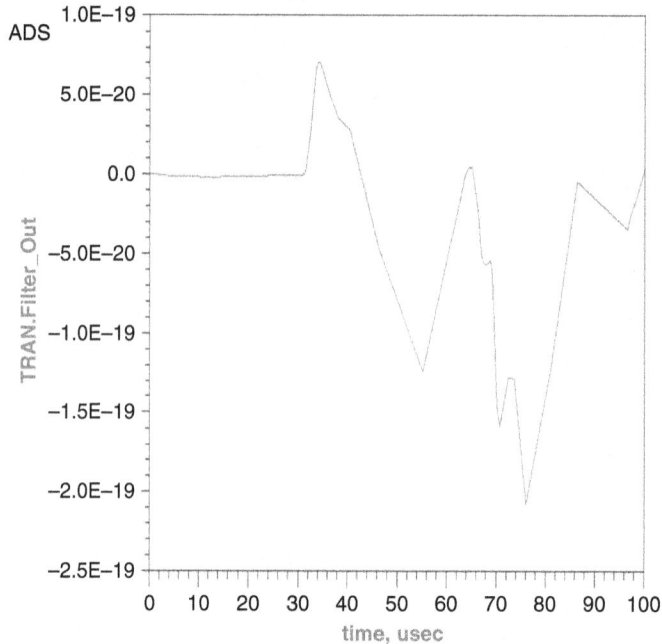

FIGURE 9.100   Filter output.

b. **Results of Filter**

Using the ADS simulation tool, the performance of the circuit is analysed and the S-parameter and the CE calculation using LISN and the input filter to modify the loss occurring in the CE are analysed using the simulation tool and the respective waveforms are obtained using the software as in Figure 9.100.

This describes the grid-to-vehicle technology. Here, the results of the system are found using the ADS simulation software and also the S-parameter measurements and the CE measurements are found and the pertinent waveforms are shown. The waveforms show that there are EMI effects present at the output so there is a need for input filter. T filter is implemented and analysed and the waveform is compared, and it is found that T filter output has less ripple when compared with other types of filter waveforms, so T filter is used. Implementation of the S-Parameter and the CE are also analysed.

9.3   MOTORING MODE FOR LOW-POWER EV USING BUCK CONVERTER

The electric vehicle requires power converter topologies to manage the flow of energy from the battery to the drive mechanism. The most recent electric vehicles are powered by a mix of two or more types of energy storage devices. The battery has a low specific power capacity and cannot power the device for a longer duration. To increase storage capacity, an ultra- or super-capacitor is paired with a battery to create a hybrid energy storage system (HESS). The HESS has a high power density, a long life, and a low price, and it is more reliable than a power bank.

In addition to the switching of a power semiconductor device, two sources run the load application. Objectifying and optimizing the parameters can result in a very efficient application process. The suggested converter has two input sources, namely, a battery and a super-capacitor. In this mode, the battery sends power to the load through the inductor. Current will flow in the forward direction, from battery to load, and motor torque will be positive. When the system experiences a sudden change in voltage, the super-capacitor will stabilize the system's voltage. The system on chip

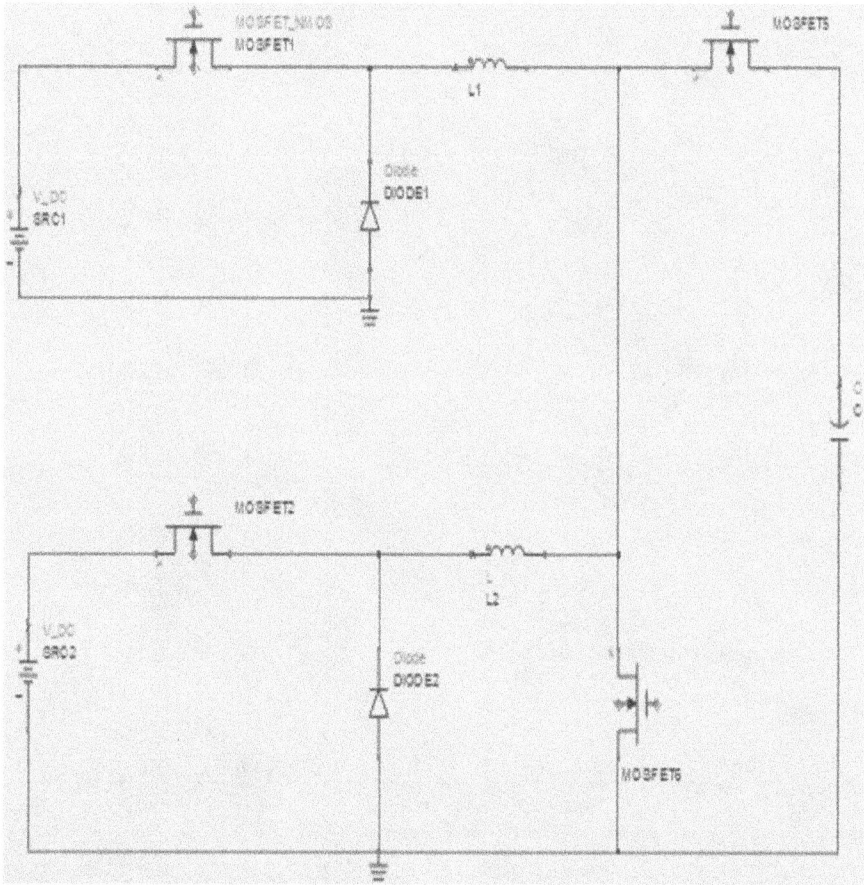

FIGURE 9.101   Buck converter circuit.

(SOC) of the battery is discharging from 100 V. Starting battery current increases, but stabilizes at 1.8 Amps after some time. The speed of the motor approaches 20 rad/s. The quick shift in load raises the armature current, causing the speed to decrease to 17 rad/s and the torque to increase to 5 Newton-meters in direct proportion to the armature current. Figure 9.101 illustrates the step-down performing circuit.

9.3.1 Simulation of Converter Circuit

The Figures 9.102 and 9.103 show the simulation of the convertor circuits.

9.3.2 System Design Calculations

Proper inductive and capacitive values have to be designed to nullify the ripple contents in the input and output sides. These values play a prominent role in the performance of the system. Table 9.5 shows the designed and simulated parameter values.

For L and C values :

$$f_s = 40 \, \text{KHz} \rightarrow T_s = 25 \, \text{us}$$

$$V_{\text{input}} = 100 \, \text{V}; \, V_{\text{output}} = 48 \, \text{V}$$

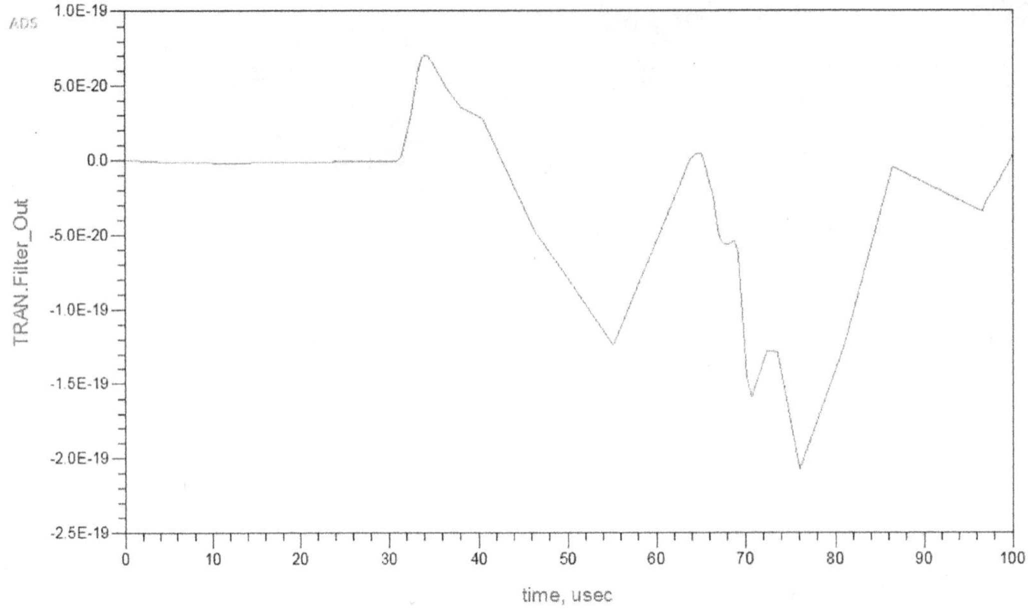

FIGURE 9.102   Converter circuit with DC block.

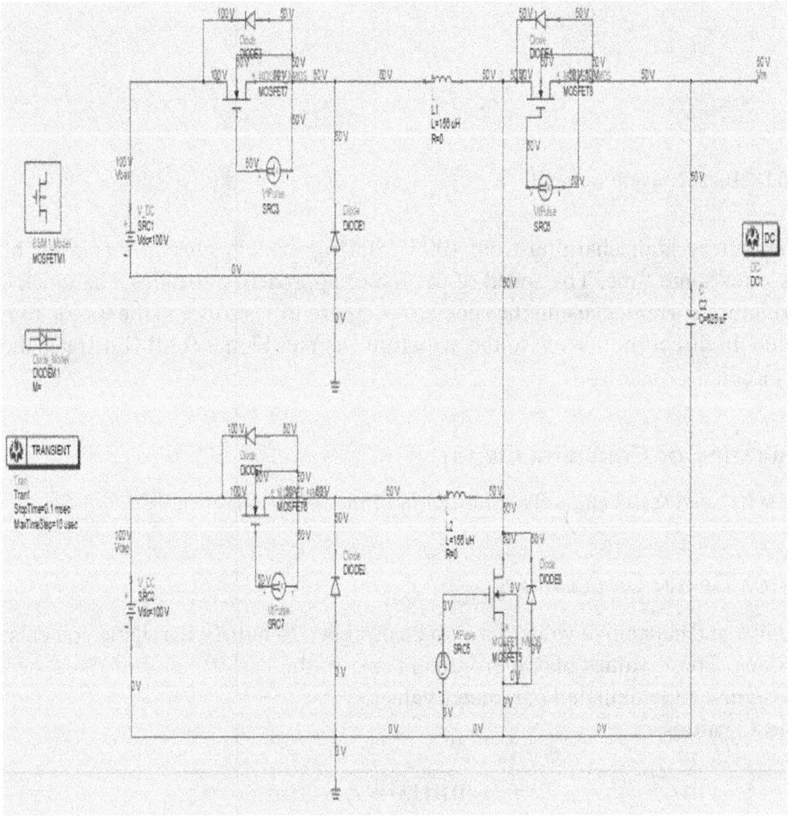

FIGURE 9.103   Converter circuit with Transient block.

TABLE 9.5
Theoretical and Simulation Values

Parameter	Designed Setup Values	Simulating Values
V_{in}	100 V	100 V
V_o	48 V	43.8 V
P_{out}	1 KW	1 KW
L	150 uH	156 uH
f_s	40 kHz	40 kHz
D	0.48	0.48
C	650 μF	625.6 μF

For Buck Converter Operation:

$$V_o = D * V_s$$

Duty Cycle, $D = 0.48$

$$L = V_o(1-D)/(f_s * I_o) \Rightarrow 156 \text{ uH}$$

$$C = I_o/(f_s * V_o) \Rightarrow 625 \text{ uF}$$

Figure 9.104 shows the instance parameter for inductor and capacitor.

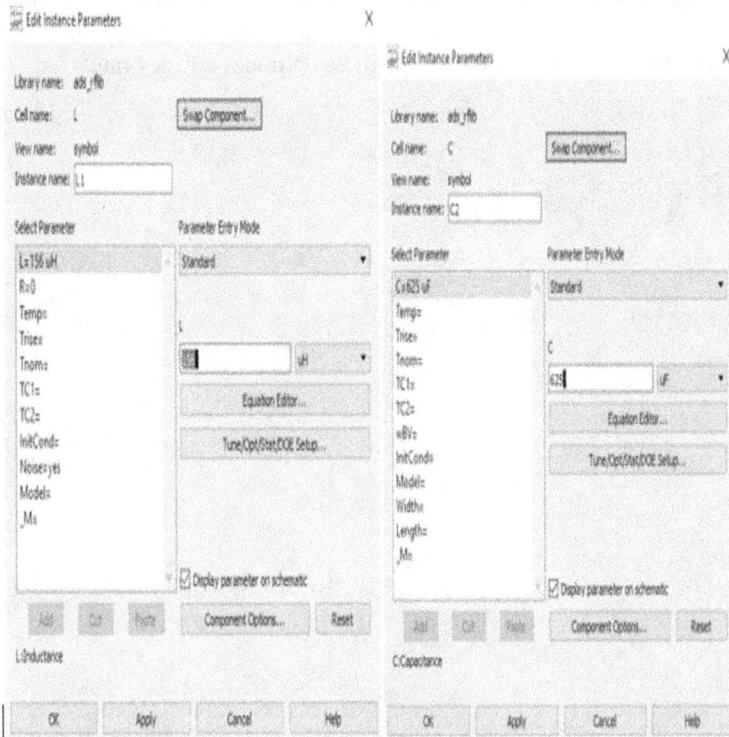

FIGURE 9.104 Instance parameter for inductor and capacitor.

9.3.3 IMPLEMENTED CIRCUIT RESULT AND INFERENCE

The above obtained output (Figure 9.106) shows that the output value gets settled down at 43.252 V. The graph intimates that there is no ripple or any high-order disturbances coming out from the converter circuit. Precisely, the converter circuit provides ripple-free output to the drive system. This is achieved only by choosing the appropriate design values of L and C values for the converter circuit. And, the required output is to be in the range of 36–48 V to drive the PMDC motor. Also, the conductive emission needs to be analysed in the system to find out the distortion level affecting the system internally or externally by having the symbol as shown in Figures 9.107 and 9.108.

S-parameter simulator [SP] is a key component to perform scattering parameter analysis for the circuit.

Clicking on the S-parameter icon, S-parameter instance tab pops up as in Figure 9.109. And then desired values can be entered.

And add the PORT IMPEDANCE TERMINATION for S-Parameter at either ends of the two-port symbolized blocks.

9.3.4 S - DOMAIN VALUES IDENTIFICATION

Click on the simulate icon [⚙] after placing the port impedance terminator on the ends (Figure 9.110).

On the whole, the circuit is tested and operated for 10 GHz frequency range (Figure 9.111).

Individual graph for S-parameter S(1,1), S(1,2), S(2,1), S(2,2) reflexes the circuit impedance and its stability shown in Figure 9.109. The overall circuit impedance value shown is 34 ohm.

The frequency-domain graph results show that the circuit is unstable and three poles lie on the right hand side of the S-Plane. So addition of zero or the filter components is a must to bring the system back to the stable region so as to operate in an optimum efficacy rate.

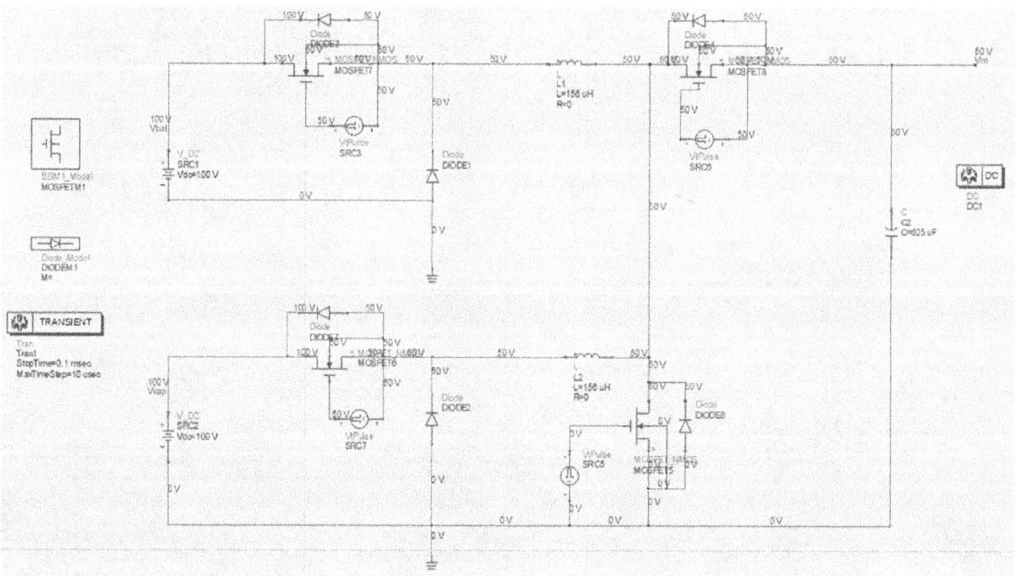

FIGURE 9.105   Implemented circuit result.

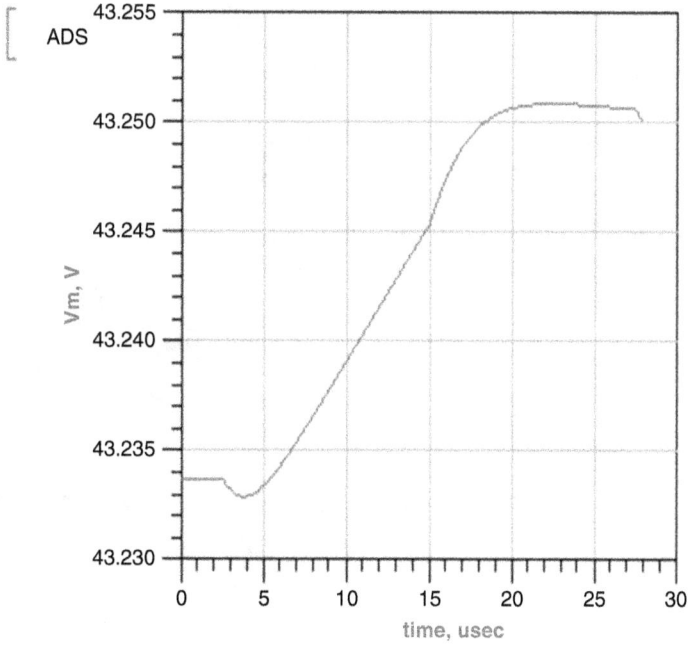

FIGURE 9.106 Output waveform.

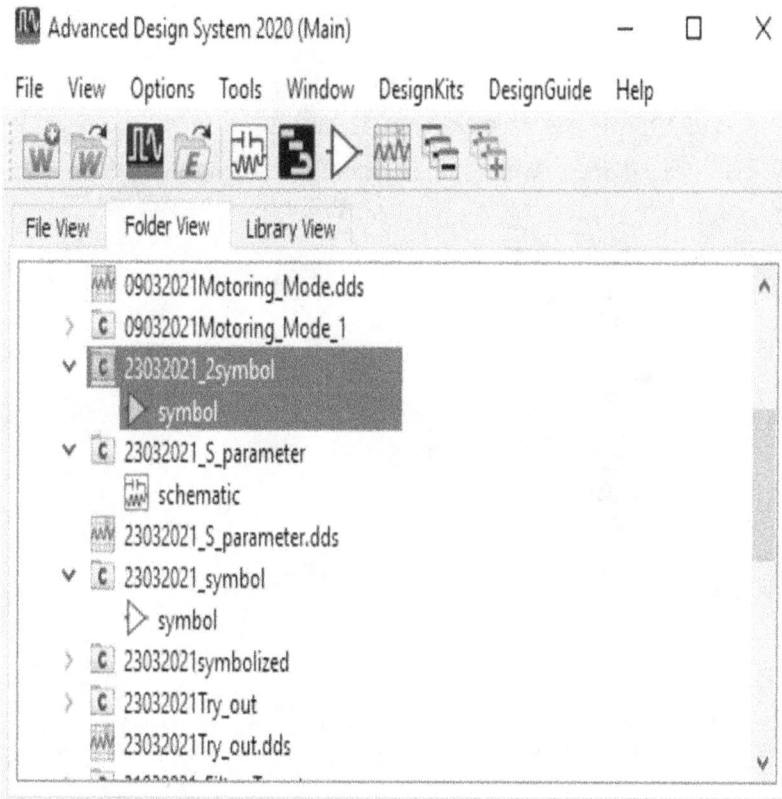

FIGURE 9.107 Advanced design system.

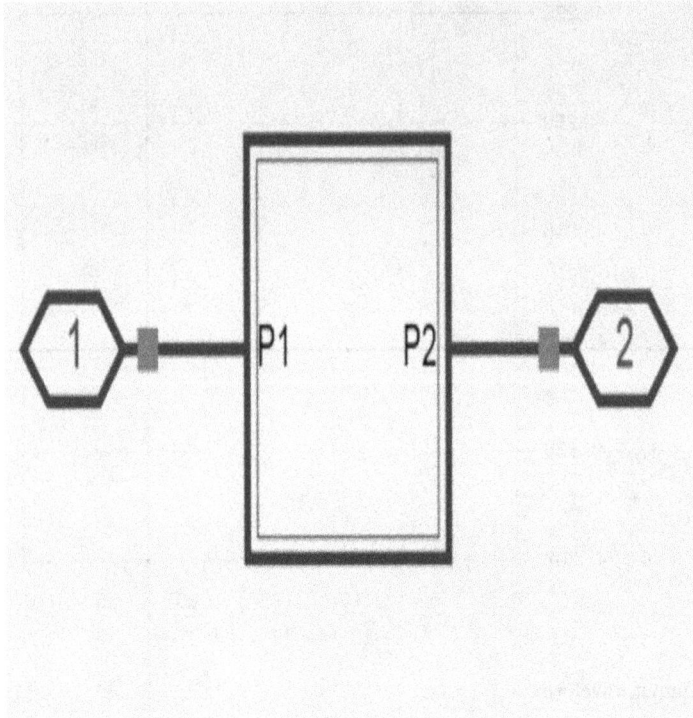

FIGURE 9.108 Port.

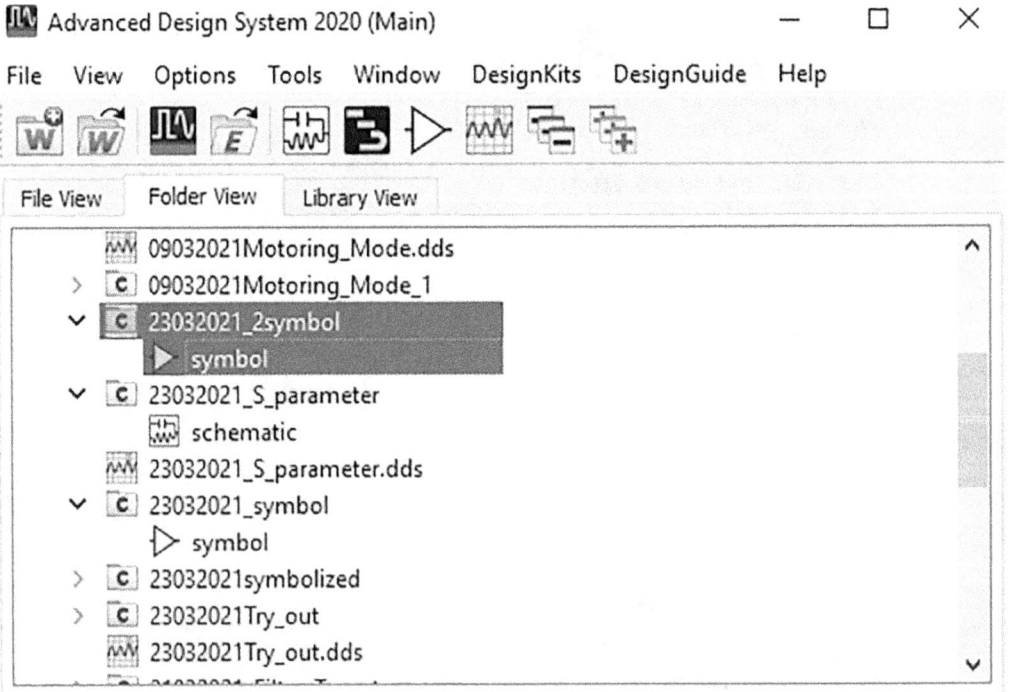

FIGURE 9.109 S-Parameter.

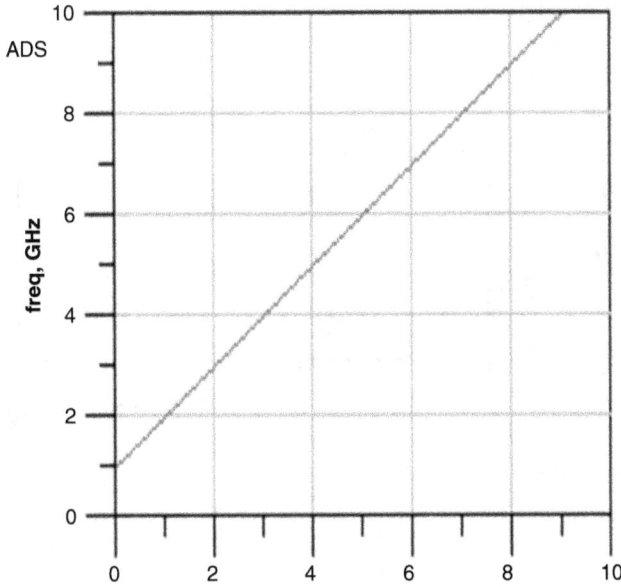

FIGURE 9.110   Waveform 1.

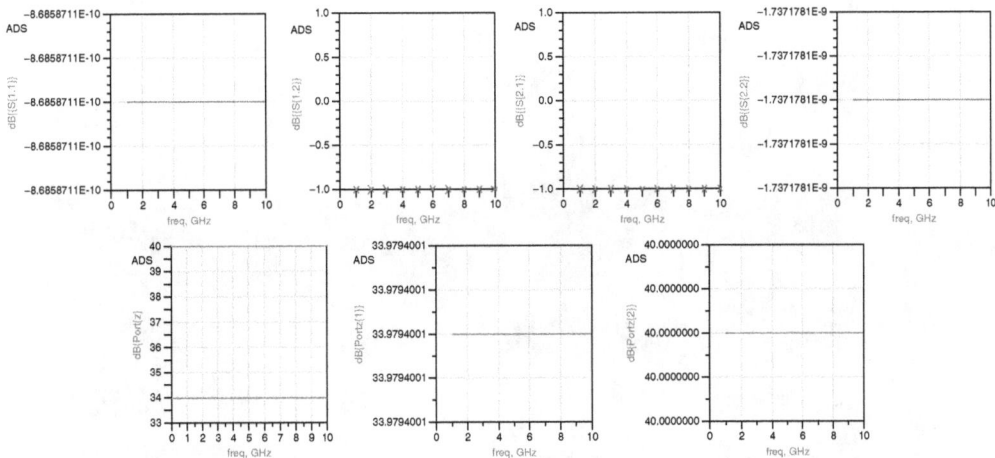

FIGURE 9.111   Waveform 2.

9.3.5   CE and Filter Spectrum Analysis

CE is the quantity of EMI that is conducted from the EUT to the outside world through the power lines. The LISN is a passive network that isolates the test system with reference impedance and provides a measuring point for monitoring CEs. The measuring bandwidth ranges from 9 kHz to 30 MHz. Since the power electronics (PE) switch (MOSFET) in the converter circuit operates at a frequency of 40 kHz, LISN is a more appropriate method for conducting emission tests on the proposed circuit. Using an RF (BNC) connector to a (test receiver) port on the LISN, a spectrum analyser or RF noise meter is utilized to examine the CEs of the EUT. As shown in Figure 9.112, a LISN is positioned between the power mains and the EUT.

As shown in Figure 9.113, this second-order low pass active filter consists of a simple passive RC filter stage that provides a low frequency channel to the input of a non-inverting operational amplifier.

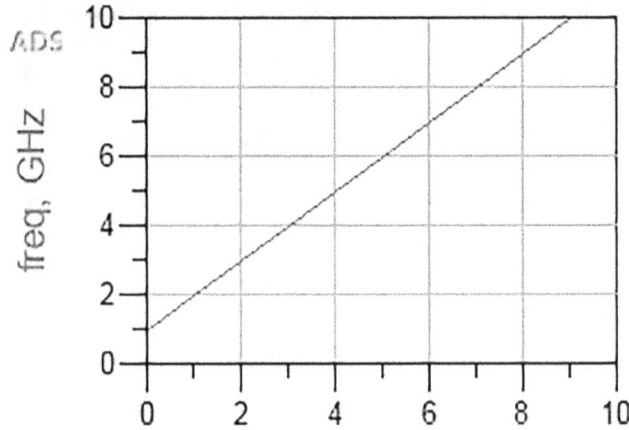

FIGURE 9.112   Circuit diagram 1.

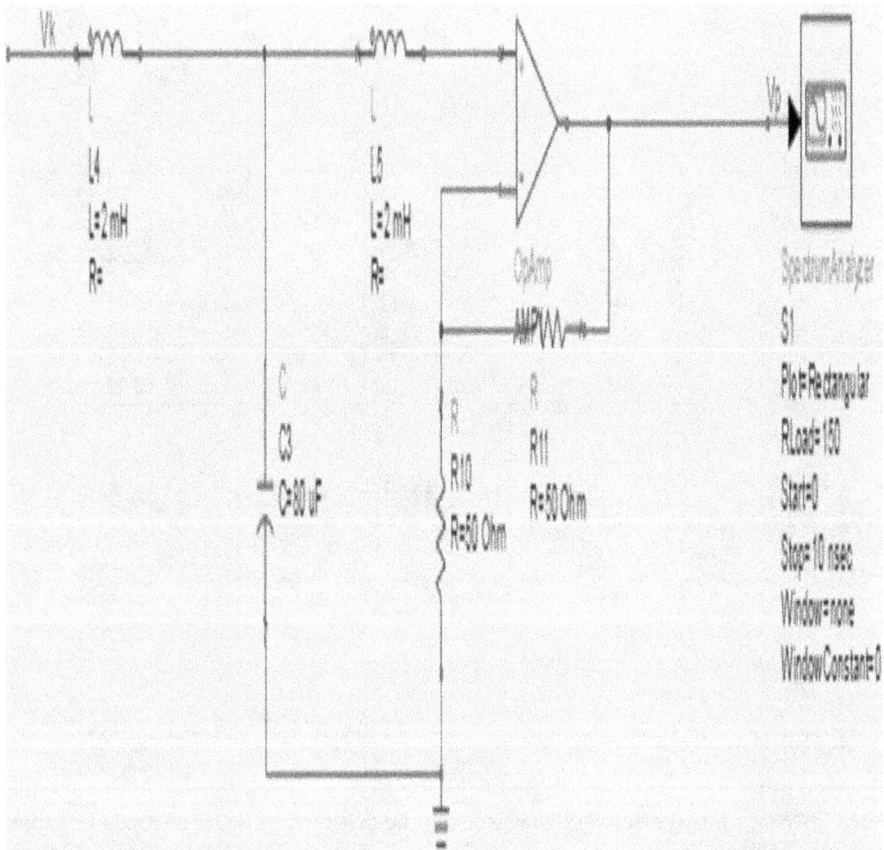

FIGURE 9.113   Circuit diagram 2.

The advantage of this setup is that the op-high amp's input impedance prevents excessive loading on the filter's output, while the op-low amp's output impedance prevents the cut-off frequency point of the filter from being impacted by variations in the load's impedance. This arrangement gives excellent filter stability. The passband gain will boost the frequency response of the circuit. For a non-inverting amplifier circuit, the size of the filter's voltage gain is a function of the feedback resistor (R11) divided by the input resistor (R10). This configuration achieves a gain value of 10.

Adding up the active filter at the $V_{measurement}$ of LISN, value of Voltage is analysed using a spectrum analyser (Figures 9.114 and 9.115).

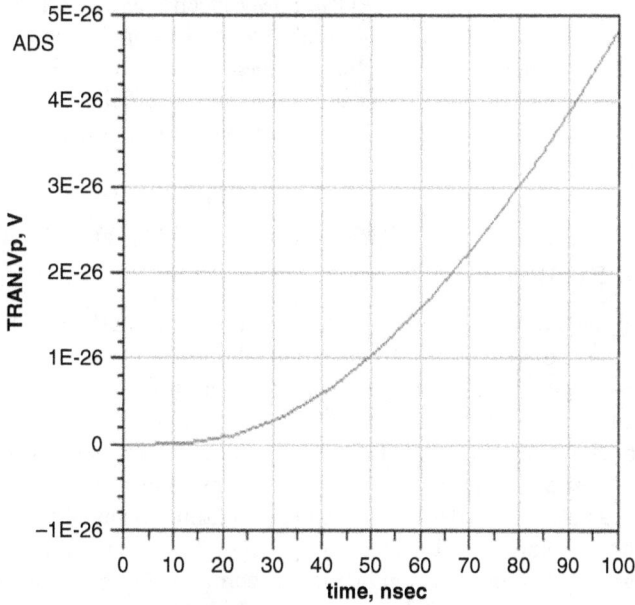

FIGURE 9.114 Waveform 3.

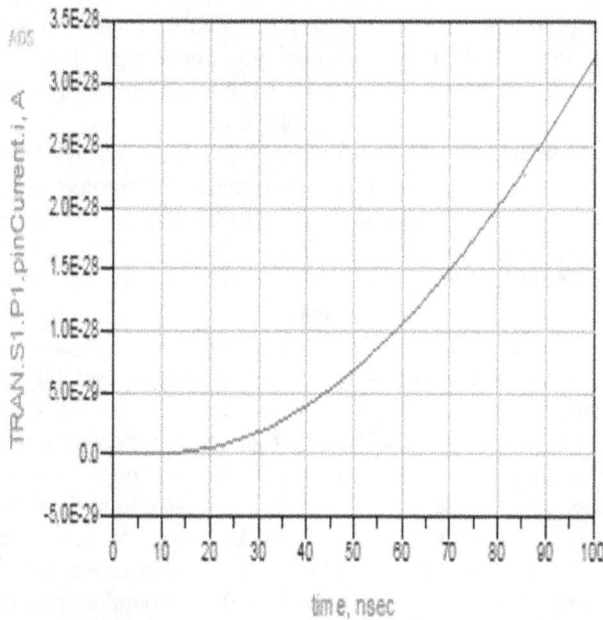

FIGURE 9.115 Waveform 4.

From the graph, it can be inferred that the distorted value is lesser than Zepto (10^{-21}) and Yocto (10^{-24}) values. And, it reveals clearly that the power conversion is good with no distortion after placing the filter.

Using a converter, the motoring function of a PMDC motor is explored here. The converter is utilized in hybrid energy storage systems with duty ratio generation to activate the switches. The DC–DC converter adjusts the output voltage and battery current of the EV or HESS power bank. The achieved efficiency of power conversion is 90.1%. Power electronic converters could significantly enhance the power density, affordability, and durability of drive applications. They can dramatically minimize switching losses at high switching frequencies, allowing PE converter topologies to increase system efficiency even further. In order to boost energy efficiency and dependability, EVs should adopt the small PE converters option. The validation of the constructed circuit is performed using ADS software.

9.4 BUCK-BOOST DC–DC CONVERTER WITH COUPLED-INDUCTOR FOR SOLAR PV APPLICATIONS

The purpose of the new coupled-inductor buck-boost converter (CIBuBoC) is to create a new configuration of a non-isolated doubled-switch buck-boost converter with coupled inductors in two cascade semi-stages. In the converter, an extremely high step-up or step-down voltage conversion ratio and step-up or step-down boundary adjustments are obtained in comparison to other types of buck-boost converters that use two power switches and a coupled inductor concurrently (CI). The suggested circuit topology contains a basic structure with two cascaded semi-stages, as well as ultra-extended output voltage, continuous input current with low ripple, positive output voltage polarity, and a common ground. These characteristics make the converter more appropriate for numerous applications, such as Solar PV systems. In addition, the voltage stress across each power switch is considerably less than other buck-boost converters leading to power MOSFETs. Consequently, the proposed converter has sufficiently high efficiency. Using the number of turns of the coupled inductor (CL) in buck-boost topology allows for excellent modification of the step-down or step-up boundary. The desirable characteristics of these converters include non-inverting wide high step-up/step-down voltage gain with enough efficiency, continuous input current with low ripple, a small number of components, and a reasonable price. Due to the significance of step-up mode operation, greater emphasis has been paid to the extension of the voltage conversion range in buck-boost converters as a quadratic or semi-quadratic coefficient.

In recent years, the electricity demand has increased dramatically, and the solution to this problem is to replace conventional power plants with distributed generations (DGs), thereby increasing the focus on the use of renewable energy sources in power generation. Renewable energy output is climate-dependent and low in consumption. To make it suitable for everyday usage, the output amplitude must be increased to a particular level. Using power converters that contain appropriate power electronic components, this can be accomplished effectively. In general, DC–DC buck-boost converters with a variable output voltage are utilized as an essential component in a variety of power electronic applications, such as renewable energy sources, portable gadgets, and automobile electronics. Nonetheless, non-isolated DC–DC converter architectures are commonly used in low-power applications due to their simple construction and less expensive implementation. These conventional converters have severe voltage gain ratio and component stress limits. To improve the performance of buck-boost converters, numerous non-isolated modified buck-boost converters are implemented. CL is increasingly used in DC–DC converters to achieve a broad voltage gain range. The suggested CIBuBoC is a modified buck-boost converter that achieves ultra-high step-up/step-down voltage conversion ratio and step-up/step-down boundary adjustment in comparison to other similar buck-boost converters. The converter employs two power switches that operate simultaneously and a CL. The voltage stress across each power switch is significantly lower than

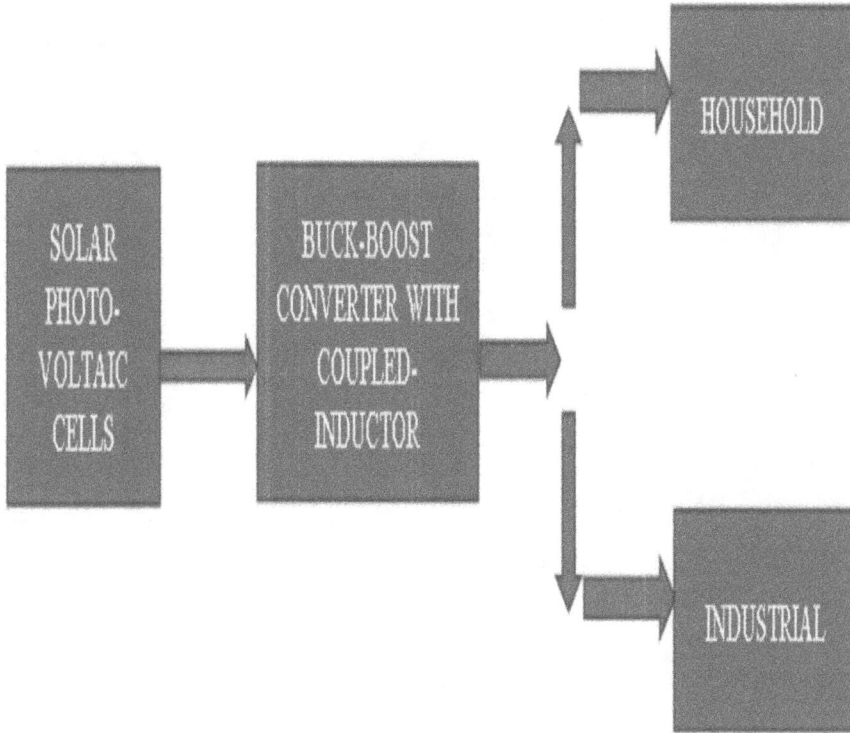

FIGURE 9.116 Block diagram of the proposed application.

in conventional buck-boost converters, which resulted in the selection of power MOSFETs with a reduced drain-source ON resistance (RDS(ON)) and, as a result, the converter operates with sufficiently high efficiency.

The primary goal of the proposed converter is to achieve a high step-up voltage conversion ratio and to increase the use of renewable energy sources in power generation in order to provide a high, constant DC output voltage to an industrial or domestic application. This high input-to-output voltage conversion ratio is obtained through the simultaneous functioning of two MOSFET switches and a linked inductor built as an ideal transformer with magnetizing and leaky inductors (Figure 9.116).

The output from the solar photovoltaic cells is given to the CIBuBoC, using which the voltage is boosted to a utility level and given for household and industrial applications.

9.4.1 CIRCUIT TOPOLOGY

Figure 9.117 shows the circuit topology of the proposed Coupled-Inductor Buck-Boost DC–DC Converter. The description about the topology is discussed below. The new CIBuBoC includes two power switches (S_1 and S_2) operating simultaneously, an input inductor (L_1), a coupled inductor (CL), three diodes (D_1, D_{01}, and D_{02}) conducting opposite to the switches, and three capacitors (C1, C01, and C02).

The coupled inductor (CL) is described as an ideal transformer with a magnetizing inductor (L_m) and a merged leaking inductor (L_k) in the primary side and a coupling coefficient (K)

$$K = \frac{L_m}{L_m + L_k}$$

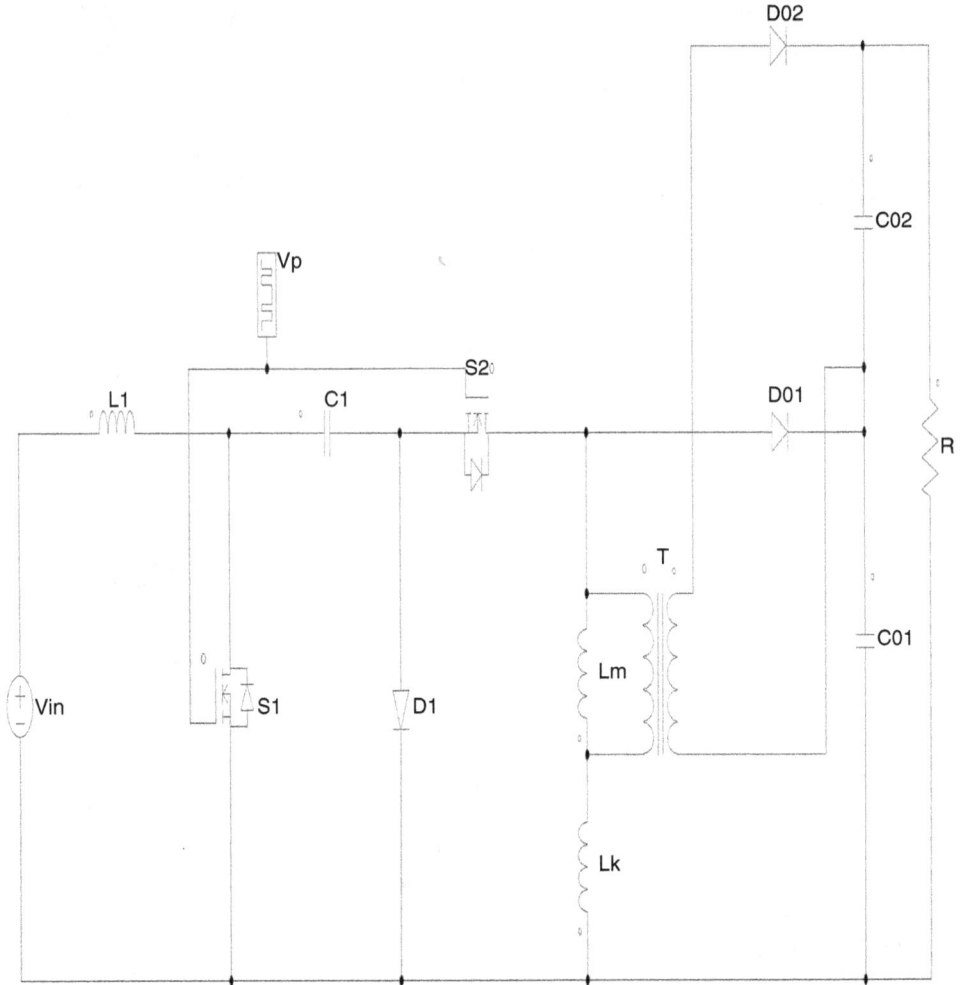

FIGURE 9.117   Circuit topology of the proposed converter.

The capacitors are large enough that the switching voltages are practically constant, and the input and magnetizing inductors are large enough that current ripples across them may be disregarded.

Each cycle consists of two operational modes. During mode I ($t=t_0$), as depicted in Figure 9.118, power switches S_1 and S_2 are simultaneously switched on, while all diodes are reverse biased and turned off. The voltages of the input source and capacitor C_1 are applied to the input inductor (L_1) and the primary side of CL, respectively, and they are energized to linearly increase the current. Thus, the load is supplied by the output capacitors C01 and C02 in series (Figure 9.118).

$$V_{L1} = V_{in}$$

$$V_{LM} + V_{LK} = V_{c1}$$

$$I_{S2} = I_{LM} = I_{LK}$$

$$I_{S1} = I_{L1} + I_{LM}$$

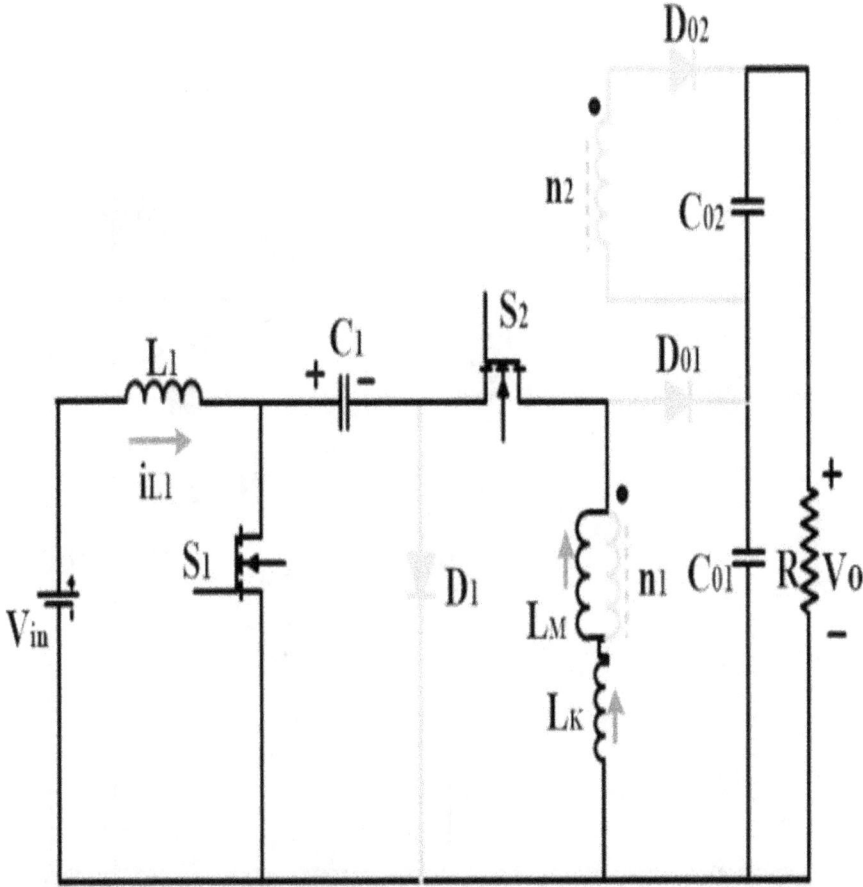

FIGURE 9.118   Operating mode I of the converter.

During mode II $(t=t_1)$, as shown in Figure 10.117, the power switches $S1$ and $S2$ are turned off and the diodes are turned-on simultaneously. During this time interval, the capacitor C1 is charged by the input inductor (L1) and the magnetizing inductor (Lm) is de-energizing to the output through the output diodes D01 and D02 so that the currents of those inductors are decreased. Since the magnetizing and the leakage inductor currents are equal in mode I, the output diode current of D02 increases into zero current condition (ZCS) (Figure 9.119).

$$V_{L1} = V_{in} - V_{C1}$$

$$V_{LM} + V_{LK} = V_{C01}$$

$$I_{D1} = I_{L1}$$

$$I_{D01} = I_{LK}$$

Thus, the block diagram, circuit topology, and the description of the proposed Coupled-Inductor Buck-Boost DC–DC Converter under two operating modes has been described clearly. In these converters, a high voltage conversion ratio using a large number of storage components has been provided. The doubled-switch topology reduces component stresses with a less number

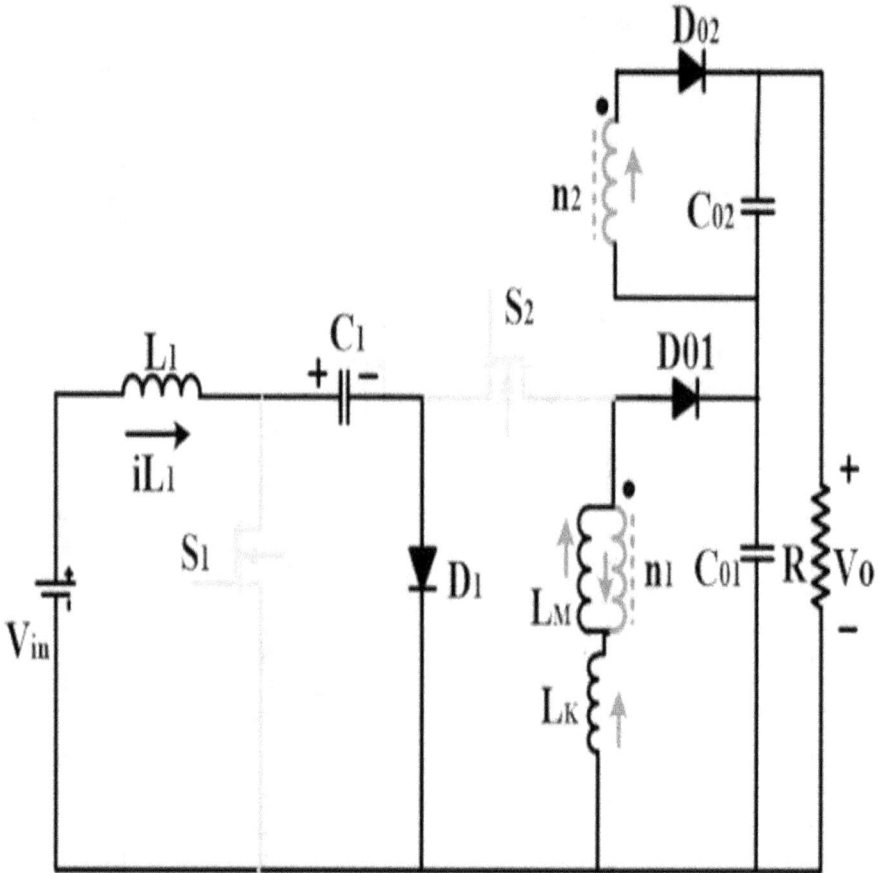

FIGURE 9.119   Operating mode II of the converter.

of storage components. The duty cycle of the converter can be reduced, which leads to decrease in the switching losses significantly and the converter has a simple structure with a common ground between the input and output voltages.

9.4.2   Design of the Converter

This section explains the designing of the proposed CIBuBoC with design formulae, design specifications, and the component model descriptions readily available in the market. Each of the energy storage elements to be connected in converter design and operational specifications has their own design formula based on which they are calculated for an efficient conversion from fixed DC input to variable DC output.

  a. **Design Formulae**

      The following are some of the design formulae derived for the design of the converter for its proposed conversion operation.

    1. Duty Cycle,

$$D = \frac{T_{ON}}{T_{ON} + T_{OFF}}$$

2. Transformer Turns Ratio,

$$n = \frac{n_1}{n_2}$$

3. Static Voltage Conversion Ratio,

$$M = \frac{(1+n)D}{(1-D)^2}$$

4. Input Inductor,

$$L1 = \frac{V_{in}D}{20\%I_{L1}f_s}$$

5. Input Capacitor,

$$C1 = \frac{M^2(1-D)^2}{1\%Rf_s}$$

6. Magnetizing Inductor,

$$Lm = \frac{RD(1-D)}{50\%M(1+n)f_s}$$

7. Leakage Inductor,

$$Lk = \frac{L_m}{K} - L_m$$

8. Output Capacitors,

$$C01 = C02 = \frac{(1+n)P_{out}}{V_{out}1\%V_{in}f_s}$$

b. **Design Specifications**

Table 9.6 shows the calculated values of the energy storage elements present in the circuit topology and other converter operational specifications for efficient conversion operation.

c. **Model Description**

Table 9.7 shows the available market model of the components based on their designed values required for the implementation of the hardware setup of the converter.

From the above given design formulae and the table on the design specifications, the proposed CIBuBoC can be implemented in any simulation software window and can be simulated and the results can be verified based on the output waveforms obtained. Also, from the model specifications table, the hardware setup of the proposed converter can also be implemented if needed.

9.4.3 SIMULATION

This section explains the simulation of the proposed Coupled-Inductor Buck-Boost DC–DC Converter using the ADS simulation software and shows the resulting waveforms obtained from simulating the converter. Figure 9.120 shows the implementation of the proposed converter in the

TABLE 9.6
Values of Energy Storage Elements

Parameter	Calculated Value
D	0.611
N	2 ($n_1 = 58$, $n_2 = 29$)
M	12.113
L1	0.27 mH
C1	123.34 μF
Lm	2.32 mH
Lk	2.32 μH (K = 0.99)
C01	134.58 μF
C02	134.58 μF

TABLE 9.7
Market Model of the Components Based on Their Designed Values

Description	Design Values	Available Market Model
MOSFET (S1, S2)	td(on) = 33 ns, td(off) = 21 ns	IRFB4227PBF
Input Inductor (L1)	0.27 mH	RLB0913-271K
Input Capacitor (C1)	123.34 μF	150 μF 450V – EC3861
Magnetizing Inductor (Lm)	2.32 mH	RLB0913-222K
Leakage Inductor (Lk)	2.32 μH	RLB0913-2R2K
Output Capacitor (C01)	134.58 μF	150 μF 450V – EC3861
Output Capacitor (C02)	134.58 μF	150 μF 450V – EC3861
Transformer (T)	$n_1/n_2 = 58/29$	EE42/21/20
Diode (D, D01, D02)	VF = 0.89 in 5A	BYV28–200-TR

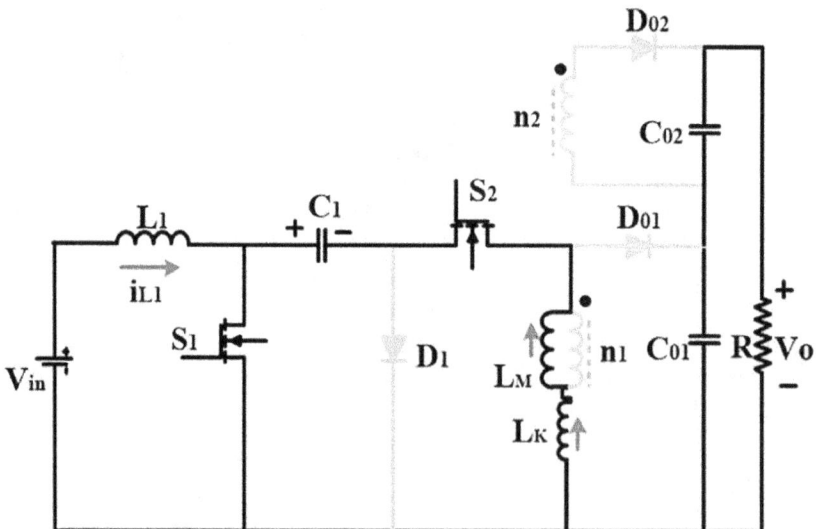

FIGURE 9.120　Schematic implementation of the proposed converter.

ADS simulation schematic window and the respective models of the components for their operation and the simulation parameters for simulating the converter.

Figures 9.121–9.127 are the resulting waveforms obtained on simulating the converter based on the design specifications.

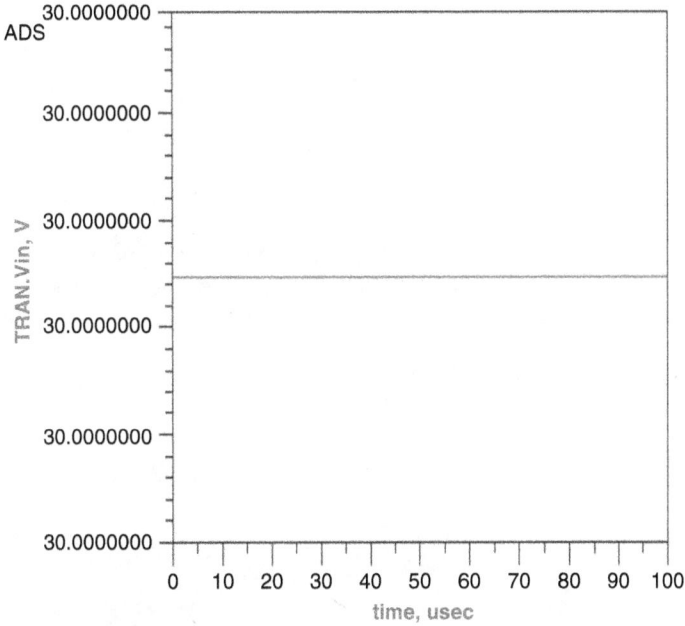

FIGURE 9.121   Input voltage to the converter.

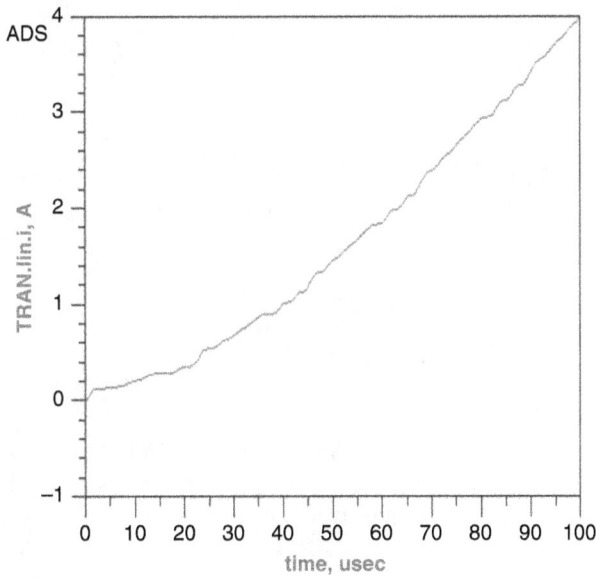

FIGURE 9.122   Input current to the converter.

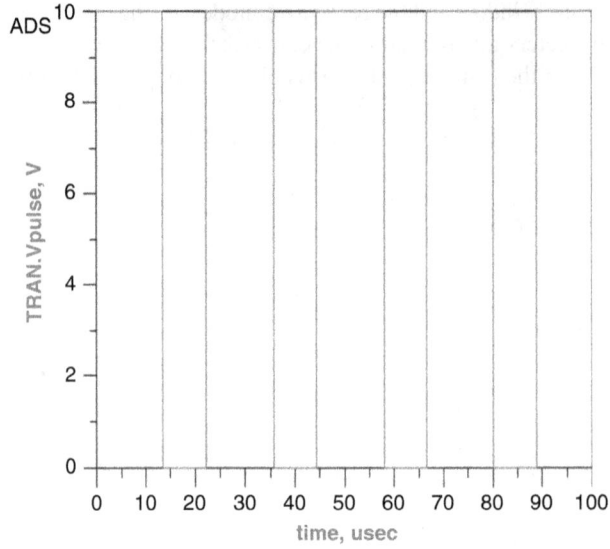

FIGURE 9.123   Voltage pulse to the MOSFET switches.

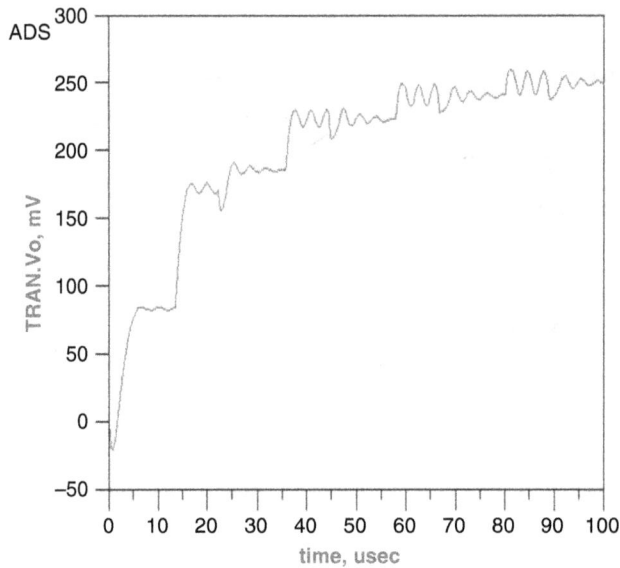

FIGURE 9.124   Output voltage from the converter.

9.4.4   S-PARAMETER AND CE

This chapter explains the calculation of the S-parameter coefficients using the sub-circuit of the converter and after that, CE is calculated from the spectrum analyser and the value is compared with the standard value and a suitable filter is designed to control the emission rate to a standard value if it exceeds the limit values.

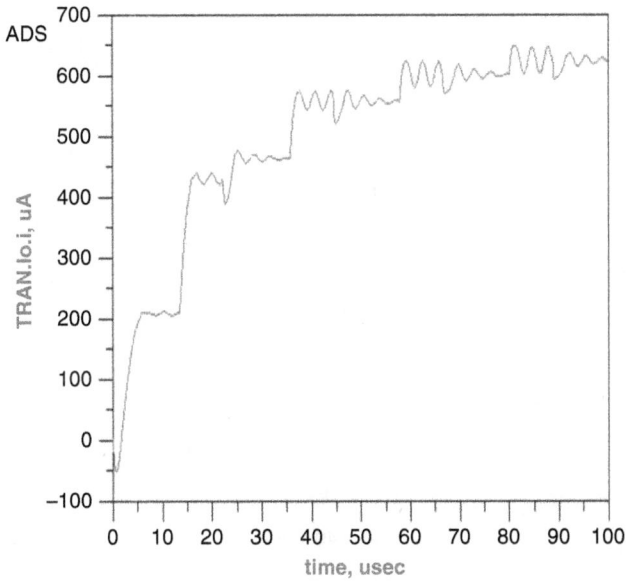

FIGURE 9.125   Output current from the converter.

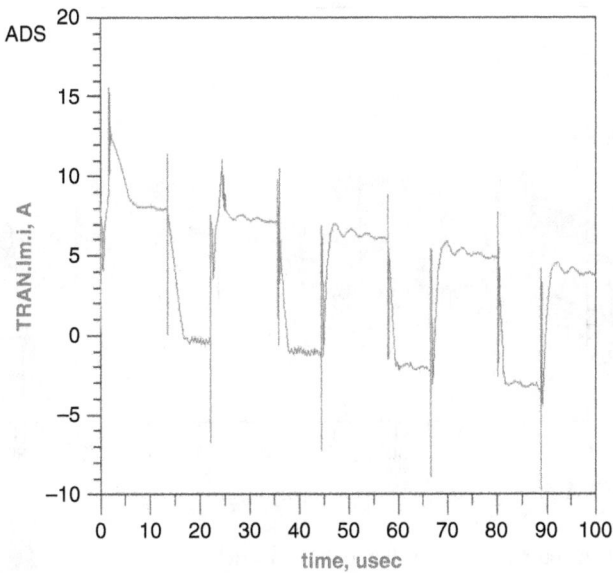

FIGURE 9.126   Magnetizing inductor current from the converter.

9.4.5   SIMULATION

Figure 9.128 shows the simulation of the S-parameter of the proposed converter by connecting an input port and an output port at the input and output sides, respectively, such that forming a sub-circuit of the converter and the Figure 9.129 shows the creation of a new symbol for the sub-circuit formed to obtain the transmission and the reflection coefficients of the S-parameters.

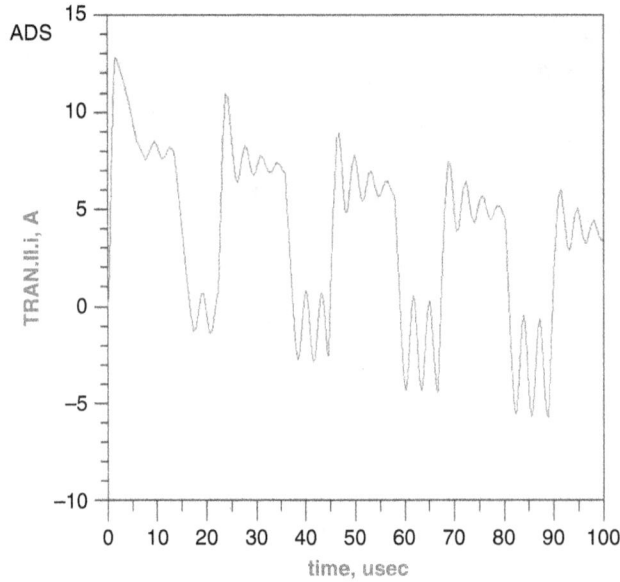

FIGURE 9.127   Leakage inductor current from the converter.

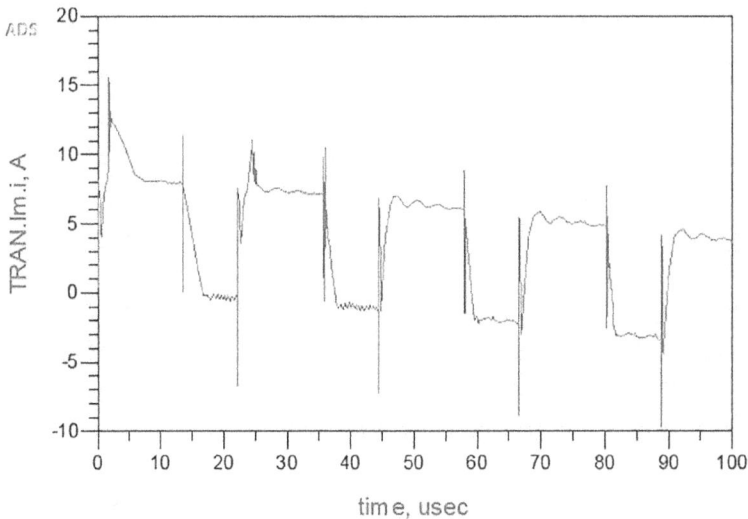

FIGURE 9.128   Input and output port insertion to the converter.

Figures 9.130–9.135 are the reflection coefficients (1,1 and 2,2) at the input and output sides and the transmission coefficients (1,2 and 2,1) between the input and output sides of the S-parameter matrix.

9.4.6   CONDUCTED EMISSION

CEs are essentially EMI or undesired noise that is generated internally by electronic or electrical components in the design. The propagation of these conducted emissions occurs along interconnecting wires. These cables may be signal ports, wired ports for telecommunications, or power conductors.

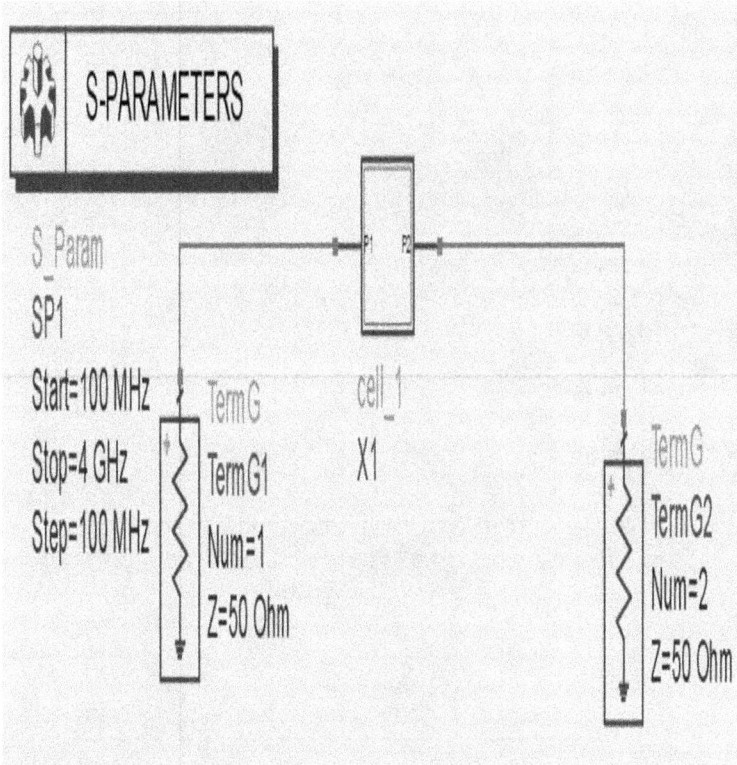

FIGURE 9.129 Two-port representation of the proposed converter.

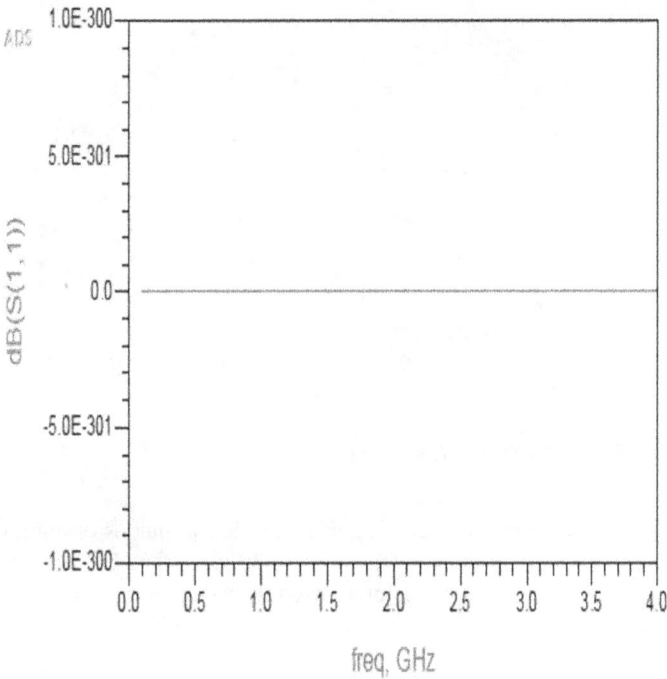

FIGURE 9.130 Reflection coefficient at the input side.

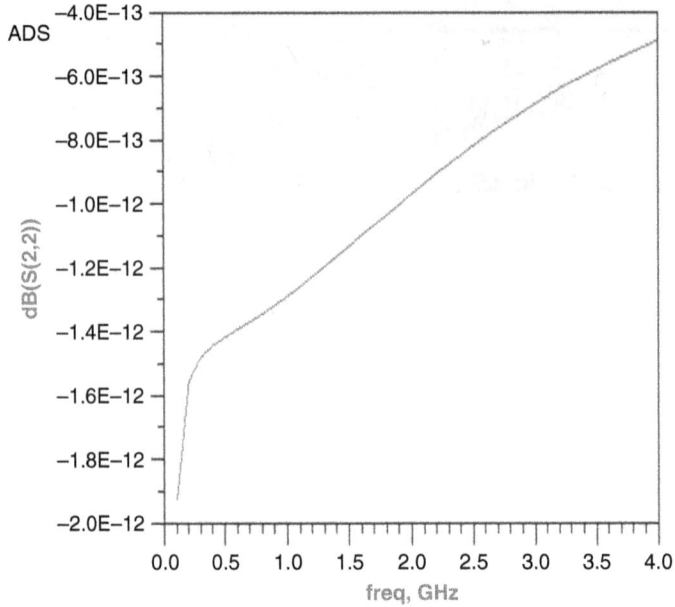

FIGURE 9.131 Reflection coefficient at the output side.

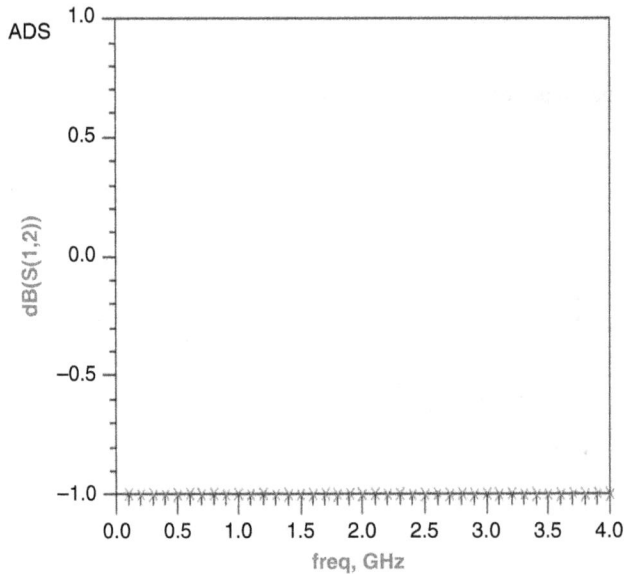

FIGURE 9.132 Transmission coefficient from input to output.

Typically, conducted emissions are classified as either discontinuous or continuous disruptions. CE requirements are defined in the range of 150 kHz to 30 MHz. The LISN is used to measure the disturbances produced by the EUT. EUT is connected to the supply mains through the LISN due the following reasons:

- LISN defines the input impedance
- LISN acts as high pass filter and an interface to EMI analyser

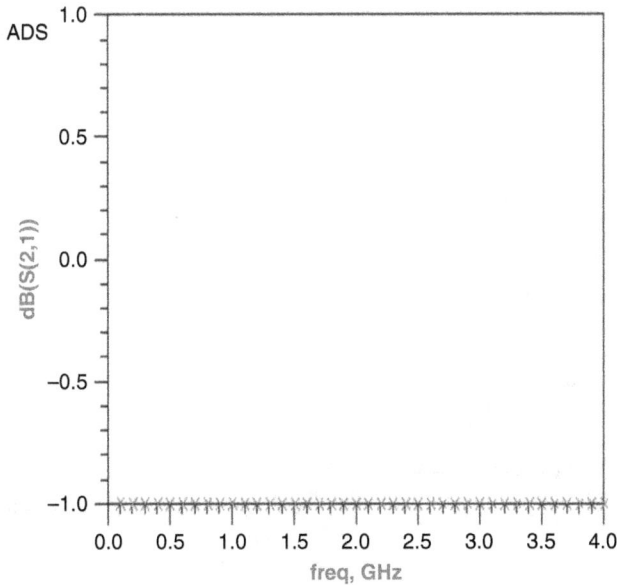

FIGURE 9.133 Transmission coefficient from output to input.

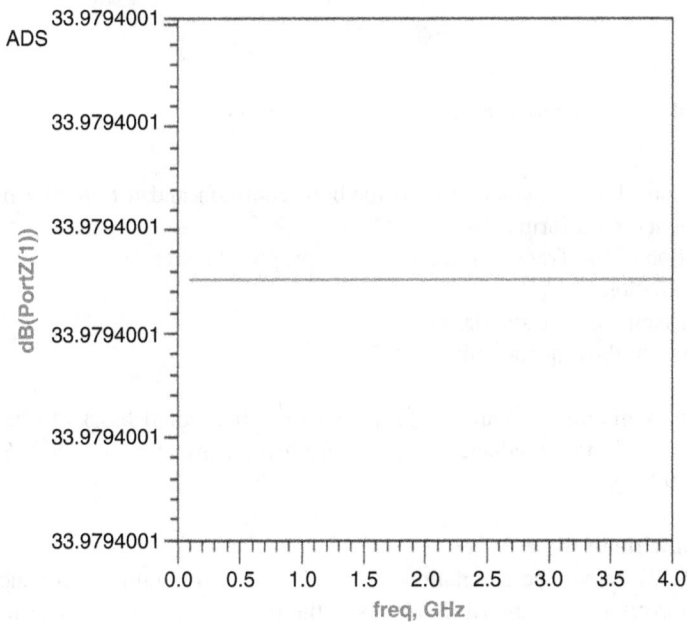

FIGURE 9.134 Input impedance.

- LISN attenuates noise coming from the supply mains
  The following are the steps involved in obtaining the conducted emission from the converter and techniques involved in the reduction of the emission rate.
  1. From the circuit model, identify the noise sources
  2. Evaluate the Main Parasitic Parameters
     a. High-Frequency Parasitic Parameter of Common Mode EMI
     b. High-Frequency Parasitic Parameter of Differential Mode EMI

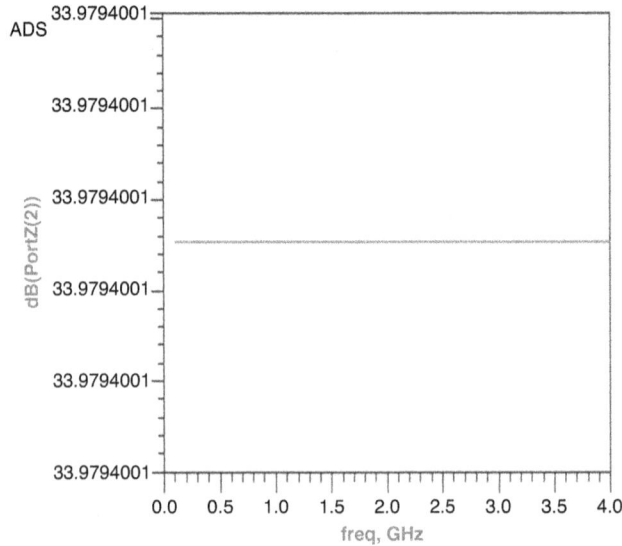

FIGURE 9.135 Output impedance.

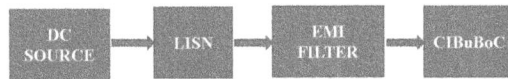

FIGURE 9.136 Block diagram for conducted emission calculation.

3. Formation of the conduction loop for both common and differential modes and its equivalent circuit formation
4. Evaluation of the Transfer Function of the proposed converter
5. CE Simulation
6. Comparison with the standards
7. Design of high-frequency filters

The following block diagram (Figure 10.134) shows the individual blocks to be connected in a cascaded form to obtain the conducted emission from the converter so as to meet the standards available (Figure 9.136).

a. **Simulation without filter**

Figure 9.137 shows the simulation of the converter to obtain the conducted emission from the converter using the LISN blocks at the positive and negative terminals, respectively, connected to a summer and a spectrum analyser.

Figure 9.138 shows the waveform of the current flowing through the default R-load inside the spectrum analyser.

9.4.7 EMI FILTER DESIGN

A simple π Current Limit Control (CLC) Filter is designed from the following L and C calculations

$$\text{I. Inductor, } L = \frac{Z_o}{nf_c} = 0.479\,\text{mH}$$

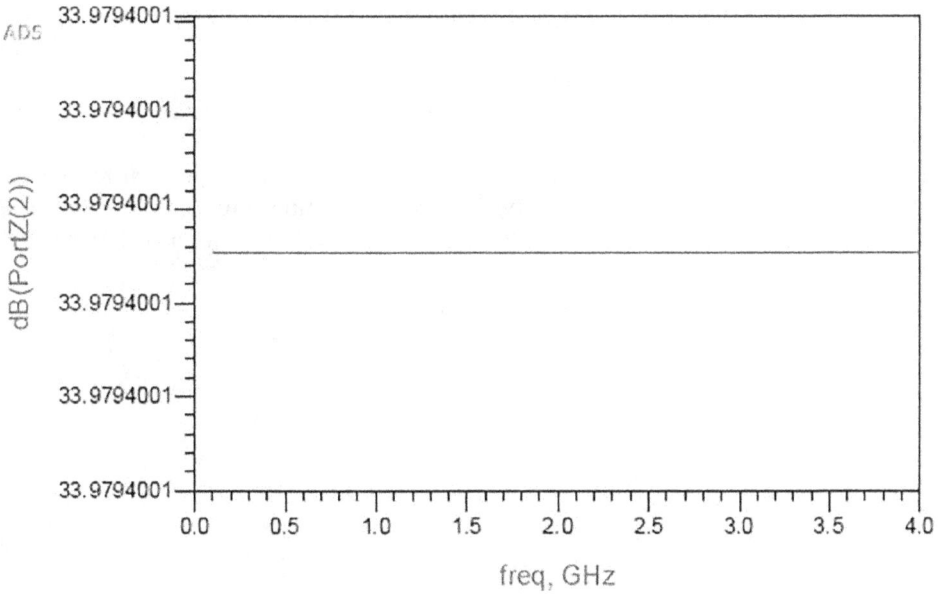

FIGURE 9.137 Schematic implementation of line impedance stabilization network (LISN) for conducted emission calculation.

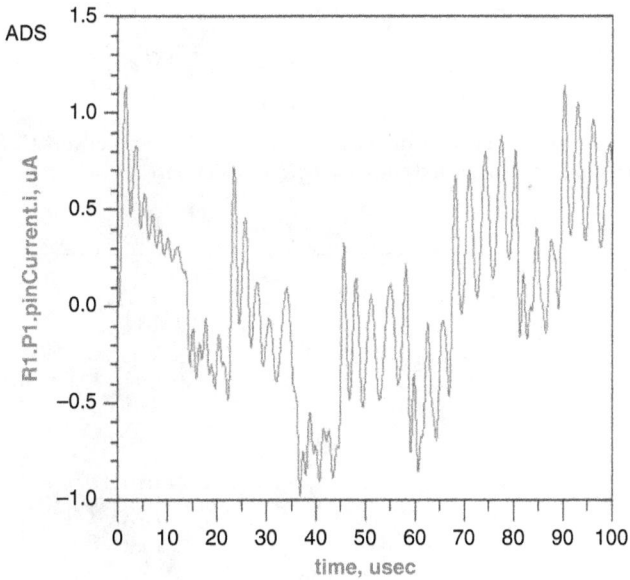

FIGURE 9.138 Output current from the spectrum analyser without filter.

where,

Z_o is the output impedance

f_c is the cut-off frequency

$$\text{II. Capacitor, } C = \frac{1}{Z_o n f_c} = 0.417 \mu F$$

Figure 9.139 shows the simulation of the converter to obtain the CE from the converter using the LISN blocks at the positive and negative terminals and an EMI filter (CLC or π) connected between the LISN summer and the spectrum analyser.

Figure 9.140 shows the waveform of the current flowing through the default R-load inside the spectrum analyser.

This CIBuBoC provides a semi-quadratic ultra-high step-up with low input current ripple and low output voltage ripple with positive polarity. With the correct ratio of turns on the coupled inductor, it also attains a high semi-quadratic voltage gain and step-up/step-down boundary adjustment.

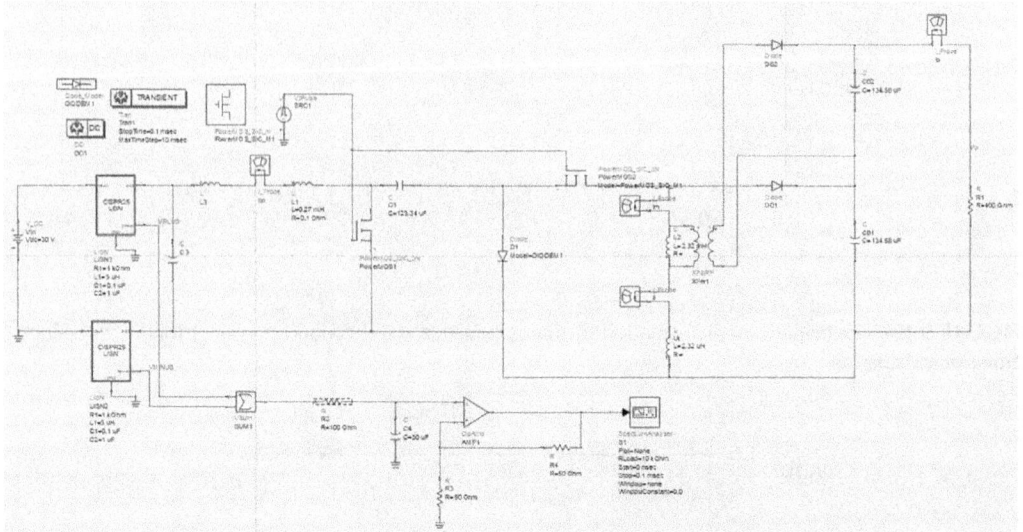

FIGURE 9.139 Schematic implementation of line impedance stabilization network (LISN) with electromagnetic interference (EMI) filter for conducted emission (CE) calculation.

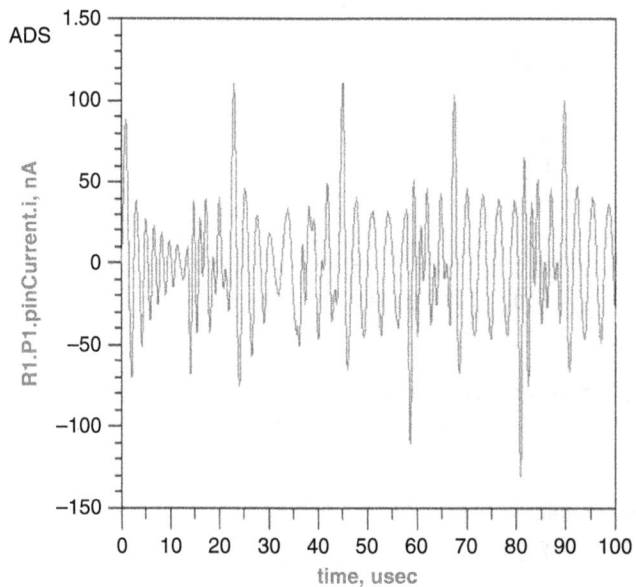

FIGURE 9.140 Output current from the spectrum analyser with filter.

The doubled-switch buck-boost topologies reduce component stresses while simultaneously increasing voltage gain and decreasing the number of storage components. Other benefits include a simple structure and the application of low voltage stress across the MOSFET switches, which reduce switching power losses and improves efficiency. For buck-boost converters, the implementation of several efficient voltage boosting approaches, such as magnetic components and switched-capacitors or switched-inductors, is restricted since these methods frequently result in the loss of step-down area. The proposed coupled-inductor converter can be used to enhance the power level of Solar PV power generation in single stage and the enhanced power can be used exclusively for residential and industrial areas.

Thus, this chapter clearly explained about the S-parameters, the design, and the implementation of the S-parameter subnetwork to obtain the reflection and transmission coefficients of the S-parameter matrix. CE is measured using LISN and a spectrum analyser. Noise reduction techniques by means of filters and the implementation of such a designed topology to reduce the CE and the resulting waveforms of all the topologies were simulated using ADS software.

BIBLIOGRAPHY

1. S. Hasanpour, A. Baghramian, and H. Mojallali, "Analysis and Modelling of a New Coupled-Inductor Buck-Boost DC/DC Converter for Renewable Energy Applications", *IEEE Transactions on Power Electronics*. December 2019. DOI: 10.1109/TPEL.2019.2962325.
2. Li Zhai, Tao Zhang, Yu Cao, Sipeng Yang, Steven Kavuma, and Huiyuan Feng, "Conducted EMI Prediction and Mitigation Strategy Based on Transfer Function for a High-Low Voltage DC-DC Converter in Electric Vehicle", Basel, MDPI, 2018. DOI: 10.3390/en11051028.
3. Uma Maheswari Yuvaraj, Amudha Alagarsamy, Ashok Kumar Loganathan, and Selvathai Thavassy, "Conducted Emission Study in Space Vector Modulated Voltage Source Inverter", *International Journal of Power Electronics and Drive Systems (IJPEDS)*, vol. 13, no. 2, pp. 988–997, June 2022. ISSN: 2088–8694. DOI: 10.11591/ijpeds.v13.i2.
4. A. Phimphui and U. Supatti, "V2G and G2V Using Interleaved Converter for a Single-Phase Onboard Bidirectional Charger", *2019 IEEE Transportation Electrification Conference and Expo, Asia-Pacific (ITEC Asia-Pacific)*, Seogwipo-si, Korea (South), 2019. DOI: 10.1109/ITEC-AP.2019.8903662.
5. R. Goswami and S. Wang, "Modeling and Stability Analysis of Active Differential-Mode EMI Filters for AC/DC Power Converters", *IEEE Transactions on Power Electronics*, vol. 33, no. 12, pp. 10277–10291, December 2018. Doi: 10.1109/TPEL.2018.2806361.
6. Mithat C. Kisacikoglu, Metin Kesler, and Leon M. Tolbert, "Single-Phase On-Board Bidirectional PEV Charger for V2G Reactive Power Operation", *IEEE Transactions on Smart Grid*, vol. 6, pp. 767–775, 2015. Doi: 10.1109/TSG.2014.2360685.
7. C. A. Bendall and W. A. Peterson, "An EV On-board Battery Charger", *Proceedings of Applied Power Electronics Conference. APEC '96*, San Jose, CA, USA, 1996, pp. 26–31 vol. 1. Doi: 10.1109/APEC.1996.500417.
8. V. Monteiro et al., "On-Board Electric Vehicle Battery Charger with Enhanced V2H Operation Mode", *IECON 2014–40th Annual Conference of the IEEE Industrial Electronics Society*, Dallas, TX, USA, 2014, pp. 1636–1642. Doi: 10.1109/IECON.2014.7048722.
9. M. Grenier, M. G. Hosseini Aghdam, and T. Thiringer, "Design of On-board Charger for Plug-in Hybrid Electric Vehicle", *5th IET International Conference on Power Electronics, Machines and Drives (PEMD 2010)*, Brighton, UK, 2010, pp. 1–6. Doi: 10.1049/cp.2010.0101.
10. Y. Uma Maheswari, A. Amudha and A. Kumar, "A Review on EMI Issues in High speed Designs and Solutions", *Journal of Electronics, Electromedical Engineering, and Medical Informatics*, vol. 4, no. 4, pp. 191–203, October 2022.
11. http://www.keysight.com

Index

For Product Safety Concerns and Information please contact our EU
representative GPSR@taylorandfrancis.com
Taylor & Francis Verlag GmbH, Kaufingerstraße 24, 80331 München, Germany

www.ingramcontent.com/pod-product-compliance
Lightning Source LLC
Chambersburg PA
CBHW080129220326
41598CB00032B/5003